STRUCTURAL DYNAMICS OF TURBO-MACHINES

STRUCTURAL DYNAMICS OF TURBO-MACHINES

Part I—*Component Dynamics, Life Cycle Evaluation*
Part II—*Structural and Operational Analysis of Engine System*
Part III—*Engine Operational Characteristics*

A S RANGWALA

Technical Director
Center for Engineering Technology
Cincinnati, USA

Consulting Engineer
Design, Stress and Dynamic Evaluation of Rotating and Reciprocating Machinery

NEW AGE REFERENCE

An Imprint of

NEW AGE INTERNATIONAL (P) LIMITED, PUBLISHERS
LONDON • NEW DELHI • NAIROBI
Bangalore • Chennai • Cochin • Guwahati • Hyderabad • Kolkata • Lucknow • Mumbai
Visit us at **www.newagepublishers.com**

Published by New Age International (P) Ltd., Publishers
First Edition: 2009
Reprint: 2017

GLOBAL OFFICES

- **New Delhi** **NEW AGE INTERNATIONAL (P) LIMITED, PUBLISHERS**
7/30 A, Daryaganj, New Delhi-110002, (INDIA)
Tel.: (011) 23253771, 23253472, **Telefax:** 23267437, 43551305
E-mail: contactus@newagepublishers.com • Visit us at www.newagepublishers.com
- **London** **NEW AGE INTERNATIONAL (UK) LTD.**
27 Old Gloucester Street, London, WC1N 3AX, UK
E-mail: info@newacademicscience.co.uk • Visit us at www.newacademicscience.co.uk
- **Nairobi** **NEW AGE GOLDEN (EAST AFRICA) LTD.**
Ground Floor, Westlands Arcade, Chiromo Road (Next to Naivas Supermarket)
Westlands, Nairobi, KENYA, **Tel.:** 00-254-713848772, 00-254-725700286
E-mail: kenya@newagepublishers.com

BRANCHES

- **Bangalore** 37/10, 8th Cross (Near Hanuman Temple), Azad Nagar, Chamarajpet, Bangalore- 560 018
Tel.: (080) 26756823, **Telefax:** 26756820, **E-mail: bangalore@newagepublishers.com**
- **Chennai** 26, Damodaran Street, T. Nagar, Chennai-600 017, **Tel.:** (044) 24353401, **Telefax:** 24351463
E-mail: chennai@newagepublishers.com
- **Cochin** CC-39/1016, Carrier Station Road, Ernakulam South, Cochin-682 016, **Tel.:** (0484) 2377303, **Telefax:** 4051304
E-mail: cochin@newagepublishers.com
- **Guwahati** Hemsen Complex, Mohd. Shah Road, Paltan Bazar, Near Starline Hotel, Guwahati-781 008,
Tel.: (0361) 2513881, **Telefax:** 2543669, **E-mail: guwahati@newagepublishers.com**
- **Hyderabad** 105, 1st Floor, Madhiray Kaveri Tower, 3-2-19, Azam Jahi Road, Near Kumar Theater, Nimboliadda Kachiguda, Hyderabad-500 027, **Tel.:** (040) 24652456, **Telefax:** 24652457
E-mail: hyderabad@newagepublishers.com
- **Kolkata** RDB Chambers (Formerly Lotus Cinema) 106A, 1st Floor, S N Banerjee Road, Kolkata-700 014
Tel.: (033) 22273773, **Telefax:** 22275247, **E-mail: kolkata@newagepublishers.com**
- **Lucknow** 16-A, Jopling Road, Lucknow-226 001, **Tel.:** (0522) 2209578, 4045297, **Telefax:** 2204098
E-mail: lucknow@newagepublishers.com
- **Mumbai** 142C, Victor House, Ground Floor, N.M. Joshi Marg, Lower Parel, Mumbai-400 013
Tel.: (022) 24927869, **Telefax:** 24915415, **E-mail: mumbai@newagepublishers.com**
- **New Delhi** 22, Golden House, Daryaganj, New Delhi-110 002, **Tel.:** (011) 23262368, 23262370, **Telefax:** 43551305
E-mail: sales@newagepublishers.com

ISBN: 978-81-224-2230-6
C-16-11-9784

Printed in India at Glorious Printers, Delhi.
Typeset at In-house, Delhi.

NEW AGE INTERNATIONAL (P) LIMITED, PUBLISHERS
7/30 A, Daryaganj, New Delhi-110002
Visit us at **www.newagepublishers.com**
(CIN: U74899DL1966PTC004618)

Preface

Considerable interest in the application of the theory of structural dynamics to the design of compressors, steam and gas turbines and pumps has existed for several years. The need for a comprehensive textbook on the dynamics of the rotating and stationary blades and vanes and the associated disks and shafts incorporating the most recent developments of the subject has been strongly felt for a long time.

Since the advent of the earliest water-driven power saw mills, problems of deformed and broken turbine blades, shafts and bearings have plagued the operators and manufacturers of the machines. Problems associated with relative motion between the rotating and stationary parts and lubrication were so extensive that little effort was expended in understanding the impact of material fatigue, elevated temperature and load cycling on the dynamic characteristics of the airfoils. Although a very extensive amount of research has been done, and an even greater number of publications exist on the subject, little effort has been spent in putting together in one concise publication topics including conceptual design, fluid flow, structural dynamic analysis, design optimization, vibration measurement and dynamic balance. Numerous other topics closely related to operation, manufacturing and materials selection of turbo-machinery components and system have been covered extensively. Special emphasis is placed on computer simulation using finite element methods, correlation of analysis with experimental test results and procedures to improve performance efficiency and structural integrity.

The basic premise in the operation of all turbo-machines calls for an interaction between the fluid medium flowing over the surfaces of the stationary and rotating airfoils. Hence, the aerodynamic and structural dynamic characteristics of the airfoils are closely intertwined. The overall profile of the blades must be contoured to maximize the aerodynamic efficiency, but at the same time the part must have adequate structural strength to withstand the many different dynamic excitations imposed on it. Dynamic loads arise from many sources, the predominant one being the source of the operating principles itself on which the machine is designed. When a rotor blade passes the stationary vanes of the nozzle, it experiences repeated fluctuating lift and moment loads at a frequency dependent on the number of vanes and the speed of the machine. The rotating airfoils are flexible members, and possess a number of natural frequencies of vibration about their torsional axis and bending in and out of the plane of rotation of the disk. In addition to the steady centrifugal forces arising from its mass, the airfoil must also withstand the dynamic loads due to the aerodynamic excitation. Although the blades are designed to avoid resonance at its design speed, resonant vibrations are still encountered. A good example is an aircraft engine as the aircraft accelerates from ground to flight idle, cruise and takeoff speeds.

This book, together with its companion book titled **Turbo-Machinery Dynamics: Design and Operation**, is written to meet the needs of students in engineering colleges and practicing engineers in a large variety of industries where turbo-machines are used. All of the material has been specifically tailored for college undergraduate and graduate level design engineering and vibration of rotating machine courses. Electronic spread-sheet type of calculations are used in example problems to calculate natural frequencies of vibration, dynamic response, fatigue life and design parameters related to fluid flow and component sizing. It is expected that the reader is familiar with basic to medium level calculus offered at the college undergraduate level.

The book is split into three parts. The first part focuses on dynamic analysis of engine components. The four chapters in the section look into the theory of vibrations, fan and compressor airfoils, turbine blade and vane and bearings and seals for various applications of turbo-machines. The second part investigates the dynamic behavior of the engine system, specifically the torsional and flexural aspects of the engine's rotor system. The third part delves into associated topics in structural vibrations, such as experimental and test measurement of operating parameters, balancing of rotor, and turbo-machinery noise. The companion volume to this book (**Turbo-Machinery Dynamics: Design and Operation**) is devoted to various applications of turbo-machines, aerodynamic design and operating features of compressor and turbine components and of the engine as a system, and also discusses other closely related topics of interest (super alloys, their characteristics, and manufacturing methods). Since the aerodynamic design of a turbo-machinery and its parts play such an overwhelming role in establishing the structural dynamic behavior of the components, module or assembly during operation, a close correlation has been maintained throughout the book between the design and dynamics disciplines.

The first chapter of part I establishes the basics of vibration theory and methods for its application to turbine parts. The second, third and fourth chapters are devoted to applications of the dynamic theory to fan and compressor airfoils, to blades and vanes in the turbine's elevated temperature regime and to the role of bearings to support the rotor and of seals to prevent the gases from escaping. The close interaction between aerodynamics and the structural integrity in the form of strength and component life management issues of the airfoils receives special attention. In the second part, shaft torsional vibrations are investigated in chapter five and flexural rotor dynamics in chapter six. In part III, chapter seven delves into instrumentation and techniques needed to analyze operating parameters during operation of the engine, chapter eight discusses procedures for static and dynamic balancing procedures for many different rotor configurations and chapter nine looks into noise generation mechanism and their suppression in turbo-machines.

The associated book (**Turbo-machinery Dynamics: Design and Operation**) considers applications of turbo-machines for air transportation and power generation and design features of engine parts in the relatively cooler zone in the compressor and in the hotter combustor and turbine flow paths. Material selection and super alloy characteristics as also manufacturing methods are also included.

A list of symbols is provided mostly to facilitate identification with commonly used parameters in the equations and the associated text. However, because of the considerable number of topics the corresponding variables are adequately defined within each section. Oftentimes it is found necessary within the sections to redefine many of the symbols for convenience and better understanding of the subject matter. Thus, the list of symbols may be used only as a general guideline.

I gratefully remember and appreciate past students of the course on this topic who have sent in comments and reported errors, and express my hope that those who work with this treatise will do likewise. I am indebted to Mr. Mark Belloni and Dr. Fred Ehrich of General Electric Company for performing a vast amount of computational work in finite element analysis and for valuable advise on the text and layout of the book. I greatly appreciate comments provided by Dr. Ahmad Kamel, Mr. George Robinson and Dr. Raj Subbiah of Siemens-Westinghouse Power Corporation for checking the problems and reading the proof.

Orlando, FL **A.S. Rangwala**

Contents

PART-II

PART-III

Introduction

Dynamic analysis of rotating machinery has come a long way since Professor Stephen Timoshenko considered the case of a uniform shaft with a disk at each end in his first technical paper "O Yavleniyakh Rezonansa v Valakh (On Resonance Phenomena in Shafts)" in the *Bulletin of the Polytechnical Institute* of St. Petersburg. Modern turbo-machines have a large number of compressor and turbine stages. The design of turbo-machinery airfoils is far more complicated due to the complex configuration. The shape of the airfoils is designed from aerodynamics consideration, but the blades must structurally withstand constant changes in the loads imposed by the flow of fluids over its surface.

When Abdulla S. Rangwala came to America in 1967 as a student he was influenced by the works of Professors S. Timoshenko and J. den Hartog, using Lord Rayleigh's method for resolving vibrations problems in engineering. His first job was with the Large Steam Turbines Department of General Electric Company in Schenectady, NY. His initial practical experience was with calculating fundamental periods of torsional and flexural vibrations of turbine rotor and journal bearing systems and in balancing of disks and rotors. At GE's Aircraft Engines Group he gradually shifted his attention to design and evaluation of compressor and turbine components. His experience with practical problems has culminated in his writing of **Structural Dynamics of Turbo-machines** and the companion treatise **Turbo-machinery Dynamics: Design and Operation**. The two complementary books represent a unique compilation of a large number of topics in an organized manner that is closely associated with the design and operation of turbo-machinery. The author presents the latest technical developments in the areas of design, development, and structural dynamic evaluation for turbine engineering, manufacturing, and operation engineers.

With the advent of computers, many important developments in the design and development of turbo machinery have occurred. While the computers do not fundamentally change the principles of fluid flow and structural vibration mechanics, it greatly influences the choice of methods of calculation that are most attractive. To uphold the technical excellence and unique appeal while keeping pace with new developments in the field is no small responsibility, and the author is to be commended for his fine works.

Brewster, OH. **Mark Belloni**

Part I

Component Dynamics, Life Cycle Evaluation

CHAPTER 1

Vibration Theory Fundamentals

1.1 INTRODUCTION

Mechanical vibrating systems consist of elements such as a spring for storing potential energy, mass and inertia for kinetic energy, and damper for dissipating mechanical energy. The vibration process alternatively converts energy between its potential and kinetic forms. In its general sense the vibration is a periodic motion that repeats itself in all its details after a certain interval of time, called the period of vibration. Some energy must be replaced in each cycle of vibration from an external source to maintain the vibration.

In a linear spring the change in length is proportional to the force acting along its length, the constant of proportionality referred to as the spring's flexibility, which is the reciprocal of its stiffness. The ideal spring has no mass, and so the forces acting at its opposite ends are equal and opposite. A mass is a rigid body; so its acceleration, according to Newton's second law of motion, is proportional to the force acting on the mass. In a viscous damper the applied force is proportional to the relative velocity of its connecting points. The damping coefficient provides the proportionality constant, and is the characteristic parameter of the damper. The ideal damper also is massless, and so the force at one end is equal to and opposite to the force at the other end.

Besides translational motion, a vibrating system may execute rotational motion. The elements of a mechanical system executing pure rotation of the parts are analogous to the elements executing pure translation. In a rotational system inertia stores kinetic energy, while stiffness and damping parameters are defined with reference to angular rotation and velocity. The analogy between translational and rotational motion of a vibrating body carries on to the mathematical equations describing the motion of the system; force used in linear motion is replaced by torque for rotational motion. A plot of the linear displacement, or angular rotation, against time may be a complicated curve. The simplest kind of periodic motion is a harmonic motion, with the displacement expressed as a harmonic function of time t and angular velocity ω, also referred to as the circular frequency, and is measured in radians per second. The maximum value of the displacement, x_0, is called the amplitude of vibration. The period of vibration, T, measured in seconds, is the reciprocal of the frequency of vibration, f, measured in cycles per second. The relation between ω, f and T are as follows. A full cycle of vibration takes place when ωt has passed through 360°, or 2π radians. Then the sine function resumes its previous path. Thus, when $\omega t = 2\pi$, the time interval t is equal to the period T, or $T = 2p/w$ seconds. Since f is the reciprocal of T, $f = \omega/2p$ cycles per second. For rotating machinery the frequency is often expressed as cycles per minute, so $f = 30\omega/p$. In a harmonic motion displacement is given by $x = x_0 \text{Sin}(\omega t)$. Velocity is determined by differentiation with respect to time, so $dx/dt = x_0\omega \cos(\omega t)$. Thus, velocity is also harmonic, with a maximum value of $x_0\omega$. Acceleration is given by the second derivative of x with respect to time t:

$$d^2x/dt^2 = -x_0\omega^2 \sin(\omega t) \tag{1.1}$$

also harmonic, with a maximum value of $x_0\omega^2$.

A vibrating system is said to have one degree of freedom if its geometrical position can be expressed at any instant in time by one parameter only. For example, a rotating shaft in rigid bearings is fully described by the angle between a reference line and the vertical axis. Another example is a piston moving in a cylinder, so the system can be specified by the distance from the cylinder end, and thus may be classified as a single degree of freedom system. In general, if it takes n numbers to specify the position of a mechanical system, the system has n degrees of freedom. A disk moving in its plane without restraint has three degrees of freedom: the x and y displacements of its center of gravity and the angle of rotation about the center of its gravity. A circular disk rolling down an inclined plane has only one degree of freedom; if it partly slides and rolls, it has two degrees of freedom, one of translation and one of rolling. A rigid body moving freely in space has six degrees of freedom, three of translation and three of rotation. Thus, it takes six coordinates to express its position, usually denoted as x, y, z, θ, φ and ψ.

A single degree of freedom system is not necessarily a simple one. For example, a gas turbine engine with a rigid shaft and rigidly attached to the mount has only one degree of freedom. This is because a single parameter, such as the angle through which the shaft has turned, determines completely the position of all its moving parts. But if the engine is mounted by flexible attachments so that it can freely move in all directions as on the mount skid, it has seven degrees of freedom, six in reference to the motion of the engine's center of gravity as a rigid body in free space and shaft angle as the seventh coordinate.

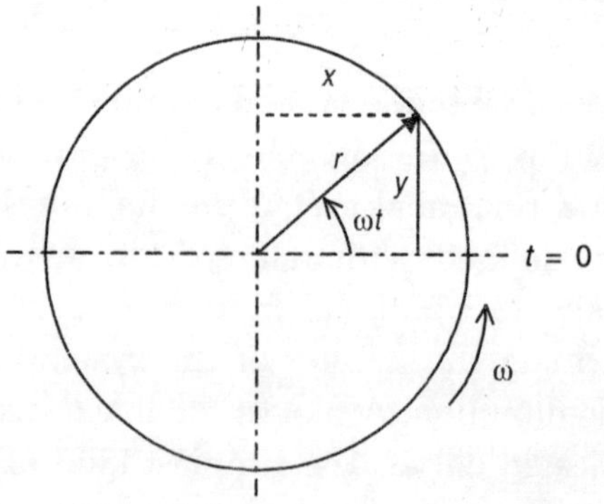

Fig. 1.1. *Horizontal Projection of a Rotating Vector*

1.2 VECTOR METHOD OF REPRESENTING DYNAMIC MOTION

The motion of a vibrating particle can be conveniently represented by means of a rotating vector, r (Fig. 1.1).

When time is reckoned from the horizontal position of a vector rotating at a uniform angular velocity, the horizontal projection of the vector is $x = r \cos(\omega t)$, and the vertical projection will be $y = r \sin(\omega t)$. Either projection can be taken to represent harmonic motion. This representation gives rise to the name circular frequency for ω.

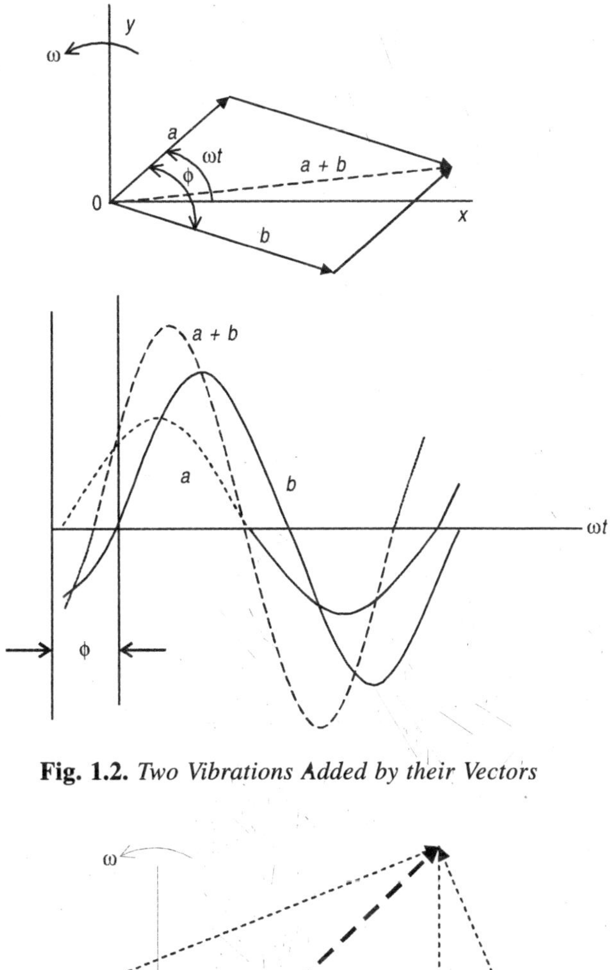

Fig. 1.2. *Two Vibrations Added by their Vectors*

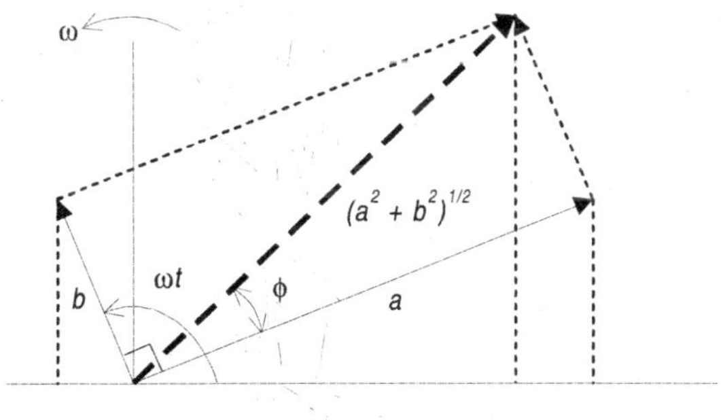

Fig. 1.3. *Addition of Sine and Cosine Waves*

This method of representing reciprocal motion is convenient. If a point is simultaneously subjected to two motions of the same frequency which differ by the phase angle ϕ, namely

a cos (ωt) and b cos (ωt - ϕ), the geometric sum of the two vectors gives the total motion (fig. 1.2). Parallelogram a, b rotates in a counterclockwise direction with uniform angular velocity ω, and the horizontal projections of all the vectors represent displacement as a function of time. The horizontal axis in the upper diagram represents a specific instant in time for which the vector diagram is drawn. Displacement of the sum is the sum of ordinates for a and b.

Addition of the two vectors is permissible only if vibrations are of the same frequency. Motions a sin (ωt) and a sin ($2\omega t$) can be represented by two vectors, the first rotating at an angular speed ω, the second at twice that speed. The relative position of these two vectors is changing continuously, thus a geometric addition of the vectors has no physical implication. A special case of vector addition often encountered is the addition of sine and cosine waves of different amplitudes, a sin (ωt) and b cos (ωt). The two vectors are perpendicular (fig. 1.3), so $a \sin(\omega t) + b \cos(\omega t) = [(a^2 + b^2)^{1/2}] \times \sin(\omega t + \phi)$, where $\tan \phi = b/a$. As an example, consider the sum of two motions 10sin ($25t$) and 20sin ($25t$ + 1). At time $t = 0$, vector a can be vertically up, with no horizontal projection. The second vector b is turned 1 radian (or $180/\pi$ degrees) in a counterclockwise direction from the first vector. The magnitude of vector sum is $\{a^2 + b^2 + 2ab \cos(180/\pi)\}^{1/2} = 26.76$ units long.

1.3 COMPLEX NUMBERS METHOD

The vector method offers visualization of harmonic motion, which is simpler than the consideration of the sine wave by itself. For numerical calculations, however, it is not well adapted. Complex numbers employing imaginary numbers offer a simpler method. A complex number can be represented graphically by a point in a plane where the real component is plotted horizontally and the imaginary part along the vertical axis. Using $j = (-1)^{1/2}$, imaginary numbers are j, $2j$, $3j$, etc. By joining that point with the origin, the complex number can be made to represent a vector (fig. 1.4). If the angle of the vector is α with the horizontal axis, and of length a, then the vector is $a(\cos \alpha + j \sin \alpha)$. Since a harmonic form of motion can be represented by a rotating vector, the variable angle ωt may be substituted for fixed angle α to obtain a rotating vector. Its horizontal projection will then be an alternating motion. The sum of the two motions considered earlier by the vector method can be solved using complex numbers. The first vector is $+10j$ and the second vector is $+20j \cos(180/\pi) - 20 \sin(180/\pi) = +10.8j - 16.8$. The sum of the two vectors is $-16.8 + 20.8j$, representing a vector of length $(16.8^2 + 20.8^2)^{1/2} = 26.76$ units.

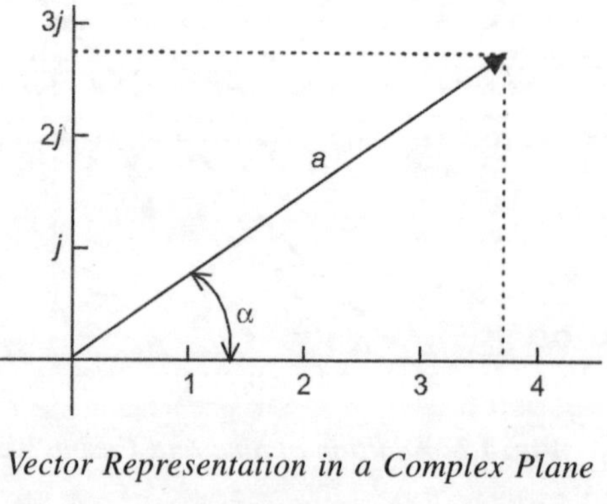

Fig. 1.4. *Vector Representation in a Complex Plane*

Differentiating $a\{\cos(\omega t) + j \sin(\omega t)\}$ gives $j\omega a\{\cos(\omega t) + j \sin(\omega t)\}$. Note that by definition, $j.j = -1.0$. Thus, differentiation of a complex number is equivalent to multiplication by $j\omega$.

In vector representation differentiation multiplies the length of the vector by ω and turns it ahead by 90°. Thus, multiplying a complex number by j is equivalent to moving the vector by 90° without changing its magnitude. In making extended calculations with complex numbers the ordinary rules of algebra are followed. Algebraic calculations are performed without much recourse to their physical meaning, only the final answer is interpreted using the real part for the vector magnitude and phase angle from both the real and imaginary parts.

A concept of importance in many applications is that of work done by a harmonically varying force on a harmonically varying motion of the same frequency. Consider a force $P = P_0 \sin(\omega t + \varphi)$ acting upon a body for which the motion is given by $x = x_0 \sin(\omega t)$. The work done by the force during a small displacement dx is Pdx, which may be written as $Pdt(dx/dt)$. During one cycle of vibration ωt varies from 0 to $2\pi/\omega$. After integration between limits 0 and 2π, the work per cycle is

$$W = \pi P_0 x_0 \sin(\varphi).$$

In order to understand this expression for work a graphical approach is useful. In fig. 1.5 the ordinates are the displacement x and the in-phase component of the force (*ref*: den Hartog, 1984). Between A and B the force is positive, say upward, and the body is moving in the same direction, so the work done is positive. Between B and C, on the other hand, the body moves downward toward the equilibrium point while the force is still positive, so that negative work is done. The work between A and B cancels that between B and C, and over a whole cycle the work done is zero. Thus, if a harmonic force acts on a body subjected to a harmonic motion of the same frequency, the component of the force in phase with the displacement does no work. Since the velocity vector is 90° ahead of displacement, it follows that force does work only with that component which is in phase with the velocity.

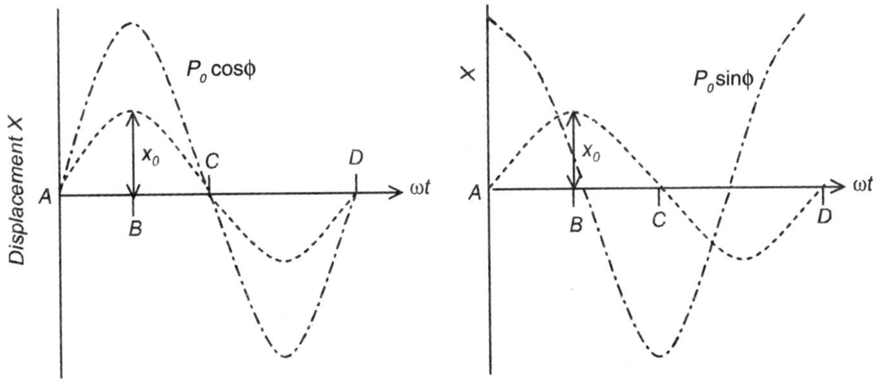

Fig. 1.5. *Work Done on Harmonic Motion: In Phase* (left) *and 90° Out of Phase* (right)

1.4 FREE VIBRATIONS WITHOUT DAMPING

Consider a mass m suspended by a spring from a fixed location, as shown in fig. 1.6. Stiffness of the spring is defined by k, the force required to extend it by 1 unit. A damper is also placed between the mass and the fixed wall. The damper does not transmit force to the mass when at rest, but as soon as it moves it exerts a damping force of cdx/dt. It is proportional to the velocity and directed opposite to the mass. c is the coefficient of viscous damping. An external alternating force $P_0 \sin \omega t$ is also acting on the mass. The problem consists in determining the motion of mass m due to this external force. The equation of motion can be obtained by applying Newtons second law of motion, which states that force is equal to mass multiplied by acceleration. All forces acting upward will be considered negative, and positive when acting downward. Spring force has the magnitude $-kx$, and is negative since it pulls upward. The damping force acting on the mass is also negative, $-c dx/dt$. The downward forces acting on the mass may then be expressed by:

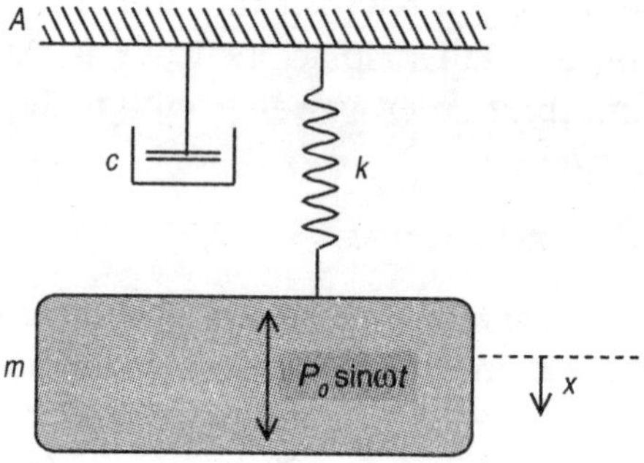

Fig. 1.6. *Single Degree of Freedom Vibrating System*

$$-kx - c\frac{dx}{dt} + P_0 \sin \omega t \tag{1.2}$$

Newton's law gives:

$$m\, d^2x/dt^2 + c dx/dt + kx = P_0 \times \sin \omega t \tag{1.3}$$

This equation is known as the differential equation of motion of a single degree of freedom system. Before developing a solution of the general equation, simplified cases will be considered first. If there is no external applied force and no damping, the equation reduces to:

$$m\, d^2x/dt^2 + kx = 0 \tag{1.4}$$

The most general solution of this equation is:

$$x = C_1 \sin \sqrt{\frac{k}{m}} t + C_2 \cos \sqrt{\frac{k}{m}} t \tag{1.5}$$

where C_1 and C_2 are arbitrary constants. Physically, the result as it stands is indefinite, since the constants can take any value. This is because the problem itself was not fully defined. If it is specified that the mass is pulled out of its equilibrium position to $x = x_0$, then released without initial velocity, two conditions are specified:

At $\quad t = 0: \quad x = x_0 \quad$ and $\quad dx/dt = 0.$

The first condition gives $C_2 = x_0$. For the second condition, (1.5) must be differentiated, giving $C_1 = 0$. Substitution into (1.5) leads to the specific solution:

$$x = x_0 \cos \sqrt{\frac{k}{m}} t \tag{1.6}$$

This represents an undamped vibration. Denoting the period of vibration by T,

$$T = 2\pi\sqrt{\frac{m}{k}} \tag{1.7}$$

Customarily, $(k/m)^{1/2}$ is denoted by ω_n, called the natural circular frequency. The natural frequency f_n is given by:

$$f_n = \frac{1}{T} = \frac{1}{2\pi}\sqrt{\frac{k}{m}} = \frac{\omega_n}{2\pi} \tag{1.8}$$

measured in cycles per second.

In the derivation of the differential equation of motion the effect of gravity has been neglected. The amplitude of vibration is measured from the position where the downward force mg is held in equilibrium by an upward force $k\delta$, δ being the extension of the spring due to gravity. If x_1 is measured from the unstressed position of the spring, then $x_1 = x + \delta$. In the differential equation of motion, then, x must be replaced by x_1, and on the right hand side of the equation a force mg must be added.

If it is assumed that the motion is harmonic, the frequency can be calculated from an energy consideration. In the center of the swing the mass has maximum kinetic energy, while at either extreme position it has no kinetic energy. At the same time the spring is stretched (or compressed) to the fullest at the extreme positions, and has elastic, or potential, energy stored in it. Between the middle and extreme positions the spring-mass system has both kinetic and potential energy, the sum of which has to be constant since there is no external work done on the system. Thus, the kinetic energy at the center position must equal elastic energy at the extreme position. For a linear spring the potential energy when stretched over a distance x is $\frac{1}{2}kx^2$. The kinetic energy at any instant is $\frac{1}{2}mv^2$. If the motion is assumed to be $x = x_0 \sin(\omega t)$, then velocity $v = x_0\omega \cos(\omega t)$. At the extreme position the potential energy is $\frac{1}{2}kx_0^2$, and the kinetic energy at the center position where the velocity is maximum is $\frac{1}{2}m\omega^2x_0^2$. Equating the two energies, $\frac{1}{2}kx_0^2 = \frac{1}{2}m\omega^2x_0^2$, from which $\omega^2 = k/m$, independent of the amplitude x_0. This method is of significance when dealing with problems of greater complexity, where the generalized energy method, known as the method of Rayleigh, leads to a result.

In still another variation of fig. 1.6, instead of the sinusoidal force acting on the mass, the upper end of the spring is made to move up and down with an amplitude a_0, the motion of A being $a_0 \sin(\omega t)$. It can be shown that this motion of the top of the spring is equivalent to a force on the suspended mass. If x is the downward displacement of the mass, spring extension will be $x - a_0 \sin(\omega t)$, spring force is $-k\{x - a_0 \sin(\omega t)\}$ and damper force is $-c\{dx/dt - a_0\omega \cos(\omega t)\}$. Newton's law then gives the following expression:

$$m\,d^2x/dt^2 + c\,dx/dt + kx = ka_0 \times \sin \omega t + ca_0\omega \times \cos \omega t \tag{1.9}$$

As mentioned earlier in section 1.2, the sum of a sine and cosine waves of the same frequency is also harmonic, so:

$$m\,d^2x/dt^2 + c\,dx/dt + kx = [(ka_0)^2 + (ca_0\omega)^2]^{1/2} \times \sin(\omega t + \varphi) \tag{1.10}$$

Thus, the motion of the top of the spring with amplitude a_0 is equivalent to a force of amplitude $\{(ka_0)^2 + (ca_0\omega)^2\}^{1/2}$ and phase $\varphi = \tan^{-1} c\omega/k$. Terms ka_0 and $ca_0\omega$ are the maximum values of the spring and damper forces, while the expression inside the parenthesis provides the peak magnitude of the force when the mass is clamped.

1.5 TORSIONAL PENDULUM

Figure 1.7 shows a disk of moment of inertia J attached to a shaft of torsional stiffness k, defined as the torque necessary to produce one radian twist at the disk. The twisting motion of the disk is under the influence of an externally applied torque $T_0 \sin(\omega t)$. Since only the torsional displacement φ is required to describe its position from the equilibrium position, this is a single degree of freedom problem (*ref*: den Hartog, 1984). For a rotating body, Newton's law states that torque is equal to moment of inertia times angular acceleration, or:

$$T = J \times d^2\varphi/dt^2 \tag{1.11}$$

The torques acting on the disk are the spring torque, damping torque and external torque. The spring torque $-k\varphi$, where φ is in radians, is negative, just as the spring force in the previous case was $-kx$. The stiffness k will depend on the shaft cross-section geometry and its material shear modulus. The damping torque is $-c\, d\varphi/dt$, caused by the shaft material's internal damping. The damping constant c results in a torque on the disk caused by an angular speed of rotation of 1 radian per second. Application of Newton's law yields the differential equation of motion:

$$J \times d^2\varphi/dt^2 + c \times d\varphi/dt + k \times \varphi = T_0 \times \sin \omega t \tag{1.12}$$

This equation is similar to the translational case, and the solution for the undamped free condition can be readily obtained by substituting J for m, T_0 for P and φ for x in the linear case.

Any system with inertia, elasticity and damping proportional to the velocity, where the displacement can be described by a single quantity, can be defined in this manner. Consider two disks of moment of inertia J_1 and J_2 joined by a shaft of torsional stiffness k, fig. 1.8. On the first disk torque $T_0 \sin \omega t$ is acting. Both disks can assume an angular position independent of the other by twisting the shaft. However, if the quantity of interest is angle of twist in the shaft, it is possible to express the motion in terms of this quantity alone. Let φ_1 and φ_2 be the angular displacements of the disks, then $\varphi_1 - \varphi_2$ is the shaft twist, $k(\varphi_1 - \varphi_2)$ is the spring torque, $c\, d(\varphi_1 - \varphi_2)/dt$ is the damping torque. Applying Newton's law to the first disk, then the equation of motion is:

$$T_0 \times \sin \omega t = J_1 \times d^2\varphi_1/dt^2 + c \times (d\varphi_1/dt - d\varphi_2/dt) + k \times (\varphi_1 - \varphi_2) \tag{1.13}$$

and to the second disk,

$$0 = J_2 \times d^2\varphi_2/dt^2 + c \times (d\varphi_2/dt - d\varphi_1/dt) + k \times (\varphi_2 - \varphi_1) \tag{1.14}$$

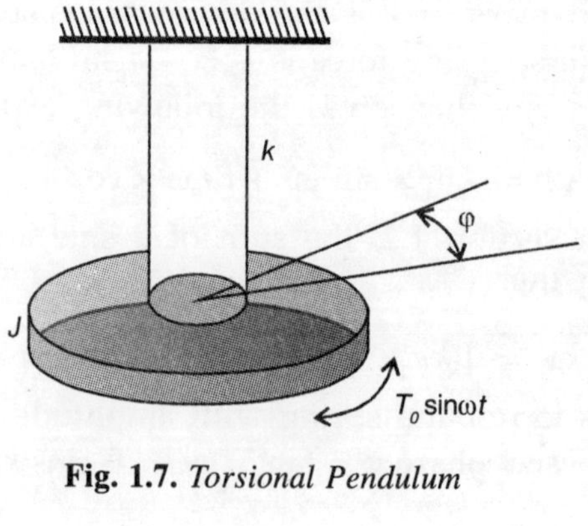

Fig. 1.7. *Torsional Pendulum*

Rearranging the variables and subtracting the results from the two equations yields:

$$(T_0/J_1)\sin\omega t = (d^2\varphi_1/dt^2 - d^2\varphi_2/dt^2) + c(1/J_1 + 1/J_2)(d\varphi_1/dt - d\varphi_2/dt) + k(1/J_1 + 1/J_2)(\varphi_1 - \varphi_2) \quad (1.15)$$

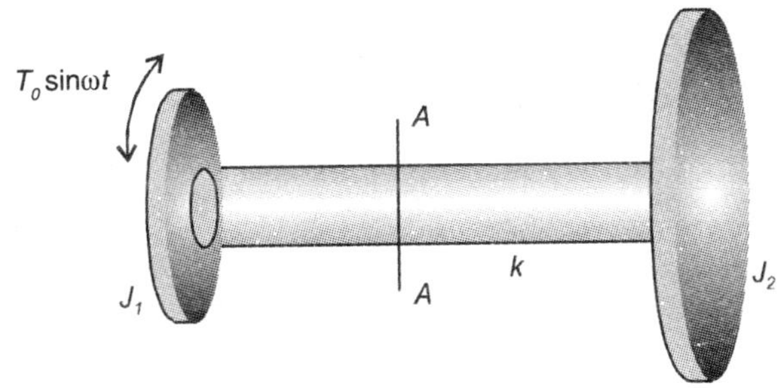

Fig. 1.8. *Two Disks Mounted on an Elastic Shaft*

Defining the angle of twist $(\varphi_1 - \varphi_2) = \psi$, then multiplying the equation by the expression $J_1J_2/(J_1 + J_2)$ gives the resulting equation of motion with a single twist angle variable ψ:

$$J_1J_2/(J_1 + J_2)(d^2\psi/dt^2) + c(d\psi/dt) + k\psi = \{T_0J_2/(J_1 + J_2)\}\sin\omega t \quad (1.16)$$

The solution of this equation gives information on the angle of twist, but angular rotation of individual disks cannot be obtained by this method. A variation is shown in fig. 1.9 where a gear and pinion system is attached to the shaft. Assume gears G and P to be without any inertia. Also assuming the gear teeth to be rigid, the torsional flexibility is then limited to the two shafts. Gear ratio is n. The differential equations of motion for this system could be derived from Newton's law, but the system could be reduced to that of fig. 1.8 by omitting the gears and replacing J_2, k_2 and ψ by equivalent parameters. If J_2 is clamped and torque T_0 is applied to J_1, then J_1 will rotate through an angle φ_0, so $k = T_0/\varphi_0$. Because of the gears the torque in the shaft k_2 is T_0/n, and the angle of twist of k_2 is T_0/nk_2. Since J_2 is fixed, this is the angle of rotation of the pinion P. Angle of the gear G is n times smaller, or T_0/n^2k_2. Adding angle T_0/k_1 for shaft stiffness k_1 gives the angular rotation of J_1. Thus, the equivalent shaft stiffness k may be obtained from the following expression:

$$\frac{1}{k} = \frac{\phi}{T_0} = \frac{1}{k_1} + \frac{1}{n^2k_2} \quad (1.17)$$

Now consider the inertia. Angular acceleration α in k_1 and G becomes $n\alpha$ in k_2 and P. Hence the torque in k_2 is $n\alpha J_2$. This is also the torque at the pinion P. Gear G sees it n times larger, so the torque at A is $n^2\alpha J_2$ and the equivalent of J_2 in the gearless system is n^2J_2. In general, then, a geared system such as that of fig. 1.9 can be reduced to an equivalent non-geared system (fig. 1.8) in the following manner.

Divide the system into separate parts each of which has the same speed within itself. Figure 1.9 has only two parts, but there may be several. Choose one of these parts as the primary and assign numbers n to each of the other parts so that n is the speed ratio with respect to the primary. $n > 1$ for speeds higher than the primary part's speed, while the speed ratio of the primary part is unity. Next, remove all gears and multiply all spring constants k and inertias J by the factors n^2. The differential equation of the reduced gearless system is then the same as that of the original geared design.

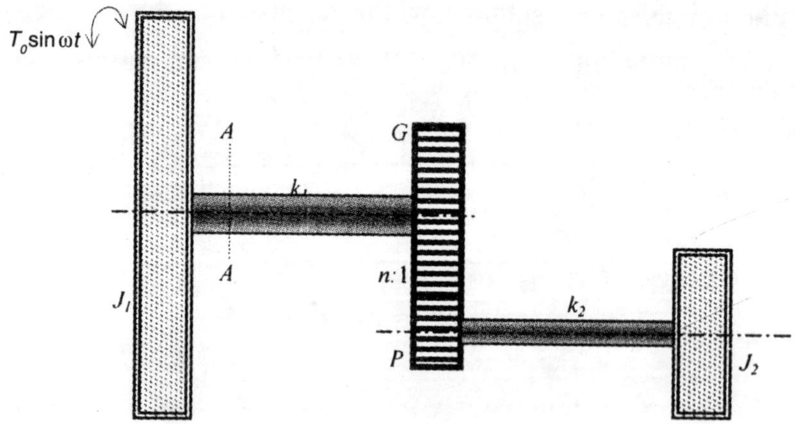

Fig. 1.9. *Geared System*

1.6 FREE VIBRATIONS WITH VISCOUS DAMPING

Undamped free vibrations persist indefinitely, which cannot happen; all free vibrations eliminate with time. The term viscous damping is associated with the expression cdx/dt since it represents the conditions of damping due to the viscosity of oil in a dashpot. Other types of damping exist, but their solution cannot be determined with as much ease. Consider the function $x = e^{st}$, where t is the time and s an unknown constant. Upon differentiation the same function multiplied by a constant results. This function when substituted in the equation:

$$m(d^2x/dt^2) + c(dx/dt) + kx = 0 \tag{1.18}$$

permits division by e^{st} and leads to an algebraic equation instead of a differential equation. Thus, assuming e^{st} as the solution, then the above equation becomes:

$$(ms^2 + cs + k).e^{st} = 0 \tag{1.19}$$

This is a quadratic in s with two values:

$$s_{1,2} = -\frac{c}{2m} \pm \sqrt{\left(\frac{c}{2m}\right)^2 - \frac{k}{m}} \tag{1.20}$$

The most general solution of the problem is:

$$x = C_1 e^{s_1 t} + C_2 e^{s_2 t} \tag{1.21}$$

where C_1 and C_2 are arbitrary constants. The physical significance of this equation will depend on whether the expressions for s are real or complex. When $(c/2m)^2 > k/m$, both values of s will be real. Also, they are both negative since the square root is smaller than the first term $c/2m$. Thus (1.21) describes a solution consisting of the sum of two decreasing exponentials, as shown in fig. 1.10. The case of $C_1 = 1$ and $C_2 = -2$ is shown as a dashed line.

This fig. shows that the motion is not oscillatory but rather a creeping back to the original position. This is due to the fact that when $(c/2m)^2 > k/m$ the damping c is large. For smaller values of c pertaining to more practical cases, (1.20) gives complex values for s. The damping c at which this transition occurs is called critical damping c_c:

$$c_c = 2m\sqrt{\frac{k}{m}} = 2\sqrt{mk} = 2m\omega_n \tag{1.22}$$

If the damping is less than critical damping, the solution consists of two factors, the first a decreasing exponential and the second a sine wave. The combined result is an exponentially decreasing sine wave, lying in the space between the exponential curve on both sides of the phase angle axis (fig. 1.11). The smaller the damping constant c, the flatter will be the exponential curve and the more cycles it will take for vibrations to be eliminated.

The rate of this dying down of the vibrations is of interest, and can be calculated by considering two consecutive maximum values of the curve, A-B, B-C, and so on. During the time interval between any two such peak values the amplitude of vibration diminishes from:

$$e^{-\frac{c}{2m}t} \text{ to } e^{-\frac{c}{2m}\left\{t+\frac{1\pi}{q}\right\}} \text{ where } q = \sqrt{\frac{k}{m} - \frac{c^2}{4m^2}} \tag{1.23}$$

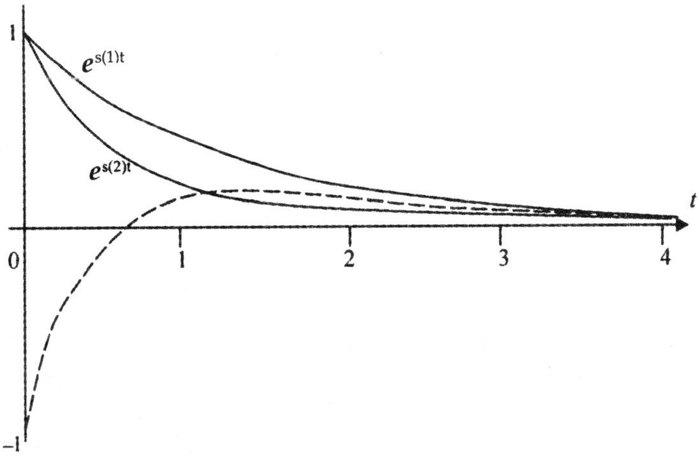

Fig. 1.10. *Motion of Single Degree of Freedom System with* $c/c_c > 1$

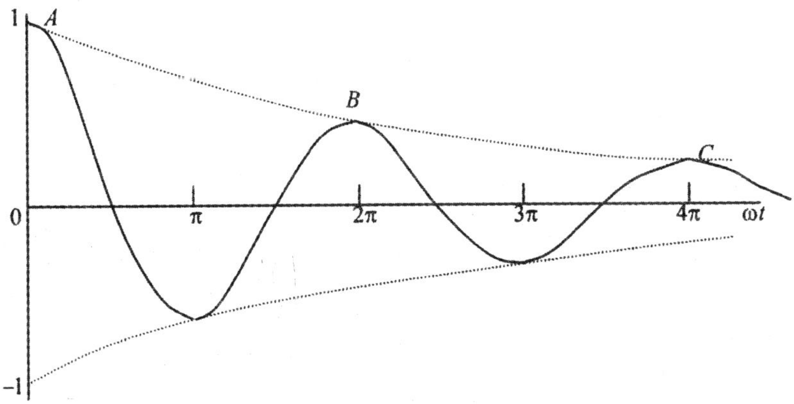

Fig. 1.11. *Free Vibration of System with* $c/c_c < 1$

The latter of these two expressions is equal to the first one multiplied by the constant factor:

$$e^{\frac{\pi c}{mq}} \tag{1.24}$$

which factor is smaller than unity. This factor is the same for any two consecutive maxima, and is independent of the amplitude or of time. Thus, the amplitudes decrease in a geometric series.

The frequency of vibration diminishes with increasing damping. If written in a dimensionless form with the help of (1.23), it becomes:

$$\frac{q}{\omega_n} = \sqrt{1-\left\{\frac{c}{c_c}\right\}^2} \tag{1.25}$$

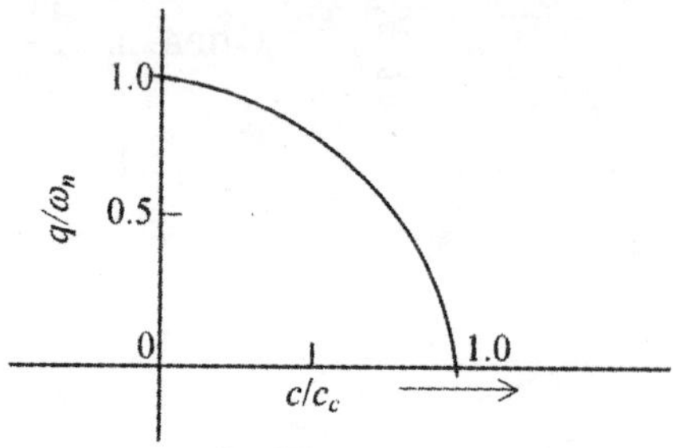

Fig. 1.12. *Natural Frequency As Function of Damping*

This equation is plotted in fig. 1.12, where the ordinate q/ω_n is the ratio of the damped to the undamped natural frequency, and the abscissa is the ratio of the actual to the critical damping. The fig. is a circle; for critical damping the natural frequency is zero. Due to the horizontal tangent of the circle at $c = 0$, the natural frequency is nearly constant and equal to $\sqrt{(k/m)}$ for all typically encountered values of damping, $c/c_c < 0.2$.

1.7 FORCED VIBRATIONS WITHOUT DAMPING

Another particular case of (1.3) is the one where the damping term is made zero:

$$m(d^2x/dt^2) + kx = P_0 \sin \omega t \tag{1.26}$$

Substituting a harmonic solution of the type $x = x_0 \sin \omega t$ into this equation yields:

$$-mw^2 x_o \sin \omega t + kx_o \sin \omega t = P_0 \sin \omega t \tag{1.27}$$

producing the result:

$$x = \frac{P_0/k}{1-(\omega/\omega_n)^2} \sin \omega t \tag{1.28}$$

The expression P_0/k has the same physical significance of static deflection of the spring under a constant load P_0. Thus, if $P_0/k = x_{st}$, the solution becomes:

$$\frac{x}{x_{st}} = \frac{1}{1-(\omega/\omega_n)^2} \sin(\omega t) \tag{1.29}$$

Including the homogeneous solution containing the two integration constants, the total solution of the problem becomes:

$$x = C_1 \sin \omega_n t + C_2 \cos \omega_n t + \frac{x_{st}}{1-(\omega/\omega n)^2} \sin \omega t \tag{1.30}$$

The first two terms represent the undamped free vibrations, the third term is the undamped forced vibration. Examining the implications of this result, x/x_{st} is a sine wave with an amplitude of $1/\{1 - (\omega/\omega_n)^2\}$. For values of $\omega/\omega_n > 1$ the amplitudes are negative, and of $\omega/\omega_n < 1$ the amplitudes are positive. When amplitudes of motion are negative, it only means that the amplitudes are positive but with a phase angle lag of 180°.

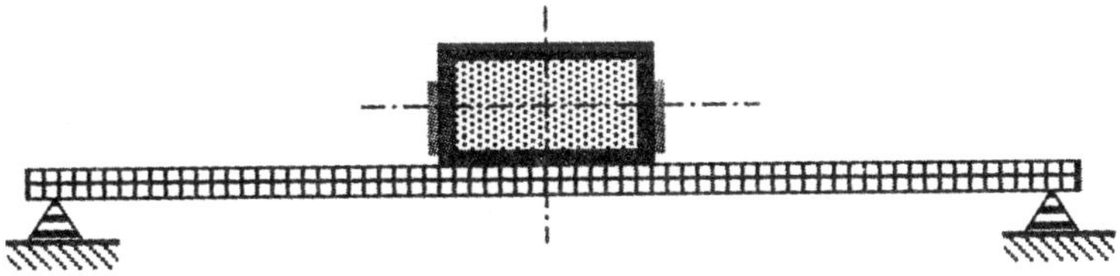

Fig. 1.13. *Unbalanced Rotating Machine on a Beam*

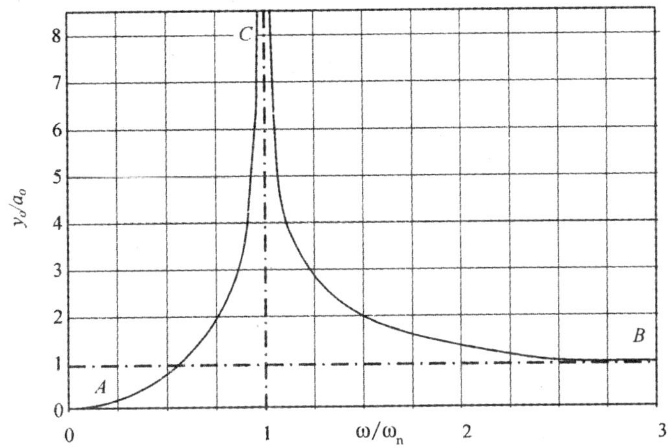

Fig. 1.14. *Resonance Diagram for Rotating Machine on a Beam Example*

It also means that for $\omega/\omega_n < 1$ the force and motion are in phase, and they are out of phase for $\omega/\omega_n > 1$. For $\omega/\omega_n < 1$ the mass is below the equilibrium position when the force pushes downward; the mass is above the equilibrium position when the force pushes downward for $\omega/\omega_n > 1$. At $\omega/\omega_n = 1$ the forced frequency coincides with the natural frequency, so the force is pushing the mass at the right time in the right direction. The amplitude then increases indefinitely. In the case of a pendulum, it is pushed slightly in the direction of motion every time it reaches the end of a swing; a small force is required to produce large amplitude. This phenomenon is known as 'resonance', and the natural frequency is called the 'resonant frequency'.

An important application of this theory is when the amplitude of the impressed force varies with ω^2 (*ref*: den Hartog, 1984). Figure 1.13 shows a beam on two supports carrying a rotating machine with an unbalanced centrifugal force of $m_1\omega^2 r$, where m_1 is the unbalance mass and r is its distance from the centerline of the shaft. This force can be resolved into a vertical component $m_1\omega^2 r$ sin (ωt) and a horizontal component $m_1\omega^2 r$ cos (ωt). If it is assumed that the beam is rigid for displacements in the horizontal direction and flexible for displacement y in the vertical direction, then it becomes a single degree of freedom system. Here the motor mass m and beam stiffness $k = 48EI/l^3$ are acted upon by the frequency dependent vertical disturbing force of amplitude $m_1\omega^2 r$. Considering the motion of the motor and beam system, it is equivalent to motor mass inertia force of $m\omega^2 a_0$ acting on the beam, where the support is moving with amplitude a_0, and the force is absent. The solution for this problem can be found directly from (1.28) by substituting $m\omega^2 a_0$ for P_0. Then displacement amplitude:

$$y_0 = \frac{m\omega^2 a_0 / k}{1-(\omega/\omega_n)^2} = \frac{(\omega/\omega_n)^2}{1-(\omega/\omega_n)^2}$$

or,

$$\frac{y_0}{a_0} = \frac{(\omega/\omega_n)^2}{1-(\omega/\omega_n)^2} \tag{1.31}$$

If y_0/a_0 is plotted as a function of ω/ω_n, the result is shown in fig. 1.14. At C there is resonance, and so spring extension is infinitely large in the absence of damping. Since this result is not in agreement with observations, it is necessary to consider damping. It shows the relative motion of a system in which the end of the spring is subjected to an alternating motion of constant amplitude a_0. The plot also shows the absolute motion of a system in which the mass experiences a force of variable amplitude $m\omega^2 a_0$. The ordinates of the three points A, B and C representing (1.31) can be physically explained as follows. At A the excitation frequency ω is nearly zero, so the top of the spring moves at a slow rate, the mass follows this motion and the spring does not stretch; hence $y_0 = 0$. At B the motion of the top of the spring is very rapid, so the mass cannot follow and stands still in space.

1.8 FORCED VIBRATIONS WITH VISCOUS DAMPING

The complete equation of motion is:

$$m(d^2x/dt^2) + c(dx/dt) + kx = P_0 \sin(\omega t) \tag{1.32}$$

The full solution consists of the complete solution of the equation with the right hand side set to zero and a particular solution of the whole equation (1.32), or:

$$x = e^{-\frac{c}{2m}t}\{C_1 \sin(qt) + C_2 \cos(qt)\} + \textit{particular solution} \tag{1.33}$$

Thus, only the particular solution needs to be determined. Assuming a solution of the type $x = x_0 \sin(\omega t - \varphi)$, the particular solution can be determined by finding x_0 and φ. Newton's law requires that the sum of all the forces be zero at all times, implying that the sum of the vertical and horizontal components must also be zero. From these conditions the unknowns x_0 and φ are solved:

$$x_0 = \frac{P_0/k}{\sqrt{\left\{1-\frac{\omega^2}{\omega_n^2}\right\}^2 + \left\{2\frac{c}{c_c}\frac{\omega}{\omega_n}\right\}^2}} \tag{1.34}$$

$$\tan(\varphi) = \frac{2\frac{c}{c_c}\frac{\omega}{\omega_n}}{1-\left(\omega^2/\omega_{n^2}\right)} \tag{1.35}$$

The expressions for the amplitude and the phase angles are in terms of ratios of frequency and damping, where c_c is the critical damping. P_0/k may be interpreted as the static deflection of the spring due to P_0. These relations are plotted in fig. 1.15. The amplitude diagram contains a family of curves, one for each value of damping. All curves lie below the one for zero damping, since the amplitude of forced vibration is reduced by the damping. The maximum value of the different curves do not occur any more at $\omega/\omega_n = 1$ but at a smaller frequency. In fact, three different frequencies need to be distinguished, all of which coincide for $c = 0$:

- $\omega_n = \sqrt{(k/m)}$ = undamped natural frequency
- $q = \sqrt{\frac{k}{m} - \left\{\frac{c}{2m}\right\}^2}$ = damped natural frequency
- The frequency of maximum forced amplitude, or the resonant frequency.

The phase angle diagram is also of interest. For no damping it was seen that the force and displacement are in phase below resonance and 180° out of phase above resonance. So the phase angle shows a discontinuous jump at the resonance point. For damping values different from zero the other curves in the phase angle diagram are plotted. In general, damping tends to smooth out the sharpness of the undamped diagrams for both amplitude and phase.

Energy relations involved in this process also give a deeper understanding. For very slow motions $\varphi = 0$, and little work is done over a whole cycle, or no mechanical energy is converted into heat. Starting from the equilibrium position, the external force moves through a certain distance before reaching the extreme position, when it does work. But that work is used to store potential or elastic energy into the spring. During the next quarter cycle the motion goes against the external force and the spring gives up its stored energy. At slow speeds, then, the work of the external force is transferred into elastic energy and nothing is converted into heat. At resonant frequency $\varphi = 90°$, and work dissipated per cycle is $\pi P_0 x_0$. The external force is equal and opposite to the damping force in that case, so energy is dissipated by the damper. The spring force and the inertia force are equal and opposite, and also in phase with displacement. Each of these forces individually perform work during a quarter cycle, but store the energy to return it during the next quarter cycle. The work is stored periodically as elastic energy in the spring and as kinetic energy in the mass.

Returning to the assertion pertaining to the particular solution, the general solution consists of the damped free vibration superposed on the forced vibration. After a short time, damped free vibration disappears and the forced vibration alone persists. For this reason the forced vibration may also be referred to as sustained vibration and the free vibration becomes transient vibration. The values of the constants C_1 and C_2 depend on the conditions at the start and can be calculated from these conditions using the analytical process described earlier. It is possible to describe the whole process by physical reasoning. For example, consider a mass suspended by a spring and acted upon by an external harmonic force of frequency eight times as slow as the natural frequency of the system. The mass is clamped while the external force is acting. Suddenly the clamp is withdrawn. Assume the damping in the system is such that the free vibration decreases by 10 percent for each cycle. Assume also that the clamp is released when the forced vibration has its maximum amplitude. From the initial conditions it follows that at the instant of the release the mass has no deflection and no velocity. The prescribed forced vibration starts with $x = x_0$ and $dx/dt = 0$. These two conditions can only be satisfied by starting a free vibration with $x = -x_0$ and $dx/dt = 0$. Then the combined motion will start at zero with zero velocity. Figure 1.16*a* shows the free vibration, 1.16*b* shows the forced vibration and 1.16*c* shows the combined motion.

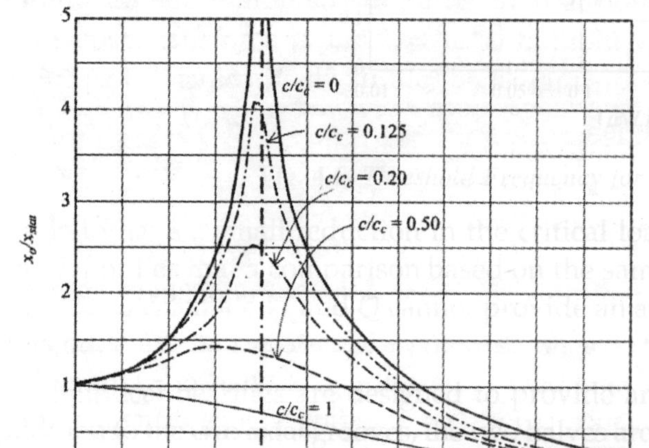

Fig. 1.15. *(a) Amplitudes of Forced Vibrations, (b) Phase Angle between Force and Displacement as a Function of Frequency for Various Damping Values*

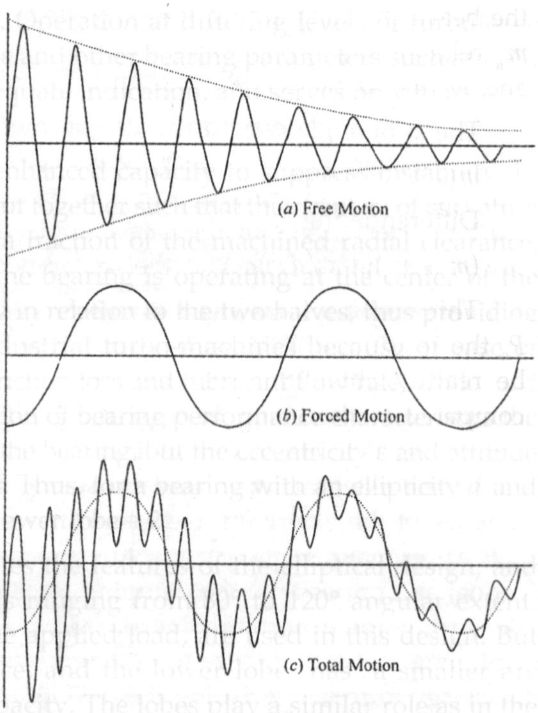

Fig. 1.16. *Starting Transient Motion of a Forced Vibration*

Note that the transient vibration disappears quickly and that the maximum amplitude at the start is nearly twice as great as the sustained final amplitude. If the difference between free and forced frequencies is small and damping is also small, the free and the forced vibration vectors would retain nearly the same relative motion during one revolution, the included angle between them changing only slightly. The vectors can then be added geometrically, and during one revolution of the two vectors the motion will practically be a sine wave of frequency $\omega_1 \approx \omega_2$ and amplitude the arithmetic sum of the two amplitudes. After a large number of cycles, the relative position of the vectors varies, and the magnitude of the sum vectors also changes. The resulting motion can then be described as an approximate sine wave with frequency ω_1 and an amplitude varying slowly between the sum and the difference of the amplitudes of the free and forced vibrations. This phenomenon is known as beats. Note that the free vibrations will be eliminated after a while due to energy dissipation from the damping. In order to have sustained beats it is necessary to have two sustained, or forced, vibrations.

1.9 VIBRATION DUE TO ROTATING ECCENTRIC WEIGHT

A classic example of forced vibrations with viscous damping is that of a rotating machine of mass m with a rotating eccentric mass m_u mounted in a shaft and bearings, as shown in fig. 1.17. The rotating eccentric mass m_u follows a circular path of radius e with respect to the bearings. Considering only the motion in the vertical direction, the displacement of mass m_u relative to m is $x_2 - x_1 = e\ \sin(\omega t)$, where x_2 and x_1 are the absolute displacements of m_u and m and ω the angular speed of the machine's rotating system (*ref:* Crede *et al.*, 1988).

The differential equation of motion of the system is:

$$m(d^2x_1/dt^2) + m_u(d^2x_2/dt^2) + c(dx_1/dt) + kx_1 = 0 \tag{1.36}$$

Differentiating $(x_2 - x_1)$ with respect to time and substituting in (1.36),

$$(m + m_u)(d^2x_1/dt^2) + c(dx_1/dt) + kx_1 = m_u e\omega^2 \times \sin \omega t \tag{1.37}$$

This equation is of the same form as (1.32): by substituting $(m + m_u)$ for m and $m_u e\omega^2$ for P_0 the resulting displacement, velocity and acceleration of the rotating machine mass m can be readily obtained. Note that the magnitude of eccentric mass m_u will be insignificant compared to the machine mass m.

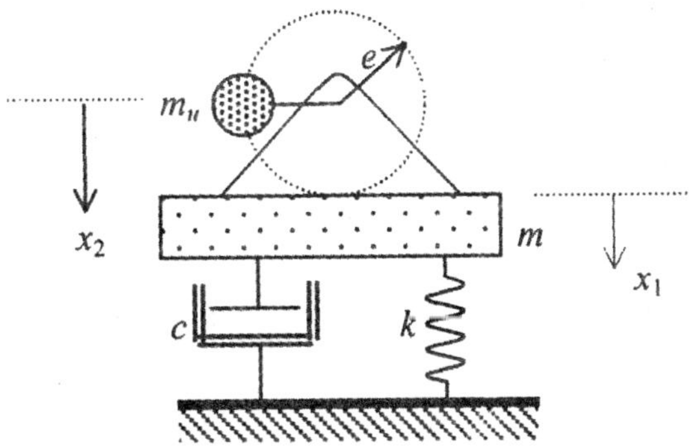

Fig. 1.17. *Forced Vibrations from Rotating Eccentric Weight*

When the machine is mounted on an elastic foundation, the motion of the foundation comes into play. As discussed earlier in section 1.7, displacement forced on an otherwise unforced system is equivalent to a system where a force is applied on that system. Therefore, a rotating machine with an eccentric mass in the rotor mounted elastic foundation behaves as if two distinct forcing functions are applied on it, one due to the eccentric mass whose frequency coincides with the angular speed of the rotor, and the other due to motion from the foundation. The frequency of the foundation motion will depend on the elastic stiffness of the foundation and the mass of the rotating machine. Damping characteristics of the foundation will control the amplitude of motion of the machinery from both sources of excitation, since the foundation will be in the direction of the load path traveling from the rotor to the foundation. Relative motion between the rotor and machine housing will be dependent on the material damping of the shaft and the bearings.

As an example of the above discussion, consider a centrifugal compressor weighing 3,000 lbs mounted on a simply supported elastic beam of negligible weight, of the type shown in fig. 1.13. The elastic beam sags 2.5 in. under the weight of the compressor, and has a damping coefficient of 28 lbs for a velocity of 1 in. per second. The elastic beam is moved up and down at resonant speed with amplitude of 1 in. Assume the center of the compressor to coincide with the center of the beam along the beam's length. The natural frequency of vibration is $(g/\delta_{st})^{1/2} = (386/2.5)^{1/2} = 12.43$ radian/second. At resonance the disturbing force is $\{(ka_0)^2 + (ca_0\omega)^2\}^{1/2}$. Here stiffness of the beam $k = 3000/2.5 = 1200$ lbs/in; $a_0 = 1$ in.; damping $c = 28$ lbs-second/in; and natural frequency $\omega = \omega_n = 12.43$ radians/second. The disturbing force is then equal to 1249.5 lbs. The amplitude of motion of the compressor then is disturbing force divided by the product of c and ω, and is found to be 3.59 in.

1.10 VIBRATION ISOLATION

An unbalanced machine has to be installed in a structure where vibration is undesirable. Prime examples of this situation are gas turbine engines attached to an airplane wing and on board a ship. The problem is one of mounting the machine in such a manner that no vibrations will appear in the structure to which it is attached.

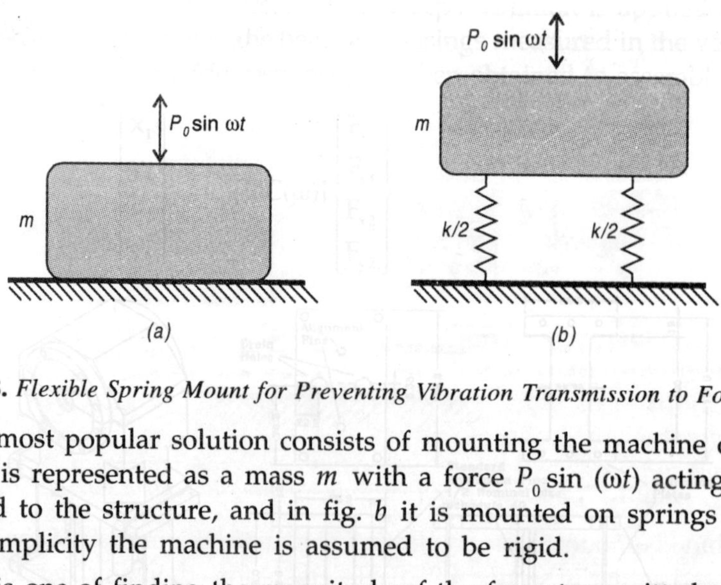

Fig. 1.18. *Flexible Spring Mount for Preventing Vibration Transmission to Foundation*

Probably the most popular solution consists of mounting the machine on springs. In fig. 1.18 the machine is represented as a mass m with a force $P_0 \sin(\omega t)$ acting on it. In fig. *a* it is solidly attached to the structure, and in fig. *b* it is mounted on springs with a combined stiffness k. For simplicity the machine is assumed to be rigid.

The problem is one of finding the magnitude of the force transmitted to the floor by the machine. Since only the springs are in contact with the foundation, the only force transmitted is the spring force, and that has the amplitude kx, damping being neglected here. If x_0 is the displacement of the mass and $x_{st} = P_0/k$ is the static displacement, then:

$$\frac{x_0}{x_{st}} = \frac{x_0}{P_0/k} = \frac{kx_0}{P_0} = \frac{\text{spring force}}{\text{impressed force}} \tag{1.38}$$

$$= \frac{\text{transmitted force}}{\text{impressed force}} = \text{Transmissibility}$$

The ideal is to have this ratio zero, but practical considerations aim to make it small. If the spring constant $k = \infty$, as in fig. *a*, the natural or resonant frequency is infinite. The operating frequency is then small compared to the natural frequency, ω/ω_n is nearly zero and the transmitted force equals the impressed force. Physically this is obvious, since the rigid floor does not permit mass m to move, and all the force P_0 is transmitted to the foundation. It is necessary to design the support springs so as to make the natural frequency of the whole machine low compared with the disturbance frequency. Thus, the springs should be soft (*ref*: Crede *et al.*, 1988).

Other interesting observations can be made from equation (1.29). If ω is smaller than $\omega_n\sqrt{2} = \sqrt{(2k/m)}$, the springs make matters worse, since the transmissibility is greater than unity. If the natural frequency is one-fifth of the disturbing frequency, the transmissibility is 1/24, a good ratio.

So far the support has been considered without damping, which is the case in steel springs. If cork or rubber padding is used for the purpose, then damping is not negligible and must be accounted for. The system is then represented by fig. 1.19.

Now the displacement curve is not directly proportional to the amplitude of the transmissibility curve. The force is made up of spring force kx_0 as well as the damping force $c\omega x_0$. As shown earlier the two forces are out of phase by 90°. Consequently, their vector sum is the total transmitted force.

The amplitude x_0 is given by the following expression:

$$\text{Transmitted force} = P_0 \frac{\sqrt{1+\left\{\frac{c\omega}{k}\right\}^2}}{\sqrt{\left\{1-\frac{\omega^2}{\omega_n^{\,2}}\right\}^2+\left\{2\frac{c\omega}{c_c\omega_n}\right\}^2}} \tag{1.39}$$

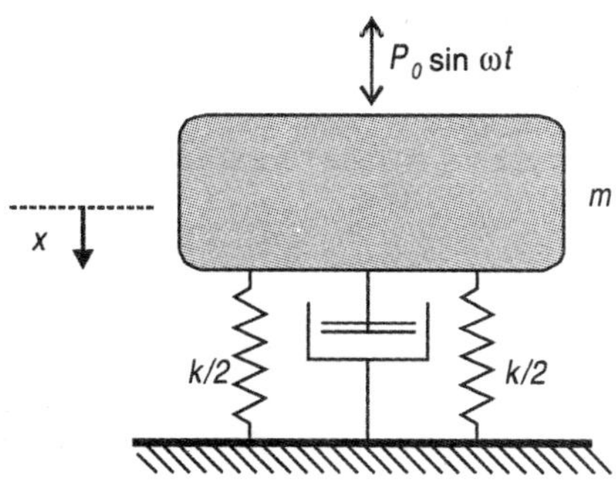

Fig. 1.19. *A Spring Support with Damping*

Transmissibility is defined as the ratio of transmitted force to impressed force. Thus, damping is advantageous only in the region $\omega/\omega_n < 1.41$, where spring mounting makes matters worse. For all values of ω/ω_n where spring mounting helps, the presence of damping makes the transmissibility worse. However, this statement is not as important as it may appear at first sight. For one thing the effect of damping is not great, and can be easily offset by making the springs weaker. Secondly, operation at resonance point $\omega/\omega_n = 1$ is never the intention, although it may sometimes unfortunately occur, and then the presence of damping is highly desirable. Thus, in spite of the dictum, some damping in the springs is generally advantageous.

Practical cases of isolation by means of springs occur in many machines. Their main application lies in apparatus that is inherently unbalanced, or as in the case of most reciprocating machines, develops a non-uniform torque.

1.11 CANTILEVER BEAM ATTACHED TO WHEEL

Consider a beam of irregular cross-section attached to a wheel of radius R rotating at constant speed ω (fig. 1.20). The beam deflects from position PQ to PQ' as a consequence of centrifugal forces acting on it (*ref*: Rao, 1991). Displacement vector QQ' may be written as:

$$\bar{u} = \overline{QQ'} = u_x\bar{e}_x + u_y\bar{e}_y + u_z\bar{e}_z \tag{1.40}$$

where $\bar{e}_x, \bar{e}_y, \bar{e}_z$ represent unit vectors in the fixed system AOB. Coordinates x and y lie in the cross-sectional plane of the beam, and z represents the polar axis. Unit vectors

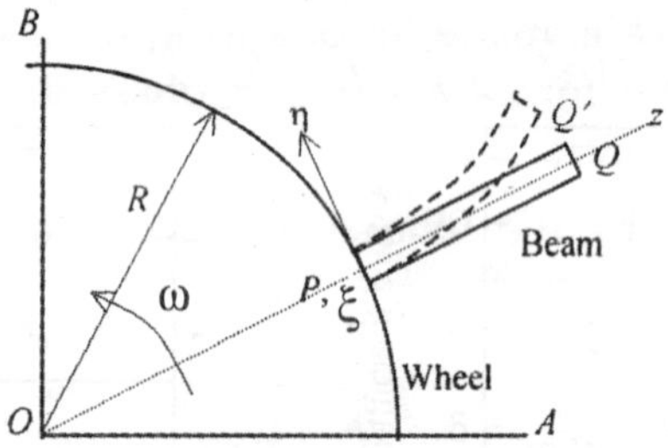

Fig. 1.20. *Cantilever Beam Attached to Rotating Wheel (Rao, 1991)*

$\bar{e}_\xi, \bar{e}_\eta, \bar{e}_z$ are in the rotating coordinate system ξ, η, z. In the rotating frame (1.40) be expressed as:

$$\bar{u} = u_\xi\bar{e}_\xi + u_\eta\bar{e}_\eta + u_z\bar{e}_z \tag{1.41}$$

Position vector between O and Q' is given by the expression:

$$= \overline{OQ} = \overline{OP} + \overline{PQ} + \overline{QQ}$$

$$= r_\xi\bar{e}_\xi + r_\eta\bar{e}_\eta + r_z\bar{e}_z \tag{1.42}$$

Differentiation of this vector with respect to time yields velocity of Q:

$$d\bar{r}/dt = (dr_\xi/dt)\bar{e}_\xi + (d\bar{e}_\xi/dt)r_\xi + (dr_\eta/dt)\bar{e}_\eta + (d\bar{e}_\eta/dt)\bar{r}_\eta + (d\bar{r}_z/dt)\bar{e}_z + (d\bar{e}_z/dt)\bar{r}_z \tag{1.43}$$

Since $d\bar{e}_z/dt = \omega e_\eta, d\bar{e}_\eta/dt = -\omega e_z$ *and* $d\bar{e}_\xi/dt = 0$, (1.43) reduces to:

$$d\bar{r}/dt = (dr_\xi/dt)\bar{e}_\xi + (dr_\eta/dt + \omega r_z)\bar{e}_\eta + (dr_z/dt - wr_\eta)\bar{e}_z \tag{1.44}$$

and velocity squared is given by:

$$v^2 = (dr/dt)^2 = (d r\xi/dt)^2 + (d r\eta/dt)^2 + (dr_z/dt)^2 + 2\omega(r_z dr_\eta/dt - r_\eta\, dr_z/dt) + \omega^2(r_\eta^2 + r_z^2) \quad (1.45)$$

Rotary inertia and Coriolis terms are not of consequence. Hence, centrifugal terms for displacement of points on the line of centroid will take the form:

$$\omega^2(r^2_\eta + r_z^2) = \omega^2[2(R+z)u + y^2\cos^2\varphi + x^2\sin^2\varphi - xy\sin 2\varphi] \quad (1.46)$$

where φ is the stagger angle about the ξ axis in the disk plane. Eliminating rigid body terms, total kinetic energy of the blade is:

$$T = \frac{1}{2}\int v^2 dm = \frac{1}{2}\int \rho v^2 dA dz$$

or

$$T = \int_0^L \{m/2[(dx/dt)^2 + (dy/dt)^2] + I_{cg}/2(d\theta/dt)^2 + \omega^2(R+z)u\,m + \omega^2/2[m(y^2\cos^2\varphi + x^2\sin^2\varphi - xy\sin 2\varphi\}dz \quad (1.47)$$

Here ρ is mass density, m is mass per unit length, I_{cg} is mass polar moment of inertia of the beam and θ is its torsional displacement. Displacement $u(z, t)$ is the total axial displacement due to bending of the beam. For a non-uniform beam cross-section, displacements along the length of the beam will vary with area, x and y coordinates and their time derivative and angle of twist. Coordinates x and y may be assumed to be quadratic functions of z, since effect on the natural frequencies is not substantial. Cross-sectional area and other mechanical properties, however, may require a polynomial of z.

Displacements in the beam may be expressed as: $u_x = x - y\theta$, $u_y = y + x\theta$ and $u_z = x(x' - y\theta') - y(y' + x\theta') + u$ (denoting differentiation with respect to z). Longitudinal strain arising from bending is:

$$e_{zz} = -xx'' - yy''$$

Strain due to axial displacement is:

$$V = \frac{1}{2}\int_0^L (EI_{yy}x''^2 + EI_{xx}y''^2 + 2EI_{xy}x''y'')dz \quad (1.48)$$

When the cross-section is not circular warping may be expected. Section properties can be expressed as a polynomial of some degree. Total strain energy is obtained from the sum of (1.47) and (1.48). Differential equations of motion can be derived from application of Hamilton's principle to the energy expressions, then solved by the method of Galerkin. Ritz's procedure may also be gainfully employed by minimizing the Lagrangian expression $\Gamma = T - V$ and obtain the eigen-value problem. Shape functions may be assumed in a form of series for beam bending and torsion:

$$x = \sum_1^n f_i(Z)q_{1i} \qquad y = \sum_1^n f_i(Z)q_{2i} \qquad \theta = \sum_1^n f_i(Z)q_{3i} \quad (1.49)$$

where $z = Z/L$ and q_{1i}, q_{2i}, q_{3i} are generalized coordinates and L is beam length. The functions f_i must satisfy the beam's boundary conditions (namely, zero deflection and slope at fixed end and zero bending moment and shear force at free end). Expressions for potential and kinetic energies can then be by a classical numerical integration procedure to set up the eigen-value problem. Eigen-values and corresponding mode shapes can be obtained through a library routine. Convergence of the lower eigen-values may be expected when using a five term solution in the shape functions for each coordinate x, y and θ.

To determine the effects of a tapered beam, mass and stiffness distribution will change along the length, thus affecting both bending and torsional modes. If a rectangular section is used, beam width and height may be written in the form:

$$b = b_0(1 - \alpha H) \quad \text{and} \quad h = h_0(1 - \beta H) \tag{1.50}$$

where b_0 and h_0 represent values at the beam root, $H = zL$ and α and β are parameter values. If a single term approximation for uncoupled bending vibration in yz plane is used, the expression for the fundamental mode takes the form (*ref:* Rao, 1965):

$$p_i^2 = \frac{12EI_0}{m_0L^4}\frac{\left(\frac{12}{5}-\frac{2\alpha}{5}-\frac{6\beta}{5}+\frac{12\alpha\beta}{35}+\frac{12\beta^2}{35}-\frac{9\alpha\beta^2}{70}-\frac{3\beta^3}{70}+\frac{2\alpha\beta^3}{105}\right)}{\left(\frac{104}{45}-\frac{584\alpha}{315}-\frac{584\beta}{315}+\frac{5353\alpha\beta}{3465}\right)} \tag{1.51}$$

By defining $\lambda_{si}^2 = \frac{\rho A_0 L^4}{EI_0} p_i^2$,, frequency parameter ratios for the first three modes using a five-term solution are given in figs. 1.21, 1.22 and 1.23. An error of under ½ percent is encountered when using this procedure for the fundamental mode of vibration. Experimental results (*ref:* Carnegie, *et al.*, 1965-66) are shown for comparison.

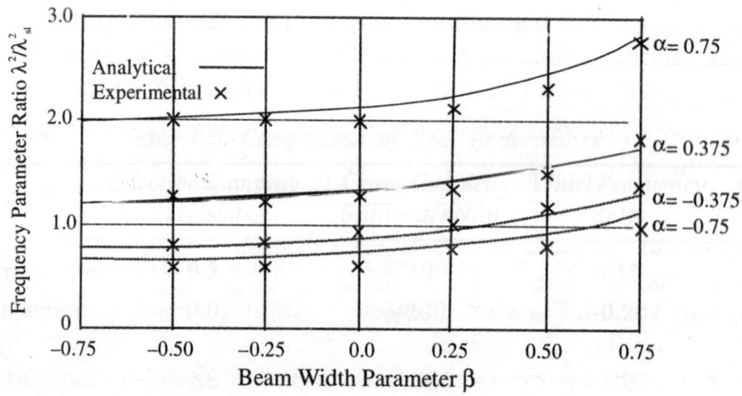

Fig. 1.21. *Effect of Beam Width Taper on Frequency – Mode # 1 (Rao, 1991)*

Close scrutiny of the plotted results indicates some interesting aspects. A reduction in the width of the rectangular section (but height held constant) increases the fundamental and higher bending mode frequencies of vibration. A similar reduction in the height of the beam (width held constant) increases the fundamental mode frequency, but decreases the higher bend mode frequencies.

Changes in torsional stiffness cannot be so readily evaluated due to the presence of hyperbolic tangent entities in the series expression. Generally, a decrease in the beam

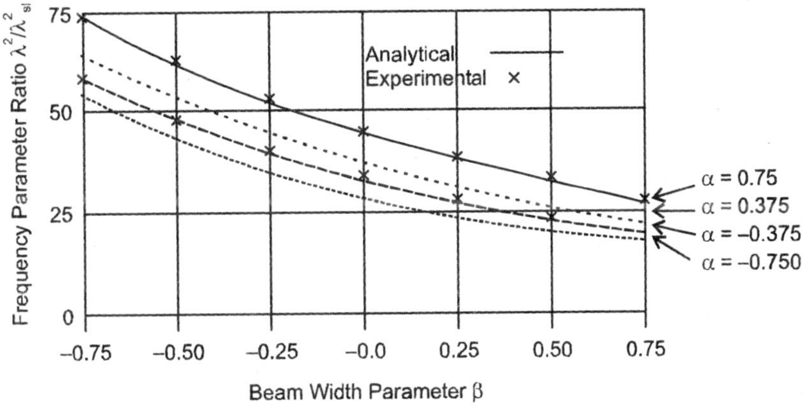

Fig. 1.22. *Effect of Beam Width Taper on Frequency – Mode # 2 (Rao, 1991)*

But a similar reduction in height, width held steady, will decrease the torsional mode frequencies.

Coupling between the bending modes in the xz and yz planes may be expected if a symmetric beam is pre-twisted. Figure 1.24 provides graphical depiction of the coupling as a function of the frequency ratio p/p_{y1}, the ratio of principal moments of area of beam section and the angle of twist γ. Three different twist angles are used. The effect of angle of twist is little on the first mode, but the second coupled mode combines the characteristics of the uncoupled first xz mode and the second yz mode, the extent of coupling depending also on numerical values of $\sqrt{(I_{xx} I_{yy})}$. The third coupled mode frequency increases with the angle of twist relative to the uncoupled frequency.

Besides translation in both lateral directions, elements in a beam rotate due to the slope at that location. This results in an associated inertia term, called the rotary inertia. Classical beam theory does not include its effects, as also the deflection due to shear. Hence, the angle of rotation of the element is the same as the slope. If the shear energy is also included in the equations of motion of the beam, correction factors derived by Sutherland *et al.*, (1954) for shear deformation and rotary inertia for various beam cross-sections must be applied. This factor depends on the radius of gyration $r_g = \sqrt{(I/AL^2)}$, dominating with larger values on higher modes of vibration.

Increased torsional stiffness is a consequence of pre-twist in a beam. The additional stiffness on top of that obtained from St. Venant's procedure arises from the inclination of the axial fibers, and is expressed by Carnegie (1959) by:

$$c_r = c_3\left(\frac{\gamma}{L}\right)^2 \tag{1.52}$$

where $C_3 = E\int_A \left(r^4 - \frac{r^2 I_p}{A}\right) dA$, r is element distance from the centroid in a symmetric cross-section and I_p defines the second polar moment of inertia. For asymmetric sections this expression takes the form:

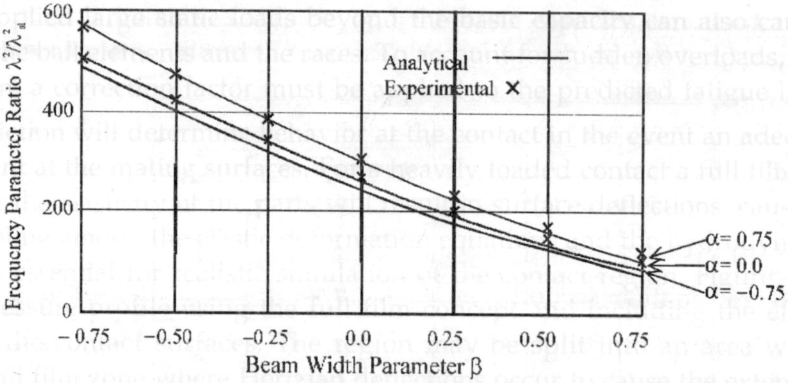

Fig. 1.23. *Effect of Beam Width Taper on Frequency – Mode # 3 (Rao, 1991)*

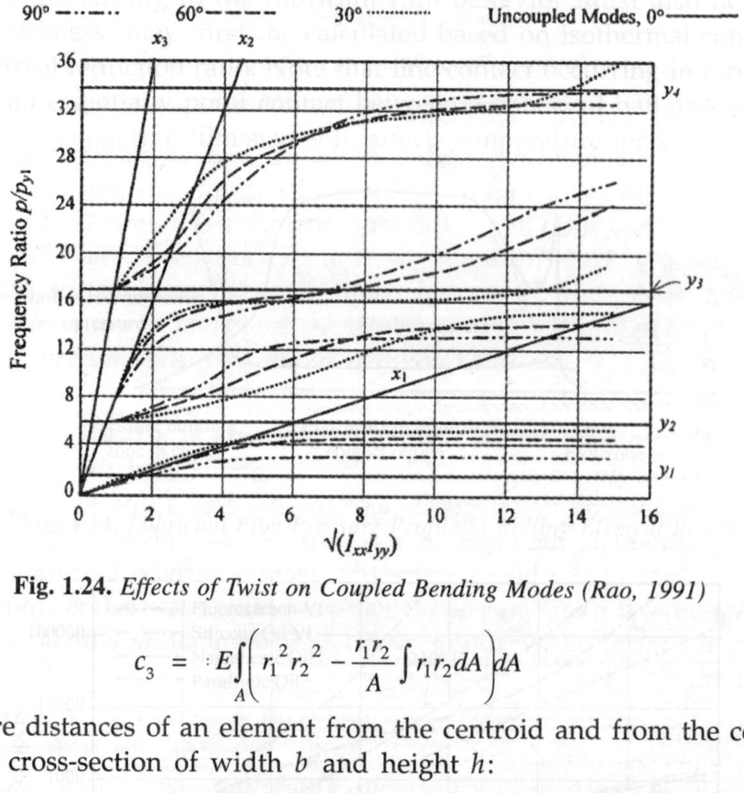

Fig. 1.24. *Effects of Twist on Coupled Bending Modes (Rao, 1991)*

$$c_3 = E\int_A \left(r_1^2 r_2^2 - \frac{r_1 r_2}{A}\int r_1 r_2 dA \right) dA$$

where r_1 and r_2 are distances of an element from the centroid and from the center of torsion. For a rectangular cross-section of width b and height h:

$$c_3 = E\frac{(b^4 + h^4)bh}{180}$$

Torsional stiffness of a pre-twisted beam may be estimated from fig. 1.25 as a function of the twist angle for a beam of rectangular section and a typical airfoil section used in turbo-machines. The values are obtained by adding c_3 to St. Venant's coefficient c_r. Torsional frequencies may then be estimated from the equation:

$$p_n = \frac{n\pi}{2L}\sqrt{\frac{C}{I_{cg}}} \text{ radian/second, } n = 1, 3, 5\ldots \tag{1.53}$$

where C is torsional stiffness and I_{cg} is polar mass moment of inertia about the centroid. Torsional motion along the length of the beam results in warped cross-sections, producing an axial displacement proportional to the warping function and the beam slope. The associated energy due to the displacement may be expressed by:

$$\nu_b = \frac{1}{2} c_1 \theta''^2$$

where $c_1 = EI_{\phi\phi} = E\int \phi^2 dA.$. The warping function mentioned above for a thin section is given by $\varphi = xy$. For a rectangular section then $c_1 = (Eb^3h^3)/144$.

The effect of additional bending on the torsional mode is obtained from Carnegie (1959) using Rayleigh's principle:

$$p_n = \frac{n\pi}{2L}\sqrt{\frac{C}{I_{cg}}}\sqrt{1+\frac{En^2\pi^2b^3h^3}{576L^2C}} \tag{1.54}$$

where n is odd.

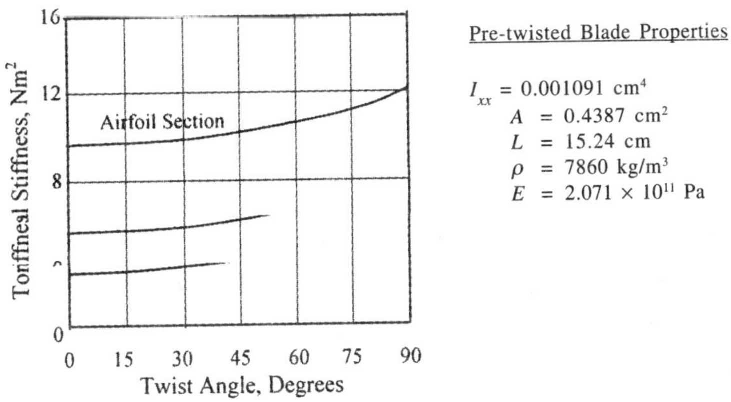

Fig. 1.25. *Torsional Stiffness of Pre-twisted Beams (Carnegie, 1959)*

1.12 BEAM SUBJECTED TO CENTRIFUGAL FORCE

Centrifugal forces acting axially in a cantilever beam of the type discussed in the previous section have the tendency of stiffening the beam. Consequently, as the rotor speed increases, bending natural frequencies are increased; torsional frequencies, however, are unaffected. The effect is similar to that observed in a piano wire; tuning frequency is modified by adjusting the tension in the wire with one end fixed and the other wrapped around a pin. Defining a non-dimensional stiffness parameter $\alpha^2 = m\omega^2L^4/EI_{xx}$, fig. 1.26 shows the impact of stiffening on eigen value $\lambda\alpha$ for the first three modes, with deflection occurring in the flap-wise bending direction at a setting angle of 0°. Disk radius is zero. Centrifugal forces have a substantial effect on the fundamental mode, reducing considerably for upper level modes. Centrifugal forces also play a lesser role on bending modes occurring parallel to the longer edge of the beam (*ref*: Rao, 1991).

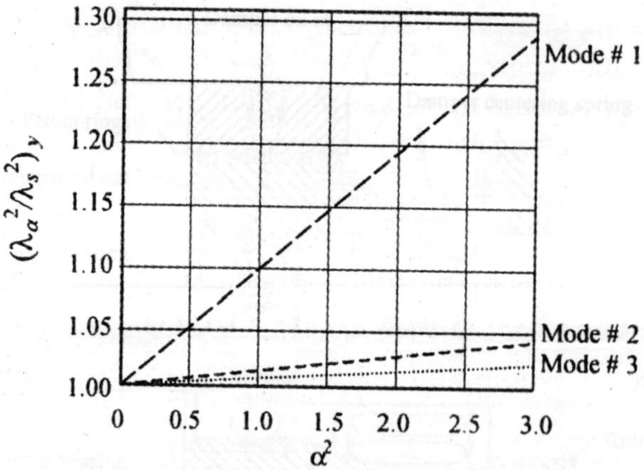

Fig. 1.26. *Centrifugal Stiffening Effect in Flap-wise Bending Cantilever Beam (Rao, 1991)*

When disk radius is non-zero, centrifugal force stiffening in the beam is increased, further raising bending mode natural frequencies. Effects of disk radius on bending modes parallel to beam's shorter edge are shown in fig. 1.27 for the first three modes, where *r* represents disk radius divided by beam length.

Because of additional stiffening disk radius is once again noted to play a dominant role on the fundamental mode frequency for motion along the blade's chord. Also, the frequency parameter ratio increases linearly with disk radius.

Next, consider the effects of setting the cantilever beam at an angle to the axis of disk, so that stagger angle φ about the ξ axis is non-zero. Defining φ′ = (φ – 90°), fig. 1.28 provides the effects of setting angle on bending mode frequencies for motion parallel to the short edge for three different values of α^2. The variation of bending mode frequencies is linearly proportional to $\sin^2 \varphi'$, with the fundamental mode being affected dominantly. Higher modes are less affected. Bending frequencies for motion parallel to the long edge are also increased when φ′ is increased, but the influence is less pronounced. Higher modes are even less affected.

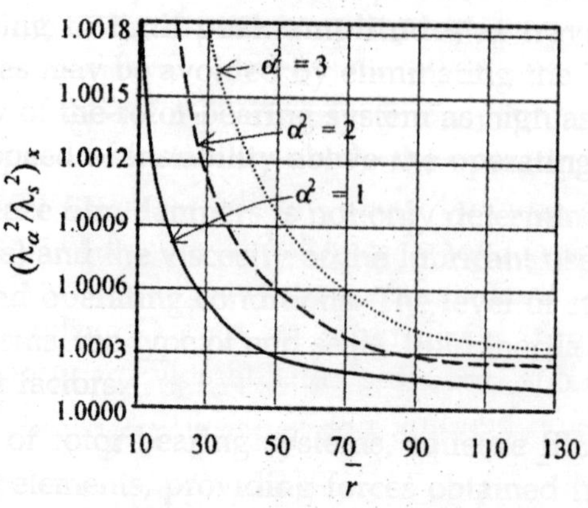

Fig. 1.27. *Effect of Disk Radius on Beam First Chord-wise Bending Frequency (Rao, 1991)*

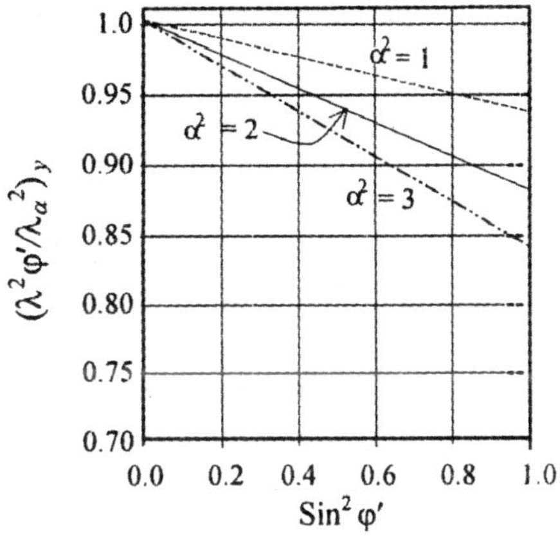

Fig. 1.28. *Beam Setting Angle and First Bending Frequency (Rao, 1991)*

1.13 FLEXIBLY MOUNTED BEAM

Practical considerations oftentimes result in a non-rigid attachment of the cantilever beam to a rotating disk. Support stiffness plays a role in bending and torsional modes of beam vibration (*ref*: Rao, 1991). A uniform cross-section of the beam will be considered to illustrate the impact of loose mounting, with the beam restrained by a spring K_z for torsional movement and by a spring K_x for bending displacements in the yz plane as shown in fig. 1.29.

Torsional vibrations are analyzed by using Hamilton's principle of variation:

$$\delta \int_{t_1}^{t_2} \left[\int_0^L \left(\frac{1}{2} C\theta'^2 - \frac{1}{2} Icg \left(d\theta \ dt \right)^2 \right) dz + \frac{1}{2} K_z \theta_0^2 \right] dt = 0 \qquad (1.55)$$

leading to:

$$\int_{t_1}^{t_2} \left[\int_0^L \left(I_{cg} d^2\theta / dt^2 - C\theta'' \right) \delta\theta dz + C\theta \big|_0^L + K_z \theta_0 \, \delta\theta_0 \right] dt = 0 \qquad (1.56)$$

Using non-dimensional terms:

$$\begin{aligned} \theta'' + \lambda^2 \theta &= 0 \\ \theta'' + \lambda^2 \theta &= 0 \qquad Z = 0 \\ \theta' &= 0 \qquad Z = 1 \end{aligned} \qquad (1.57)$$

where

$$\alpha = \frac{K_z L}{C}, \ \lambda^2 = \frac{I_{cg} L^2 p^2}{C}, \ Z = z/L \text{ and } ' = d/dZ.$$

Equation (1.57) may be solved to yield: $\theta = A \cos \lambda Z + B \sin \lambda Z$, and application of boundary conditions provides:

$$\alpha \cot \lambda - \lambda = 0$$

and:

$$\theta = \cot \lambda \cos \lambda Z + \sin \lambda Z$$

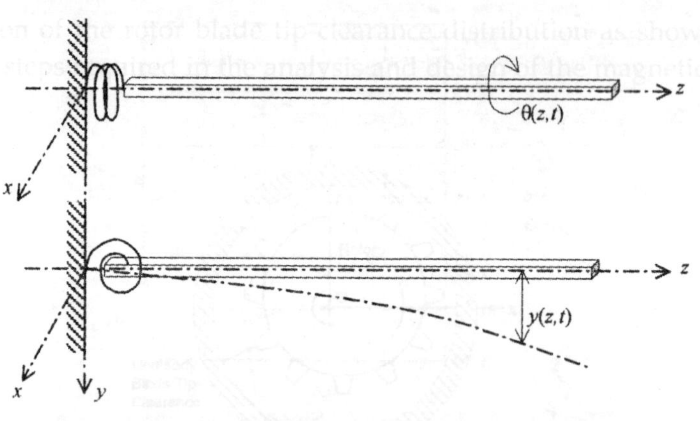

Fig. 1.29. *Flexible Beam Support – Torsion and Bending*

When $\alpha = 0$ the frequency equation takes the form: $\sin\lambda = 0$, and for $\alpha = \infty$ it becomes: $\cos\lambda = 0$. Torsional frequencies as a function of root stiffness are shown in fig. 1.30.

Bending of the beam is affected by an additional strain due to stiffness of the support, given by: $V_s = \frac{1}{2}K_x(y_o')^2$. Application of Hamilton's principle yields:

$$\delta\int_{t_1}^{t_2}\left[\int_0^L EI_{xx}y''^2 - \frac{1}{2}m(dy/dt)^2\,dz + \frac{1}{2}K_x y_0'^2\right]dt = 0 \tag{1.58}$$

Applying the variation,

$$\int_{t_1}^{t_2}\left[\int_0^L \{md^2y/dt^2 + EI_{xx}y''''\}\delta y dz + EI_{xx}y'\delta y'|_0^L - EI_{xx}y'''\delta y|_0^L + K_x y'_0\,\delta y'_0\right]dt = 0$$

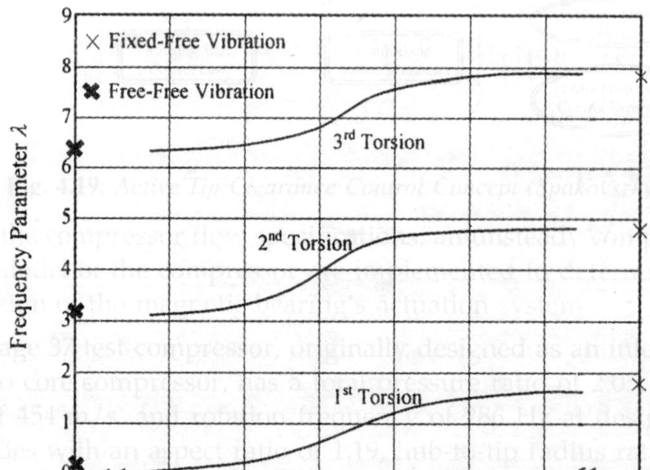

Fig. 1.30. *Beam Support Stiffness and Torsional Frequencies (Rao, 1991)*

This equation provides differential equations and boundary conditions of a non-dimensional form:

$$Y'''' - \lambda^4 Y : = 0$$
$$Y = 0;\ Y'' - \beta Y' = 0;\ Z = 0 \qquad (1.59)$$
$$Y'' = 0;\ Y''' = 0;\ Z = 1$$

where: $Y = y/L$; $Z = z/L$; $' = d/dz$; $\lambda^4 = m\dfrac{L^4\omega^2}{EI_{xx}}$; $\beta = \dfrac{K_x L}{EI_{xx}}$.. λ and β are non-dimensional frequency and support stiffness parameters, respectively. Solution of (1.59) is of the form:

$$Y = A \cosh \lambda Z + B \sinh \lambda Z + C \cos \lambda Z + D \sin \lambda Z \qquad (1.60)$$

Using the boundary conditions of (1.57), this leads to:

$$\begin{bmatrix} \dfrac{\beta}{\lambda}(Ch+C)+Sh & \dfrac{\beta}{2\lambda}(Ch+C)-S \\ \dfrac{\beta}{\lambda}(Sh+S)+Ch & \dfrac{\beta}{2\lambda}(Sh+S)-C \end{bmatrix} \begin{Bmatrix} B \\ D \end{Bmatrix} = 0 \qquad (1.61)$$

where $C = \cos \lambda$, $Ch = \cosh \lambda$, $S = \sin \lambda$, $Sh = \sinh \lambda$. Expansion of the determinant yields the frequency equation:

$$(1 + C \times Ch)(\beta/\lambda) + Sh \times C - Ch \times S = 0 \qquad (1.62)$$

Corresponding mode shapes are:

$$Y(Z) = \sin\lambda Z - \frac{\beta}{2\lambda}\left(1-\frac{b}{a}\right)\cos\lambda Z - \frac{b}{a}\ \sinh\lambda\ Z + \frac{\beta}{2\lambda}\left(1-\frac{b}{a}\right)\cosh\lambda Z \qquad (1.63)$$

where $a = \dfrac{\beta}{2\lambda}(Ch+C)+Sh$; $b = \dfrac{\beta}{2\lambda}(Ch+C)-S$. When $\beta = 0$, then support stiffness $K_x = 0$, and boundary conditions are for a hinged-free beam. The frequency equation is then:

$$Sh \times C - Ch \times S = 0 \qquad (1.64)$$

If $\beta = \infty$, then support stiffness $K_x = \infty$, boundary conditions for the cantilever are:

$$1 + C \times Ch = 0$$

Figure 1.31 provides the functional relationship between frequencies and b.

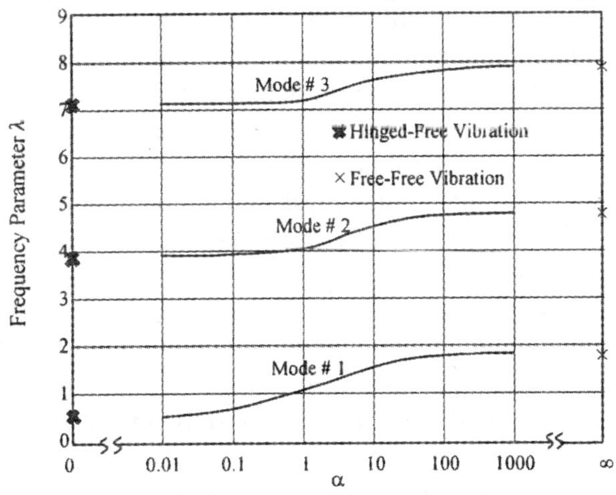

Fig. 1.31. *Beam Support Stiffness and Bending Frequencies (Rao, 1991)*

1.14 ROW OF CONNECTED BEAMS

Combining a number of cantilever beams at their free ends to form an assembly considerably alters the dynamic pattern of behavior when compared with a beam standing alone. Adjacent beam's modes and the connecting band's elastic and inertia characteristics introduce fixed-pinned modes in between the cantilever modes of vibration. The concept has practical applications in many long fan and turbine blades connected by shrouds at the mid-span and at the outer tip (*ref:* Rao, 1991).

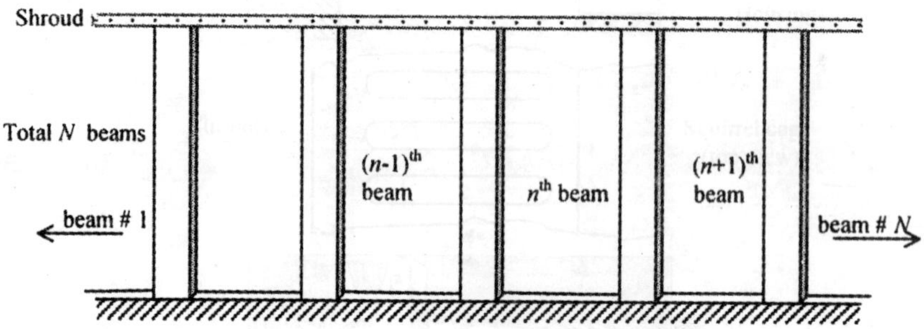

Fig. 1.32. *Row of Connected Beams*

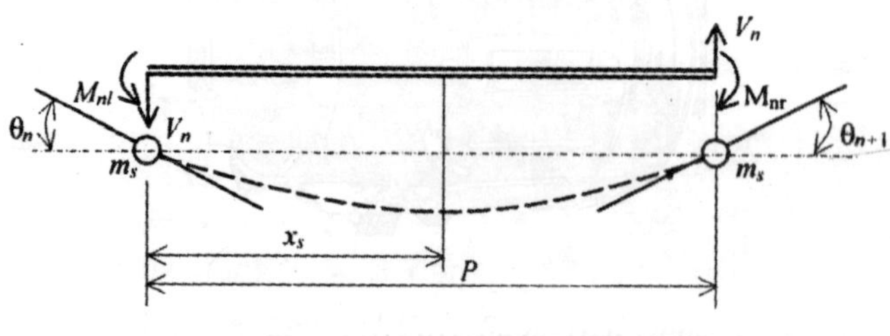

Fig. 1.33. *Shroud Pitch Length*

Figure 1.32 shows a row of N beams of uniform length L, mass per unit length m_b and E_bI_b flexural rigidity. A shroud (shown in fig. 1.33) of mass per unit length m_s, flexural rigidity E_sI_s and pitch P connects the free end of the cantilever beams. Shear force at the two ends of the shroud segment may be expressed by:

$$V_n = \frac{M_{nr} - M_{nl}}{P} \tag{1.65}$$

Bending moment and slope at the two ends of the shroud are related:

$$M_{nl}\frac{P}{E_sI_s} + 4\theta_n + 2\theta_{n+1} = 0 \tag{1.66}$$

$$M_{nr}\frac{P}{E_sI_s} - 2\theta_n + 4\theta_{n+1} = 0 \tag{1.67}$$

Then: $$V_n = \frac{6E_sI_s}{P^2}(\theta_n + \theta_{n+1}) \tag{1.68}$$

Bending moment at any distance x_s of the nth shroud will then be:

$$M_{xs} = M_{nl} + V_n x_s$$

$$= -\frac{E_s I_s}{P}(4\theta_n + 2\theta_{n+1}) + \frac{6E_s I_s}{P^2}(\theta_n + \theta_{n+1})x_s \tag{1.69}$$

Expression for total strain energy in the shroud is:

$$U_s = \sum_{n=1}^{N-1} \frac{2E_s I_s}{P}\left\{\theta_n^{\ 2} + \theta_{n+1}^{\ 2} + \theta_n \theta_{n+1}\right\} \tag{1.70}$$

Strain energy of the beams as a consequence of bending is:

$$U_b = \sum_{n=1}^{N} \int_0^L \frac{E_b I_b}{2}(y_n'')^2 dx_b \tag{1.71}$$

Kinetic energy of the complete shroud band is:

$$V_s = \sum_{n=1}^{N} \frac{1}{2} m_s (dy_{nl} / dt)^2 \tag{1.72}$$

and kinetic energy of the blades is: $V_b = \sum_{n=1}^{N} \int_0^L \frac{1}{2} m_b (dy_n / dt)^2 dx_b$ (1.73)

Equations (1.68) to (1.71) may be used to obtain Lagrange's function:

$$\Gamma = U_b - V_b + U_s - V_s \tag{1.74}$$

So from Hamilton's principle: $\delta \int_{t_1}^{t_2} \Gamma dt = 0$

and since $\theta_n = y'_{nl}$, the following restraint conditions and differential equations are obtained for the nth blade:

$$\left[E_b I_b y_n''\right]'' + m_b (d^2 y_n / dt^2) = 0, \; n = 1, 2, \ldots N \tag{1.75}$$

at $\qquad x_b = 0 \qquad E_b I_b \left[y_n' \delta y_n' - y_n''' \delta y_n\right] = 0, \qquad n = 1, 2, \ldots N$ (1.76)

at $\qquad x_b = L \qquad \delta y_n' \left[E_b I_b y_n'' + \frac{2E_s I_s}{P}(y_{n-1}' + 4y_n' + y_{n+1}')\right] = 0$ (1.77)

$$\delta y_n \{-E_b I_b y_n''' + m_s (d^2 y_n / dt^2)\} = 0, \; n = 1, 2, \ldots N \tag{1.78}$$

Since all beams of the group are fixed at the base, $y = y' = 0$. Boundary conditions at the tip will control the cantilever, fixed-hinged and fixed-fixed modes of vibration. When the shroud band is not substantial the fixed-fixed modes will not occur. For uniform beams the solution of (1.75) and satisfying the boundary conditions of (1.76) at $x_b = 0$ provides the cantilever modes equations of the form:

$$y_n = A_n (\cos mx_b - \cosh mx_b) + B_n (\sin mx_b - \sinh mx_b) \tag{1.79}$$

where $m^4 = (m_b\omega^2)/E_bI_b$. From (1.79) the derivatives of y_n at $x_b = L$ are obtained:

$$\begin{aligned} y'_{nl} &= [-A_n(S+Sh)+B_n(C-Ch)]m \\ y''_{nl} &= [-A_n(C+Ch)-B_n(S+Sh)]m^2 \\ y'''_{nl} &= [A_n(S-Sh)-B_n(C+Ch)]m^3 \end{aligned} \tag{1.80}$$

where $S = \sin mL$, $Sh = \sinh mL$, etc. The second boundary condition provides:

$$B_n = RA_n \tag{1.81}$$

where
$$R = \frac{(S-Sh)+\beta(C-Ch)}{(C+Ch)-\beta(S-Sh)}$$

and
$$\beta = \frac{m_S\omega^2}{E_bI_bm^3} \tag{1.82}$$

Using (1.81) and (1.77), the frequency determinant is obtained:

$$\Delta = \begin{vmatrix} t_o & t_1 & 0 & 0\ldots & 0 & 0 \\ t_1 & t_2 & t_1 & 0\ldots & 0 & 0 \\ 0 & t_1 & t_2 & t_1\ldots & 0 & 0 \\ \ldots & \ldots & \ldots & \ldots & \ldots & \ldots \\ 0 & 0 & 0 & 0\ldots & t_2 & t_1 \\ 0 & 0 & 0 & 0\ldots & t_1 & t_o \end{vmatrix} = 0 \tag{1.83}$$

where: $t_1 = \dfrac{2r}{mL}[-(S+Sh)+R(C-Ch)]$, $t_2 = 4t_1 - [(C+Ch)+R(S+Sh)]$,

$$t_0 = 3t_1 - [(C+Ch)+R(S+Sh)], \quad r = \frac{E_sI_sL}{E_bI_bP} \tag{1.84}$$

For fixed-supported modes the secondary boundary conditions (1.77 and 1.78) will take the form $Y_{nL} = 0$. This leads to: $R = \dfrac{(C-Ch)}{(S-Sh)}$, with the frequency determinant of (1.83) remaining unchanged.

Smith (1948) provides for groups of 6 beams. The ratio of frequency of the group and the fundamental freestanding beam frequency is plotted in fig. 1.34 against the ratio of flexural rigidity r of one shroud pitch and the beam for varying shroud and beam mass. Frequencies are noted to increase with flexural rigidity of the shroud, while deceasing with shroud mass. Five fixed-pinned modes appear in the group (one less than the number of beams in the group), three of them displaying odd symmetry and two showing even symmetry.

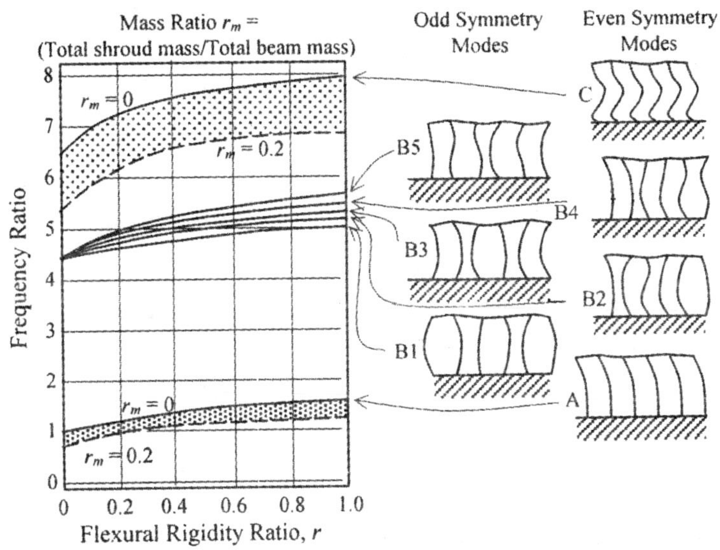

Fig. 1.34. *Vibration Characteristics of Group of Six Beams (Smith, 1948)*

The procedure may be extended to any number of beams in the group; results may be expected to follow the same trend as shown in this example.

1.15 ENERGY METHODS IN SHELL VIBRATION

When turbine and compressor blades are long, their dynamic characteristics are adequately simulated by elastic beam. Shorter length and wider chord blades, on the other hand, behave more like a plate or a shell. Kinetic and potential energy expressions for a pre-twisted shell need to be established prior to application of the Ritz procedure to obtain natural frequencies of vibration (*ref:* Rao, 1991).

Differential geometry of the shell must be first defined to obtain strain displacement equations. Figure 1.35 shows the central plane of a rectangular and uniformly pre-twisted plate of length l and width b using rectangular coordinates xyz and curvilinear coordinates $\alpha\beta$. Plate aspect ratio is then $r = l/b$. Position vector of P on curve α is:

$$\bar{M}(\alpha,\beta) = (\alpha \cos \Gamma\beta\ \bar{i} + (\alpha \sin \Gamma\beta\ \bar{j} + (\beta)\bar{K}) \tag{1.85}$$

where $\bar{i}, \bar{j}, \bar{k}$ are unit vectors and $\Gamma = \gamma/l$ is the unit pre-twist angle. Coefficients of the first quadratic form of the undeformed central plane may be obtained from the relations (*ref:* Veizer, 1961):

$$A = |\bar{M},\alpha| = 1,\ B = |\bar{M},\beta| = [1+\Gamma^2\alpha^2]^{1/2}$$

and

$$X = \text{Cos}-1\ \frac{\bar{M}_{,\alpha}}{A}\frac{\bar{M}_{,\beta}}{B} = \frac{\pi}{2} \tag{1.86}$$

Basic vectors in the central plane are unit vectors $\bar{M}_{,\alpha}/A$, $\bar{M}_{,\beta}/B$, and $\bar{n}$, where

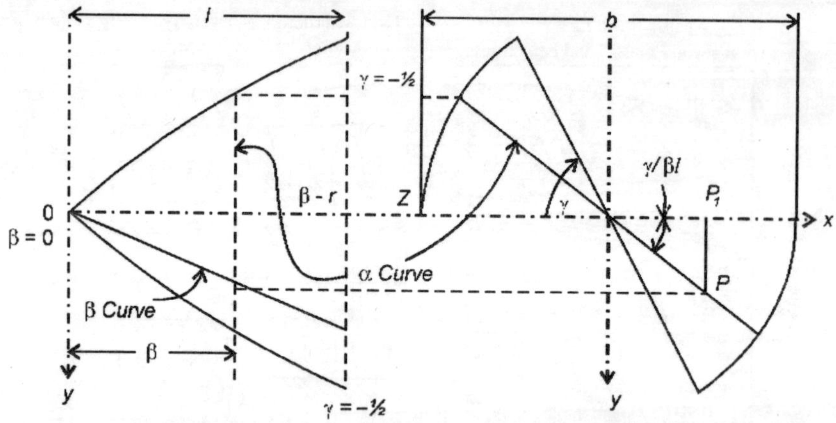

$\Gamma = \gamma/l$; $r = l/b$; $OP = \alpha$; $OP_1 = \alpha \mathrm{Cos}(\gamma/l)$; $PP_1 = \gamma \mathrm{Sin}(\gamma\beta/l)$

Fig. 1.35. *Central Surface of Uniformly Twisted Plate (Rao, 1991)*

$\bar{n} = \dfrac{\bar{M}_{,\alpha}}{A} \times \dfrac{\bar{M}_{,\beta}}{B}$. Coefficients of the second quadratic form are:

$$L = \bar{n} \cdot \bar{M}_{,\alpha\alpha} = 0,$$

$$M = \bar{n}.\bar{M}_{\alpha\beta} = -\frac{\Gamma}{B}, \quad N = \bar{n}.\bar{M}_{\beta\beta} = 0 \tag{1.87}$$

Curvature of radii R_1 and R_2 in α and β directions and radius of torsion R_{12} are:

$$\frac{1}{R_1} = \frac{L}{A^2} = 0, \qquad \frac{1}{R_2} = \frac{N}{B^2} = 0, \qquad \frac{1}{R_{12}} = \frac{M}{AB} = \frac{\Gamma}{B^2} \tag{1.88}$$

Curvatures along the α and β directions are zero, hence coordinates α and β coincide with the real direction of asymptotes, and the shell is of negative Gaussian curvature, given by the expression:

$$K = \frac{LN - M^2}{A^2 B^2 \sin^2 x} = -\frac{\Gamma^2}{[1 + \Gamma^2 \alpha^2]^2} \tag{1.89}$$

Deformed and undeformed shell surfaces are shown in fig. 1.36. The coordinates of a surface equidistant z from the middle surface are non-orthogonal. Longitudinal strains ε_α and ε_β and shear strain $\omega_{\alpha\beta}$ of the middle surface are:

$$\varepsilon_\alpha = \frac{\bar{M}_{,\alpha}}{A} \cdot \frac{\bar{U}_{,\alpha}}{A}, \quad \varepsilon_\beta = \frac{\bar{M}_{,\beta}}{B} \cdot \frac{\bar{U}_{,\beta}}{B}$$

$$\omega_\alpha = \frac{\bar{M}_{,\beta}}{B} \cdot \frac{\bar{U}_{,\alpha}}{A}, \quad \omega_\beta = \frac{\bar{M}_{,\alpha}}{A} \cdot \frac{\bar{U}_{,\beta}}{B} \tag{1.90}$$

$$\omega_{\alpha\beta} = \omega_\alpha + \omega_\beta$$

and the associated bending strain components of the middle surface are:

$$\chi_\alpha = \frac{\bar{\Omega}_\alpha}{A}.\frac{\overline{M}_{,\beta}}{B}, \quad \chi_\beta = \frac{\bar{\Omega}_\beta}{B}.\frac{\overline{M}_{,\alpha}}{A}$$

$$\tau_\alpha = \frac{\bar{\Omega}_\alpha}{A}.\frac{\overline{M}_{,\alpha}}{A}, \quad \tau_\beta = \frac{\bar{\Omega}_\beta}{B}.\frac{\overline{M}_{,\beta}}{B}$$

$$2\tau_{\alpha\beta} = \tau_\alpha - \tau_\beta - \frac{\varepsilon_a + \varepsilon_\beta}{R_{12}} \tag{1.91}$$

where $\overline{\Omega}$ is a vector of elastic rotation.

The displacement vector of the middle surface has components u, v and $-w$ in the $\overline{M}_{,\alpha}, \overline{M}_{,\beta}$ and $\bar{n}$ directions. The stretching and bending strains and rotations for the shell are obtained from Vezier (1961).

$$V = \frac{E}{2(1-v^2)}\int_V \left[\varepsilon_\alpha + \varepsilon_\beta + 2v\varepsilon_\alpha\varepsilon_\beta + \frac{1-v}{2}\omega_{\alpha\beta}\right]ABd\alpha d\beta dz \tag{1.92}$$

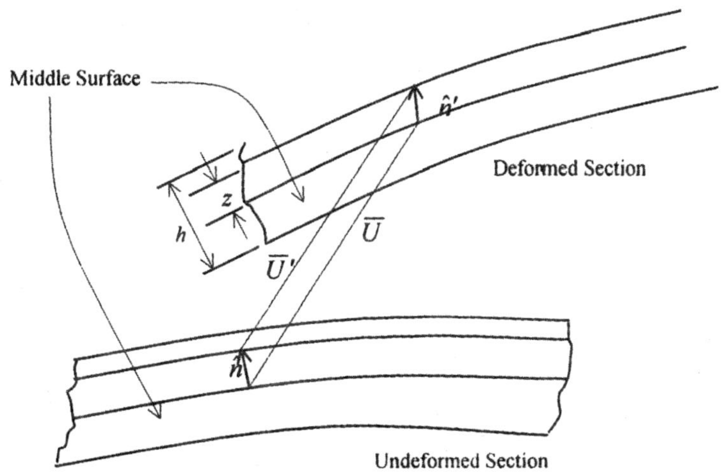

Fig. 1.36. *Plate Undeformed and Deformed Sections (Rao, 1991)*

Application of thin shell theory implies neglecting powers of z and that of z/R_{12} greater than or equal to 2; integrating over the thickness of the shell, the resulting expression for the potential energy is given by equation (1.91). Note that strain energy due to stretching of the middle surface is given by the first terms proportional to h, while the second and third terms proportional to h^3 denote strain energy due to bending. The first and second set of terms are the conventional strain energy encountered in shell analysis (*ref*: Love, 1944). The third set of terms is due to the retention of z/R_{12} when compared with unity. The relative order of the third set of terms, when compared with the first or second sets, is h/R_{12} (*ref*: Novozhilov, 1959). Hence, the third set of terms in the expression may be retained. The potential energy for the shell may then be expressed in the form:

$$V = \frac{E}{2(1-v^2)}\iint \left[\begin{array}{l} h\left\{(\varepsilon_\alpha + \omega_\beta)^2 - 2(1-V)\left(\varepsilon_a\varepsilon_\beta - \frac{\omega^2\alpha\beta}{2}\right)\right\} \\ +\frac{h^3}{6R_{12}}\{(\chi_\alpha+\chi_\beta)^2 - 2(1-v)(\chi_\alpha\chi_\beta - \tau_{\alpha\beta})^2\} \\ +\frac{h^3}{6R_{12}}\{v-2)\omega_{\alpha\beta}(\chi_\alpha+\chi_\beta)+(\varepsilon_\alpha\tau_\beta - \varepsilon_\beta\tau_\beta) \\ +v(\varepsilon_\beta\tau_\alpha - \varepsilon_\alpha\tau_\beta)-(6-2v)\tau_{\alpha\beta}(\tau_\alpha+\varepsilon_\beta)\} \end{array}\right] ABd\alpha d\beta \tag{1.93}$$

Kinetic energy of the vibrating shell is:

$$T = \frac{1}{2}\int\{(du/dt)^2 + (dv/dt)^2 + (dw/dt)^2\}dm$$

Since $dm = AB\rho dz\, d\alpha d\beta$, then:

$$T = \iint \frac{\rho h}{2}\rho^2\left[(u^2+v^2+w^2)\frac{h^2}{12}(v^2{}_\alpha + v^2{}_\beta)\right]ABd\alpha d\beta \tag{1.94}$$

where the first set represents energy due to translation and the second set due to rotation.

A gain in potential energy due to twisting of the plate as a consequence of centrifugal force may be expected when the part is mounted on a rotating wheel (see fig. 1.37). Gain in element strain energy due to deflections η_D and z_D is:

$$dU_\eta = -\int_0^{\eta_D} dF_\eta d\eta_D = -ABd\alpha d\beta\rho h\omega^2\left\{\eta_D\alpha\sin(\phi+\Gamma\beta)+\frac{\eta_D^2}{2}\right\} \tag{1.95}$$

$$dU_z = -\ ABd\alpha d\beta\rho h\omega^2\left\{(R+\beta)z_D + \frac{z_D^2}{2}\right\} \tag{1.96}$$

Net gain in potential energy of the plate due to rotation is given by the sum of above two expressions after integration. By using the Hamilton principle, differential equations of motion are set up, with the expressions giving rise to a combination of linear, non-homogeneous and non-linear terms.

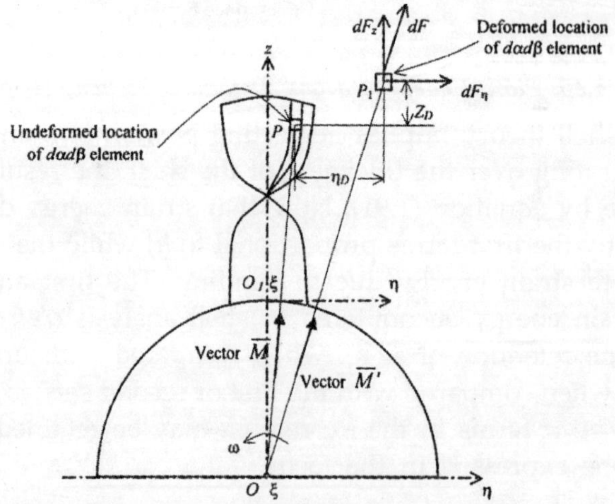

Fig. 1.37. *Plate Twist due to Centrifugal Force (Rao, 1991)*

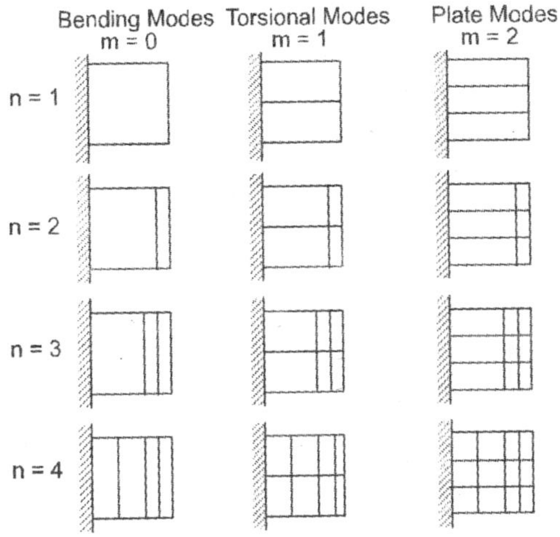

Fig. 1.38. *Plate Vibration Modes (Grinsted, 1952)*

Displacements may be assumed in the form of a combination of functions of α and β, with the equations satisfying geometric boundary conditions $u = v = w = 0$ on the fixed edge. Figure 1.38 provides nodal patterns for the different modes of vibration. Bending modes correspond to $m = 0$, with no nodal line in the direction of $\beta = 0$. Thus, $0/n$ indicates symmetrical bending modes about the line of axis $\alpha = 0$. Torsional modes are anti-symmetric, having one nodal line along $\alpha = 0$, and are indicated by $1/n$. $m > 1$ represents plate modes, with $2/n$, $4/n$, etc. being symmetric and $3/n$, $5/n$, etc. being anti-symmetric modes.

Figure 1.39 provides following solutions for the fundamental bending frequency of a stationary pre-twisted plate. Petricone *et al.* (1970) use the Rayleigh-Ritz method, indicating frequency increases with pre-twist. Flat plate triangular elements are used in Rawtani *et al.* (1969). McBain (1975) employs quadrilateral plate elements of the NASTRAN computer program, and also provides experimental results. Holographic experimental results by Gupta (1979) show a consistent decrease in frequency with the angle of pre-twist. Second and third bending mode frequencies are also provided in the same references.

Swaminadham *et al.*, (1975) obtained natural frequencies for a rotating plate (4.57 × 2.54 × 0.3175 cm mild steel, with pre-twist of 30° mounted on a 4.445 cm radius disk at a stagger angle of 0°), and used the beam theory and test experiments.

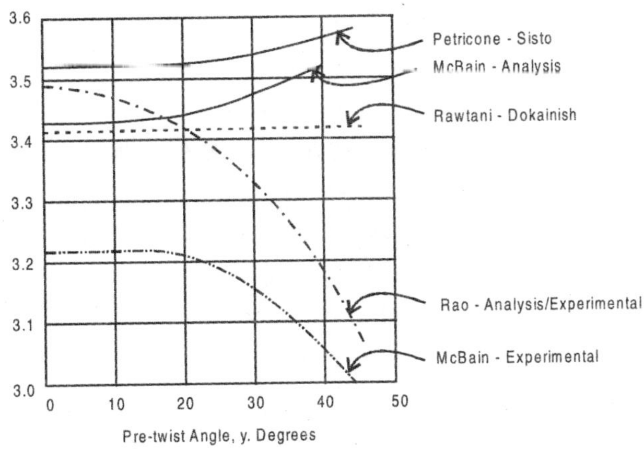

Fig. 1.39. *Primary Bending Mode Frequency of Pre-twisted Plate*

Table 1.1 compares the frequencies with the Ritz method. Higher natural frequencies are calculated by the beam theory, the plate theory predicting results closer to experiments. Predictions, from finite element calculations using parametric elements (*ref*: Thomas *et al.*, 1976-1977) are shown in fig. 1.40.

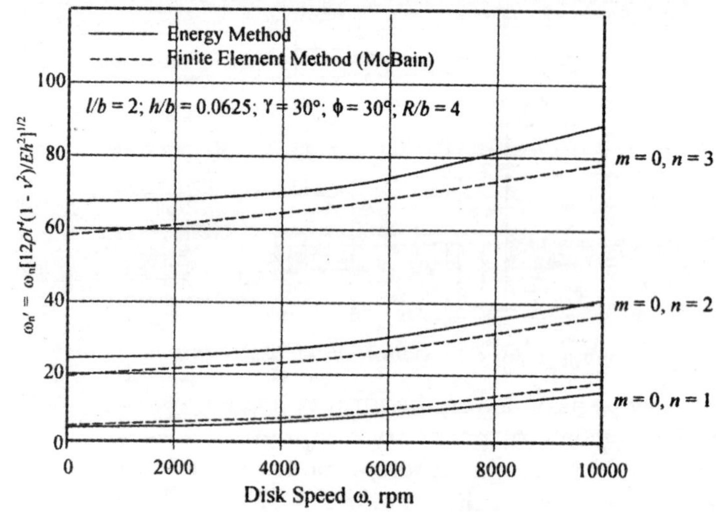

Fig. 1.40. *Rotating Plate Frequencies – Bending Mode*

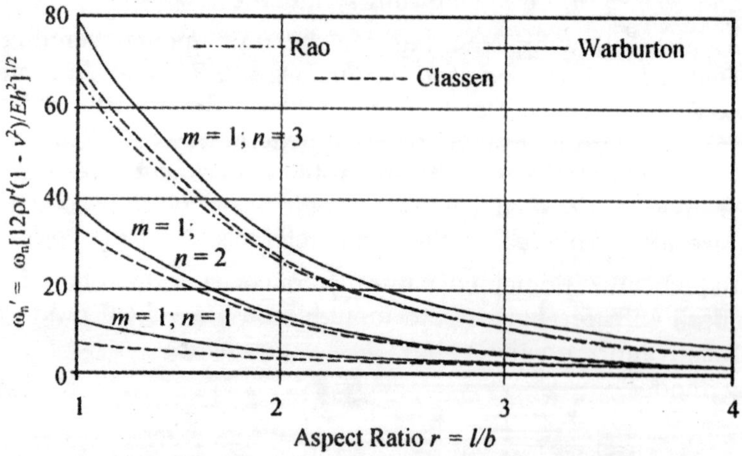

Fig. 1.41. *Torsional Frequencies of Flat Plate*

Gupta *et al.*, (1978) use the flat plate theory to determine effects of aspect ratio on torsional modes of vibration of a stationary blade. Figure 1.41 provides the results, together with results from Warburton (1954). Warburton assumed a displacement field consisting of a beam function in both directions of the plate. Aspect ratios greater than 4.0 have little effect on the non-dimensional frequency.

Pre-twist in a plate increases the natural frequency of torsional vibrations, just as in the vibration of beams. Rotary inertia also has an impact on torsional frequencies. In the case of

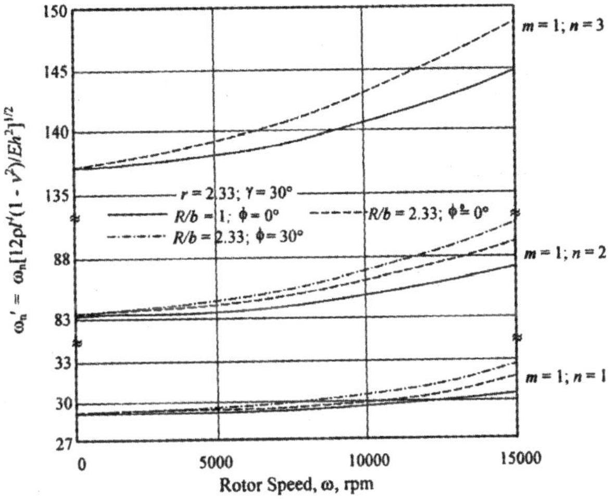

Fig. 1.42. *Torsional Frequencies of Rotating Pre-twisted Plate (Rao, 1991)*

Table 1.1. ***Fundamental Bending Mode Frequency of Rotating Pre-twisted Plate***

$$\omega_0' = \omega_0[12\rho l^4(1-\nu^2)/Eh^2]^{1/2}$$

Speed, rpm	*Beam Theory*	*Ritz Method*	*Experimentally Measured*
0	3.3961	3.3178	3.3190
2000	3.4226	3.3262	3.3535
6000	3.4705	3.3931	3.4226
10000	3.5820	3.5225	3.5449

centrifugal forces arising from mounting on a rotating disk, however, the beam theory predicts little change in torsional modes, but the plate theory has an influence, albeit to a lesser extent than in bending modes.

Theoretical frequencies for a plate of $r = 2.33$, $R/b = 1$, $\phi = 0°$ and with pre-twist of 0° and 12° are shown in fig. 1.42. Effects of stagger angle, dominant only at high rotor speeds, and disk radius are seen. Relative increase in natural frequencies is less conspicuous for torsional modes than for bending modes.

1.16 VIBRATION IN CIRCULAR DISK

A disk rotating about its central axis experiences axial vibrations at the rim, with it's geometry and method of attachment to the drive shaft controlling the natural frequencies. Lower order modes are primarily affected by thickness at the center hub and rim radius, while higher order modes are controlled by thickness at the rim. The vibration pattern comprises of waves rotating forward and backward in the disk relative to the direction of rotation; expressed mathematically, the terms include integral nodal diameters and nodal circles (*ref:* Rao, 1991).

The equation of motion of a circular plate vibrating at frequency ω with out-of-plane displacements w in the axial direction is expressed as:

$$\nabla^4 w(r,\theta) - \beta^4 w(r,\theta) = 0 \tag{1.97}$$

where $\nabla^2 = \frac{\delta^2}{\delta r^2} + \frac{1}{r}\frac{\delta^2}{\delta\theta^2}$ and $\beta^4 = \frac{\omega^2\rho}{H}$. Here, ρ is plate mass per unit area, r is disk radius and $H = \frac{2Eh^3}{3(1-\nu^2)}$, with h as disk half thickness and ν as the Poisson's ratio. Solution of the equation of motion is:

$$\begin{aligned} w = &\{A_{1n}J_n(\beta r) + A_{3n}Y_n(\beta r) + B_{1n}I_n(\beta r) + B_{3n}K_n(\beta r)\} \sin n\theta \\ &+ \{A_{2n}J_n(\beta r) + A_{4n}Y_n(\beta r) + B_{2n}I_n(\beta r) + B_{4n}K_n(\beta r)\} \cos n\theta \end{aligned} \quad (1.98)$$

Y and K are Bessel functions; when the argument is zero the functions become infinite. For finite deformations in a circular plate this equation takes the form:

$$w(r, \theta) = L_n\{J_n(\beta r) + \mu_n I_n(\beta r)\}\cos(n\theta - \varepsilon_n) \quad (1.99)$$

where: $\cos(n\varepsilon_n) = \frac{A_{1n}}{L_n} = \frac{B_{1n}}{\mu_n L_n}$, $\sin(n\varepsilon_n) = \frac{A_{2n}}{L_n} = \frac{B_{2n}}{\mu_n L_n}$. The frequency equation takes the form:

$$\frac{p^2}{2n}\begin{bmatrix} \{p^4 + (1-\nu)^2 n^2 (n^2 - 1)\}(J_{n-1}I_{n+1} + J_{n+1}I_{n-1}) \\ -2p^2(1-\nu)n\{(n-1)I_{n-1}J_{n-1} + (n+1)I_{n+1}J_{n+1}\} \end{bmatrix} = 0 \quad (1.100)$$

where $p = \beta q$, $I_n\pm_1$ and $J_n\pm_1$ are $I_n\pm_1(p)$ and $J_n\pm_1(p)$ respectively, and q is the disk radius. For n=0, or no nodal diameters, (1.100) becomes:

$$p(J_0I_1 + I_0J_1) - 2(1 - \nu)I_1J_1 = 0 \quad (1.101)$$

Using p_{ns} (n = 0, 1, 2..., s = 0, 1, 2...) as roots of (1.100) and (1.101) for n nodal diameters and s nodal circles, the natural frequencies may be expressed by:

$$\omega_{ns} = p_{ns}^2\sqrt{\frac{2Eh^3}{3\rho(1-\nu^2)q^4}} \quad (1.102)$$

Corresponding mode shapes are given by:

$$\varphi_{ns}(r,\theta) = \left[J_n\left(p_{ns}\frac{r}{q}\right) + \mu_{ns}I_n\left(p_{ns}\frac{r}{q}\right)\right]\cos(n\theta - \varepsilon_n) \quad (1.103)$$

where $\mu_{ns} = \frac{p^3J'_n + (1-\nu)n^2(pJ'_n - J_n)}{p^3I'_n - (1-\nu)n^2(pI'_n - I_n)}$, J' and I' are derivatives with respect to the radius r.

1.17 COUPLED DISK-BLADE SYSTEM

The Bernoulli-Euler theory of beam bending provides the differential equation of motion in the form:

$$\frac{\partial^4 w}{\partial l^4} + \frac{A\rho}{EI}\frac{\partial^2 w}{\partial t^2} = 0 \quad (1.104)$$

If $w = W(l)e^{i\omega t}$, then:

$$w = \{A\cos(\lambda l) + B\sin(\lambda l) + C\cosh(\lambda l) + D\sinh(\lambda l)\}e^{i\omega t} \quad (1.105)$$

where $\lambda^4 = \dfrac{\omega^2 A\rho}{EI}$. For a free beam the frequency equation is:

$$\cos(\lambda L) \times \cosh(\lambda L) = 1$$

where l is distance measured along beam axis and L represents beam length. Principal modes are available from the characteristic equation:

$$\phi_r(l) = \cos(\lambda_r l) + \cosh(\lambda_r l) - \sigma_r\{\sin(\lambda_r l) + \sinh(\lambda_r l)\} \qquad (1.106)$$

where $\sigma_r = \dfrac{\cosh(\lambda_r, L) - \cos(\lambda_r, L)}{\sinh(\lambda_r, L) - \sin(\lambda_r, L)}$. Application of a force $Fe^i\omega^t$ or a couple $Me^i\omega^t$ results in a displacement and slope. Use of influence coefficients (defined as ratio of load and deformation, both measured at the beam's root, $l = 0$) facilitates evaluation of vibration characteristics of attached components such as a blade and a disk by applying dynamic equilibrium and compatibility between the attached components with varying displacement fields (*ref*: Rao, 1991). A circumferentially varying shear force per unit length, $P_n \cos(n\theta)e^i\omega^t$, with zero bending moment is considered. Next, a circumferentially varying moment per unit length, $M_n \cos(n\theta)e^i\omega^t$, with zero shear force is applied at the rim. The procedure yields the influence coefficients listed below.

Direct coefficient between force and deflection: $${}_N\beta_\phi = -\frac{K_5}{EI\lambda^3 K_3} \qquad (1.107)$$

Cross coefficient between force and slope: $${}_N\beta_\phi = -\frac{K_1}{EI\lambda^2 K_3} \qquad (1.108)$$

Cross coefficient between moment and deflection: $${}_C\beta_w = -\frac{K_1}{EI\lambda^2 K_3} \qquad (1.109)$$

Direct coefficient between moment and slope: $${}_C\beta_\phi = -\frac{K_1}{EI\lambda K_3} \qquad (1.110)$$

In a similar manner coefficients between the root and tip of the attached beam may be expressed.

$${}_N\beta'_w = \frac{K_8}{EI\lambda^3 K_3} \qquad (1.111)$$

$${}_N\beta'_\phi = -\frac{K_{10}}{EI\lambda^2 K_3} \qquad (1.112)$$

$${}_C\beta'_w = -\frac{K_{10}}{EI\lambda^2 K_3} \qquad (1.113)$$

$${}_C\beta'_\phi = \frac{K_7}{EI\lambda K_3} \qquad (1.114)$$

In the expressions for coefficients, $K_1 = \sin\lambda L \sinh\lambda L$,

$$K_3 = \cos\lambda L \cosh\lambda L - 1$$

$$K_5 = \cos\lambda L \sinh\lambda L - \sin\lambda L \cosh\lambda L$$

$$K_6 = \cos\lambda L \sinh\lambda L + \sin\lambda L \cosh\lambda L$$
$$K_7 = \sin\lambda L + \sinh\lambda L,$$
$$K_8 = \sin\lambda L - \sinh\lambda L$$
$$K_{10} = \cos\lambda L - \cosh\lambda L \qquad (1.115)$$

Using a two degree of freedom model for a disk D with one blade B, let F and M represent the force and couple with response x and θ. Applying the corresponding coefficients, the displacement field takes the form:

$$X_D = {}_F\alpha_w F_D + {}_M\alpha_w M_D$$
$$\theta_D = {}_F\alpha_\phi F_D + {}_M\alpha_\phi M_D$$
$$X_B = {}_F\beta_w F_D + {}_M\beta_w M_D$$
$$\theta_B = {}_F\beta_\phi F_D + {}_M\beta_\phi M_D \qquad (1.116)$$

Assume F and M to be the force and moment between the mating disk and beam. Deflections at the attachment point are equal. Then:

$$F = F_D + F_B$$
$$M = M_D + M_B \qquad (1.117)$$
$$X = X_D = X_B$$
$$\theta = \theta_D = \theta_B \qquad (1.118)$$

and since no loads are applied,

$$F_D = -F_B \text{ and } M_D = -M_B. \qquad (1.119)$$

Conditions (1.117) and (1.119) may be used in (1.116) to give the frequency equation:

$$\begin{bmatrix} ({}_F\alpha_w + {}_F\beta_w)({}_M\alpha_w + {}_M\beta_w) \\ ({}_F\alpha_\phi + {}_F\beta_\phi)({}_M\alpha_\phi + {}_M\beta_\phi) \end{bmatrix} \begin{Bmatrix} F_B \\ M_B \end{Bmatrix} = 0 \qquad (1.120)$$

Next, consider Q identical beams attached to the disk, with the assembly vibrating in a mode with n nodal diameters. At each disk-beam attachment point three coordinates for translation and three for rotation will be used to obtain a 6×6 matrix of influence coefficients. Reaction forces and moments assume a harmonic distribution around the disk rim of period π/n, so the force on j^{th} blade will be:

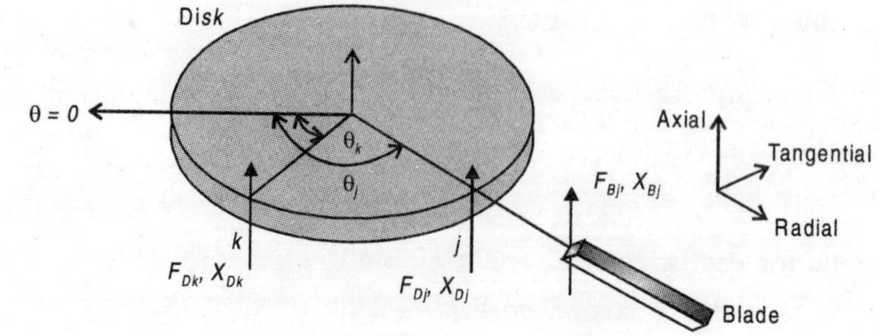

Fig. 1.43. *Reaction Loads and Deflections at Disk-Beam Attachment (Rao, 1991)*

$$f_j = -2\varepsilon A_o \cos\frac{2\pi nj}{Q} \qquad (1.121)$$

where A_o is force amplitude and 2ε is angular width of the beam. Expression for the couple takes a similar form. In fig. 1.43 vectors $\{F_D\}$ and $\{F_B\}$ are reaction forces and moments acting on the disk and the beams, and vectors $\{X_D\}$ and $\{X_B\}$ are the corresponding displacements and rotations. The components are:

$$\{F_D\} = \begin{Bmatrix} N_{ax} \\ N_{\tan} \\ N_{rad} \\ R_{ax} \\ R_{\tan} \\ R_{rad} \end{Bmatrix}; \qquad \{X_D\} = \begin{Bmatrix} W_{ax} \\ W_{\tan} \\ W_{rad} \\ \varphi_{ax} \\ \varphi_{\tan} \\ \varphi_{rad} \end{Bmatrix}; \qquad \{F_B\} = \begin{Bmatrix} N_{ax} \\ N_{\tan} \\ N_{rad} \\ R_{ax} \\ R_{\tan} \\ R_{rad} \end{Bmatrix};$$

$$\{X_B\} = \begin{Bmatrix} W_{ax} \\ W_{\tan} \\ W_{rad} \\ \varphi_{ax} \\ \varphi_{\tan} \\ \varphi_{rad} \end{Bmatrix} \qquad (1.122)$$

Rim displacement at beam j will be the sum of displacements from all Q forces:

$$\{X_D\}_j = \sum_{k=1}^{Q} (\alpha)_{jk} \{F_D\}_k \qquad (1.123)$$

where $(\alpha)_{jk}$ is the coefficient matrix for response at beam j due to force at beam k. When the beam attached at point j is considered:

$$\{X_B\}_j = [\beta]_j \{F_B\}_j \qquad (1.124)$$

If the beams are rigidly attached to the disk, then:

$$\{X_B\}_j = \{X_D\}_j\ ,\ j = 1, 2, \ldots Q \qquad (1.125)$$

$$\{F_B\}_j = -\{F_D\}_j\ ,\ j = 1, 2, \ldots Q \qquad (1.126)$$

Combining the above equations yields:

$$\sum_{k=1}^{Q} (\alpha)_{jk} \{F_D\}_k + [\beta]_j \{F_D\}_j\ ;\ j = 1, 2, \ldots Q \qquad (1.127)$$

This leads to the frequency determinant:

$$[\Delta]_{6Q \times 6Q} \{F_D\}_{6Q \times 1} = 0 \qquad (1.128)$$

For a frequency value satisfying (1.128), setting $F_{D1} = N_{ax1} = 1$ unit determines the remaining parameters in $\{F_D\}$ by simultaneously solving the $(6Q - 1)$ equations; mode shapes are obtained from (1.124).

1.18 EXAMPLE PROBLEMS

Example Problem # 1.1: A force 5 sin ($2\pi 60t$) acts on a displacement of (1/10) × sin ($2\pi 60t - \pi/4$). What is the work done during the first second? (Units: inch, pound, seconds).

Solution: The force is $\pi/4$ radians (45°) out of phase with the displacement; so it can be resolved into two components, each of magnitude $5/\sqrt{2}$ lbs in phase and 90° out of phase with displacement. The component in phase with the displacement does no work. Work done by the 90° out of phase component per cycle is: $\pi P_0\, x_0 = \pi.(5/\sqrt{2}).(1/10) = 1.11$ in-lbs. In the first second there are 60 cycles, so the work performed is determined to be 60 × 1.11 = 66.6 in-lbs.

Example Problem # 1.2: In the above example what is the work during the first 0.01 second?

Solution: In the first 0.01 second there will be 60 × 0.01 = 0.6 cycle, or 0.6 × 360° = 216°. For part of the cycle, work is determined by integration.

$$W = \int P dx = \int P_o \sin(\omega t).x_o \omega.\cos(\omega t - \varphi)dt$$

$$= 5(1/10)\int_0^{216°} \sin(\omega t).\cos(\omega t - 45°)d(\omega t)$$

$$= (1/4)\cos(45°).\sin^2(\omega t) + (1/2)\,\sin(45°)[(\omega t/2 - (1/4)\sin(2\omega t)]\}_0^{216°}$$

$$= (1/4) \times (0.707) \times (-0.588)^2 + [216 \times 0.707/(4 \times 57.3)] - 0.707 \times 0.951/8$$

$$= +\ 0.643 \text{ in-lbs.}$$

Example Problem # 1.3: In the system shown in fig. 1.44 the mass weighs 2 lbs, spring stiffness is 25 lbs/in.; l = 8 in. and $a = b$ = 4 in. The system damping constant acting on the mass c = 0.005 lbs/in/sec. What is the damper critical damping?

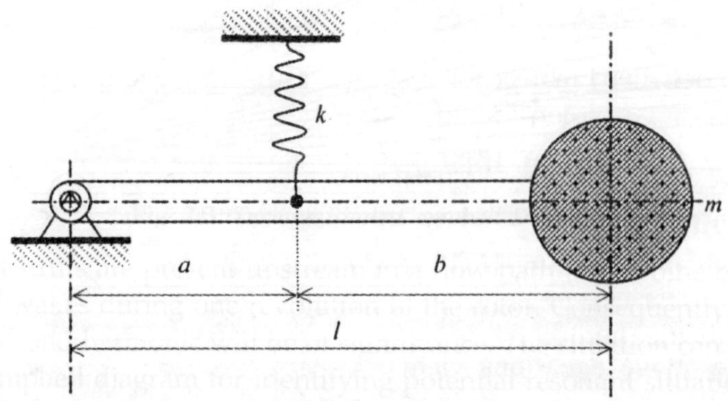

Fig. 1.44.

Solution: The undamped natural frequency is ω_n = Ö(k/m). The equivalent spring constant is $k.a^2/l^2$, or 25 ´ (16/64) = 6.25 lbs/in. Then:

$$\omega_n = \sqrt{(k/m)} = \sqrt{(6.25 \times 386/2)} = 34.73 \text{ radians/sec.}$$

The critical damping constant is:

$$c_c = 2m\omega_n = 2 \times (2/386) \times 34.73 = 0.36 \text{ lbs/in/sec.}$$

Example Problem # 1.4: Find the rate of decay of free vibration in the example problem # 1.3.

Solution: The rate of decay is:

$$\Delta x/x = \delta = 2\pi \times (c/c_c) = 2\pi \times (0.005/0.36) = 0.087$$

Example Problem # 1.5: A variable length cantilever beam consists of a strip of spring steel 0.20 in. wide and 0.020 in. thick. It carries a weight of ¼ oz. at its free end. What should be the free length of the strip if it is required to have frequencies from 6 cycles/sec to 60 cycles/sec?

Solution: The spring constant of a cantilever beam is $3EI/l^3$. The moment of inertia of the cross-section is:

$$I = (1/12)bh^3 = .2 \times (.02)^3 / 12 = 1.33 \times 10^{-7} \text{ in}^4.$$

Bending constant $EI = 30 \times 10^6 \times 1.33 \times 10^{-7} = 4.0$ lb s-in^2, so spring constant $k = 12/l^3$ lbs/in. At the end, mass $m = 1/(4 \times 16 \times 386) = 4.05 \times 10^{-5}$ lbs-in^{-1}-sec^2.

Mass per inch length of strip $\mu = 0.2 \times 0.02 \times 0.28/386 = .29 \times 10^{-5}$ lbs-in^{-1}-sec^2. So total mass $= (4.05 + 0.29l) \times 10^{-5}$ lbs-in^{-1}-sec^2.

Maximum length frequency is 6 cycles/sec, or, $\omega^2 = (2\pi \times 6)^2 = 1421$ (radians/sec)^2.

Natural frequency: $\omega^2 = k/m$, or $1421 = 12 \times 10^5/\{(4.05 + 0.29l)l^3\}$

By a method of trial and error, length of the beam is: $l = 5.33$ in.

Example Problem # 1.6: A flywheel may be represented approximately by a solid steel disk of 4.5 in. diameter and 2 in. thickness. The shaft on which it is mounted has a torsional stiffness of 9.5 in-lbs/radian. Find the natural frequency of vibration of the system.

Solution: Weight of the flywheel is:

$$(\pi/4) \times (4.5)^2 \times 2.0 \times .28 = 8.9 \text{ lbs.}$$

and its moment of inertia is:

$$I = \tfrac{1}{2}(mr^2) = \tfrac{1}{2}(8.9/386)(2.25^2) = 0.0584 \text{ lbs-in-second}^2.$$

The natural frequency thus is:

$$\omega_n = \sqrt{(k/I)} = \sqrt{(9.5/0.059)} = 12.75 \text{ radians/second.}$$

or,

$$f_n = 12.75/2\pi = 2.03 \text{ cycles/second.}$$

Example Problem # 1.7: A stroboscope is a device for producing intermittent flashes of light with which rapid vibratory motions can be made to appear to stand still, or to move very slowly. The flashes of light are of extremely short duration. When a vibrating object illuminated with this light is adjusted to the same frequency as the vibration, the object will be seen in a certain position; then it will be dark, and the object is invisible while traveling through its cycle. When it returns to the first position after one cycle another flash of light occurs. Thus the object appears to stand still. Now consider a point located 4 in. from the axis of a machine rotating at 10,000 rpm. If it is required to see the point with a blurring of less than 1/32 in., what should be the duration of the light flashes?

Solution: The point in question travels at:

$$10{,}000 \times 2\pi \times 4/60 = 4{,}189 \text{ in/second}$$

Flash time duration multiplied by blur velocity of the point will give blur size, which is $4189t = (1/32)$, so $t = 1/(4189 \times 32) = 1/134{,}000$ second.

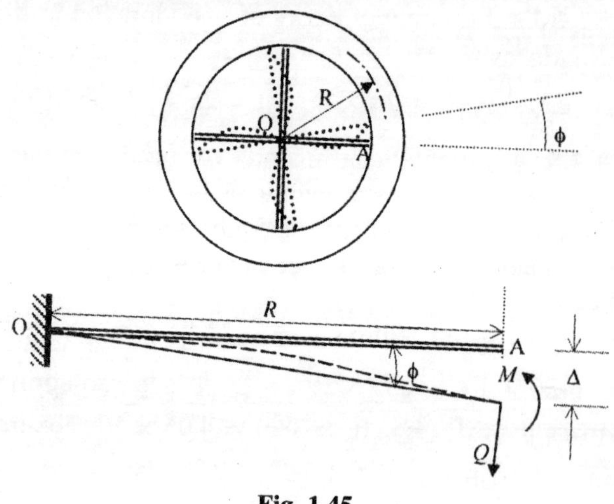

Fig. 1.45.

Example Problem # 1.8: A flywheel consists of a heavy rim of weight W = 475 lbs and mean radius R = 22.75 in. attached by four flexible prismatic spokes (fig. 1.45). If the hub is held fixed, derive expressions for loads in terms of deflection and slope in the spokes and find the period of free rotational vibration of the rim about its central axis O. Neglect the mass of the spokes, and assume spoke bending stiffness B = 2,250 lbs/in.

Solution: Consider the rim to have a small angle of rotation ϕ from its equilibrium position as shown. Each spoke behaves as a beam built in at the hub and constrained to move with the rim at the other end. At the outer end of the spoke shear force Q and bending moment M act, which may be determined from formulas for beam stiffness:

$$Q = \frac{12B\Delta}{R^3} - \frac{6B\phi}{R^2} = \frac{12\times 2250\Delta}{22.75^3} - \frac{6\times 2250\phi}{22.75^2}$$

$$= 2.293\Delta - 26.08\phi$$

$$M = \frac{6B\Delta}{R^2} - \frac{4B\phi}{R} = \frac{6\times 2250\Delta}{22.75^2} - \frac{4\times 2250\phi}{22.75}$$

$$= 26.08\Delta - 395.6\phi$$

If the rim is assumed to be rigid, the tangent to the elastic line at the outer end of each spoke will be radial. Then deflection $\Delta = R\phi$, giving:

$$Q = 6B\phi/R^2 = 26.08\phi \text{ and } M = 2B\phi/R = 197.8\phi$$

Then the total moment on the rim will be:

$$M_{total} = 4QR - 4M = 16B\phi/R = 1582\phi$$

and the rotational spring constant is:

$$k_r = \frac{M_{total}}{\phi} = 1582 \text{ in.-lbs/radian.}$$

Since the mass moment of inertia of the flywheel rim is:

$$J = WR^2/g = (475 \times 22.75^2)/386 = 636.9 \text{ in.-lbs/second}^2,$$

then the period of free natural vibration of the rim is:

$$\tau = 2\pi\sqrt{\frac{WR^3}{16gB}} = 2\pi\sqrt{\frac{JR}{16B}} = 2\pi\sqrt{\frac{636.9 \times 22.75}{16 \times 2250}} = 3.986 \text{ second}$$

Example Problem # 1.9: Using Rayleigh's method for calculating the angular frequency of the fundamental mode of vibration of the beam with two masses, shown in fig. 1.46. Assume $W_1 = W_2 = W$, flexural rigidity of the beam as EI and neglect the mass of the beam.

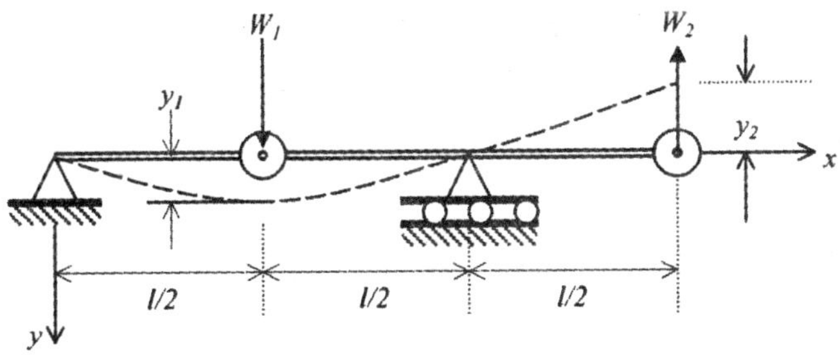

Fig. 1.46.

Solution: From Rayleigh's method, for n masses on a beam the formula for angular frequency is:

$$p^2 = \frac{g\sum_{j=1}^{n} W_j y_j}{\sum_{j=1}^{n} W_j y_j^2}$$

It will be assumed that during vibration the beam maintains a shape similar to the static deflection curve due to the two weights acting in opposite directions. The corresponding static deflections are:

$$y_1 = \frac{W_1 l^3}{48EI} + \frac{W_2 l^3}{32EI} = \frac{5Wl^3}{96EI}$$

$$y_2 = \frac{W_1 l^3}{32EI} + \frac{W_2 l^3}{8EI} = \frac{5Wl^3}{32EI}$$

Substituting these values into the expression for angular frequency gives:

$$p = \sqrt{\frac{192EIg}{25Wl^3}} = 2.77128 \times \sqrt{\frac{EIg}{Wl^3}}$$

Example Problem # 1.10: If $W_1 = 3W_2$ in the above problem, what percentage change in the frequency is obtained?

Solution: Since $W_1 = 3W_2 = 3W$ and $W_2 = W$, static deflections are:

$$y_1 = (3/32) \times (Wl^3/EI) \text{ and } y_2 = (7/32) \times (Wl^3/EI)$$

Substituting in the expression for frequency:

$$p = 2.59554 \times \sqrt{\frac{EIg}{Wl^3}}$$

Hence the frequency is reduced by (2.7713 – 2.5955)/2.7713 = 6.34%

Example Problem # 1.11: A simply supported beam is impacted by a free falling sphere of weight W from height h, as shown in fig. 1.47. Determine an expression for maximum momentary deflection and stress produced in the beam.

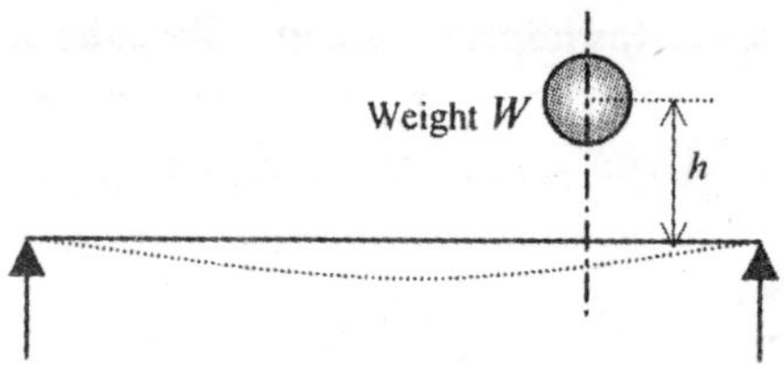

Fig. 1.47.

Solution: When a load W falls a distance h on the beam, it causes a peak instantaneous deflection δ in the beam where the ball strikes. The total work done by the ball is $W(h + \delta)$. Assume a static equivalent load W_{eq} that produces an equal deflection δ at the same point on the beam, with the load applied gradually. The work done by this equivalent load is $\frac{1}{2}(W_{eq}\delta)$. If the supports at the two ends of the beam are rigid, the work done by the falling and the stationary loads are the same, or:

$$W(h + \delta) = \tfrac{1}{2}(W_{eq}\delta)$$

The static deflection δ may be expressed in terms of the equivalent static load W_{eq}. For example, if the load is applied centrally on the beam, $\delta = (W_{eq}l^3/48EI)$, where l is the length of the beam and EI represents the beam stiffness in flexure.

Example Problem # 1.12: If the ball weighs 8000 lbs, produces a static deflection of 0.75 in. and a maximum bending stress of 5500 lbs/in² in the above problem, from what height can a ball of 4000 lbs be dropped at the same point in the beam if the maximum permissible bending stress is 22000 lbs/in²?

Solution: The required static load for a stress of 22000 lbs/in² is:

$$W_{eq} = (22000/5500) \times 8000 = 32000 \text{ lbs}$$

Since static stress and deflection are proportional to the load causing them, the corresponding deflection is: $\delta = (32000/8000) \times 0.75 = 3.0$ in. Substituting the values in the expression obtained from the previous example problem,

$$4000(h + 3.0) = \tfrac{1}{2}(32000 \times 3.0)$$

Hence, $h + 3.0 = 12.0$ and the required height is 9.0 in.

Example Problem # 1.13: Consider a prismatic bar rotating about an axis normal to the beam, resulting in a state of centrifugal stress in the bar (fig. 1.48). If angular speed is 32,500 rpm, outer and inner radii are 5.5 in. and 2.5 in. and material density is 0.28 lbs/in³, determine the maximum stress.

Solution: At any radius in the bar stress due to centrifugal force is:

$$\sigma = \int \rho\omega^2 r dr = (\rho\omega^2/2)(r_o^2 - r_i^2)$$

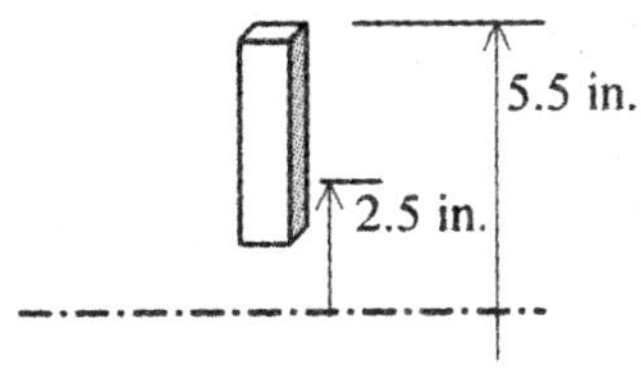

Fig. 1.48

Stress at the inner radius near the axis of rotation is:

$$\sigma = [(0.28/386) \times (2\pi \times 32500/60)^2](5.5^2 - 2.5^2)/2 = 100827 \text{ lbs/in}^2.$$

REFERENCES & BIBLIOGRAPHY

Anwar, I., "Computerized Time Transient Torsional Analysis of Power Trains", ASME paper # 89-DET-74, New York, 1989.

Carnegie, W., "Vibrations of Pre-twisted Cantilever Blading", Proceedings, Institute of Mechanical Engineers, Volume 173, Page 343, 1959.

Carnegie, W., Dawson, B., Thomas, J., "Vibration Characteristics of Cantilever Blading", Proceedings, Institute of Mechanical Engineers, Volume 180 (31), Page 71, 1965-1966.

Choi, Y.S., Noah, S.T., "Forced Periodic Vibration of Unsymmetric Piecewise-Linear Systems", Journal of Sound and Vibration, Volume 121(3), Page 117-126, 1988.

Classen, R.W., Thorne, C.J., "Vibrations of a Rectangular Plate", Journal of Aerospace Science, Volume 29, Page 1300, 1962.

Crede, C.E., Ruzicka, J.E., "Theory of Vibration Isolation", Shock & Vibration Handbook, McGraw-Hill, New York, 1988.

Den Hartog, J.P., "Mechanical Vibrations", Dover Publications, Inc., New York, 1984.

Deutschman, A.D., Michels, W.J., Wilson, C. E., "Machine Design – Theory & Practice", MacMillan Publishing Co., New York, 1989.

Ehrich, F. F., "Rotor Dynamic Response in Nonlinear Anisotropic Mounting Systems", Proceedings of the 4th International Conference on Rotor Dynamics, IFTOMM, 1-6, Chicago, September 7-9, 1994.

Ehrich, F.E., "Handbook of Rotor Dynamics", Krieger Publishing Co., Malabar, FL, 1999.

Eshelman, R.L., "Torsional Vibrations in Machine Systems", Vibration, 3(2): 3, 1987.

Gale, N.F., Widener, S.K., Sui, P.C., Zhang. H., "Analytical Engine Design Methods: A Review", SAE paper # 950806, 1995.

Goldstein, H., "Classical Mechanics", Addison-Wesley Publishing Co. Inc., Mass, 1987.

Grinsted, B., "Nodal Pattern Analysis", Proceedings, Institute of Mechanical Engineers, Volume 166, Page 309, London, 1952.

Gupta, K., "Vibration of Small Aspect Ratio Blades", Ph.D. Thesis, Indian Institute of Technology, New Delhi, 1979.

Gupta, K., Rao, J. S., "Flexural Vibrations of Pre-twisted Cantilever Plates", Journal of Aeronautical Society of India, Volume 30, Page 131, 1978.

Hildebrand, F.B., "Advanced Calculus for Applications", Prentice-Hall, New Jersey, 1983.

Jeon, H., Tsuda, K., "Theoretical Analysis of the Undamped, Coupled Torsional – Axial Vibration of Marine Diesel Engine Shafting", Bulletin of the M.E.S.J., Volume 4, Number 7, Page 41, 1989.

Law, B., "Prediction of Crankshaft and Flywheel Dynamics", Institute of Mechanical Engineers, Paper # C382/046, Pages 487-498, Perkins Technology, London, 1989.

Lee, H.C., Bishop, R.E.D., "Applications of Integral Equations to Flexural Vibrations of a Wedge with Rotary Inertia and Shear", Journal of Franklin Institute, 277, Page 327 (1964).

Love, A.E.H., "A Treatise on the Mathematical Theory of Elasticity", Dover Publications, 1944.

Mabie, H. H., Reinholtz, C.F., "Mechanism and Dynamics of Machinery", John Wiley & Sons, 1987.

MacBain, J.C., "Vibratory Behavior of Twisted Cantilever Plates", Journal of Aircraft, Volume 12, Page 343, 1975.

McLachlan, N.W., "Ordinary Nonlinear Differential Equations", Oxford University Press, New York, 1956.

Meirovitch, L., "Analytical Methods in Vibration", Macmillan Publishing Co., Inc., New York, 1984.

Novozhilov, V.V. "The Theory of Thin Shells", P. Nordhoff Ltd., Netherlands, 1959.

Petricone, R., Sisto, F., "Vibration Characteristics of Low Aspect Ratio Compressor Blades", ASME paper # 70-GT-94, New York, 1970.

Rangwala, A.S., "Reciprocating Machinery Dynamics: Design and Analysis", Marcel Decker, New York, 2001.

Rao, J.S., "The Fundamental Flexural Vibration of a Cantilever Beam of Rectangular Cross-Section with Uniform Taper", Aero Q., Volume 16, Page 139, 1965.

Rao, J.S., Jadvani, H.M., "Free and Forced Vibration of Turbine Blades", Vibration of Bladed Disk Assemblies, ASME Publication, Page 11, New York, 1983.

Rao, J.S., "Mechanical Vibration", Addison-Wesley Publishing Co., Reading, Mass., 1990.

Rao, J.S., "Turbo-machine Blade Vibration", Wiley Education Ltd., John Wiley & Sons, New Delhi, India, 1991.

Rawtani, S., Dokainish, M.A., "Bending of Pre-twisted Cantilever Plates", Transactions, CASI, Page 89, 1969.

Rayleigh, "The Theory of Sound", Volume # I, Macmillan & Co. Ltd., London, 1894.

Roark, R. J., Young, W.C., "Formulas for Stress and Strain", 5th edition, McGraw-Hill, New York, 1985.

Robeson, G., "Cosworth – The Search for Power", Patrick Stephens Limited, London, 1991.

Schroen, B.G., "Die Dynamik der Verbrennungskraftmaschine", Springer-Verlag, Vienna, 1982.

Shemeld, D.E., "A History of Development in Rotor Dynamics - A Manufacturer's Perspective", CP2443, NASA, Washington, D.C., 1986.

Smith, D.M., "Vibration of Turbine Blades in Packets", Proceedings, 7th Conference for Applied Mechanics, Volume 3, Page 148, London, 1948.

Subramanyam, K.B., Kulkarni, S.V., Rao, J.S., "Coupled Bending-Torsion Vibrations of Rotating Blades of Asymmetric Aerofoil Cross-section with Allowance for Shear Deflection and Rotary Inertia by Use of the Reissner Method", Journal of Sound and Vibrations, Volume 75(1), Page 17, 1981.

Sutherland, R.L., Goodman, L.E., "Vibration of Prismatic Bars Including Rotary Inertia and Shear Deflections", University of Illinois, 1954.

Swaminathan, M., Rao, J.S., "Vibrations of Rotating Pre-twisted and Tapered Blades", Mechanical Machine Theory, Volume 12(4), Page 331, 1977.

Thomas, J., Mota Soares, C.A., "Dynamic Analysis of Rotating Turbine and Compressor Blading", Vibrations in Rotating Machinery Conference Proceedings, Institute of Mechanical Engineers, Page 231, 1976-77.

Timoshenko, S.P., "On the Correction for Shear of the Differential Equations for Transverse Vibrations of Prismatic Bars", Phil. Mag., 41, Page 566 (1929).

Timoshenko, S., Young, D.H., Weaver, W., "Vibration Problems in Engineering", John Wiley & Sons, Inc., New York, 1984.

Vezier, G.A.L., "Theory of Elastic Thin Shells", Pergamon Press, 1961.

Wachel, J.C., Tison, J.D., "Vibrations in Reciprocating Machinery and Piping Systems", Proceedings of the 23rd Turbo-machinery and Piping Systems, Texas A & M University, College Station, TX, 1994.

Warburton, G.B., "The Vibration of Rectangular Plates", Proceedings, Institute of Mechanical Engineers, 168, Page 371, 1954.

Watanabe, I., "Natural Frequencies of Flexural Vibrations of Tapered Cantilevers with Uniform Thickness", Proceedings, 1st Japan National Congress Applied Mechanics, Page 547, 1951.

Wehrli, V.C., "Uber Kritische Drezahlen unter Pulsierender Torsion", Ingenieur Arch, 33:73-84, 1993.

CHAPTER 2

Fan and Compressor Airfoils

2.1 INTRODUCTION

Compressor blades are designed from aerodynamic considerations to maximize the development of air pressure, but must be capable of withstanding the fluctuating lift and moment loads repeatedly at a frequency dependent on the number of stationary nozzle vanes and the speed of the machine. The considerable dynamic loads imposed on them as they traverse through the varying pressure field past the non-rotating vanes inherently cause failures of blades. The airfoils are flexible structural members possessing a large number of natural frequencies, and the potential to interact with vane excitation frequencies in the region is present. Even in power generating machines operating at a near fixed speed it is difficult to avoid resonance during start-up and shutdown operations, when the instantaneous rotor speed leads to a coincidence of the excitation and blade frequencies. Aircraft engines, on the other hand, operate through a large operating speed range as the engine accelerates from ground to flight idle, cruise and takeoff speeds. The problem of ensuring the integrity of the airfoils then assumes even greater proportions.

Air flowing through the annulus of an axial compressor performs the relatively more difficult task of going against an adverse pressure gradient. As the compression ratio increases, the danger of stall troubles rises correspondingly, which is all too prevalent in axial compressors. In individual airfoils stalling is encountered when the difference between the flow direction and the blade's angle of incidence becomes excessively large. The stability of the flow is jeopardized by the very fact that the movement is taking place in the same direction as that of the increase in pressure. Flow reversals occur under the right conditions of mass flow and rotor speed, and which vary from those for the blade's design.

Air is first imparted acceleration by the rotating blades, then decelerated in the stator passages, where the kinetic energy is converted to static pressure. A series of diffusions take place in both the rotor and stator passages. Absolute velocity of the working fluid is increased in the rotor, but the relative velocity will decrease. It is imperative for flow area to diminish moderately in a diffusing flow, hence a single compressor stage can only deliver a small increment in the pressure, considerably less then what a turbine can advantageously use with its efficient pressure gradient, converging blade passages and accelerating flow. This explains why a few turbine stages can power many more stages of the compressor. As fluid density increases with progression of the flow, axial velocity does not change appreciably. During operation at lower speeds density in the latter stages deviates further from the design point, causing the flow velocity to reach a point where the blades may stall and the compressor to surge.

Airfoil sections are designed to maximize efficiency, hence the need to pass greater flow of air at high pressure ratio calls for increasing the Mach number. In the first stage the larger tip radius increases peripheral speed, posing a significant issue. Oftentimes compressor blades are required to operate over a transonic range, meaning that flow over a portion of the blade exceeds unity Mach number. Blade sections based on circular arcs, sometimes referred to as bi-convex blades, have been found to be effective in the transonic mode of operation. With even greater Mach numbers parabolic based sections are required. Mach numbers higher than 1.5 are now employed in compressors of industrial gas turbines and for fans of high bypass ratio turbofan engines. At entry velocity below the critical Mach number the performance of the blades does not exhibit much variation with the speed. Above the critical speed the losses mount rapidly to a point where an appreciable increase in pressure increment is not experienced, and the blade loses the capacity to provide diffusion for the flow. At zero incidence the Mach numbers are in the region of 0.7 to 0.85. Increased Mach numbers also reduce the range of incidence angle for which losses may be at an acceptable level.

At the inlet the air temperature, and hence the acoustic velocity, is lowest, so compressibility effects play a major role in the front stages of the compressor. Blade tip velocity is also the highest at the first stage, and is significant if shock losses and noise are to be controlled. Because of the increased flow whirl and the speed required to obtain constant work input at all heights, Mach number in the stator at the hub radius will be the maximum. In the first stator stage Mach number may vary from 0.55 at the root to 0.45 at the tip, while in the rotor the corresponding values are 0.70 and 1.2, and 0.9 at the mean radius. Diffusion factors generally correlate well with measured losses for subsonic conditions, and are substantially higher with supersonic losses.

Even at low spin speed the fan tip velocity tends to be high in modern turbofan engines, mainly because of the large diameter required, and may be in the range of 1.4 to 1.6 Mach number. Double circular arc based airfoil profiles do not perform satisfactorily under these circumstances, and must be specially developed. Supersonic diffusion must then be provided to a Mach number of 1.2 prior to the normal shock at entry to the passage between the blades. For diffusion to occur above the acoustic speed, it interestingly calls for a reduction in the flow area, and may be accomplished either by decreasing the annulus area or by making the suction side slightly concave. A combination of the two methods may add to the benefit. A fan blade section with little curvature is shown in fig. 6.2. Losses as a consequence of shock are only a part of the problem, since interaction between the boundary layer and shock waves magnify the viscous losses (*ref*: Saravananamuttoo, 2001).

A damper at part span is required in long and flexible fan blades to control torsional and flexural motion, and also proves advantageous in the event of ingestion of a foreign object during the take-off roll of the aircraft. However, performance of the portion of the blade in the proximity of the damping device is diminished, especially if it is located where the Mach number is high. Wide chord fan blade development has eliminated the need for the dampers, but improvements in manufacturing and stress analysis techniques have played no small role in bringing about this progress. A fan rotor with integral wide-chord blades machined from a single forging has been developed by some aircraft engine manufacturers. A hollow core with stiffeners helps to cut the weight of some very large fan blades.

2.2 FINITE ELEMENT METHOD IN BLADE VIBRATIONS

A number of difficulties are encountered in analyzing shells shaped in the form of turbomachinery airfoils, and the problem is compounded due to varying thickness and substantial deformation. Finite element programs developed for this application go a long way in the design process. Classical shell theory plays a major role in the formulation of shell elements (*ref*: Rao, 1991). Three-dimensional elements modified to satisfy the underlying shell characteristics may also be employed. Flat shells are oftentimes substituted for curved ones for simplification of geometry and the analytical process. Triangular and iso-parametric quadrilateral shell elements curved in two planes are also available. But most of them suffer from limitations in applicability, arising from assumptions within the underlying shell theory. Problems come up when working in the three-dimensional realm in the form of large stiffness coefficients for displacements along an edge when all degrees of freedom are retained at each node.

A form of three-dimensional shell element for use in shell structures is illustrated in fig. 2.1 (*ref*: Ahmad *et al.*, 1970).

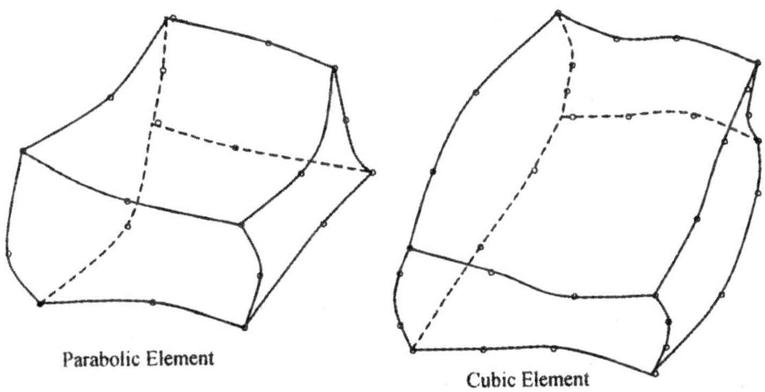

Fig. 2.1. *Hexahedral Three Dimensional Elements*

The original normal to the mid-surface is assumed as inextensible and straight, while the strain energy corresponding to stresses perpendicular to this surface is ignored. Hence, elastic modulus in the normal direction becomes zero. Shear deformation is important in thick shell elements; this is achieved by permitting the normal to remain perpendicular to the deformed middle surface.

A typical thick shell element is shown in fig. 2.2. External faces are curved, but straight lines define sections across the thickness. A pair of i_{top} and i_{bottom} describes the geometry in Cartesian coordinates. Assume ξ and η are curvilinear coordinates in the middle plane and ζ a linear coordinate along the thickness of the shell, their values changing between +1 and –1 on the faces of the shell. The Cartesian and curvilinear coordinates are related by:

$$\begin{Bmatrix} x \\ y \\ z \end{Bmatrix} = \sum_{i=1}^{8} N_i(\xi,\eta)\frac{1+\varsigma}{2}\begin{Bmatrix} x_i \\ y_i \\ z_i \end{Bmatrix}_T + \sum_{i=1}^{8} N_i(\xi,\eta)\frac{1-\varsigma}{2}\begin{Bmatrix} x_i \\ y_i \\ z_i \end{Bmatrix}_B \tag{2.1}$$

Here $N_i(\xi,\eta)$ is a function assuming unity value at node i and zero at other nodes. N_i are shape functions of a square parent element with 8 nodes. The functions are parabolic in ξ and η, so the curved shape of the element can take a parabolic form. With additional mid-side nodes the shape functions take the form:

$$N_i = \frac{1}{4}(1+\xi_0)(1+\eta_0) - \frac{1}{4}(1-\xi^2)(1+\eta_0) - \frac{1}{4}(1-\eta^2)(1+\xi_0) \tag{2.2}$$

where $\xi_o = \xi\xi_i$, $\eta_o = \eta\eta_i$, ξ_i and η_i being +1 and –1 and i = corner node, with the mid-side node $\xi_i = 0$.

$$N_i = \frac{1}{2}(1+\eta_0)(1-\xi^2) \tag{2.3}$$

and where $\eta_i = 0$, $\quad Ni \;=\; N_i = \frac{1}{2}(1+\xi_0)(1-\eta^2)$ (2.4)

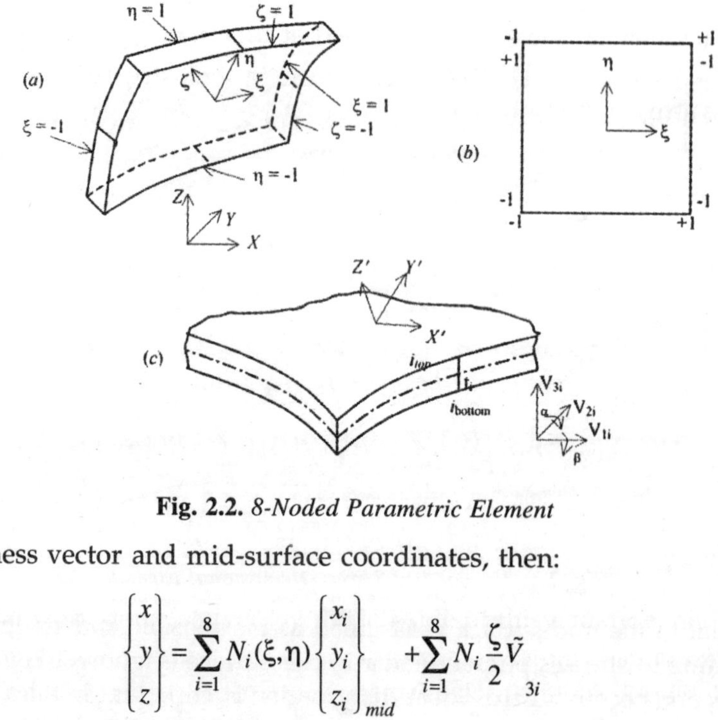

Fig. 2.2. *8-Noded Parametric Element*

Using a thickness vector and mid-surface coordinates, then:

$$\begin{Bmatrix} x \\ y \\ z \end{Bmatrix} = \sum_{i=1}^{8} N_i(\xi,\eta)\begin{Bmatrix} x_i \\ y_i \\ z_i \end{Bmatrix}_{mid} + \sum_{i=1}^{8} N_i \frac{\zeta}{2}\tilde{V}_{3i} \tag{2.5}$$

where x_i, y_i, z_i are average values of the coordinates at the two surfaces. Then:

$$\begin{Bmatrix} u \\ v \\ w \end{Bmatrix} = \sum_{i=1}^{8} N_i \begin{Bmatrix} u_i \\ v_i \\ w_i \end{Bmatrix} + \sum_{i=1}^{8} N_i \frac{\zeta t_i}{2}[\tilde{V}_{1i} - \tilde{V}_{2i}]\begin{Bmatrix} \alpha_i \\ \beta_i \end{Bmatrix} \tag{2.6}$$

where $\tilde{V}_{1i}, \tilde{V}_{2i}$ are unit orthogonal vectors with displacements in global axes x, y, z. u_i, v_i, w_i are displacements at the mid-surface nodes and β, α are rotations about $\tilde{V}_{1i}, \tilde{V}_{2i}$, providing a total of 5 degrees of freedom at each node. In matrix notation, $\{U\} = [N]\{a\}$, where $\{a\}$ is a column vector and $[N]$ is obtained by expanding equation 6.18.

With the displacements available, element properties, strains and stresses need to be defined. From fig. 2.2 the strain components are:

$$[\varepsilon'] = \begin{Bmatrix} \varepsilon_{x'} \\ \varepsilon_{y'} \\ \gamma_{x'y'} \\ \gamma_{x'x'} \\ \gamma_{y'x'} \end{Bmatrix} = \begin{Bmatrix} u'_{,x'} \\ v'_{,y'} \\ u'_{,y'} + v'_{,x'} \\ w'_{,x'} + u'_{,x'} \\ w'_{,y'} + v'_{,x'} \end{Bmatrix} = [L]\{U'\} \tag{2.7}$$

where:

$$[L] = \begin{bmatrix} ,x' & 0 & 0 \\ 0 & ,y' & 0 \\ ,y' & ,x' & 0 \\ ,z' & 0 & ,x' \\ 0 & ,z' & ,y' \end{bmatrix} \text{ and } \{U'\} = \begin{Bmatrix} u' \\ v' \\ w' \end{Bmatrix}$$

Hence: $\{U'\} = [N]\{a'\}$ and $\{\varepsilon'\} = [B]\{a'\}$ where $[B] = [L][N]$.

Corresponding stresses in matrix form are:

$$\{\sigma'\} = [D]\{\varepsilon'\} \tag{2.8}$$

where elasticity matrix $[D]$ is:

$$[D] = \frac{E}{1-\nu^2}\begin{bmatrix} 1 & 0 & 0 & 0 & 0 \\ 0 & 1 & 0 & 0 & 0 \\ 0 & 0 & \frac{1-\nu}{2} & 0 & 0 \\ 0 & 0 & 0 & \frac{1-\nu}{2\kappa} & 0 \\ 0 & 0 & 0 & 0 & \frac{1-\nu}{2\kappa} \end{bmatrix} \tag{2.9}$$

where E and ν are Young's modulus and Poisson's ratio. Factor κ approximates displacement due to shear. Properties of elements call for integration over their volume, and take the form of:

$$\int [S]dxdydz \tag{2.10}$$

where $[S]$ is a function of the global coordinates x, y, z, and $[S] = [B]^T[D][B]$, with strain defined by:

$$\{\varepsilon\} = [B]\{a\}; \quad \{a\} = \begin{Bmatrix} a_1^e \\ a_2^e \\ \cdot \\ \cdot \\ a_8^e \end{Bmatrix}$$

Thus, $[B]$ is defined in terms of displacement derivatives in the local Cartesian coordinates x', y', z' and $\{a\}^e$ is the displacement field. Integration of the element in the curvilinear coordinates can be performed after transformations from the local to the global system and then to the curvilinear ξ, η, ζ coordinates.

Equation (6.13) relates global displacements u, v, w to the curvilinear coordinates. Derivatives of these displacements with respect to ξ, η, ζ may be obtained by the Jacobian matrix, so derivatives of displacements in global coordinates are given by: $[U_d]_{global} = [J]^{-1}[U_d]_{curvilinear}$. Matrix terms for the derivatives of the displacements in the curvilinear coordinates may then be written. Also, components of the Jacobian matrix in curvilinear coordinates can be written using equations (6.8). Further transformation from global to local Cartesian coordinates allows establishment of the strains. Direction of the local orthogonal axes can be ascertained by obtaining a vector normal to the surface ζ = constant by taking a vector product of any two vectors tangent to the surface. Global derivatives of displacements u, v, w are next transformed to local derivatives of local orthogonal displacements to explicitly obtain displacement derivatives at any ξ, η, ζ in the element as also the components of strain.

The infinitesimal volume in curvilinear coordinates is:

$$dxdydz = \det[J]d\xi d\eta d\zeta$$

So the stiffness matrix is: $$[K]^e = \iiint_{-1 to +1} [B]^T [D][B]\det[J]d\xi d\eta d\zeta \tag{2.11}$$

and the mass matrix is: $$[M]^e = \iiint_{-1 to +1} [N]^T \rho[N]\det[J]d\xi d\eta d\zeta \tag{2.12}$$

To obtain the equations of motion, the principle of virtual work can be gainfully employed to derive expressions for equilibrium of the body as an assembly of m finite elements. According to this principle, the total internal virtual work done by compatible small virtual displacements applied to a body must equal the total external virtual work in order to maintain equilibrium of the body. Mathematically the requirement may be given by the expression:

$$\sum_m \int \{\bar{\varepsilon}\}^T [\sigma]dV = \sum_m \int \{\bar{U}\}^T f_B dV - \sum_m \int \{\bar{U}\}^T \{dU/dt\} f_F dV$$

$$= -\sum_m \int \{\bar{U}\}^T \rho\{d^2U/dt^2\}dV \tag{2.13}$$

Internal virtual work done is given by the left side of this equation, and equals the stresses going through the virtual strains corresponding to the virtual displacements $\{\bar{U}\}$. External work is given on the right side. It equals work done by body forces f_B, friction force $\{dU/dt\}f_F$ and inertia force $\rho\{d^2U/dt^2\}$ going through the virtual work. Virtual displacements in global coordinates in x, y, z directions are given by:

$$\{\bar{U}\} = \begin{Bmatrix} \bar{u} \\ \bar{v} \\ \bar{w} \end{Bmatrix} = \delta\{U\}, \qquad \{[\sigma] = [D]\{\varepsilon\}, \qquad \{\bar{\varepsilon}\} = \delta\{\varepsilon\} \tag{2.14}$$

Substitution in equation (6.20) yields equilibrium requirements. Variation in strain energy V of the elastic continuum and in the potential energy W of the applied loads may then be used in Hamilton's equation $\delta\pi = 0$, where $\pi = V - W$, or may be expressed as:

$$\delta\pi = \sum\delta\left(\frac{1}{2}\int\{a\}^T[B]^T[D][B]\{a\}dV + \int\{a\}^T[N]^T\rho[N]\{d^2a/dt^2\}dV\right)$$

$$+\sum\delta\left(\frac{1}{2}\int\{a\}^T[N]^T[N]\{da/dt\}f_F dV - \int\{a\}^T[N]^T f_B dV\right) = 0 \quad (2.15)$$

Simplification yields: $[M]\{d^2a/dt^2\}+[C]\{da/dt\}+\{a\}=\{F_B\}$ (2.16)

where: $[M] = \sum_m \int [N]^T \rho[N]dV$

$$[C] = \sum_m \int [N]^T[N]f_F dV$$

$$[K] = \sum_m \int [B]^T[D][B]dV$$

$$\{F\} = \sum_m \int [N]^T f_B dV$$

Individual element matrices given by (2.11) and (2.12) may be used for numerical integration:

$$[K]^e = \sum_{i=1}^{n}\sum_{j=1}^{n}\sum_{k=1}^{n} H_i H_j H_k [B]^T[D][B]\det[J] \quad (2.17)$$

$$[M]^e = \sum_{i=1}^{n}\sum_{j=1}^{n}\sum_{k=1}^{n} H_i H_j H_k [N]^T\rho[N]\det[J] \quad (2.18)$$

where H provides weight coefficients of the Gaussian quadrature and n is order of integration. Reduced integration technique may be used, with the order of integration ξ, followed by η and finally ζ. Total stiffness and mass matrices, assembled as a summation for all elements in the system, may be obtained using the frontal housekeeping algorithm (*ref*: Irons, 1970). The eigenvalue problem is solved by a determinant search. As an illustration of this procedure, see example problem # 6.5 using a computer program developed by Gupta (1984) for determining natural frequencies of vibration and their modes.

2.3 INDIVIDUAL BLADE VIBRATION

An in-depth study of the dynamic characteristics of a turbine blade is necessary in evaluating its capability to withstand the assigned loads. What may appear at first sight a relatively simple task of interpreting the modal characteristics of the blade is oftentimes hampered by the geometric configuration; the root structure is thick, with a complex fir tree form for attachment to the disk; airfoil trailing edge is thin; cross-section profiles are varying at different radial heights; the blade is twisted and leaned. Initial blade geometry is controlled by aerodynamic and performance considerations. Within the engine, however, it must meet fluid flow and structural criteria. Chord-wise motion of the airfoil is coupled with the flap, or perpendicular to the chord, motion, primarily because of the twist. Total bending or torsional deformation is not encountered in the blade. Varying degrees of the two movements exist in each mode of vibration as a direct consequence of the coupling between the degrees of freedom. Aerodynamic and mechanical loads also sometimes tend to dramatically change the blade's profile. The blade's design may also go through modifications, such as replacing a shrouded configuration with a wide chord or large sweep blade without the shroud.

A gas turbine's rotating components are subjected to resonant vibration and flutter phenomena. Resonant vibrations result from a coincidence in the blade's natural frequency and exciting force frequency. Exciting forces arise from uneven pressure distributions around obstructions in the gas flow path. For example, a prescribed number of support frame struts, nozzle vanes, combustor burners and bleed ports will result in a corresponding number of wakes in the gas stream past those obstructions. Rotating stall in a compressor, rubbing contact between rotor and stator and meshing of gears are examples of non-integral orders of obstructions in the gas stream. Stationary components such as nozzle vanes and support struts, on the other hand, will experience excitation caused by the rotating blades or by compressor rotating stall. Dynamic stresses, thus, must be adequately taken into consideration in the design of vanes and struts, but the absence of centrifugal force poses less stringent requirements than for rotating blades.

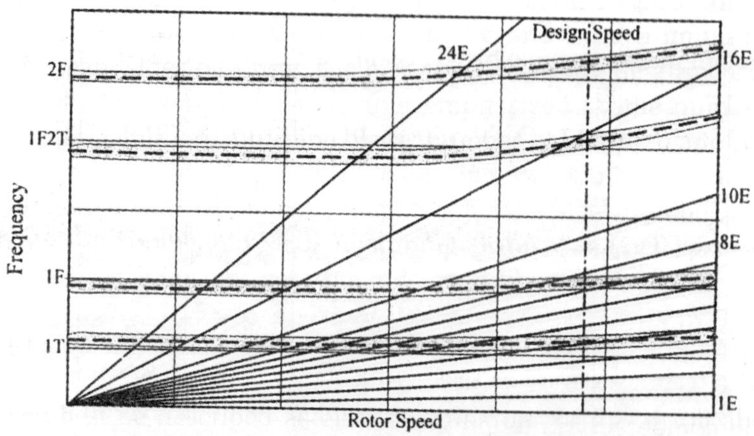

Fig. 2.3. *Turbine Blade Campbell Diagram*

When 8 support struts are present upstream in a flow path the turbine blades will encounter an equal number of wakes during one revolution of the rotor. Consequently, 8E, or 8 times rotor frequency, and its second harmonic will be of significance. The situation can be well summarized in the form of a Campbell diagram for identifying potential resonant situations, as shown in fig. 2.3. The fig. graphically depicts the extent of interference between the blade's natural frequency and the different excitation frequencies for an engine operating at an essentially fixed speed. Note that the blade's natural frequencies, especially the bending mode frequencies, increase with operating speed. This observation is consistent with the blade experiencing increased stiffness as a direct consequence of the higher centrifugal force as rotor speed increases. The example shows the blade's 1F2T (combined first flexure/second torsion) and 2F modes to be above the operating speed for 8E stimulus. The first flexural mode (1F) also has adequate margin for 16E excitation.

However, the 1F mode for the 8E and the 1F2T mode for the 16E stimulus are too close to the design speed line, which may be considered unacceptable for the blade design. The 1F mode of vibration is next to the first torsional (1T) mode, but a higher frequency level also implies increased excitation levels. Blade tuning through modification of the blade's taper characteristics may permit setting of the problem modes to above the rotor speed. The 24E stimulus intersects all of the blade modes at a low rotor speed, hence it is not of consequence.

Note that this simplified illustration only considers the eighth multiple for excitation frequency, and with the engine operating at a constant speed. Other excitation sources and frequencies will be generally present. Aircraft engines operate in a wide operating speed range as throttle movement goes from ground idle to flight idle, cruise and take off conditions. Accuracy in the calculation sequence is also a problem for the blade's higher modes of vibration. Fan blades are often subjected to distortion in the airflow from the inlet section. Low engine order excitations are generally the result, with force magnitude reducing for increased orders. To take care of the uncertainties one option is to use a ±10 to 15 percent margin for the blade's fundamental and second modes (*ref*: Rangwala, 2004). Increased engine operating speed range by a similar margin for the two modes may also be used to take care of uncertainties.

Individual blades also need to be tested using modal analysis with impulse, white noise and sinusoidal excitation applied at the leading edge and at the trailing edge. Optical measurement systems based on Electronic Speckle Pattern Interferometry may also be employed to determine natural frequencies, and even more important, mode shapes. Very often each test program produces modes that may show small differences to totally new shapes, leading to ambiguities that cannot be easily resolved. Interpretation and classification of modes in the familiar flap, edge, torsional and stripe modes of vibration is not always feasible. Finite element analysis may help to sort the problem. Elements with one or two mid-side nodes and an appropriate master degrees of freedom can help in the identification, and may also reveal new modes among those measured. The blade may be assumed fully fixed at the root.

As a rule, finite element procedures are reliable, and often used in the design process for optimization studies. In rare cases a combination of analytical and test may be necessitated. Not so negligible differences can sometimes be explained by the method of excitation, accelerometer mass, height location for analysis or inappropriate boundary conditions. Modal analysis results may be deficient in content from motion perpendicular to a sensitive direction, as may be the case in movement along an edge. Location of an anti-node may be at fault if flap, or out of plane, motion takes place where the edge shows most of the activity. Closely spaced modes can combine to appear as a natural mode, but in reality combine elements of the closely spaced modes. Instead of a cursory look at measured or calculated results, it pays to be attentive to the details. In this regard, note that when complete assemblies of bladed disks are analyzed and tested in an engine or in a rig, the amount of data to be evaluated is considerably more.

Suppression of flutter in an airfoil provides the primary inducement in introducing an elastic support in the form of a shroud. When operating at speed the protrusions on the row of blades lock to form a continuous ring to couple the blades, resulting in a stiffer assembly. Since the protrusions are in contact as the blades go through a certain amount of untwist under centrifugal action, the shrouds rub against each other. The extent of the rub and the consequent damping has been examined extensively to determine suitable boundary conditions, with assumptions ranging from a fully locked to a freely slipping condition. Precision in this matter comes into focus when analytically predicted natural frequencies of vibration do not agree with measured data. Slipping at shroud contact points assumes that the degrees of freedom related to the motion occurs without an accompanying force at the interface. A ball joint form of condition permits translational motion at the interface, but rotation can take place without the associated moments. Variation in the mode shapes obtained from the different sets of boundary conditions play a substantial role in aero-elastic stability calculations, because the numbers are used directly in work/cycle relations to predict flutter susceptibility.

Z-shaped shrouds on many low-pressure gas turbine blades also perform the function of sealing while maintaining a pre-load along the circumference during operation. As the shroud contact edges wear, a change in the dynamic characteristics of the rotating assembly may be expected. Coupled modes in a tight shroud condition degenerate into first flap and first edge-wise cantilever type modes at lower frequencies when the shroud slackens (*ref*: Hall, *et al.*, 1976). Placing frequencies of integral order vibration in the low engine orders (2E, 3E, and occasionally 4E and 5E) outside the operating range or at low shaft speeds finds favor among many gas turbine manufacturers. Higher engine order resonance reaching in the 30 to 40E may also need scrutiny if excitation is of sufficient strength and blade damping is inadequate.

Wakes, potential pressure disturbances, circumferential flow distortions and shocks in passage and secondary flows produce pressure variations that result in time varying forces on a rotating blade. Flow of gas inside a turbine is inherently unsteady, and is far from uniform on both the upstream and downstream sides of a row. Kielb *et al.*, (1992) provide a fine discussion on this subject of blade stimuli. The complexity is further emphasized by the number of parameters affecting the aero-elastic stability of a turbine blade: number of blades, blade chord, blade twist and geometry, aspect ratio, stagger, hub/tip ratio, shroud location and angle, incidence angle, blade load, tip speed, pressure distribution, shock position, inlet Mach number, natural frequency, mode shape, mechanical damping and blade mistuning. Srinivasan (1997) explores the extent of influence of the parameters at specific aerodynamic conditions.

2.4 BLADE FLUTTER

Blade flutter may best be described as elastic oscillations of the airfoil about its torsional axis as a consequence of the interaction of aerodynamic characteristics of gases flowing pass a cascade of rotating blades and the cascade's own geometry. A combination of experimental test data and computational fluid dynamics methods are required to reliably predict flutter behavior in blades. Flutter is more commonly encountered in long and slender low-pressure turbine and compressor blades, which lack adequate torsional stiffness.

The most important contributor in the flutter mechanism is the relative location of the airfoil's torsion axis, followed by reduced natural frequency of the fundamental modes and steady state blade loading. Here the fundamental modes are defined as bending along and normal to the chord and torsion about the blade's leading edge. Dynamic mode shapes play a considerable role in the flutter mechanism, while the effect of inlet flow incidence angle is minimal. Flutter frequency where the modes become unstable has been observed in some studies to be between 0.1 and 0.4 of each of the fundamental mode's frequency. Unsteady pressures arising on the surface of a blade during its vibratory motion are a consequence of the contributions from all of the blades in the cascade. To maximize stability and cascade unsteady response, factors that contribute to aerodynamic damping must be understood. Following Bolcs *et al.*, (1986), the damping coefficient for a purely bending mode is related to dimensional work by the relation:

$$\Xi_h = -\frac{w}{\pi q c^2 h^2} \tag{2.19}$$

where h is the plunging amplitude normalized by the chord, Ξ is damping coefficient, q is dynamic pressure, w is work in traveling wave domain and c is the blade chord. The damping coefficient for torsion is:

$$\Xi_\alpha = -\frac{w}{\pi q c^2 \alpha^2} \tag{2.20}$$

with α representing the torsional amplitude.

Flutter in airfoils is perhaps the most serious cause of blade failures. A principal reason is that flutter is an aero-elastic instability, and once initiated, cannot be stopped readily. In an unstable condition aerodynamic forces induced due to blade vibrations feeds energy into the structure, thus escalating stresses with each cycle of vibration. Flutter resulting from vibrations is non-integral, and so the frequency-speed characteristic does not follow the engine order line (fig. 2.4). This feature must be closely watched for during engine or rig tests. When the frequency speed characteristic rides along an engine order line, it is watched for resonant conditions. When it departs from the engine order line, the potential for aero-elastic instability is present. However, low level flutter type stresses have been reported to occur that do not necessarily build up (*ref*: Lubomski, 1980).

An important parameter that is also a measure of unsteadiness is called the reduced frequency, k, and is defined by $b\omega/V$, where b is the semi-chord, ω is the frequency of vibration and V is flow velocity relative to the blade. The parameter indicates the ratio of time taken for a sinusoidal wake imbedded in the free stream moving at velocity V to travel the semi-chord length to the time required for the airfoil to complete one cycle of vibration. At low reduced frequency ($k \sim 0.05$), wake wavelength $\lambda = VT = 2\pi V/\omega$ is large relative to the chord, and is small for $k \sim 1$ (*ref*: Platzer, *et al.*, 1988).

Panovsky *et al.*, (1998) and Nowinski *et al.*, (1998) identify the mode shape as the most important design parameter affecting the flutter characteristics of a blade through experimental correlation. A key to conducting a complete study of the mode shapes is the ability to superpose the three fundamental mode shapes to get the equivalent of the torsion at any axis. The fundamental modes are defined as bending along and normal to the chord, and torsion about the blade leading edge. In fig. 2.5 bending normal to the chord in the η-direction is referred to as the 'flex' mode and bending along the chord in the ξ direction is the 'axial' mode.

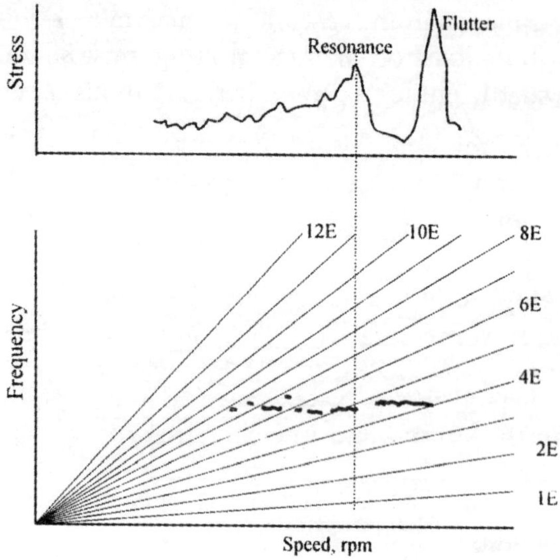

Fig. 2.4. *Blade Campbell Diagram from Engine Test*

By exploiting linearity, the unsteady solution for torsion of amplitude α about any location (ξ, η) can be determined by the superposition method of the modes in question. Using vector relations, it can be shown (*ref*: Panovsky, 1997) that the amplitude of the fundamental modes is given by:

$$\{a(\xi,\eta;\alpha)\} = \begin{Bmatrix} h_\xi \\ h_\eta \\ \alpha \end{Bmatrix} = \alpha \begin{Bmatrix} \eta \\ -\xi \\ 1 \end{Bmatrix} \tag{2.21}$$

Work about a torsion axis (*ref*: Whitehead, 1987) at point (ξ, η) for the modes is:

$$w_l = \{a\}^T[b_l]\{a\} \tag{2.22}$$

where matrix $[b_l]$ is the work matrix and $[a]$ is the modal amplitude vector. Each entry of the (3×3) matrix gives work done by combining pressures with displacements in the modes. Influence coefficient matrix $[B_k]$ for the fundamental modes for the kth blade may be determined similarly:

$$W_k = \{a\}^T[B_k]\{a\} \tag{2.23}$$

The procedure calls for running the unsteady code in the three modes with unit amplitudes for selected values of the inter-blade phase angle (IBPA). This angle provides a phase relationship that represents the motion of a blade with respect to other blades. For example, in a well-defined traveling wave mode, the motion of a blade can be shown to be identical to the motion of its neighbor except for a phase angle between them. Thus, each blade experiences exactly the same motion but at a slightly different time. An intermediate step is to determine damping values as a function of torsion axis location for each value of IBPA, and is repeated over a range of torsion axis locations. Negative values of the damping coefficient indicate instability. To determine overall stability, the damping must be checked for all considered values of IBPA and torsion axis locations. These results can be used in conjunction with the value of reduced frequency for which each torsion axis location becomes unstable. Relatively small changes in the location of torsion axis can have considerable effects on the stability. The contours of flutter-reduced frequency are roughly aligned normal to the blade mean line at all locations along the blade surface. Consequently, changes in torsion axis location along the mean line are more significant than a shift normal to the surface.

An example pertaining to this type of problem is the T55-L-11 in the 1970's that led to a progressive failure of the power turbine second stage rotor blades (*ref*: Leyes *et al.*, 1997). The blades were long and had interlocking tip shrouds, which in time wore out due to rubbing, causing the blades to be effectively free standing. Vibratory resonance eventually caused the blades to fail. Sufficient torsional load was generated to tear loose the complete power turbine and shaft. The over speeding turbine not only tore off the tail pipe but peppered the aircraft with shrapnel. In addition, the combustor forward of the power turbine had a shield around it that was insufficient to withstand the ballistic force of a 4 in. long turbine blade. A number of failed blades went through the shroud and into the aircraft.

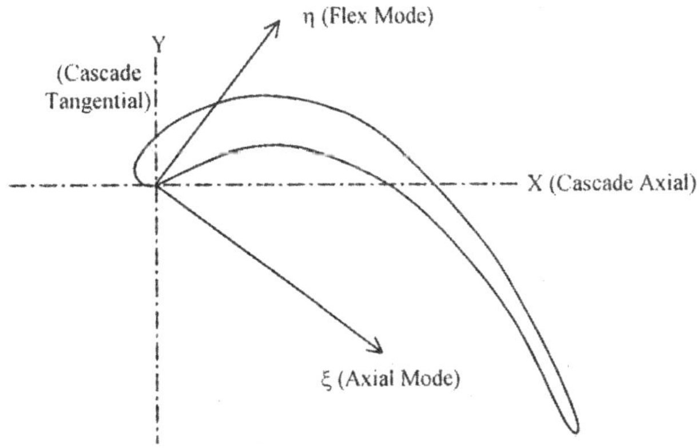

Fig. 2.5. *Fundamental Mode Shape Definition*

2.5 CASCADE FLUTTER ANALYSIS

Susceptibility of rotor blades to flutter instabilities is a major consideration in the design of turbo-machines. The flutter is caused by an interaction between the vibratory motions of an assembly of blades and the aerodynamic forces resulting from these motions. Resonant vibration is caused when a vibratory pattern in the rotating system of the blades matches both in time and space a distortion pattern in the stream.

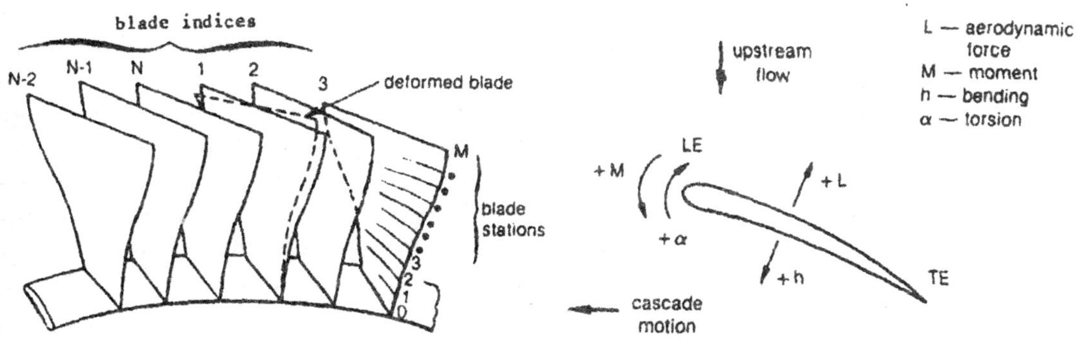

Fig. 2.6. *Cascade of Cantilevered Blades (Srinivasan, 1984)*

Experimental data obtained from tests conducted on an advanced engine over a wide range of realistic operating conditions have revealed the complex nature of cascade flutter modes. The latter are characterized by the presence of several harmonics at the flutter frequency, underlining the influence of non-uniformities of the blade properties around the rotor. The resulting system modes of the assembly depend on several important factors arising from the extent of aerodynamic and mechanical coupling.

Modern fan and compressor blades resemble highly twisted and bent plate-type configurations, whose cross-sections vary from root to tip. An assembly of such blades with or without part-span shrouds mounted on a support structure such as a hub or disk and subjected to a centrifugal and aerodynamic environment leads to an aeromechanical system of extreme complexity. The aerodynamic modeling of a vibrating cascade of blades must be analyzed under flow regimes typical of several operating conditions.

The cascade flutter problem may be formulated by using the modal characteristics of individual blades (frequencies, mode shapes, and damping) that may be available from a detailed finite element analysis or from direct experimental measurement (*ref*: Srinivasan *et al.*, 1984). Flow conditions may also vary along the span. The formulation includes considerations important in design so that parametric studies of real blade profiles can be made. Thus, coupling between the torsional and bending degrees of freedom at each span wise location of the blade and coupling among blade modes through either aerodynamic or mechanical means are inherent in the development. Numerical results are obtained for an assembly of 12 aerodynamically coupled blades on a rigid disk (fig. 2.6). A parametric study is made in which the strength of coupling (between bending and torsional degrees of freedom) in a given blade mode, mass ratio, reduced frequency, and mechanical damping are varied. Studies of flutter modes involving more than one blade mode are conducted to investigate the strength of coupling between the participating blade modes. Both selective and random mistuning procedures are used with and without mechanical damping to establish trends of behavior in regard to flutter.

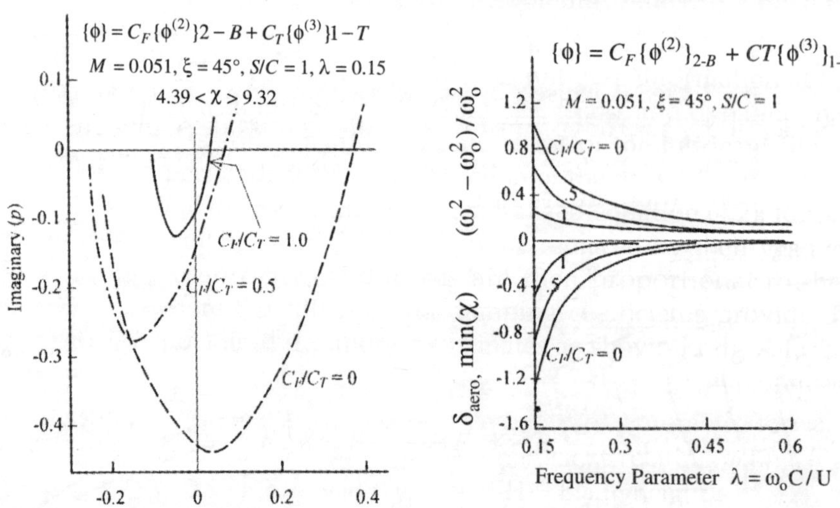

Fig. 2.7. *Influence of Coupling in the Same Mode (Srinivasan, 1984)*

The blade profile chosen represents an advanced research fan whose mode shapes are used in the baseline calculations as well as in forming new modes to vary the bending-torsion coupling.

Figure 2.7 shows the results of computation for determining the influence of introducing more or less coupling between bending and torsion in a blade mode. The plot of imaginary part of the span-wise blade mode number $p = (\lambda_0^2 - \omega_b^2)/\omega_0^2$ and the complex number λ_0 are shown as a function of the real part of p. Here λ and ω are the reduced and circular frequencies of vibration. Three sets of calculations are made for a fixed frequency parameter $\lambda = 0.15$ and varying amounts of bending and torsion loads. Either method can be used in practice to improve stability conditions. A combination of these, i.e., both mistuning and damping arranged in a random manner increases, as expected, the stability even more. A set of calculations is made for the same system to compare the improvement in stability between using random arrangement of blade frequencies and a selective arrangement of the frequencies. The latter is chosen to be a condition in which four blades in a system of 12 blades had frequencies of 15 and 30 coefficients ($C_F/C_T = 0$, 0.5 and 1). The calculated points in the left diagram represent eigen-values associated with a distinct circumferential distribution of amplitudes and phases around the periphery of the assembly, and the most negative value of Im(p) pertains to the most unstable mode for that value of C_F/C_T. A striking feature of this analysis is that the stability characteristics show a definite improvement when some bending deformation accompanies torsion deformation. The implication is that blade designs that are prone to flutter in an essentially torsion mode could be improved by a design modification which leads to introduction of bending deformation into the vibratory mode. In the figs, α = blade rotation, δ = modal damping, η = eigen-vector, c = chord, h = blade translation, l = length of blade segment, r = circumferential mode number, L = unsteady aerodynamic lift, M = unsteady aerodynamic moment, C = matrix of unsteady aerodynamic coefficients, D = hysteretic damping and S = gap. In the right diagram the flutter

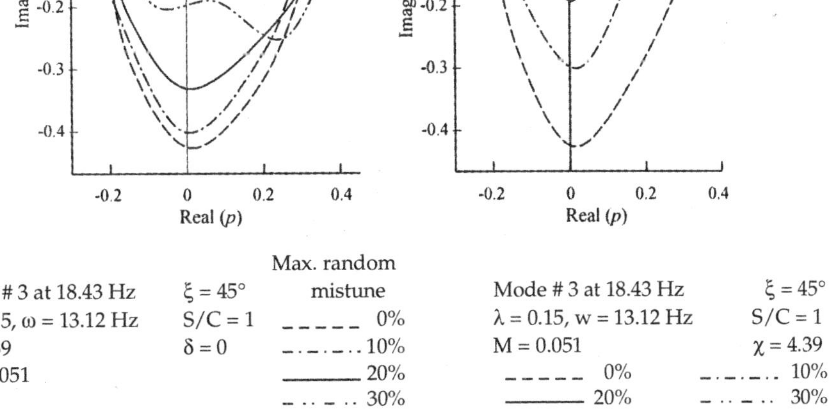

Fig. 2.8. *Effects of Random Mistuning and Damping (Srinivasan, 1984)*

characteristics are for several values of λ with C_F / C_T. At any given frequency parameter a purely torsion mode tends to become stable with any addition of bending to that mode. Since aerodynamic damping tends to increase as C_F / C_T increases, no instability is predicted for a pure bending mode, as expected. Further, the frequency of flutter ω_0 is close to the frequency of the blade by itself for large values of the λ, irrespective of the strength of coupling. The implication for an essentially aerodynamically coupled system is that for low Mach numbers and large reduced frequencies the influence of aerodynamic coupling is negligible, therefore, if flutter does take place, it will occur at a frequency close to the blade frequency.

Figure 2.8 shows the eigen-value loops resulting from a random arrangement of blade frequencies arising from mistuning and damping. The indications are clear that either method can be used in practice to improve stability conditions. The blades are placed symmetrically around the rotor. While the mistuned systems show a decided improvement over the corresponding tuned system, results indicate that the stability characteristics of the selectively mistuned system can be slightly better than the randomly mistuned system.

2.6 PREDICTION OF FORCED RESPONSE

Forced response problems are often associated with turbine blades, but fan assemblies during the development phase suffer due to their special geometry. A primary mechanism of failure in fan blades is high cycle fatigue resulting from vibrations at levels exceeding material endurance limits. In the forced response context, periodic upstream obstacles such as variable angle inlet guide vanes and struts give rise to excitation arising from blade passing frequency. Flow distortion due to non-symmetric intake duct geometry gives rise to low engine order response. The former type may be dealt with from the order of excitation deduced from the number of blades. Straightforward methods for the latter are not available because determination of low order harmonics requires detailed knowledge of the inlet flow. Both forms of excitation coexist, hence prediction of absolute vibration levels under the combined effects is necessary for establishing fatigue life. Actual vibration levels depend on unsteady aerodynamic loading and total damping in the mode of interest, determination of both of which is fraught with major difficulties. Modeling of unsteady fluid flow loading is a formidable challenge in high-speed transonic conditions. Also, accurate structural damping prediction methods under operating conditions are not available. Aerodynamic damping may also interact non-linearly with the structural motion (figure 2.9).

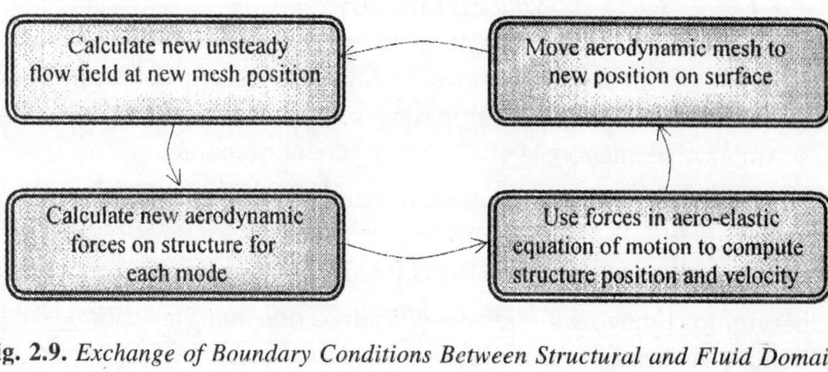

Fig. 2.9. *Exchange of Boundary Conditions Between Structural and Fluid Domains*

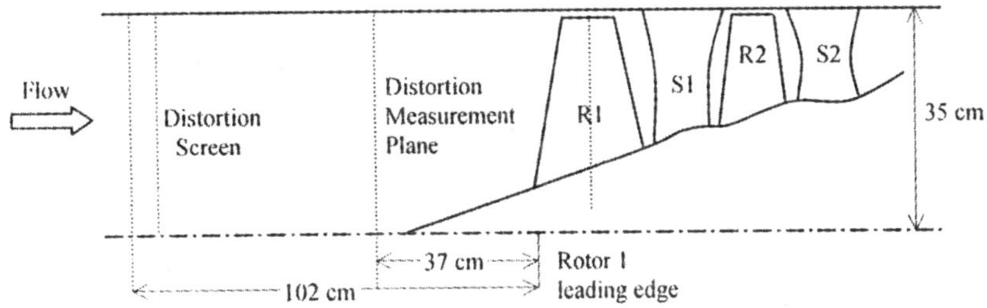

Fig. 2.10. *Test Fan Configuration (Breard, 2000)*

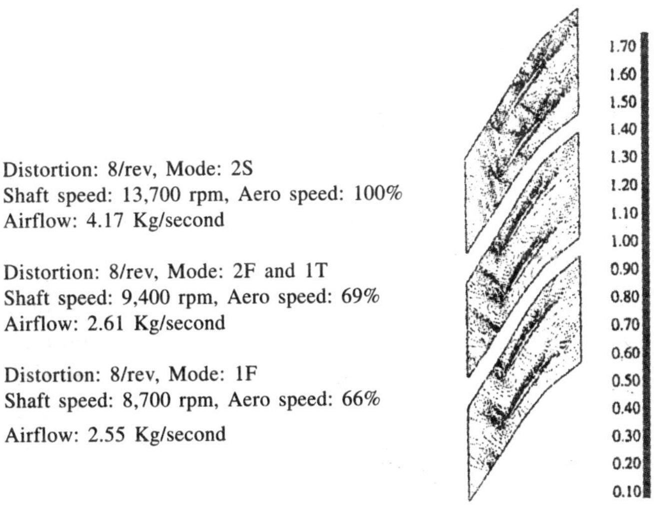

Fig. 2.11. *Steady State Mach Number Contours at 85 percent Span (Breard, 2000)*

Computational methods for unsteady aerodynamic loads arising from rotor/stator interaction use Euler or Navier-Stokes equations. Each blade row is considered separately, the separate analyses being linked through inflow boundary conditions obtained at every time from the

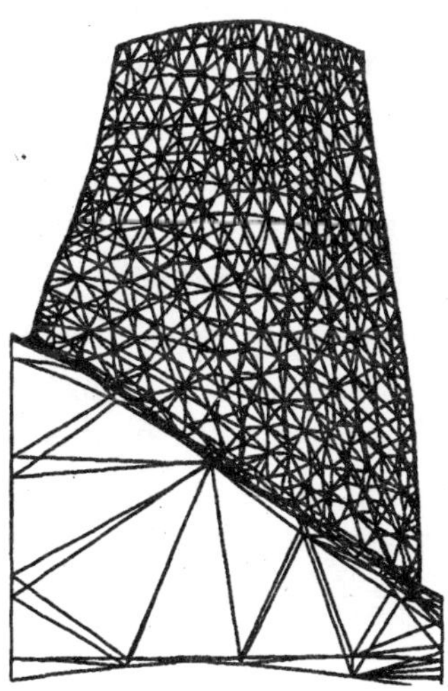

Fig. 2.12. *Structural Analysis Mesh of Fan and Disk (Breard, 2000)*

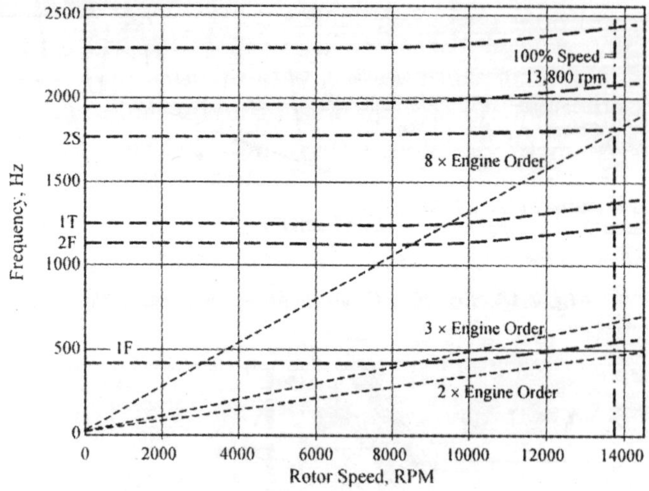

Fig. 2.13. *Campbell Diagram (Breard, 2000)*

Table 2.1. ***Blade Natural Frequencies of Vibration***

Shaft Speed (RPM)	*Aero Speed (RPM)*	*Frequency (Hz)*	*Adjusted (Hz)*	*Mode*
62%	66%	412.09	434.5 (+5.44%)	1F
68%	69%	1148.45	1251.7 (+9.03%)	2F
68%	69%	1227.86	1251.7 (+1.94%)	1T
69%	100%	1770.72	1800.0 (+1.65%)	2S

flow solution and imposed as a boundary condition to the structural model to compute the new blade position. The aerodynamic model then moves to follow the structural motion. Final operation in the cycle is the determination of the new unsteady flow solution about the new position so that the unsteady pressures become available as boundary conditions for the next time step. The origin of this integrated method can be traced back to wing flutter (*ref*: Ballhaus, *et al.*, 1978).

The numerical study is focused on a benchmark case from an experimental program known as ADLARF (Augmented Damping of Low Aspect Ratio Fans) between the US Air Force and GE Aircraft Engines. The basic objective of the research program is to evaluate the state of the art for avoiding high cycle fatigue using experimental and analytical methods (*ref*: Breard, *et al.*, 2000). The twin stage test configuration of the fan is shown in fig. 2.10. Analysis is concentrated on the first stage rotor, which has 16 low aspect ratio blades. Instrumentation measures unsteady blade surface pressures along the 85 percent span. Pressure distortions are created by screens of varying porosity mounted 1.5 × diameter upstream of the first rotor, and are measured by rakes inside the intake duct. Typical steady state results are shown in fig. 2.11 for the three speed points considered.

A typical semi-structured mesh has 180,000 points per passage, and extends to include the spinner and the intake. Because of blade untwist a new grid is generated for each speed point of interest. Boundary conditions correspond to the working line along which unsteady flow measurements are recorded.

The mesh used for cyclic symmetry structural analysis is shown in fig. 2.12, and the Campbell Diagram in fig. 2.13. Table 2.1 provides the natural frequencies of vibration.

Table 2.2. *Forced Response Computation Results*

Aero Speed (RPM)	*Mode*	*Modal Force ($N\text{-}kg^{1/2}$)*	*Aero Q-factor*	*Max. Displacement Peak to Peak, mm*
66%	1F	70.0	60	4.0
69%	2F/1T	27.5/86.0	44/80	0.14/2.11
69%	2F/1T	41.0/87.0	52/98	0.25/2.58
100%	2S	60.0	100	0.37

The resonance characteristics of interest correspond to the following four crossings: 1F/3EO, 1T/8EO, 2F/8EO and 2S/8EO. Because of their proximity modes 1T and 2F are excited together. Detailed mode shapes are shown in fig. 2.31.

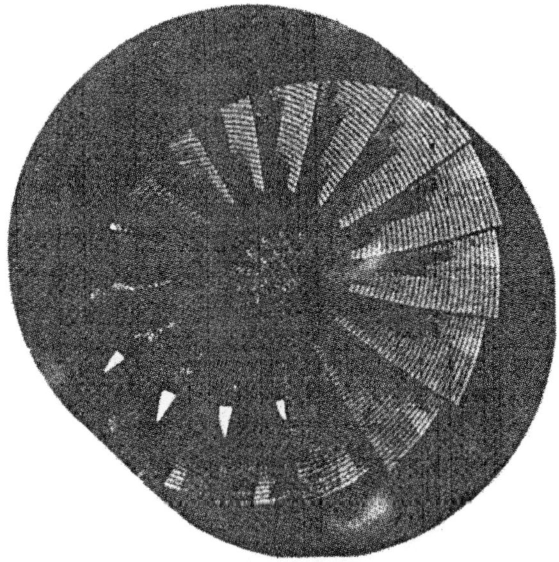

Fig. 2.14. *Full Assembly Mesh (Breard, 2000)*

Forced response computations are initiated by applying a prescribed total pressure distribution to the steady state solution. Modeling of the 3/rev distortion requires a full assembly representation, while only two blade passages are necessary for an 8/rev disturbance. To assess the effects of blade flexibility, computations with and without blade motion may be considered. A suitable number of time steps (say about 150) are needed per cycle of vibration with negligible mechanical damping.

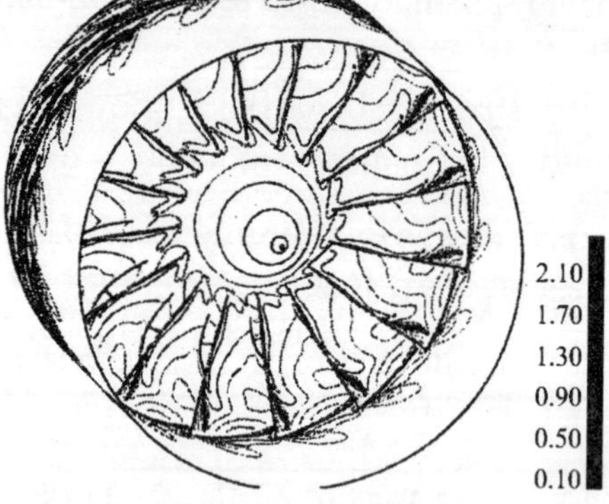

Fig. 2.15. *Steady State Mach Number Contours for 1F/3EO Crossing (Breard, 2000)*

Modal forces, displacements, aerodynamic *Q* factors and maximum actual displacements are given in table 2.2 for all resonant conditions of interest. The *Q* factor is obtained by matching modal displacements with and without blade motion:

$$Q = \frac{\omega_0^2 A_0}{F_0} \tag{2.24}$$

where ω_0, A_0, F_0 represent angular frequency of vibration, modal displacement amplitude and modal force amplitude.

Distribution of the first harmonic of the pressure across the blade along the chord at 85 percent blade span and comparison with experimental data is considered next (*ref*: Rangwala, 2004). The mesh and steady state Mach numbers are shown in figs 2.14 and 2.15. The input total pressure distortion is shown in fig. 2.16. In the case of the 1F/3EO crossing maximum computed blade vibration amplitude is close to 4.0 mm. Computed damping (ζ) is 0.0083, which agrees well with the measured damping of 0.008 for one blade and 0.010 for the adjacent blade. The time history of the envelope of modal displacement amplitude agrees well with the measured results of the blade motion, indicating blade flexibility is important for this case.

Crossings for the 2F and 1T/8EO modes indicate uncertainties in the aerodynamic conditions, so unsteady pressure variation loads are likely to appear for more than one mass flow rate. Results from the lowest mass flow rate show a better agreement with measured data.

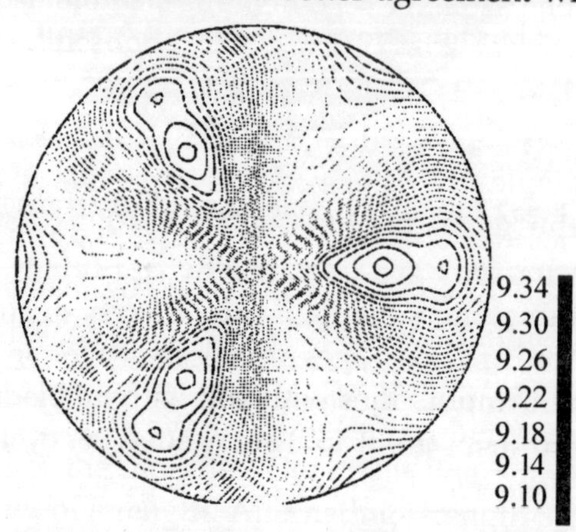

Fig. 2.16. *Total Pressure Distribution – 1F/3EO Crossing (Breard, 2000)*

The reduction in mass flow rate moves the shock upstream, consequently the unsteady pressure amplitude close to leading edge is predicted better. Measurements show significant blade-to-blade differences for both amplitude and phase, indicating a certain amount of mistuning. The amplitude of forcing increases by 50 percent for the 2F mode over the 1T mode, but aerodynamic damping decreases by 20 percent for both modes. Computed maximum vibration amplitudes for the 2F mode is 0.14 mm and 2.11 mm for the 1T mode.

Unsteady loading with and without blade motion is similar for the 2S/8EO crossing, and the measured data tends to confirm it. Unsteady loading at resonance is similar to that at off-resonance. This feature is due to small amplitudes of blade motion, and is confirmed by the low vibration signal from the strain gauges.

2.7 ASPECTS OF BLADED DISK ASSEMBLY DYNAMICS

Engine manufacturers expend considerable amount of effort to analyze and test bladed disk assemblies of fan, compressor and turbine rotors. A grasp of the dynamics of the rotating assembly of blades and disk is of fundamental importance in understanding flutter and resonance phenomena. Coupling among the blades due to shrouds and in the dovetail support will substantially alter the characteristics from those of individual freestanding blades. Vibration modes of bladed disk assemblies are affected by design features of blades, extent and nature of variation of these features among the blades, support in the dovetail and at tip or mid-span shrouds, damping, speed and of the disk on which the blades are mounted (*ref*: Srinivasan, 1997).

Natural frequencies of vibration and their associated mode shapes must be accurately known for all modes in the operating speed regime because of their prime role in forced and flutter response. Because of the nature of the assembly, the dynamic modes experience variation in amplitudes in the blade, which in turn vary around the disk's perimeter. This interaction between the disk and the blades may be observed from holograms for 5 and 2 nodal diameters of a stationary shrouded fan rotor in fig. 2.17 (*ref*: Mikolajczak *et al.*, 1975). The modes represent a standing wave, with blades at or near the nodal diameters subjected to mostly torsional motion and those at and close to anti-nodes going through predominantly bending deflection. Nodal circle modes at some location along the blade span may also be experienced by the assembly at higher frequencies. Oftentimes observed during the operation of an engine, the vibratory pattern may present itself either as a standing wave or as a traveling wave (*ref*: Rangwala, 2004). As engine speed increases, a backward traveling wave will become stationary when rotor speed reaches resonance.

A Campbell diagram displaying vibratory features in terms of frequency and speed is necessary to evaluate the performance of the bladed disk assembly. Calculated or measured natural frequencies as a function of the rotor speed for each circumferential wave number, or nodal diameter, and radial wave number, or nodal circle, are plotted on the diagram. Intersection of the characteristics with engine order lines defines the resonant speeds where operation of the engine is to be avoided. The confluence of two and three nodal diameter pattern modes is shown in the diagram of fig. 2.18 (*ref*: Pfeiffer, 1985). The style in which the frequencies for a given vibratory pattern change with engine speed are represented by the horizontal lines in this diagram. Slight variations in the otherwise circular symmetry provide the reason for the two closely spaced lines for each vibratory mode.

Torsional vibration frequencies of blades generally do not change substantially with engine speed, as opposed to bending modes. In a study conducted on an MTU blisk (where the blades are machined integrally with the disk), the Campbell diagram indicated a mere 2 percent difference in frequencies of the second bending and the first torsion modes (*ref*: Frischbier *et al.*, 1996).

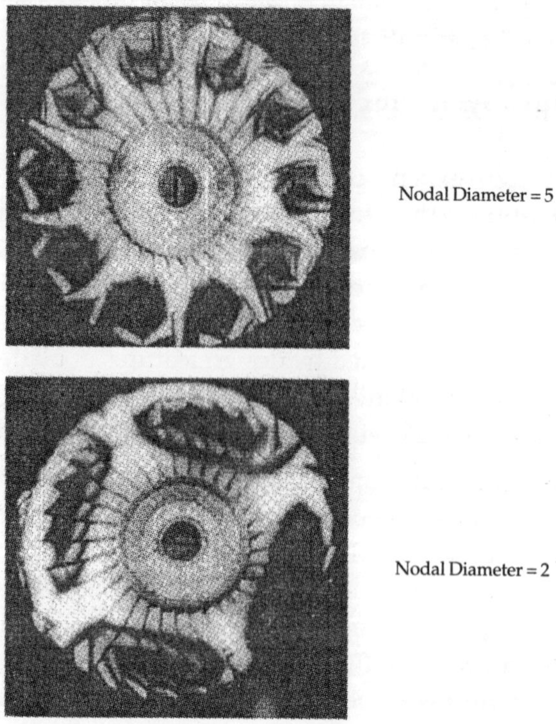

Fig. 2.17. *Vibratory Pattern of Shrouded Fan and Disk Assembly (Mickolajczak, 1975)*

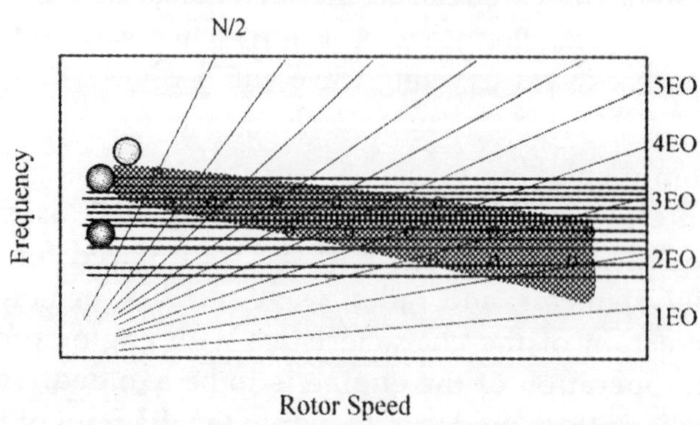

Fig. 2.18. *Bladed Disk Assembly Campbell Diagram (Pfeiffer, 1985)*

At higher speeds the bending mode frequency showed a general upswing, but the torsion mode was little affected. Aero-elastic instability was recognized to arise from the interaction of the two modes, and the rotor was redesigned through a numerical optimization technique.

Uniformity in the dynamic patterns is based in large part on the extent of similarity in the geometric features determining the stiffness and inertia characteristics, or mistuning, of individual blades in the rotor. When the blades are mostly uniform, the vibratory pattern may be identified without difficulty as corresponding to a certain nodal diameter or nodal circle mode. As variations between the blades increase, the mistuned assembly's vibration modes cannot be easily described. Non-uniform blade chord, clearance gap and angle of incidence among others affect cascade aerodynamics, and in conjunction with structural irregularities around the rotor, lead to changes in natural modes, forced responses and flutter.

Shrouds are employed to provide elastic support of fan blades to limit flutter in the blades. The protuberances on fan and low-pressure steam turbine blades interlock at speed with those on adjacent blades to form a nearly continuous band of coupled blades, resulting in a stiffer assembly. Modified shroud designs in the form of snubbers are also extensively used by turbo-machinery manufacturers at the tip or mid-span location of the blade. Stability of the stage is enhanced by locating the nearly continuous ring formed by the shrouds at approximately two-thirds of the distance between the blade's root to the tip. Uncertainties are present about the extent of gaps introduced in the mostly continuous ring as the blade untwists at speed, and lead to rub between the mating faces at the shroud. The extent of rub and consequent damping has received considerable attention over the years, but acceptable methods to analyze and control the problem remain elusive. Close attention to the boundary conditions at the shroud is necessary when the natural frequencies of a rotor fail to match measured data, as seen in the diagram of fig. 2.19, and the influence on mode shapes shown in fig. 2.20 (*ref*: Srinivasan *et al.*, 1978). In the Campbell diagram agreement is indicated with the freely slipping shroud, or no friction condition.

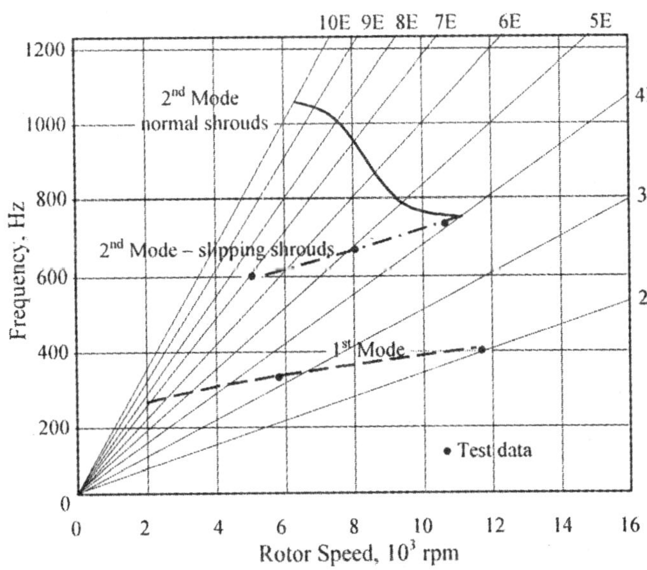

Fig. 2.19. *Shrouded Fan Campbell Diagram (Srinivasan, 1978)*

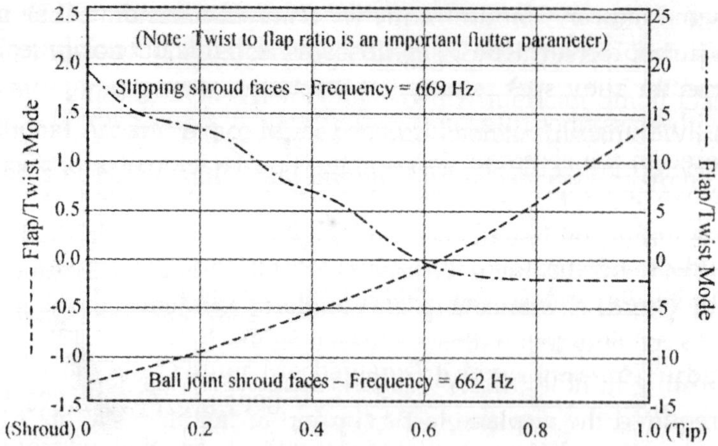

Fig. 2.20. *Effect of Shroud Boundary Conditions on Mode Shapes (Srinivasan, 1978)*

But correctness of the corresponding boundary conditions is not to be assumed, especially if mating surfaces at the shroud display signs of rub after engine test. A slipping shroud assumes the degrees of freedom to be capable of slip without an accompanying force at the location. A ball joint condition assumes translational motion is not allowed at the interface, but rotation may occur without an accompanying moment. The consequent change in mode shapes is of special interest in checking the aero-elastic stability, since the values are directly employed in the work/cycle computations for prediction of flutter.

Besides providing a preload along the circumferential direction during engine operation, interlocking shrouds at the tip of low-pressure turbine blades in aircraft engines provide the added benefit of sealing the gases past the blade tip. Z-shaped shrouds on turbine blades differ from part-span shrouds in the fan, but perform the identical function of added support. Wear and subsequent slack at the shroud contact face causes degeneration of the coupled modes of a tight shroud configuration into first flap and first edge-wise cantilever form of modes.

From practical considerations, frequencies of integral order vibration of low engine orders (2E up to 5E) need to be placed outside of, or at low speeds in, the operating speed range. The significance of higher engine orders, sometimes exceeding 30E, can only be answered by examining specific cases, and is dependent on the level of excitation, magnification factors and damping.

2.8 PROPELLER BLADE SEPARATION INCIDENT

In August 1995 an Embraer airplane on a scheduled passenger flight experienced the loss of a propeller blade from the left engine propeller while climbing though 18,000 feet at 160 knots. The aircraft crashed during an emergency landing half an hour after departing from Atlanta, Georgia. The aircraft was destroyed by ground forces and subsequent fire, with the pilot and seven passengers succumbing to fatal injuries.

Government authorities determined the probable cause of the accident as in-flight fatigue fracture and separation of the propeller (*ref:* NTSB/AAR-96/06, 1996). The resulting distortion in the engine's nacelle caused excessive drag, loss of lift on the wing and diminished control of the airplane. The fracture was attributed to a fatigue crack from multiple corrosion pits in the propeller. The manufacturer, Hamilton Standard, provides a family of composite propeller blades for use on turboprop airplanes. Figure 2.21 illustrates the design of the blade. A central solid forged aluminum alloy spar is used to carry the primary load. The airfoil shape of the blade is formed with glass fiber filled epoxy and foam, bonded to the spar with an adhesive. A tapered hole is bored in the center of the spar from the inboard end for installation of balance weights, while providing some reduction in weight. Earlier production runs of this design called for shot peening of the taper bore, but was deemed unnecessary and was discontinued. But the manufacturer's statistical data from field service experience indicate that blades without shot peening of the bores are susceptible to earlier corrosion and cracking. The blade involved in this accident was not shot peened.

It is also interesting to note that prior to certification the Federal Aviation Administration of United States required the airplane to be capable of successfully completing a flight during which likely structural damage may occur as a result of a propeller blade impact. However, the agency exempted the airplane manufacturer from complying with this requirement on condition that all practical precautions are taken in the design phase to take account of all features of the propeller to reduce the hazard, which might arise from the failure of a hub or a blade. Embraer's analysis indicated that the nacelle would not withstand the loss of a half or a full blade segment.

To comply with airworthiness requirements the propeller manufacturer is required to demonstrate vibration characteristics of the assembly to ensure resonant frequencies responsible for producing critical vibration stresses do not occur within the normal operating range of use (*ref.* Rangwala, 2004).

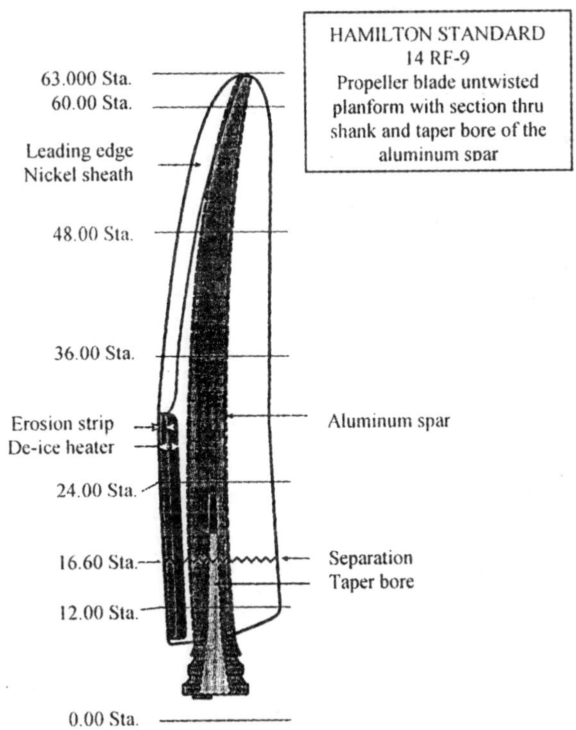

Fig. 2.21. *Propeller Blade Design (NTSB/AAR-96/06, 1996)*

Each propeller must be shown to have vibration stresses that do not exceed values that have been shown by the propeller manufacturer to be safe for continuous operation (see fig. 2.22). This must be demonstrated by measurement of stresses through direct testing, comparison with similar installations for which measurements have been made or service experience that proves the safety. A fatigue evaluation must be made and limits established for each metallic hub and blade. All reasonably foreseeable vibration load patterns must be considered. Fatigue limits must account for permissible service deterioration such as nicks, grooves, galling, bearing wear and variations in material properties. Variable pitch propellers (while engine is operating) must be subjected to a 100 hour test with the same power and speed settings when the blade experiences severe vibration characteristics. Flight tests are recommended if propeller diameter exceeds 13 feet.

Visual examination of the failed blade revealed a portion of the spar fracture was on a flat transverse plane, and contained crack arrest positions typical of fatigue cracking. Fracture faces were examined with a scanning electron microscope before the faces were cleaned.

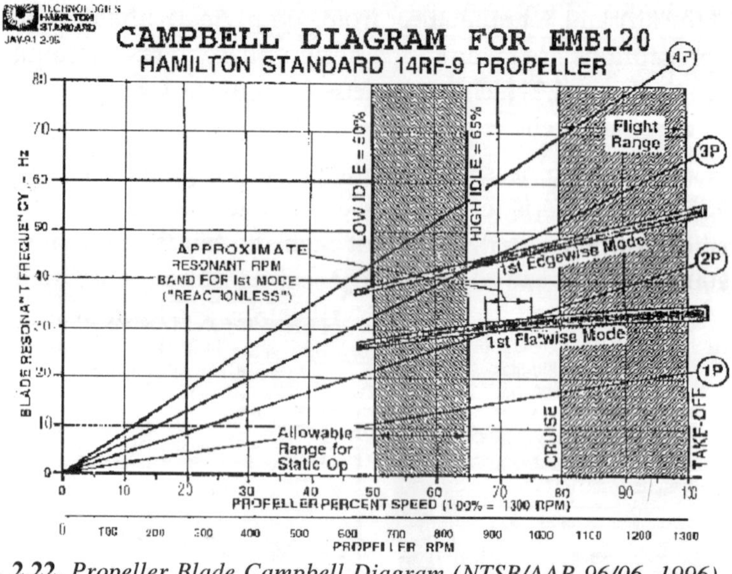

Fig. 2.22. *Propeller Blade Campbell Diagram (NTSB/AAR-96/06, 1996)*

Near the origin area of the fracture a layer of heavy oxide deposits extending to a depth of 0.049 in. from the surface was observed. Fatigue crack initiated around at least two adjacent locations on the taper bore surface. Below the surface the cracks merged to form a single crack that propagated towards the face side of the blade, then progressed circumferentially around both sides of the taper bore. The crack spread 75 percent of the spar's cross-section. In areas beyond the terminus of the fatigue rough features with a matte appearance indicated typical overstress separation. After cleaning of the fracture surface additional examination revealed fatigue crack initiating from several corrosion pits in a line of pits extending over a distance of 0.070 in.

Prior to this accident, two other incidents of failures of the composite propeller blades from this manufacturer have been recorded. On both occasions the cracks originated from inside the tapered bore. In the first case analysis indicated forces induced from the rotation of the other three blades resulted in propeller assembly imbalance and loads on the engine's forward mount that exceeded the ultimate limits. This resulted in separation of the propeller and the reduction gearbox assembly from the airplane. The flight's crew was able to accomplish a safe landing, and injuries to passengers were avoided. In the second case causal findings were similar in nature. The three remaining blades and the fourth blade stub had moved to the feathered position, with the propeller and gearbox assembly remaining within the nacelle area, partially attached to the airframe. Laboratory examination of the failed blades indicated presence of corrosion pits in the tapered bore in both instances.

2.9 ACOUSTIC RESONANCE IN MULTI-STAGE COMPRESSOR

Vibration in blades can be excited by a number of mechanisms, of which blade row interaction resulting from wakes and potential fields of adjacent blade rows and flutter are well known. Another source of observed blade excitation is associated with a resonant acoustic condition of the compressor annulus. The link with the acoustics of the system is revealed by measurements of a fluctuating pressure field, which rotates about the compressor axis at a speed close to the local speed of sound (*ref.* Rangwala, 2004). The phenomenon is sensitive to the annulus geometry and to compressor inlet conditions in a manner that is consistent with an acoustic signal. Frequencies of the fluctuating pressure signals are not necessarily integrals of shaft frequency, and are not related to the stiffness of the blade. High amplitudes of fluctuating pressure and blade stress occur in regions of irregularities in the flow field such as bleed slots or blade rows operating at extreme incidence.

Experimental measurements of the resonant acoustic feature have been made in a low speed compressor at the Whittle Laboratory of Cambridge University (*ref.* Camp, 1997 and 1997*a*). In high speed machines experimental studies of this nature are difficult to perform because of difficulties in detection and faster failures in blades due to stresses. This compressor is equipped with a variable stator mechanism to permit inlet guide vanes and stator vanes to be easily restaggered. Outer radius of the compressor is constant at 254 mm and a hub to tip ratio of 0.75. When operating at high flow coefficients with the variable stators at certain off-design settings, the compressor emits tones at a series of discrete frequencies that can be heard above the blade passing frequency and the background broadband noise.

The structure of acoustic signals in the annular ducts can be calculated by using the homogeneous wave equation in cylindrical coordinates (*ref.* Tyler *et al.*, 1962). In a stationary fluid the equation is:

$$\frac{\partial^2 p}{\partial t^2} - c^2 \nabla^2 p = 0 \tag{2.25}$$

where t is time and c is speed of sound in free space. For a narrow duct radial direction variations may be ignored, so the solution is static pressure $p(\theta, t, x) = \Theta(\theta) T(t) X(x)$.

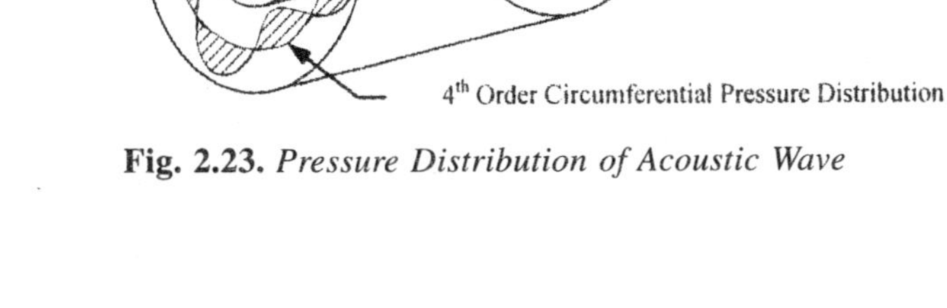

Fig. 2.23. *Pressure Distribution of Acoustic Wave*

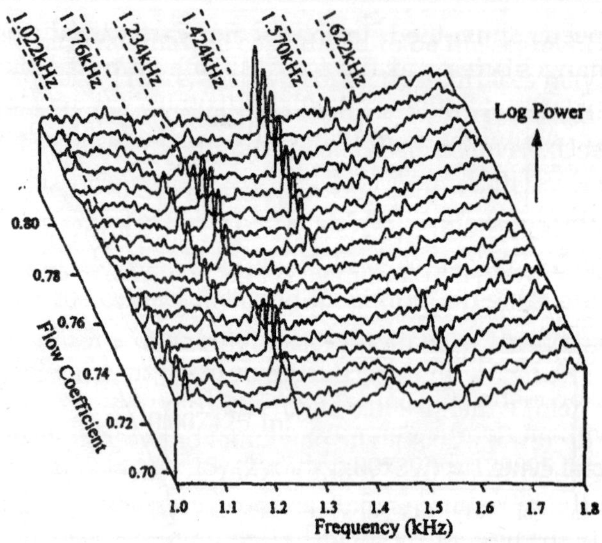

Fig. 2.24. *Pressure Amplitude Spectra as Function of Flow Coefficient (Camp, 1997)*

If the time dependent function T has frequency ω and the circumferentially varying function Θ is of order m, amplitude a_m and phase ϕ_m, then for $x = 0$ the solution is:

$$p(\theta, t, 0) = a_m \exp i(m\theta - \omega t + \phi_m) \tag{2.26}$$

where the real part is implied. Since m is an integer, variation of pressure in the θ direction is continuous and repeats around the circumference, as shown in fig. 2.23. This distribution of the wave front traces out m helices while traveling at ω/m radians per second, generating a point fluctuating pressure at frequency ω.

A source of excitation is the shedding of vortices from struts and blade rows at high incidence (*ref.* Parker *et al.,* 1985) that interacts with the flow field. Acoustic resonance is set up when the frequency of vortex shedding is close to the acoustic resonant frequency of the duct, when the vortex shedding frequency locks on to the acoustic frequency (*ref.* Cumpsty *et al.,* 1971). If now this frequency also approaches one of the natural frequencies of the blade, acoustic signal amplitude and blade mechanical stresses are reinforced. Blade stresses are particularly high if the vibrating blade also happens to be the source of shed vortices because of close interaction between the two features. The coincidence of the three frequencies is not so rare as might appear at first sight when the number of blade rows with blades vibrating in several flexure and torsion modes and considerable operating speed range of many turbo-machinery compressors is kept in mind. Acoustic resonance of the duct may also occur at many frequencies corresponding to different circumferential and radial orders.

The frequency at which vortices are shed by a body can be related to its dimensions and the velocity of the fluid by a nearly constant value of the Strouhal number, defined by $St = fD/v$, where f is the shedding frequency, D is the characteristic dimension of the body and v is flow velocity. Based on trailing edge thickness of an isolated strut, measurements indicate a value of 0.20 for this number when the natural frequency of vortex shedding is determined. At resonant conditions, however, the value of the Strouhal number may change significantly. For a compressor blade row operating at constant flow, the vortex shedding speed is proportional to blade speed, and therefore, shaft speed. Thus, when frequency of vortex shedding locks on to different acoustic modes, step changes occur in the frequency of the acoustic signal.

Figure 2.24 shows the measured spectra for frequency components between 1.0 and 1.8 kHz, indicating peaks at six frequencies. The most prominent and audible signal is at 1.424 kHz. Also, four of the frequencies (1022, 1234, 1424 and 1622 Hz) coincide with the fifth, sixth, seventh and eighth orders of 204 Hz. A Mach number based on circumferential phase velocity at tip radius is calculated from $M_\theta = (2\pi r_{tip} f)/mc$. If $m = 1, f = 204$ Hz, $r_{tip} = 254$ mm, speed of sound c = 344 m/second, then $M_\theta = 0.95$. Thus, the four modes in question are close to M_θ approaching unity value at the tip radius. The remaining two signals at 1176 Hz and 1570 Hz are the 23rd and 31st harmonics of shaft frequency (49.1 Hz).

A characteristic of acoustic resonance observed in high-speed compressors is that the frequency associated with any particular wave structure is approximately independent of shaft speed. A series of spectra have been obtained on the experimental compressor using a Kulite transducer mounted at the IGV inlet plane while the shaft speed is decelerated. The results are provided in fig. # 2.25. In the first two seconds of deceleration the signal at 1234 Hz is prominent and remains at a constant frequency, while frequencies at surrounding shaft orders fall. This characteristic is consistent with acoustic resonance in high-speed compressors.

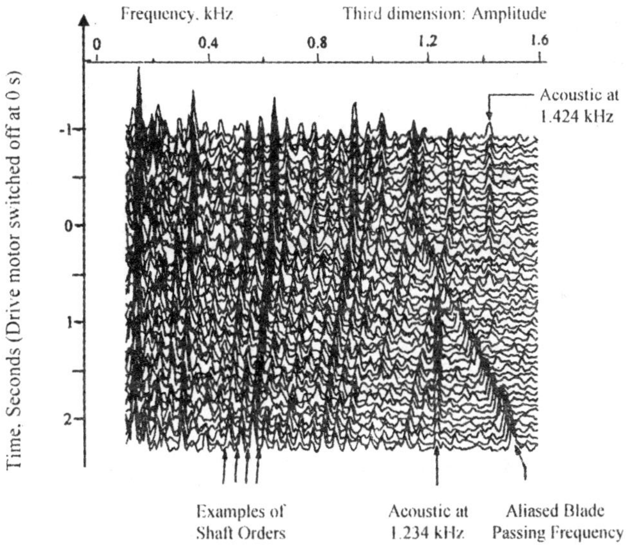

Fig. 2.25. *Pressure Amplitude Spectra as Function of Deceleration Time (Camp, 1997)*

The phase and amplitude structure of the acoustic signal is investigated using Kulite transducers in the axial and circumferential directions of the compressor and downstream duct. The 1424 Hz signal of order 7 and the gradient of its phase distribution indicate that the pressure field rotates in the opposite direction of the rotor, and may be described as a backward traveling wave. Both forward and backward traveling waves have been detected in high-speed compressors. More than one acoustic mode may sometimes exist at this frequency. Similar phase changes in the axial direction may not permit easy interpretation and identification of the waves. Acoustic waves generally propagate downstream, and travel in a direction against the rotor.

Frequencies of acoustic signals are not dependent on axial length of the compressor system. This aspect is verified on the test setup, where three different lengths of ducting have been employed on the compressor and the downstream throttle.

Axial variation of acoustic amplitude usually increases with the distance through the compressor, reaching a maximum close to the exit. The amplitudes reduce when the acoustic waves pass over support frame struts and bearing housing. It might be surmised that the acoustic wave energy does not propagate beyond the throttle, and may be reflected by the relatively flat surfaces of the support struts to get trapped within the system.

Strouhal numbers for struts and blade rows in the compressor are all potential sources of vortex shedding. For the vortex shedding frequency of 1424 Hz at a flow coefficient of 0.85 the support struts have a Strouhal number of 0.195 based on the trailing edge thickness. From the experience of gas turbine manufacturers, Strouhal numbers based on chord for blade rows that excite acoustic resonance approximate 0.6 when vortex shedding is present.

2.10 WHIRL INDUCING FORCES IN AXIAL COMPRESSOR

An increase in the radial clearance often results from rubs between rotating and stationary aircraft engine components operating at tight levels of clearance. The most frequent causes are force induced by rotor unbalance, lateral deceleration during a hard landing, lateral forces induced by high acceleration and turning maneuvers, thermal bowing and/or asymmetric ovalization of the casing, especially for fans. However, a potentially much more destructive mechanism for inducing rubs is whirl instability. Any radial deflection of the rotor relative to the stator creates circumferentially non-uniform clearances and unsteady aerodynamic forces on the rotor as each blade traverses the varying clearance gap. These unsteady forces are orthogonal to the deflection, and therefore are a significant driver of rotor whirl instabilities. The forces increase in magnitude as the deflection increases so that above the onset speed, where destabilizing forces overwhelm the stabilizing damping forces, the deflections are ultimately limited only by damage to the interacting parts or by damping forces. Consequently, an accurate determination of their magnitude and direction is of major importance in the design of safe, stable turbo-machinery components. Hysteretic whirl, whirl associated with fluids trapped within cylindrical rotor cavities and plain journal bearings (see section 13.12) are some typical examples.

Unsteady and destabilizing aerodynamic cross-axis stiffness forces that promote rotor whirl were first postulated by Thomas (1958) and Alford (1965) to explain rotor whirl instability mechanisms. In a deflected turbine rotor the airfoils in the closure zone are more highly loaded by aerodynamic forces than the airfoils in the open clearance zone because the former are operating more efficiently. This situation is shown schematically in fig. 2.26(*a*) for a turbine rotor whose centerline has been displaced upward along the ordinate by an amount $+Y$. This gives minimum clearance at the top of the turbine and maximum clearance at the bottom. The forces at these two locations are the vector sum of the mean blade force F_m and the unsteady blade force resulting from the centerline offset F_u. Summing the forces perpendicular to the axis of displacement results in a net force $F_X = F_m + F_u$, due to the difference in airfoil loading. Since F_X acts normal to the axis of displacement, it is a cross-coupled stiffening force, with its direction acting to drive the rotor in an orbital whirling motion about the centerline in the same direction as rotor rotation. Thus, F_X promotes forward whirl for turbines, with the cross-coupled aerodynamic stiffness coefficient relating acting torque T and coefficient β by the expression (D_p = mean blade diameter, H = blade height):

$$K_{XY} = F_X/Y = T\beta/(D_pH) \tag{2.27}$$

As the rotor whirls the circumferential variation in radial tip clearance causes a circumferential variation in efficiency, so that the blading with the smallest clearance is the most efficient. Alford hypothesized that the compressor would pump to a circumferentially uniform exit static pressure and therefore the more efficient blading at tight clearance would have a lower loading than the blading with larger clearance 180° away.

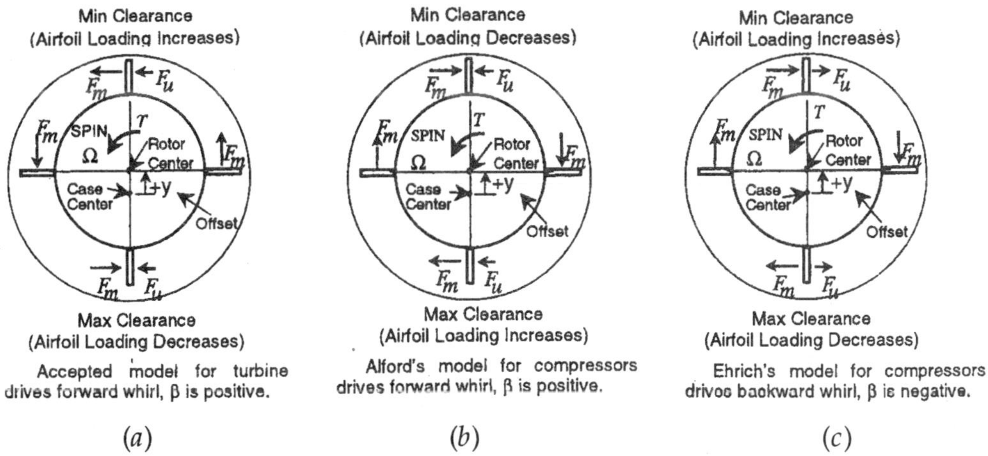

Fig. 2.26. *Mechanisms for Whirl Inducing Forces (Storace, 2000)*

This situation, illustrated in fig. 2.26(*b*), shows that the net force, $F_{X'}$ tends to cause forward rotor whirl. Thus, Alford concluded that compressors have positive β's so that whirl-inducing forces for both compressors and turbines are in the same direction.

Ehrich (1993) hypothesized differently from Alford. He reasoned that compressor airfoils with the smaller clearance would sustain a higher static pressure differential across their tips and would therefore be more highly loaded than the airfoils with larger clearance 180° away. As shown in fig. 2.26(*c*), this dictates that the net destabilizing force in compressors, $F_{X'}$ tends to produce rotor whirl counter to the direction of rotation. Thus, Ehrich concluded that compressors tend to have negative β coefficients so that the direction of whirl-inducing forces for compressors would be opposite to those for turbines.

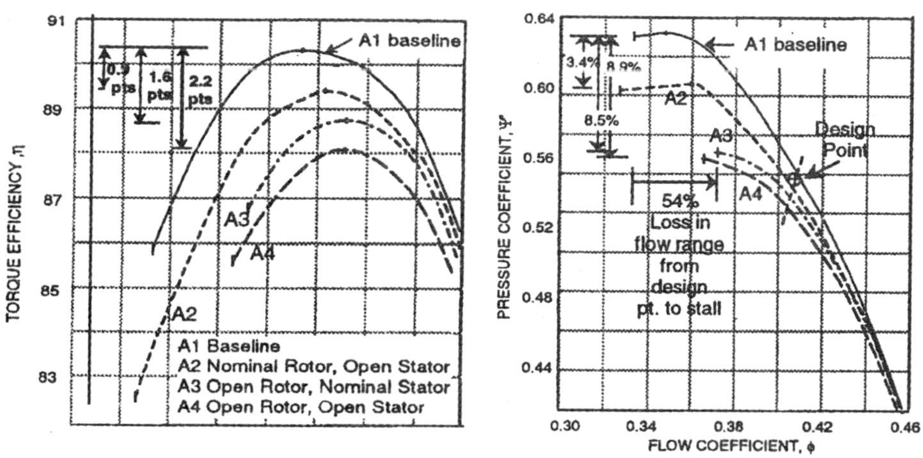

Fig. 2.27. *Effects of Variation of Clearance on Performance of Compressor A (Storace, 2000)*

In view of the disparity between Alford's and Ehrich's conjectures and the absence of a decisive resolution of rotor whirl issues, General Electric's Aircraft Engines formulated a research program (*ref*: Storace *et al.*, 2000 and Ehrich *et al.*, 2000). The test is conducted on GE's Low Speed Research Compressor to simulate the eccentricity of a whirling rotor and to measure the developed non-uniform and unsteady flow field. The test rig, which has a constant casing diameter of 1.524 m, is set up with four identical stages in order to simulate the repeating stage environment. The third stage is the test stage. The blading is representative of current design practice. Three different low-speed blading configurations are tested, being typical of modern designs, and have high hub/tip ratios of 0.85 with low aspect ratio, high solidity blading and shrouded stators. These compressors are low-speed, aerodynamic models of the middle and rear block of highly loaded, high-reaction 65 to 70 percent HP compressors in commercial gas turbine engines currently in service. Another design, also in commercial engine service, has cantilevered stators and blading with a lower hub/tip ratio of 0.70, lower reaction of 0.55, and higher aspect ratios.

Tests were performed for the configuration with two displacements of the casing centerline: a displacement of 0.1905 cm and a smaller one of 0.0965 cm. The rotor tip and the stator shroud seal are ground to allow both to run at the absolute minimum safe clearance of 0.051 cm, permitting the vehicle to enter a rotating stall. High-resolution pressure transducers, accurate to 0.01 percent of the full-scale values of either 0.068 or 0.136 bar (1 or 2 psi), are used to record steady state static and total pressures for determining both overall compressor performance and the static pressures on the stator airfoil surfaces. High response Kulite pressure transducers, having a frequency response of 20 kHz, are imbedded inside the rotor airfoil surfaces to measure the unsteady static pressures acting on the suction and pressure surfaces. The signals from the pressure transducers are digitized and ensemble-averaged. The analog data are low-pass filtered at 1 kHz to avoid aliasing using sufficient ensemble averages to reduce the effects of time-unresolved unsteadiness. A once-per-revolution pulse from an optical encoder in the casing senses the trigger airfoil on the rotor blade and initiates the sampling for the data windows for each rotor revolution. The raw pressure data for each transducer are processed using Fast Fourier Transform (FFT) methodology to give a filtered waveform consisting only of the first harmonic. The first harmonic is dominant. The 54th and 74th harmonics observed in the pressure patterns are associated with 54 rotors and 74 stators. Unsteady pressures measured at the discrete locations on the airfoil surfaces are bi-directionally curve-fit along the radial and chord wise directions to obtain continuous pressure distributions on the suction and pressure surfaces. By using the unsteady pressure distributions computed above and the geometric orientation of the airfoil surfaces in the compressor, integration over the airfoil surfaces obtains the unsteady forces acting on the various airfoils around the circumference. The unsteady forces are resolved into their tangential and radial components in the blade's coordinate system.

The component of force acting perpendicular to the direction of the rotor offset (F_X, the Thomas/Alford force) is obtained by transforming the tangential and radial blade forces to the global coordinate system fixed to the bladed disk. The β coefficient is calculated from equation (2.27).

The effects of clearance variation and centerline offset on the overall compressor performance is observed as a four-stage average of pressure coefficient and efficiency plotted as a function of the flow coefficient, referred to as the pressure characteristic. Stall margin is computed for these low-speed tests in terms of throttle margin.

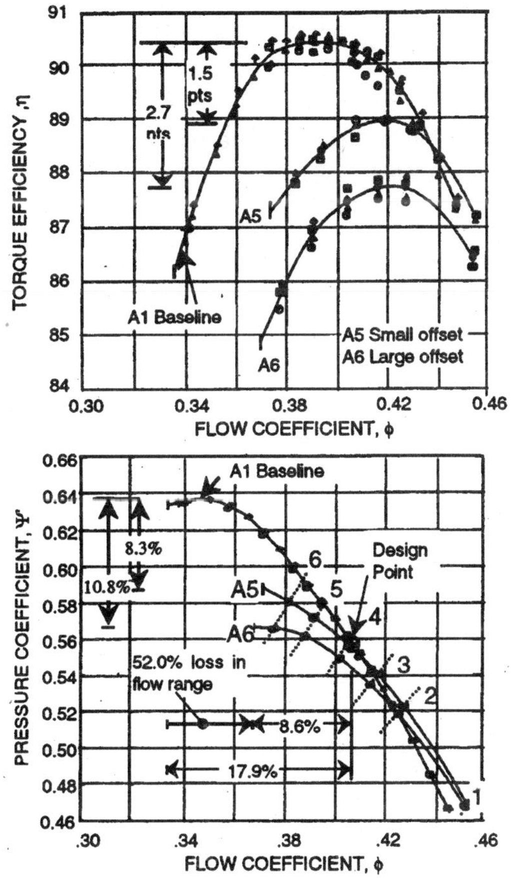

Fig. 2.28. *Effects of Rotor Centerline Offset on Performance of Compressor A (Storace, 2000)*

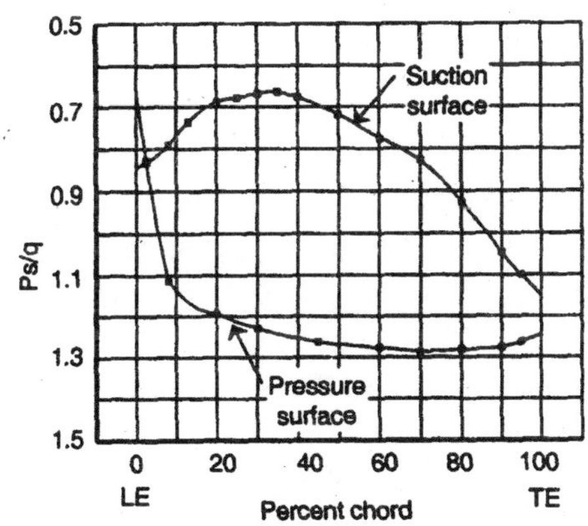

Fig. 2.29. *Steady State Static Pressure Distribution on Airfoil (Storace, 2000)*

Variations in the loading levels from high flow (low-loading) to stall were achieved by varying mass flow rate through the compressor using a discharge throttle. Lines of constant throttle setting are shown in the figs. 2.27 and 2.28 to indicate the different loading levels along the pressure characteristic. The tests are run at the design tip speed of 64.0 m/s, requiring a rotational speed of 804 rpm. This gives a Reynolds number of 3.6×10^5, representative of engine conditions. The baseline performance of compressor A, shown as curves Al, is established with no centerline offset and with circumferentially uniform, nominal levels of rotor-tip clearance and stator shroud-seal clearance. The design point is shown in both figs. The negative slope of the baseline pressure characteristic over all of the flow range from high flow to near peak pressure provides stable operation over this range, after which it begins to roll over. Stall occurs at a flow coefficient of about 0.335, as indicated by the short vertical line at the low-flow end of the pressure characteristic. The baseline compressor A has high efficiency that peaks at 90.4 percent. It also has a good throttle margin of 30.2 percent, as indicated by the 17.9 percent flow range from the design point to stall. This baseline is the performance standard against which all of the other configurations are to be compared.

The effects of varying axi-symmetric (circumferentially uniform) clearance on the performance of the compressor are shown relative to the baseline performance in fig. 2.27. The clearance variation is obtained without centerline offset and thus is typical of what occurs in a uniform rotor tip rub. Both rotor tip clearance and stator shroud-seal clearance are varied independently to separate their effects. Detailed surveys of flow properties show that when rotor tip clearance alone is increased, the dominant effect is seen in the outer 25 percent of span. Similarly, when the stator seal clearance alone is increased, the dominant effect is seen in the inner 25 percent of the span.

The effect of variation in the centerline offset on the performance is observable from the pressure and efficiency characteristics for the two levels of casing centerline offset, and are compared to those for the baseline configuration. As expected, the offset configurations with their larger average and maximum clearances have lower efficiencies and lower peak pressure rise than those for the baseline case. There is an 8.3 percent and 10.8 percent loss in peak pressure and 1.5 point and 2.7 point loss in peak efficiency for the small and large offsets. Throttle margin is 11.7 percent, which is 39 percent of the baseline value. Thus, stall margin has suffered considerably. Note that there is little change in loss of stalling flow range between small and large offsets.

Unsteady pressures on the rotor airfoils are measured as the rotor travels through varying levels of clearances caused by the centerline offset. Suction surface pressures are subtracted from pressure surface pressures to give unsteady pressure difference loading on the airfoil. Representative results showing this loading for three clearance levels around the circumference are presented in fig. 2.30 as contours of differences in unsteady static pressure, with the rotor at near minimum, nominal and near maximum clearance. The measured steady state surface static pressure distribution at 80 percent span for the rotor airfoil is shown in fig. 2.29. Airfoil unsteady loading across the span increases in the region of minimum clearance and decreases in the region of maximum clearance relative to the mean loading at nominal clearance. It is also clear that near the rotor tip in zones 1 and 2 of fig. 2.30 the airfoil unsteady tangential loading increases at near minimum clearance and decreases at near maximum clearance.

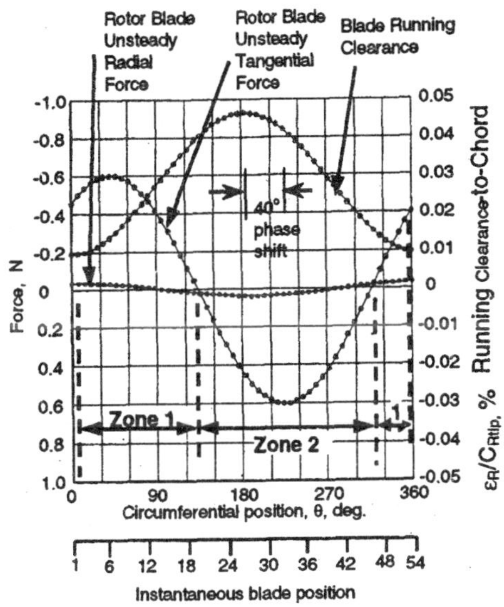

Fig. 2.30. *Variation of Whirl Inducing Forces with Tip Clearance (Storace, 2000)*

This finding confirms the correctness of Ehrich's hypothesis in fig. 2.26(*c*) about the nature of rotor whirl-inducing forces in compressors.

The unsteady pressures are reduced to unsteady blade forces as described earlier. A representative result for the large offset configuration running at high compressor loading (point 5 in fig. 2.28) is shown in fig. 2.30. The tangential and radial components of the unsteady, whirl-inducing blade force are plotted as a function of circumferential position. The circumferential variation of rotor tip clearance is also plotted. The forces need to be understood relative to the direction of rotation and the direction of the driving torque, which are counterclockwise. The centerline is offset upward, which places the minimum clearance at the top dead center. Two forces act on each airfoil, a mean force, F_m, and the tangential component of the unsteady force due to the offset, F_u. The magnitude and direction of the unsteady force will determine the direction of rotor whirl. The radial component of unsteady force is very small compared to the tangential component. The maximum tangential force occurs near the minimum clearance and the minimum tangential force occurs near the maximum clearance, again confirming the correctness of Ehrich's hypothesis. Careful use of the sign convention leads to the conclusion that in the region of negative forces shown as zone 1 in fig. 2.30 the net forces acting on the airfoils increase because the unsteady force, F_u, adds vectorially to the mean force, F_m. This occurs around the minimum clearance. In the region of positive forces shown as zone 2 in fig. 2.30 the net forces on the airfoils decrease as the unsteady force opposes the mean force. This occurs around maximum clearance near the bottom dead center. The net effect of the force distribution tends to drive the offset rotor shaft counter to the direction of rotation to drive a backward whirl.

2.11 EXAMPLE PROBLEMS

Example Problem # 2.1: A cantilever plate with a rectangular cross-section is to be analyzed using the energy procedure of sections 1.15 and 2.6. The plate is 2.0 in. long, 2.0 in. wide, 0.40 in. thick and is provided with a pre-twist angle of 20°. Material density may be assumed to be $\rho = 0.285$ lbs/in^3 and Young's modulus is $E = 27.5 \times 10^6$ lbs/in^2 and Poisson's ratio $\nu = 0.3$.

Solution: Non-dimensional frequencies of the plate are given by the expression (see table 1.1):

$$\bar{\omega} = \omega_i \sqrt{\frac{12\rho l^4 (1-\nu^2)}{Eh^2}} \tag{2.28}$$

Using the energy method based computer program (*ref*: Gupta, 1984), the finite element method calculates the first two non-dimensional frequencies at 3.2 and 21.5, which compares with the energy method values of 3.4 and 21.8.

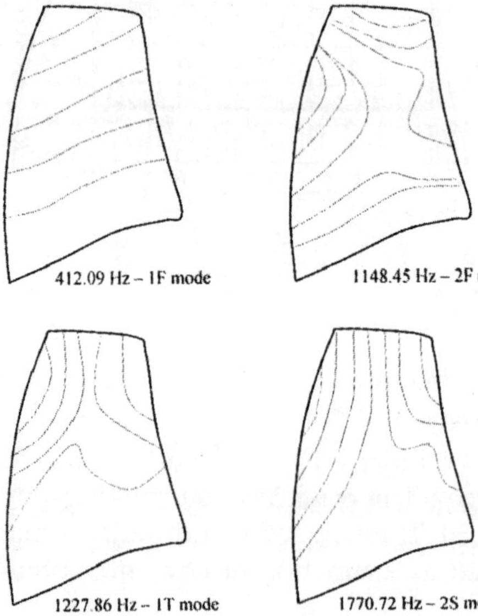

Fig. 2.31. *Mode Shapes of Fan Blade of Fig. 2.12*

Mode shapes from the finite element analysis of the fan blade shown in fig. 2.12 are provided in fig. 2.31.

Example Problem # 2.2: A bar is subjected to a tensile preload of 12500 lbs and a varying tensile load from 0 to 21250 lbs. If a stress concentration factor of k_t = 1.86 is to be applied, determine a suitable diameter for an infinite life and a factor of safety of k_s = 2.0 from the following data: yield strength S_y = 87.5 kpsi, ultimate strength S_{ut} = 112 kpsi, endurance strength $S'_e = 0.5 \times S_{ut}$ = 56.0 kpsi, surface finish modification factor k_a = 0.77, size factor k_b = 0.65, notch sensitivity q = 0.85.

Solution: Fatigue stress concentration factor $k_f = 1 + q(k_t - 1) = 1.731$. Hence endurance factor $k_e = 1/\ k_f = 1/1.731 = 0.578$, and endurance strength:

$$S_e = k_a k_b k_e S'_e = 0.77 \times 0.65 \times 0.577 \times 56.0 = 16.192 \text{ kpsi}$$

The static stress from the preload is $\sigma_{static} = 12500/\{\pi d^2/4\} = 15.916/d^2$ kpsi. The dynamic load stress range is $\sigma_{dynamic} = 21250/\{\pi d^2/4\} = 27.056/d^2$ kpsi. Alternating stress $\sigma_{alt} = 27.056/(2d^2) = 13.528/d^2$ kpsi. Mean stress $\sigma_m = \sigma_{static} + \sigma_{alt} = 29.444/d^2$ kpsi. Hence, $\sigma_{alt}/\sigma_m = 0.459$.

The fatigue diagram of fig. 2.32 provides the relation between the stress and strength characteristics. The intersection of the modified Goodman line with another line at a slope of σ_{alt}/σ_m = 0.459 defines two values of strength. Alternating strength S_{alt} corresponds to stress σ_{alt}

and strength S_m corresponds to stress σ_m. For a factor of safety of 2.0, $\sigma_{alt} \leq S_{alt}/2.0$. Thus, $13.528/d^2 \leq (29.44 \times 0.459/2.0)$, or $d = 1.414$ in.

Choose $d = 1.5$ in.

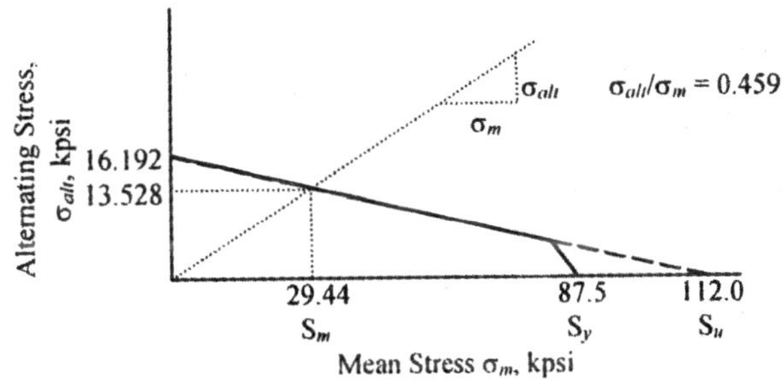

Fig. 2.32.

Example Problem # 2.3: The material for a part has $S_y = 82$ kpsi, ultimate strength $S_{ut} = 105$ kpsi, and endurance strength $S_e = 27.0$ kpsi with all corrections included. The mean and alternating stress components are $\sigma_m = 32/d^2$ kpsi and $\sigma_{alt} = 17/d^2$ kpsi. If the diameter of the part is 1.75 in., find the fact of safety using the Gerber theory.

Solution: The modified Goodman theory includes a number of conservatisms in order to account for unknown material characteristics. If the strengths are known accurately, the Gerber theory based on a parabolic relation is more extensively applied in the form:

$$(S_{alt}/S_e) + (S_m/S_{ut})^2 = 1$$

In this example, $\sigma_m = 32/1.75^2 = 10.45$ kpsi and $\sigma_{alt} = 17/1.75^2 = 5.55$ kpsi. Thus, $S_m = n\sigma_m = 10.45n$ and $S_{alt} = n\sigma_{alt} = 5.55n$. Substitution into the non-linear equation gives:

$$(5.55n/27) + (10.45n/105)^2 = 1$$

The factor of safety n is obtained by solving the equation, yielding $n = 4.067$.

Example Problem # 2.4: In example problem # 6.7 find the diameter of the component if the factor of safety is reduced to 3.25.

Solution: $S_{alt} = n\sigma_{alt} = 3.25(17/d^2) = (55.25/d^2)$ kpsi and $S_m = n\sigma_m = 3.25(32/d^2) = (104/d^2)$ kpsi. Substitute into the Gerber relationship:

$$55.25/(27 \times d^2) + \{104/(105 \times d^2)\}^2 = 1$$

or $$d^4 - 2.046d^2 - 0.981 = 0$$

A root for d in the range $1 \leq d \leq 2$ is $d = 1.564$ in.

Example Problem # 2.5: In example problem # 2.— find the diameter of the component when multiple factors of safety are a requirement: $n_1 = 2.25$ for stress amplitudes, $n_2 = 1.55$ for mean stress, $n_3 = 1.29$ for the endurance limit, $n_4 = 1.22$ for the yield strength and $n_5 = 1.20$ for the ultimate strength.

Solution: The Soderberg theory will be used for this example, which takes the form:

$$(S_{alt}/S_e) + (S_m/S_y) = 1$$

Combining the safety factors to the strength characteristics gives: minimum $S_e = S_e/n_3 = 27/1.29 = 20.93$ kpsi, minimum $S_y = S_y/n_4 = 82/1.22 = 67.21$ kpsi, minimum $S_{ut} = S_{ut}/n_5 = 105/1.20 =$

87.5 kpsi. The permissible alternating stress $\sigma_{alt,p} = n_1\sigma_{alt} = 2.25(17/d^2) = 38.25/\ d^2$ kpsi and the permissible mean stress $\sigma_{m,p} = n_2\sigma_m = 1.55(32/d^2) = 49.6/\ d^2$ kpsi. Letting $S_{alt} = \sigma_{alt,p}$ and $S_m = \sigma_{m,p}$ and substituting into the Soderberg expression gives:

$\{38.25/(20.93 \times d^2)\} + \{99.6/(67.21 \times d^2)\} = 1$

This gives $d = 1.602$ in.

■■■

REFERENCES & BIBLIOGRAPHY

Ahmad, S., Irons, B.M., Zienkiewicz, O.C., "Analysis of Thick and Thin Shell Structures by Curved Finite Elements", International Journal of Numerical Methods in Engineering, Volume 2, Page 419, 1970.

American Petroleum Institute, "Centrifugal Compressors for General Refinery Service", API STD 617, 5th edition, Washington, D.C., 1988.

Ballhaus, W.F., Goorjian, P.M., "Computation of Unsteady Transonic Flows by the Indicial Method", AIAA Journal, Volume 16, Pages 117 – 124, 1978.

Bolcs, A., Fransson, T.H. "Aero-elasticity in Turbo-machines: Comparison of Theoretical and Experimental Cascade Results, Appendix A5: All Experimental and Theoretical Results for the 9 Standard Configurations", Communication du Laboratoire de Thermique Applique et de Turbo-machines, EPF-Lausanne, No. 13, 1986.

Breard, C., Vahdati, M., Sayma, A.I., Imregun, M., "An Integrated Time-Domain Aero-elasticity Model for the Prediction of Fan Forced Response due to Inlet Distortion", ASME paper # 00-GT-373, New York, 2000.

Camp, T.R., "A Study of Acoustic Resonance in a Low Speed Multi-Stage Compressor", ASME paper # 97-GT-259, New York, 1997.

Camp, T.R., Day, I.J., "A Study of Spike and Modal Stall Phenomena in a Low Speed Axial Compressor", ASME Journal of Turbo-machinery, Volume 120, Pages 293 – 401, New York, 1997a.

Cumpsty, N.A., Whitehead, D.S., "The Excitation of Acoustic Resonance by Wake Shedding", Journal of Sound and Vibration, Volume 18, No. 3, Pages 353–369, 1971.

Ehrich, F.F., Spakovszky, Z.S., Sanchez, M.M., Song, S.J., Wisler, D.C., Storace, A.F., Shin, H.W., Beacher, B.F., "Unsteady Flow and Whirl Inducing Forces in Axial Flow Compressors: Part II - Analysis", ASME paper # 2000-GT-566, New York, 2000.

Ehrich, F.F., "Rotor Whirl Forces Induced by the Tip Clearance Effect in Axial Flow Compressors", ASME Journal of Vibration and Acoustics, Volume 115, Pages 509 – 515, New York, 1993.

Frischbier, J., Schulze, G., Zielinski, M., Ziller, G., Blaha, C., Hennecke, D.K., "Blade Vibrations of a High Speed Compressor Blisk-Rotor – Numerical Resonance Tuning and Optical Measurements", International Gas Turbine and Aerospace Congress, Birmingham, UK, June, 1996.

Gupta, D.K., "Vibration Analysis of Fluid Submerged Blades", Ph.D. Thesis, Indian Institute of Technology, Delhi, India, 1984.

Hall, R.M., Armstrong, E.K., "The Vibration Characteristics of an Assembly of Interlock Shrouded Turbine Blades", Structural Dynamics Aspects of Bladed Disk Assemblies, Srinivasan, A.V., editor, ASME, New York, 1976.

Irons, B.M., "A Frontal Solution Program for Finite Element Analysis", International Journal of Numerical Methods in Engineering, Page 275, 1970.

Kielb, R.E., Chiang, H.D., "Recent Advances in Turbo-machinery Forced Response Analyses", AIAA Joint Propulsion Conference Proceedings No. 28, 1992.

Leyes, R., Flemin, W., editors, "The History of North American Small Gas Turbine Aircraft Engines", National Air and Space Museum, Smithsonian Institute, Washington, DC, 1997.

Lubomski, J.F., "Status of NASA Full Scale Engine Aero-elasticity Research", NASA TM 81500, 1980.

Mikolajczak, A.A., Snyder, L.E., Arnoldi, R.A., Stargardter, H., "Advances in Fan and Compressor Blade Flutter Analysis and Predictions", AIAA Journal of Aircraft, Volume 12, number 4, Pages 325–332, 1975.

Nowinski, M.C., Panovsky, J., "Flutter Mechanisms in Low Pressure Turbine Blades", ASME paper # 98-GT-573, New York, 1998.

Nowinski, M.C., Panovsky, J., Bolcs, A., "Flutter Mechanisms in Low Pressure Turbine Blades", ASME Gas Turbine Conference and Exhibition, Sweden, June, 1988.

National Transportation Safety Board, "In Flight Loss of Propeller Blade – Aircraft Accident Report ", NTSB/AAR-96/06, Washington, DC, 1996.

Panovsky, J., "Flutter of Aircraft Engine Turbine Blades", Ph.D., Thesis, University of Cincinnati, 1997.

Panovsky, J., Nowinski, M.C., Bolcs, A., "Flutter of Aircraft Engine Low Pressure Turbine Blades", 8th International Symposium of Unsteady Aerodynamics and Aero-elasticity of Turbo-machines, Sweden, 1997.

Panovsky, J., Kielb, R.E., "A Design Method to Prevent Low Pressure Turbine Blade Flutter", ASME paper # 98-GT-575, New York, 1998.

Parker, R., Stoneman, S.A.T., "An Experimental Investigation of the Generation and Consequences of Acoustic Waves in an Axial Flow Compressor: Large Axial Spacing Between Blade Rows", Journal of Sound and Vibration, Volume 99, Number 2, Pages 169 – 182, 1985.

Pfeiffer, R., "Blade Vibrations of Continuously Coupled and Packed Steam Turbine LP Stages", Vibrations of Blades and Bladed Disk Assemblies, Kielb, R.E., and Rieger, N.F., editors, ASME, New York, 1985.

Platzer, M.F., Carta, F.O., "AGARD Manual on Aero-elasticity in Axial Flow Turbo-machines, Structural Dynamics and Aero-elasticity", Volume 2, 1988.

Rao, J.S., "Turbo-machine Blade Vibration", Wiley Eastern Limited, John Wiley & Sons, New Delhi, 1991.

Rangwala, A.S., "Turbo-machinery Dynamics: Design and Operation", McGraw-Hill, New York, 2004.

Saravananamuttoo, H.I.H., Rogers, G.F.C., Cohen, H., "Gas Turbine Theory", Prentice Hall, Harlow, England, 2001.

Srinivasan, A.V., Lionberger, S.R., Brown, K.W., "Dynamic Analysis of an Assembly of Shrouded Blades Using Component Modes", ASME Journal of Mechanical Design, Volume 100, Pages 520–527, New York, 1978.

Srinivasan, A.V., Fabunmi, J.A., "Cascade Flutter Analysis of Cantilevered Blades", ASME Journal of Engineering for Gas Turbines and Power, Volume 106, Pages 34–43, New York, 1984.

Srinivasan, A.V., "Flutter and Resonant Vibration Characteristics of Engine Blades", ASME paper # 97-GT-533, New York, 1997.

Storace, A.F., Wisler, D.C., Shin, H.W., Beacher, B.F., Ehrich, F.F., Spakovszky, Z.S., Sanchez, M.M., Song, S.J., "Unsteady Flow and Whirl Inducing Forces in Axial Flow Compressors: Part I - Experiment", ASME paper # 2000-GT-565, New York, 2000.

Tyler, J.M., Sofrin, T.G., "Axial Flow Compressor Noise Studies", Transactions, Society of Automotive Engineers, Volume 70, Pages 309–332, 1962.

Vahdati, M., Imregun, M., "A Non-linear Aero-elasticity Analysis of a Fan Blade Using Unstructured Dynamic Meshes", Journal Mechanical Engineering Science, Part C, 210, Pages 549–563, 1996.

Whitehead D.S., "Flutter of Turbine Blades", Unsteady Aerodynamics and Aero-elasticity of Turbo-machines and Propellers, Proceedings, 4th International Symposium, Aachen, Germany, Pages 437–452, 1987.

CHAPTER 3

Turbine and Combustor Parts

3.1 INTRODUCTION

Extensive analytical predictions and rig testing notwithstanding, blade failures in engines due to excessive vibrations still occur during final test phases. Even worse is the occurrence of failures after an engine has successfully passed a number of rigorous qualification and certification tests and goes into regular service. The failures may point to a lack of adequate design tools, but may indicate another issue arising from the business perspective: in the ongoing search for efficiency improvement and better thrust to weight ratio, engine designers and researchers are pursuing higher speeds and temperatures, radical new blade profiles, higher stage loads and lower aspect ratios. More exacting consideration of blade flutter, resonance, cooling, material characteristics and manufacturing methods is required than hitherto.

Low cycle fatigue failures of turbine blades are of great significance in aircraft engines. Thermal gradients and mechanical stress from centrifugal loading rapidly increase as the engine is started and goes to full speed during aircraft take-off from the ground. Figure 3.1 shows a typical stress and temperature distribution in a free standing and in shrouded turbine blade. The reverse pattern is repeated during landing and engine shutdown. When blade rows in a turbine are operating close to resonance conditions, they are prone to failure from high cycle material fatigue. In power generation turbines operating at a near constant speed blade resonance can occur during engine startup and shutdown conditions. Excitation is encountered from two primary sources: (*i*) due to flow path interference between stator and rotor blade rows at nozzle passing frequencies and (*ii*) due to manufacturing and assembly errors at per revolution harmonics. Interference between the stationary and moving blades is an aerodynamic phenomenon, and is based on potential interaction, wake interaction and viscous effects. Mechanical excitation from manufacturing errors and mounting of stationary diaphragms cannot be readily simulated using mathematical formulations.

Calculation of blade resonant frequencies and mode shapes of a bladed disk system calls for definition of boundary conditions at shroud and dovetail interfaces. Maximum operating speed

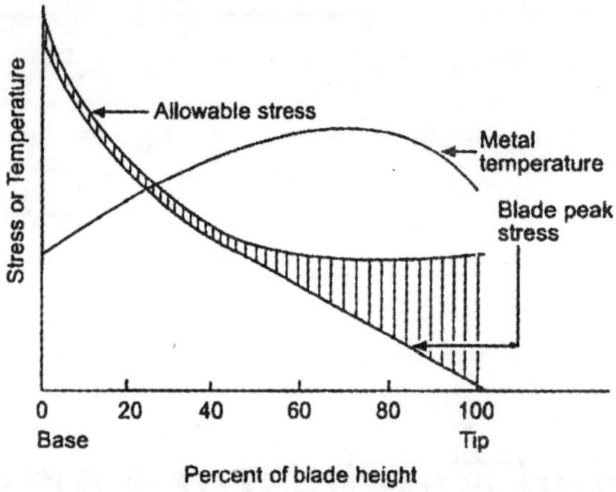

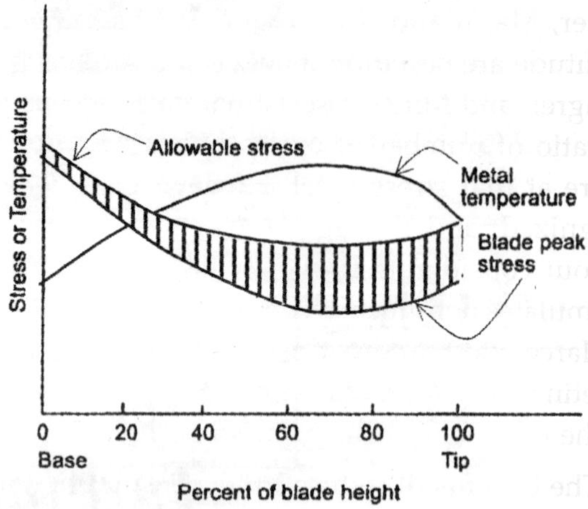

Fig. 3.1. *Stress and Temperature Distribution in Free Standing (upper) and Shrouded (lower) Turbine Blades*

of the engine cannot be established unless vulnerable frequencies and response amplitudes at resonance are known to a satisfactory degree of confidence. Research efforts have focused on the need to develop a fundamental understanding of topics such as Coulomb and viscous damping, coolant passage turbulators and unsteady viscous flow in turbo-machines. Without adequate knowledge of damping, for example, there is no alternative but to guess values for the coefficient in determining the response from a particular excitation. Note that damping may arise from material characteristics, friction, aerodynamic flow and possibly from impact.

Shrouds in the form of a protrusion are used in turbine blades to alleviate problems arising from dynamic motion in the blade. Long and slender turbine blades in the last stages of gas turbines can take advantage of the support provided by mating surfaces of shrouds on adjacent blades to reduce flexural and twisting motion at the tip. Even longer blades in low-pressure steam turbines may be equipped with some form of dampening mechanism at mid-span and tip locations. However, the shroud also imposes penalties in the form of a mass at the tip of the blade, which requires body loads due to the centrifugal force field to be carried by the rest of the airfoil. It also adds to the manufacturing cost of the blade. In a shrouded blade system the protrusion constrains blade motion not only along the contact plane between the shrouds of adjacent blades but also along the normal direction of the plane. In-plane tangential relative motion is mostly two-dimensional. On the other hand, normal relative motion can cause variation of the normal contact load and, in extreme cases, separation of the contact interface.

Turbine blades and vanes constitute a considerable portion of the total cost of the equipment, given the fact that thousands of airfoils are used in any turbo-machine. Accurate determination of operating life of turbine and compressor blades plays a central role in the design of aircraft power plants. The rotating parts must be retired prior to failure, but must still possess adequate life to be commercially acceptable to airline operators. Life estimation using stress based theories is a multi-faceted technology, and calls for calculation of mean steady stresses, dynamic stresses, failure surface, load history and cumulative damage. The influence of mean stress may be described by a number of linear and non-linear relations, for example, Goodman, Soderberg,

Gerber, Marin and Kececioglu. Several cumulative theories for alternating stress of variable amplitude are based on damage accumulated during the load cycles. A common theory due to Palmgren and Miner asserts that damage fraction at any stress level is linearly proportional to the ratio of number of cycles of operation to the total number of cycles that would produce failure at that stress level. However, the order of application of different stress levels is not recognized, and damage is assumed to accumulate at the same rate at a given stress level without consideration of past history. Experimental evidence indicates that fatigue damage accumulates non-linearly, depending on the alternating stress level. Non-linear theories proposed by Marco and Starkey, Corten and others rely on some exponent of the same ratio. A problem sometimes encountered with the application of non-linear theories is the lack of material data for the exponent at different stress levels.

The benefits of reduced fuel consumption and increased power arising from increasing the turbine inlet temperature have been clearly brought out in chapters 2 and 3. In spite of losses experienced in cooling of blades and vanes, the gains are still considerable. Methods to cool the blades receive serious research attention. Cooling of blades with liquids is difficult because of practical problems associated with delivery and retrieval of the coolant in the primary cooling system in the forced or free convection modes, or in a closed secondary system. Difficulties are also encountered due to corrosion and deposits in open systems. In closed systems it is difficult to obtain sufficient secondary surface cooling area at the base of the blade. Internal air cooling in the forced convection mode is more practical in engines. Turbine blade metal temperatures may achieve a reduction of 200° to 300°C by channeling 1.5 – 2.0 percent of the airflow for cooling for each blade row. The blades may be cast with internal cooling passages in the core or forged and drilled with holes of any required shape and size using electro-discharge, electro-chemical or laser drilling process. Outer surface cooling is achieved by pushing cooling air out of holes in the blade walls. In this process heat is extracted more uniformly from the surface and at the same time provide a layer of cooler air isolating the metal from the hot gases of the main stream. The concept of transpiration cooling requires porous blade walls for the cooler internal air to ooze out from the internal blade cavity, but successful application will depend on the availability of appropriate porosity in the skin material and manufacturing methods.

Cooling of the rotating airfoils still represents unusual difficulties from engineering and manufacturing considerations. At elevated gas and metal temperatures oxidation and creep impose limitations on the blade's capabilities. Unlike rotating airfoils, nozzle vanes do not experience high stress levels. In spite of it, cooling of annulus walls and stator vanes require special attention.

Thermal barrier coats represent perhaps the most promising and exciting development in super alloy coatings. Any mechanism by which the temperature limits can be raised by overcoming hot-section material restraints is of significant interest, and thermal barrier coatings offer this potential. Coatings of ceramic or metallic or a combination of the two are applied on the substrate of a super alloy to preclude or inhibit direct interaction between the substrate and a potentially damaging environment. This damage can be either metal recession due to oxidation/corrosion or a reduction in the mechanical properties of the substrate due to diffusion of harmful species into the alloy at elevated temperature.

3.2 INTEGRITY EVALUATION OF TURBINE BLADE

High cycle fatigue of rotating turbine components is a serious problem since it has potential to cause substantial damage. Highly loaded blades experience alternating stresses from aerodynamic excitation. The blades are subjected to phenomena such as stator wake, blade flutter, rotating stall and acoustic resonance, but the link between fluid dynamics and structural mechanics must be established.

Turbocharger turbine axial blades operating at variable speeds are exposed to unsteady dynamic forces σ. The forces are set up by the engine stroke, charging system, gas entry at inlet and nozzle vanes. A research program initiated by ABB Turbo Systems of Switzerland aims to take a combined computational fluid dynamics and finite element analysis approach to the problem (*ref*: Filsinger *et al.*, 2001). As a first step the transient flow behavior in the turbine cascade is simulated using time-dependent temperature and pressure inlet boundary conditions. Forced blade response u due to the pulsating pressure distribution $p(t)$ is then obtained. Coupling between the two numerical methods is achieved by a Fourier decomposition (amplitude P_k, phase delay β_k) of the time resolved excitation forces $F(t)$ acting on the rotating bladed disk.

The fluid dynamics program is based on a two-dimensional time accurate multi-block Euler/Navier-Stokes solver. The integrated post-processing offers a close link to mechanical integrity codes by determining blade forces, with calculations done in the absolute frame of reference and using a moving grid for the rotor. At the intersection between stator and rotor grids the cells overlap, and an interpolation technique is used. Two-dimensional Euler equations are valid for the flow simulation on a circumferential stream plane of a selected radius with constant radial thickness.

Fig. 3.2. *Computational Fluid Dynamics Grid (Filsinger, 2001)*

Figure 3.2 shows details of the grid per block. Unsteady inlet boundary conditions are determined by simulating the diesel engine's behavior with respect to the exhaust pipe system and the turbocharger's related performance. In this procedure, items pertaining to the varying ambient conditions, turbocharger specifications and load acceptance must be addressed. Total pressure and temperature values are unsteady due to the pulsating nature of the engine's exhaust flow. In the finite element routine disk assemblies containing N symmetric blades coupled tangentially through the rotor are analyzed. The disk assembly is a rotationally periodic structure with identical blades, and hence the cyclic wave theory may be applied. Static and dynamic deformations for the whole disk can then be represented by a single blade with the application of complex circumferential boundary conditions.

In the computation sequence for free vibrations, harmonic vibration of a single coupled blade (without damping) is represented by:

$$\left[M(e^{jn\phi})\right]\{d^2q/dt^2\}+\left[K(e^{jn\phi},\Omega)\right]\{q\}=\{0\} \tag{3.1}$$

where $j = (-1)^{1/2}$ and $\varphi = 2\pi/N$ is the circumferential periodicity of the disk sector. Nodal diameter n assumes values 0, 1, 2, . . up to $N/2$ for even N and $(N\text{-}1)/2$ for odd N. $[M]$ and $[K]$ represent complex inertia and non-linear stiffness matrices with respect to rotational speed Ω, and depend on the nodal diameter n. Quantities$\{q\}$ and $\{d^2q/dt^2\}$ are complex displacement and acceleration vectors of the blade. Kinematic cyclic constraints of the form:

$$\begin{aligned}\{q\}_{right} &= \{q\}_{left}\,e^{jn\phi}\\ \{d^2q/dt^2\}_{right} &= \{d^2q/dt^2\}_{left}\,e^{jn\phi}\end{aligned} \tag{3.2}$$

are applied between nodes on the right and left sector sides. Rewriting the Euler function in trigonometric notation, eigen frequencies of the cyclic finite element model can be computed in the real domain. Hence, the cyclic model has to be represented by two identical finite element meshes, with nodal boundarics on the sector sides (fig. 3.2) constrained as given in equation (3.2). For each mode i and nodal diameter n (n = 0 to $N/2$), two identical eigen-frequencies are computed, which refer to two possible mode shapes of the bladed disk assembly. Static calculations may be readily performed by substituting n = 0 in (3.1) and (3.2), so the static equation of the assembly rotating at Ω becomes:

$$[K(\Omega)]\{q\} \;=\; \{P_0\}+\{T\}+\{F(\Omega)\} \tag{3.3}$$

where $\{P_0\}$, $\{T\}$ and $\{F(\Omega)\}$ are stationary gas pressure, thermal and centrifugal forces. For the given value of Ω, eigen-frequencies of the bladed disk may then be computed.

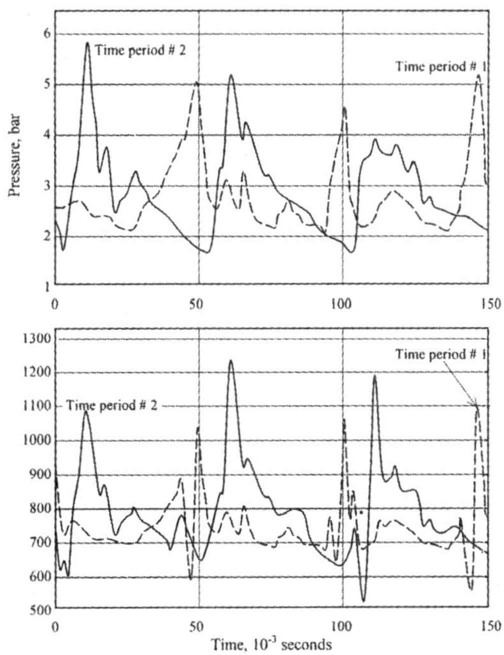

Fig. 3.3. *Boundary Conditions at Inlet: Pressure (Upper), Temperature (Lower) (Filsinger, 2001)*

Gust response and motion dependent unsteady aerodynamics assure the calculations of aero-damping. Total blade damping arising from material, micro-frictional and aero-damping may be evaluated from measured resonance peaks with the aid of fractional power bandwidth method (*ref*: Harris *et al.*, 1995). In this example values of fraction of critical damping range from 0.15 to 0.40.

Excessive computational time may be avoided by performing the finite element calculations in the frequency domain. As a first approach a single excitation force split into axial and tangential directions from the CFD calculated pressure distribution maybe applied assuming a uniform radial distribution along the blade height. This simplification is valid to some extent, since the emphasis is on vibrations caused by pressure pulses of the diesel engine's exhaust, which mostly excites low blade bending modes (*ref*: Szwedowicz *et al.*, 1996).

Forced vibration of a single coupled blade may be expressed in the Cartesian system by:

$$\left[M(e^{jm\phi})\right]\{d^2q/dt^2\}+[D]\{dq/dt\}+\left[K(e^{jm\phi},\Omega)\right]\{q\}=\{P\} \tag{3.4}$$

where damping matrix $[D]$ is determined by the Rayleigh dissipation model as a linear combination of mass and stiffness matrices. Generalized vector $\{P\}$ describes the non-uniform pressure distribution along the circumference. For a blade rotating at constant angular speed Ω, pressure P is a periodic excitation function with period $\tau = 2\pi/\Omega$. After applying the computed eigen frequencies $\omega_{i,n}$ and mode shapes $\phi_{i,n}$, the steady response of the disk is given for each nodal diameter n in the rotating frame as:

$$m_{i,n}d^2u_{i,n}/dt^2 + 2\omega_{i,n}\xi_i du_{i,n}/dt + k_{i,n}u_{i,n} = \{\phi\}_{i,n}^T\{P_k\}e^{j(k\alpha_l-\beta_k)}e^{jk\Omega t} \tag{3.5}$$

where k describes the engine order, i indicates the number of considered mode shapes and α_l denotes the tangential position of the excited mode. Amplitudes P_k and phase delay β_k are obtained from the Fourier decomposition of the non-uniform pressure P. In the twin finite element models excitation loads of each engine order k must be imposed simultaneously on node l of the real and imaginary blade:

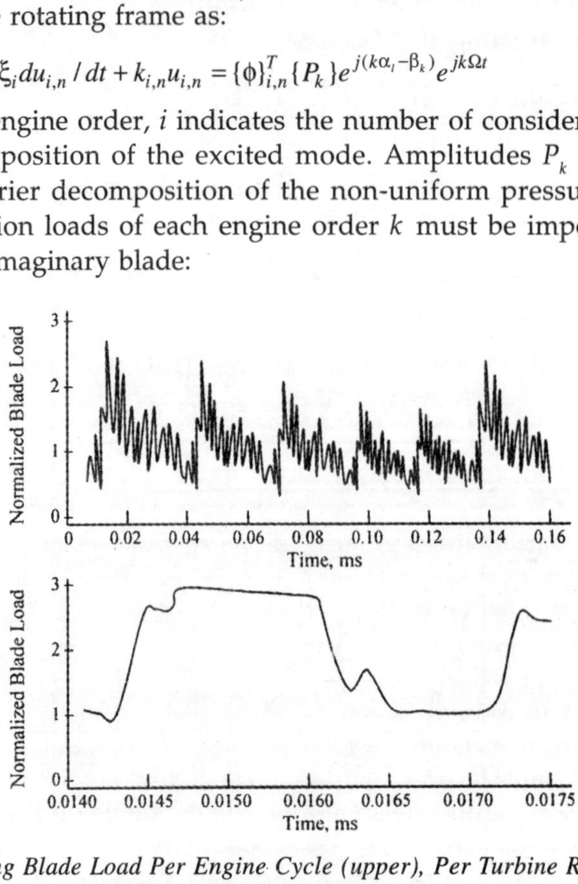

Fig. 3.4. *Resulting Blade Load Per Engine Cycle (upper), Per Turbine Revolution (lower) (Filsinger, 2001)*

$$F_{l,\chi}^{real} = P_{k,\chi}[\mathrm{Cos}(k\alpha_l - \beta_{k,\chi}) + j\mathrm{Sin}(k\alpha_l - \beta_{k,\chi})] \tag{3.6}$$

$$F_{l,\chi}^{imag} = P_{k,\chi}[\mathrm{Sin}(k\alpha_l - \beta_{k,\chi}) - j\mathrm{Cos}(k\alpha_l - \beta_{k,\chi})] \tag{3.7}$$

where χ refers to both axial and tangential directions. Due to orthogonality conditions between the disk modes and the circumferential pressure distribution in (8.37), the disk assembly oscillating at nodal diameter n is in resonance if the excitation order $k = \kappa \cdot N \pm n$ or $k = \kappa \cdot (N + 2) \pm n$ where $\kappa = 0, 1, 2\ldots \infty$ for even or odd number of blades in the rotor (*ref*: Rangwala, 2004).

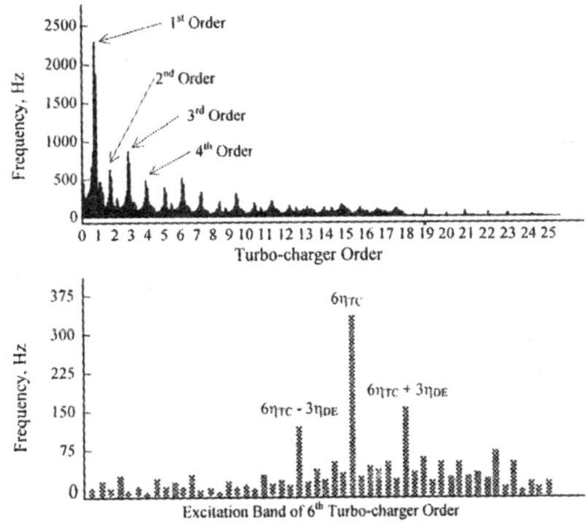

Fig. 3.5. *Blade Load Excitation* (upper) *Full Spectrum,* (lower) *Sixth Engine Order Spectrum (Filsinger, 2001)*

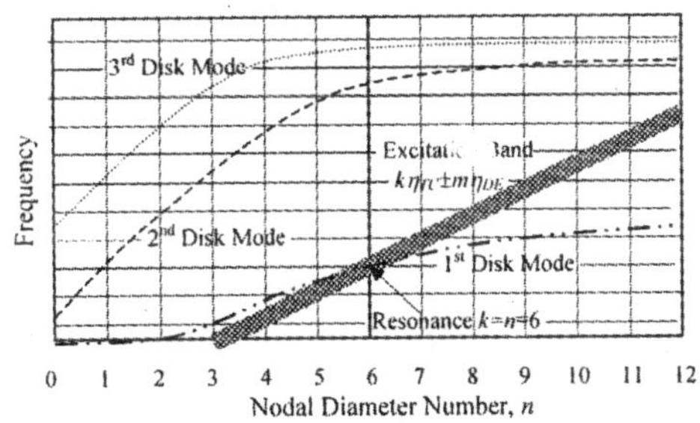

Fig. 3.6. *Nodal Diameter Diagram of Bladed Disk (Filsinger, 2001)*

Excitation pulse P_k varies with time, hence it must be transformed into the frequency domain. This is achieved by employing an ordinary Fourier decomposition. Note that both amplitude and phase relationships must be accurate in order to get the right load. With the help of these numerical tools the bladed disk assembly may be analyzed for any operating condition. Load from a three-pulse charged diesel engine is considered here, with inlet casing split into two separate segments and resulting in a total of six pulses per engine cycle. Figure 3.3 shows the pressure and temperature variations. Each inlet segment is then connected to an exhaust pipe serving three cylinders, and the pressure pulses are then directly led to the turbocharger turbine. This results in considerable differences between the entrance conditions at the two inlets, hence high dynamic loads may be expected to act on the blades.

Only one resulting force acts on each blade. Figure 3.4 shows the calculated force for a complete engine cycle, which equals two revolutions of the four-stroke diesel engine. The turbine wheel's own rotation must also be recognized, and the second half of fig. 3.4 represents loads during one turn.

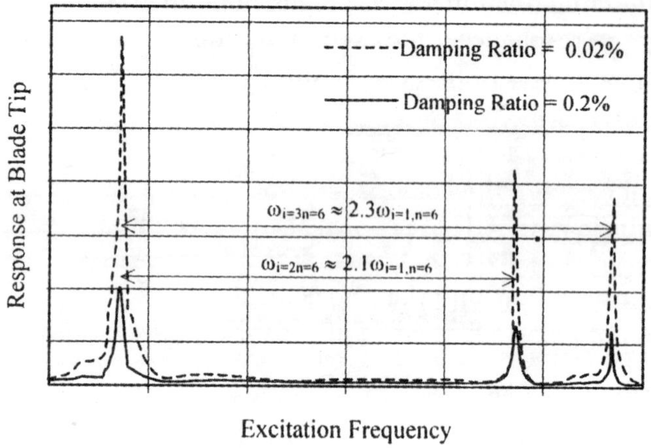

Fig. 3.7. *Blade Resonant Amplitudes (Filsinger, 2001)*

Resonance occurs when excitation order k equals nodal diameter number n. Intersections between the excitation bands and eigen-frequencies, defining possible resonance conditions may be identified with the aid of a Campbell diagram in combination with the nodal diameter diagram shown in fig. 3.5, where the sixth order is chosen for illustration. Nodal diameters for the bladed disk assembly are shown in fig. 3.6.

In the CFD calculations the rotor speed may be adjusted to resonance of the first blade mode for the sixth engine order. In the finite element code forced response is then performed according to the excitation band of the same order. If only material damping is to be considered (as may be determined from a hammer test), the value may be less than 0.02 percent. Figure 3.7 provides a comparison of resonance amplitudes of the lowest blade modes excited by the sixth engine order for damping ratios of 0.02 percent and 0.2 percent.

3.3 RESPONSE DUE TO BLADE-VANE INTERACTION

A complete and accurate definition of the resonant vibration response of turbine bladed disks includes characterization of the structural dynamics and the forcing function. The dominant source of excitation can be attributed to the presence of circumferentially oriented flow distortions created by stator vanes, struts, bleed manifolds, rotating stall and inlet configurations. Fourier analysis of the distortions yields the significant coefficients (*ref*: Jay *et al.*, 1983). These components have the potential to excite a bladed disk resonance at speeds corresponding to the intersection of the excitation frequency of the particular Fourier component K, ($K \times$ rotational speed), with the natural frequencies of the bladed disk system. In general, when a sinusoidal distortion field with K patterns acts on the tuned disk, response in the Kth diametral nodal pattern results. In addition, other excitations can also be obtained (*ref*: Ewins, 1968).

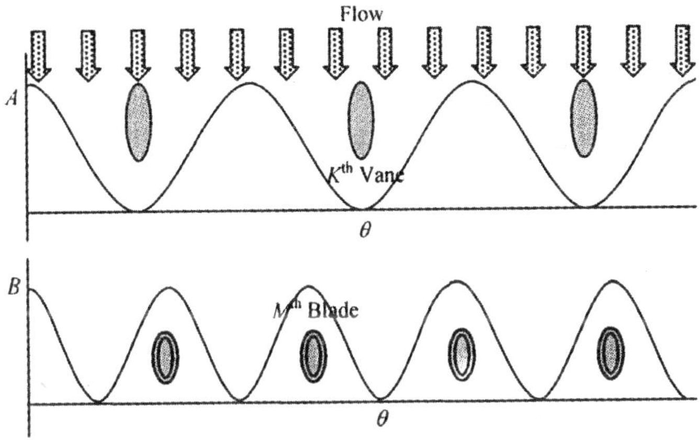

Fig. 3.8. *Flow Past Cascade of Stationary and Moving Airfoils*

A bladed disk with M number of blades can respond to a Kth component of excitation in harmonic or diametral patterns, which are functions of the sum and difference in multiples of M and K. To illustrate this ($M - K$) type excitation, an analytical model assuming sinusoidal representations for the blades and vanes and using dual (orthogonal) mode definition of the diametral modes are made (*ref*: Stange *et al.*, 1981). The model yields the mechanism of the ($M - K$) type response for both tuned and mistuned systems. The validity of the model is verified from experimental data.

Blade vibration due to an upstream vane row results when a uniform flow impinges on the set of K vanes, creating K distortions through which the set of M blades pass during rotation. The primary concern in this analysis is the fundamental harmonic resulting from these K distortions. In fig. 3.8 the airfoils of a bladed disk, rotating in the $+\theta$ direction at a speed of Ω are assumed to present a sinusoidally varying obstruction to the incoming distorted flow. Considering the bladed disk response with respect to the disk fixed coordinate system, the sinusoidal excitation pattern created by the vane row can be regarded as moving in the $-\theta$ direction at a rotational speed Ω. This excitation pattern creates a force on the blades, which can be expressed as:

$$V(\theta,t) = \frac{A}{2}[1+\cos K(\theta+\Omega t)] \tag{3.8}$$

where A is as indicated in the fig. 3.8. The presence of the blade cascade in the flow field can be represented by a Fourier series in the circumferential direction. Considering the fundamental component of the series, the blades will be represented as:

$$B(\theta) = \frac{1}{2}[1+\cos M\theta] \tag{3.9}$$

So the force created on the blades then can be written as:

$$F(\theta,t) = V(\theta,t)\times B(\theta) \tag{3.10}$$

The effect of the forcing function on the cos $n\theta$ and sin $n\theta$ modes can be computed through the generalized forces, N_{nc} and N_{ns}, from:

$$N_{nc} = \int_0^{2\pi} F(\theta,t)\times \text{Cos}(n\theta)d\theta \tag{3.11}$$

$$N_{ns} = \int_0^{2\pi} F(\theta,t)\times \text{Sin}(n\theta)d\theta \tag{3.12}$$

Substitute the expression for force into the integrals for N_{nc} and N_{ns}. For vibratory response the terms independent of time are not of interest. The remaining terms are zero, unless $n = K$, $n = M - K$, or $n = M + K$. Three cases can be distinguished:

$$n = K \qquad N_{nc} = \frac{A\pi}{4}\cos K\Omega t \tag{3.13}$$

$$n = M - K \qquad N_{nc} = \frac{A\pi}{8}\cos K\Omega t \tag{3.14}$$

$$n = M + K \qquad N_{nc} = \frac{A\pi}{8}\cos K\Omega t \tag{3.15}$$

Similarly for the Sine participation factors:

$$n = K \qquad N_{ns} = \frac{A\pi}{4}\sin K\Omega t \tag{3.16}$$

$$n = M - K \qquad N_{ns} = \frac{A\pi}{8}\sin K\Omega t \tag{3.17}$$

$$n = M + K \qquad N_{ns} = -\frac{A\pi}{8}\sin K\Omega t \tag{3.18}$$

With non-zero participation factors, response of the bladed disk in diametral nodal lines described in the three cases is possible when occurring at the intersection of the n diametral node family modes and the vane generated frequency $K\Omega$.

Response in the disk due to the forcing function of Kth order for the three given cases $n = K$, $n = M - K$ and $n = M + K$ is:

$$W(\theta,t) = \frac{\cos(n\theta)}{M_{nc}D_{nc}}\left(\frac{A\pi}{4}\right)\cos(K\Omega t - \varphi_{nc}) + \frac{\sin(n\theta)}{M_{ns}D_{ns}}\left(-\frac{A\pi}{4}\right)\sin(K\Omega t - \varphi_{ns}) \tag{3.19}$$

$$W(\theta,t) = \frac{\cos(n\theta)}{M_{nc}D_{nc}}\left(\frac{A\pi}{8}\right)\cos(K\Omega t - \varphi_{nc}) + \frac{\sin(n\theta)}{M_{ns}D_{ns}}\left(\frac{A\pi}{8}\right)\sin(K\Omega t - \varphi_{ns}) \tag{3.20}$$

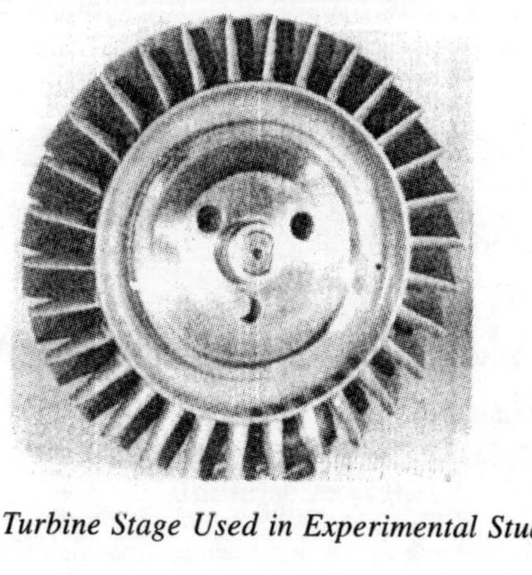

Fig. 3.9. *Turbine Stage Used in Experimental Study (Jay, 1983)*

$$W(\theta,t)=\frac{\cos(n\theta)}{M_{nc}D_{nc}}\left(\frac{A\pi}{8}\right)\cos(K\Omega t-\varphi_{nc})+\frac{\sin(n\theta)}{M_{ns}D_{ns}}\left(-\frac{A\pi}{8}\right)\sin(K\Omega t-\varphi_{ns}) \quad (3.21)$$

where M_{nc} and M_{ns} are generalized modal masses, W_{nc} and W_{ns} are modal natural frequencies, $D_{nc,ns}=\{[W^2_{nc,ns}-(K\Omega)^2]+4\beta^2_{nc,ns}(K\Omega)^2\}^{1/2}$ and β_{nc} and β_{ns} are modal damping factors for the cosine and sine modes.

In a perfectly tuned bladed disk $W_{nc}=W_{ns}=W_n$, $M_{nc}D_{nc}=M_{ns}D_{ns}=M_nD_n$ and $\varphi_{nc}=\varphi_{ns}=\varphi_n=\pi/2$.

Then for $n = K$,

$$W(\theta,t)=\left(\frac{A\pi}{4M_nD_n}\right)\sin(K\Omega t+n\theta) \quad (3.22)$$

and for $n = M - K$,

$$W(\theta,t)=\left(\frac{A\pi}{4M_nD_n}\right)\sin(K\Omega t-n\theta) \quad (3.23)$$

With K greater than M, the n diametral pattern travels opposite the direction of rotation. Thus, the number of vanes and blades determine the direction of travel of the n lobed pattern relative to the disk (*ref*: Rangwala, 2004). In these traveling patterns, equal blade amplitudes would be expected with an inter-blade phase angle dependent on the number of blades M and the diametral pattern n.

The bladed disk assembly chosen for this investigation is shown in fig. 3.9. The aerodynamic characteristics of this turbine stage are typical of advanced turbine designs.

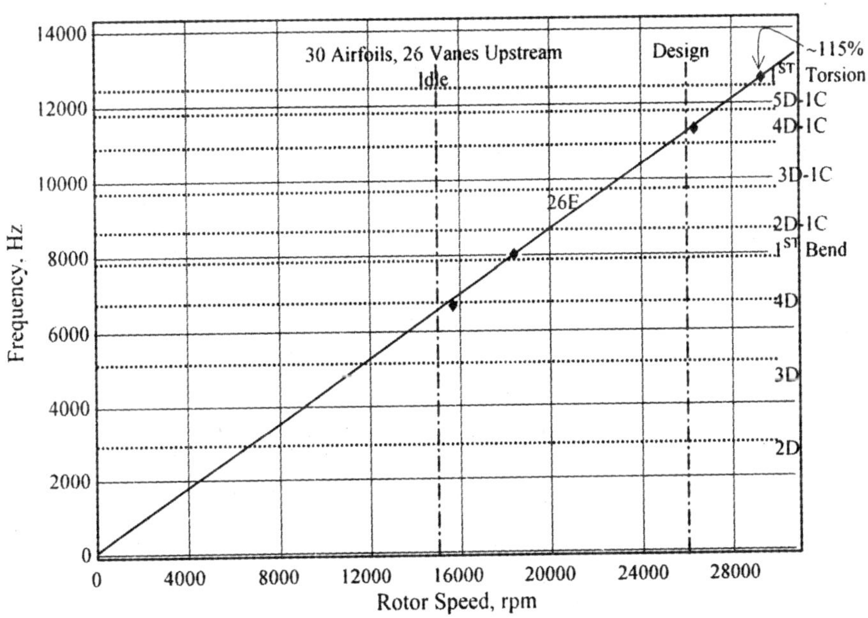

Fig. 3.10. *Bladed Disk Assembly Resonance Diagram (Jay, 1983)*

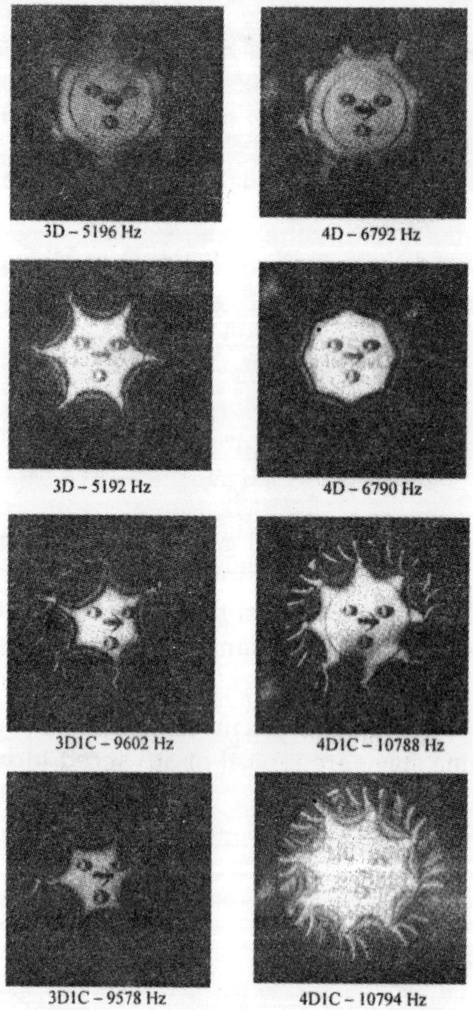

Fig. 3.11. *Holograms for Third and Fourth Split Mode Diametral Patterns (Jay, 1983)*

Thirty blades are machined integrally with the disk as shown. Definition of the blade's first bending and first torsional modes is accomplished to indicate the degree of frequency mistuning in the blading. Two methods are used. In the first, each blade is individually excited while the others remain unconstrained. In the second method, all blades except the excited blade are damped and mistuned with the addition of a sealing material. Overall agreement between the two methods is excellent. The frequency range measured for first bending mode is from 8059 to 8323 Hz and for the first torsional mode, 12074 to 12502 Hz. The average first bending frequency is 8191.5 Hz and the average first torsional frequency is 12279 Hz. The frequency-speed diagram based on the bench tests is presented in fig. 3.10. The expected engine order line of maximum response corresponds to the 26 upstream stators.

Special emphasis is focused on the third and fourth diametral patterns, holograms of which are presented in fig. 3.11. Orthogonal or dual modes are found for the modes as indicated. In the zero circumferential modes, the modal split for the third diametral pattern is only 4 Hz or .07 percent of the mean frequency, while the fourth diametral pattern modal split is 2 or less

Hz. Orthogonality of the modes can be clearly seen in the one circumferential family of modes. Modal split factors are 0.25 percent for the third diametral modes and 0.06 percent for the fourth diametral modes. These results indicate a very closely tuned rotor, the largest mistuning being noted for the third diametral, first circumferential mode. Strain gages mounted on each airfoil during this testing confirm the holographic studies.

The dynamic tests consist of slow rotor acceleration between 14,000 and 30,000 rpm at varying aerodynamic conditions. During the testing period, dynamic signals from eight adjacent blades are observed and recorded. The resonant data defined using the standard three-axis waterfall chart is shown in fig. 3.12, which present dynamic stress amplitude and resonant intersections as a function of rotor speed, with the dynamic behavior of blade # 12 indicated. The largest response, indicated by number 1, occurs at a rotor speed of approximately 25,000 rpm, consistent with the excitation created by the 26 inlet vanes.

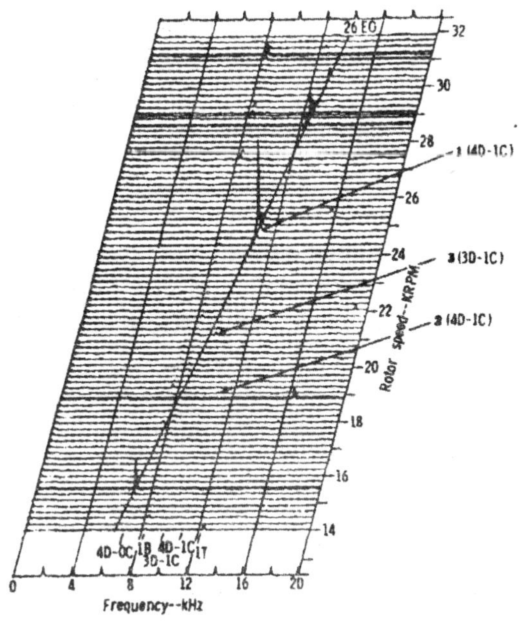

Fig. 3.12. *Waterfall Chart for Dynamic Response of Blade # 12 (Jay, 1983)*

Examination of this resonance by using the Campbell diagram (fig. 3.10) would indicate a four diametral, onc circumferential nodal pattern. Thus, the 26 vanes and 30 blades configuration yield a four diametral pattern response.

The highest stress measured on the blade hub exceeded 40000 psi (275.79 MPa). Two other resonances are significant in this discussion. The smal1 resonance denoted by 3 in the waterfall chart is identified as a three diametral, one circumferential nodal pattern excited by the 27th engine order. The resonance denoted by 2 corresponds to the four diametral, one circumferential nodal pattern excited by the 34th engine order. The source of the 27 and 34 engine orders are most probably due to the interaction of the 26 vanes and upstream total pressure and temperature instrumentation. More detailed evaluations were performed using a dual channel FFT analyzer with the capability to perform cross-spectrum analyses at expanded frequency resolution about the resonant frequency. Figure 3.13 presents the amplitude content at various frequencies for strain gage signatures from blades 11 and 12 obtained during acceleration.

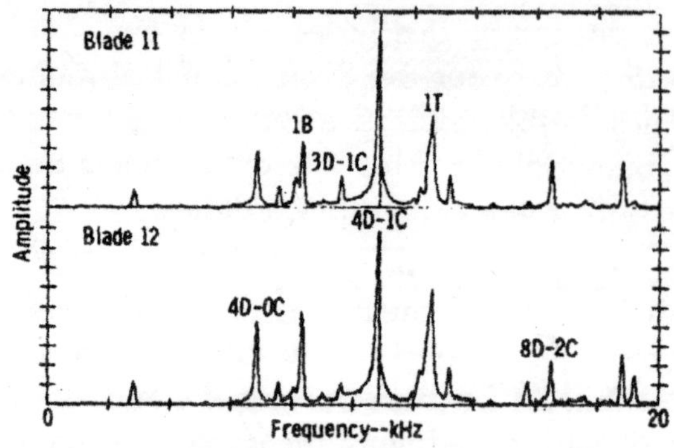

Fig. 3.13. *Overall Response Spectrum During Acceleration (Jay, 1983)*

3.4 PERIODIC RESPONSE OF SHROUDED BLADES

Protruding shrouds on a blade constrain its motion not only along the contact plane but also along the normal direction of the plane, thus resulting in a complex contact kinematic system. The in-plane tangential motion induces stick-slip frictional forces (*ref*: Menq, *et al.*, 1998), and normal motion causes variation of the contact normal load, sometimes even resulting in separation of the contact interface. A two-dimensional model of the contact kinematics can retain the normal component of the relative motion that causes normal load variation, while in-plane relative tangential motion will reduce the linear motion. If radial components are to be included in a three-dimensional model, in-plane relative tangential motion may be assumed to be elliptical, composed of motions along the principal major and minor axes. A variable normal friction force may then be applied separately to each individual linear motion to obtain equivalent stiffness and damping characteristics of the shroud contact (*ref*: Chen, *et al.*, 1999).

Figure 3.14 shows a shrouded blade system of two neighboring blades contacting each other through protruding shrouds. Blade coordinate system is defined by x (tangential), y (axial) and z (radial) directions. Shroud coordinate axis v is along normal direction, and u and w-axes are on the contact plane. The contact plane of the three-dimensional shroud contact is defined by angles Ψ (shroud angle) and Φ (inclination angle). The two coordinate systems are related to each other by a coordinate transformation matrix as a consequence of the rotations. When subjected to cyclic excitation the motion of the blades may be assumed to be periodic, so the relative motion across the shroud contact is also periodic in three-dimensional space. A substructure may be used to model the shroud contact, representing the contact plane friction interface and portions of mating shrouds. The substructure will then have two mass-less elastic beams from A to B, held together by a preload n_O. The difference in the motions of A and B then describes the relative three-dimensional motion of the two adjacent shrouds. Friction force may then be analyzed by reducing the motion into periodic motions in the contact plane and normal to the contact plane.

Constraint forces on the contact plane have two components, the variable normal contact force and the induced friction force. Depending on the amplitude and phase of the in-plane relative motion, and also the normal load from the vibratory motion, friction contact force will cause stick, slip or separation during a full cycle of motion.

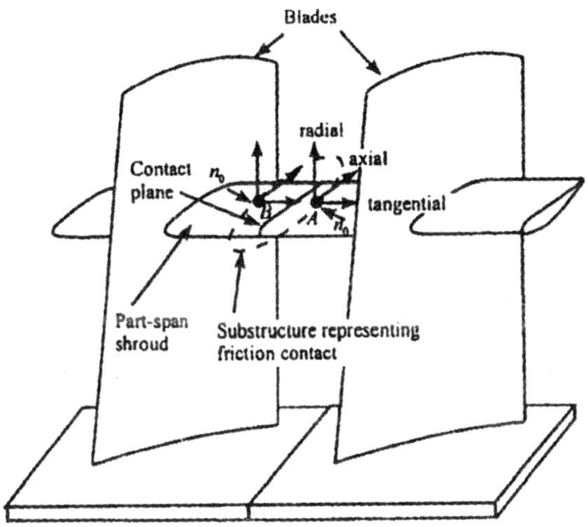

Fig. 3.14. *Shroud Contact Between Neighboring Blades (Chen, 1999)*

Separation occurs when the applied normal load vanishes at an instant. Transition from stick to slip occurs when friction force on the plane equals the varying slip load, or stick condition:

$$\left|\bar{f} = \bar{K}_u(\bar{u} - \bar{u}_o) + \bar{f}_o\right| < \mu n, dw/dt = 0 \tag{3.24}$$

slip condition:
$$\bar{f} = \mu n \frac{d\bar{w}/dt}{|\bar{w}|}, d\bar{w}/dt \neq 0 \tag{3.25}$$

where $\bar{u}, \bar{w}$ are the input relative tangential motion and contact point slip motion vector $[u\ w]^T$, $\bar{f}$ is induced friction force vector, v, u, w are relative motions in the shroud coordinate system, μ is the coefficient of friction, n is contact load and $\bar{K}_u$ is stiffness in the u direction. Subscript o refers to initial values at the beginning of the stick state.

To understand the transition from slip to stick, Coulomb friction law calls for velocity of relative motion to reach zero; friction force can then be obtained by solving the initial value problem since initial friction force at the beginning of slip is known. A numerical integration method such as the fourth order Runge-Kutta method may be used to obtain the friction force.

Transition from stick/slip to separation occurs when the normal load vanishes. Transition from slip condition to one of stick occurs when the velocity of relative motion reaches zero, implying:

$$\bar{f}^T \bar{K}_u d\bar{u}/dt - \mu^2 n(dn/dt) = 0 \tag{3.26}$$

Separation to stick/slip condition transition calls for development of normal load on the contact plane. Mathematically the criterion may be expressed as $n = 0$, $dn/dt \geq 0$.

Similarly, stick/slip to separation transition may be expressed by $n = 0$, $dn/dt < 0$. These criteria may be used to simulate hysteresis loops of the friction contact when experiencing periodic relative motion. The resulting constrained force can then be characterized by the relative motion between adjacent shrouds. By employing fast Fourier transform the constrained force can be approximated by a series of harmonic functions in a multi-harmonic balance method to solve for periodic response.

Analysis of a shrouded blade system is simplified if the blades are tuned, so each blade then has identical dynamic characteristics. Also, the excitation of interest is the one induced by the blades rotating through circumferential variations in the flow field. Effectively, then, each blade is then exposed to a periodic excitation possessing the same amplitude but differing by a phase angle proportional to the location of the blade on the disk. Forced response of the system may then also be considered to be periodic of the same fundamental period as the excitation. This allows representation of the blade's motion and the non-linear constrained force by infinite Fourier series. Truncating the series after m terms provides an approximate solution, with each blade vibrating in the same manner but with the appropriate inter-blade phase difference $k\Psi$ for the kth harmonic component from its adjacent blades. The inter-blade phase angle $\phi = 2\pi E/N$, where N is the number of blades and E is the engine order of excitation.

In a tuned system the condition of cyclic symmetry can be applied in deriving the relative motion of a contact shroud between points B and A. If the motions of the two contact points of the blade at left are defined as $[x_r\ x_l]^T$, then motion of point B is given as x_r, and the motion of point A differs from x_l with the corresponding inter blade phase angle. Both x_r and x_l are periodic motion, hence the relative motion w_r is also a periodic trajectory in three-dimensional space, which may then be transformed to the shroud coordinate system.

After the decomposition the u and w components can induce stick-slip friction, while the v component can cause normal load across the interface to vary dynamically. u and w motions are coupled together when inducing stick-slip friction. Using fast Fourier transforms, the constraint force can be approximated by a series of harmonic functions using complex Fourier coefficients of the kth harmonic component along u, v, w axes.

Predicted response of a shrouded blade system is shown in fig. 3.15, where the frequencies are normalized with respect to the first mode natural frequency. The first twenty vibration modes are employed. As the contact pre-load changes, two limit cases occur, namely the fully separate and the fully stuck cases. Both cases are linear problems since a non-linear force term does not appear. The fully separate case occurs when the interface has such a large initial gap that the vibrating neighbor shrouds cannot contact each other. On the other hand, when the pre-load of the interface exceeds a level depending on external excitation, the shroud contact interface remains fully stuck. Then the shroud does not dissipate energy, but provides additional stiffness to cause higher resonant frequencies.

In between the two linear cases the constraint force consists of non-linear friction force and the variable normal load. The significance of the variation of the contact normal load depends on the direction of the resulting relative motion at the shroud contact and the orientation of the contact plane. If the variation is not large, the effect of the shroud constraint is similar to that a platform damper. The variable normal load can also directly impose non-linear stiffness into the system due to the intermittent separation of the contact surface. This can result in a multi-valued response, leading to a jump phenomenon (*ref*: Thomson, 1988).

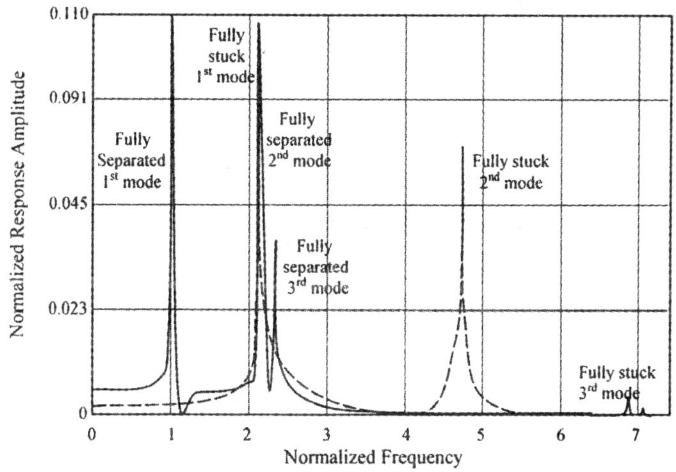

Fig. 3.15. *Periodic Response of Shrouded Blade System (Chen, 1999)*

The increase in the resonant amplitude causes the pre-load interface to separate, so the interface temporarily does not provide stiffness. The overall effect is similar to that of a softened spring that gives rise to the response with a resonance peak bending towards the lower frequencies.

3.5 CYCLICALLY SYMMETRIC STRUCTURES

The geometry of an impeller or a bladed disk is cyclically symmetric about its center axis. Thus, an identical sector comprising of a portion of the disk and a blade repeats itself as many times as the number of blades in the rotating component to form the complete part. A disk sector and a blade also interface with neighboring components. Advantage of this fact is taken to calculate natural frequencies and mode shapes by mathematically modeling just one sector. The feature, known as modal cyclic symmetry, saves significant amount of computational time. There are two principal benefits in treating the disk as a cyclic assembly rather than an assembly of a number of arbitrary components. First, the degrees of freedom at each interface between adjacent disk sectors are eliminated by cyclic constraints, thus yielding a smaller model. Second, disk component mode shapes bear a greater resemblance to system modes, which improves modal convergence. Furthermore, the mode shapes of the complete structure can be viewed from the analysis of the single segment.

As a component of a rotating system a typical disk may participate in the vibratory motion of the assembly, depending upon flexibility of the disk. Understanding of the procedure for modal cyclic symmetry is facilitated by the concept of nodal diameters and circles, where the word nodal is referred to in the context of vibrations. Consider the appearance of a flat disk vibrating in a certain mode. Many mode shapes will contain straight-line contours of zero displacements out of the plane, crossing the entire disk. These modes may be fixed relative to the disk or may travel in the same or opposite direction to that of the rotor speed.

As component geometry increases in complexity the zero displacement contours are not clearly defined in a mode shape. Nodal diameters may not always correspond to the number of lines of zero displacement, hence the definition is more general. The number of nodal diameters is an integer, and determines the variation in the value of displacement in any direction at points spaced at a circumferential angle equal to the angle of the base sectors. The normal displacement w for a stationary disk vibrating in its nth diametrical mode shape is defined by:

$$w(\theta_0, t) \;=\; A \times \sin(n_s \theta_0) \times \cos(p_n t) \tag{3.27}$$

where p_n is the resonant frequency of the nth mode, θ_o is the absolute angular coordinate and A is the resonant amplitude, a function of radius r and proportional to the magnitude of the excitation force. For a disk rotating at angular velocity Ω, θ_o may be expressed in terms of angular position of the disk $\theta_o = \theta - \Omega t$, hence:

$$w(\theta,t) \;=\; \frac{A}{2}\sin[n\theta-(p_n+n\Omega)t]+\frac{A}{2}\sin[n\theta+(p_n-n\Omega)t] \tag{3.28}$$

Modal cyclic symmetry may be performed for a stress free component when the part is not rotating, or in a pre-stressed condition when the part is rotating and subjected to centrifugal forces. Tensile forces in the disk will alter the natural frequencies of vibration. The stress-free procedure calls for performing the modal analysis after definition of the base sector and definition of boundary conditions. In the definition of the cyclically symmetric basic sector model in the cylindrical coordinate system, care must be exercised to ensure the angle spanned equals 360° divided by the number of blades. Internal coupling and constraints are permissible. Radial impellers sometimes have one or more splitter blades between the main blades. The splitter blades must then be included in the base sector. The base sector must have matched sets of nodes on the lower and higher angle edges, offset geometrically by the sector angle. The edges do not need to be flat in the cylindrical sense, and may take any shape. For the pre-stressed modal cyclic symmetry analysis a static solution is required to define the stresses in the base sector. Stress results are stored in the model, summary boundary conditions used in the static solution are removed, and then cyclic symmetry modal analysis is performed.

The combination of lightweight disks and low aspect ratio blades considerably increase the risk for highly coupled blade-disk vibrations, an oftentimes complex dynamic response that cannot be identified using standard strain gauge methods. Strain gauge data is localized with the available number and area of coverage of the rotating assembly, restricting the observation of mostly complex and non-symmetric vibration modes. Holographic interferometry imaging method, consisting of optically suppressing the rotational motion of the wheel by passing the image through a prism rotating at half the rotor speed, is used in many instances to study the dynamic response. The prism causes the disk image to rotate in the opposite direction at the same speed as the rotating test piece to cancel out the rotational motion. The disk's resonant response may then be recorded using double-pulsed holographic interferometry equipment (*ref*: Wadell, 1973 and Stetson *et al.*, 1977).

In an experiment conducted by US Air Force an annular disk is mounted at the bore and rotated at speeds ranging from 2,000 to 8,000 rpm, with its resonant modes excited by an oscillator driven electromagnet (*ref*: MacBain, *et al.*, 1978). The de-rotated image is examined by a double-pulsed ruby laser.

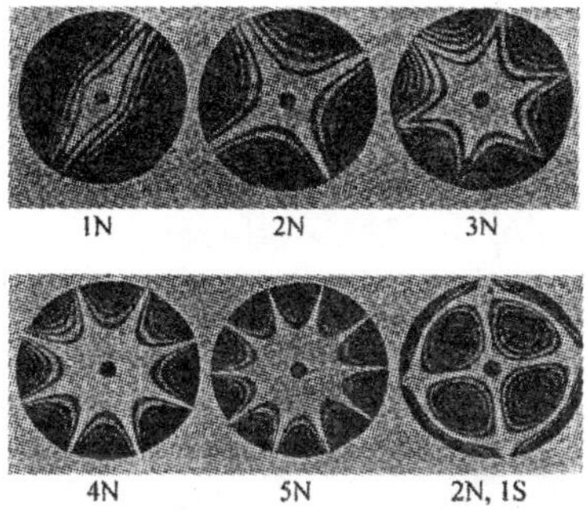

Fig. 3.16. *Resonant Mode Shapes for Stationary Disk (MacBain, 1978)*

For this study the disk, made of AISI 1018 cold rolled steel, is of 10.0 in. diameter, 0.625 in. thickness and has a 1.0 in. center hub diameter of 1.0 in. thickness with a 0.5 in. bore. Attention is focused on the first five diametrical modes and one complex mode of vibration in the speed range.

Figure 3.16 provides details of the resonant modes using siren excitation. Because of the high degree of radial symmetry, the orientation of the nodal line is governed by the exciter's location and geometry. Each mode of vibration consists of a pair of natural frequencies and mode shapes, only one mode of each pair being shown in the fig. (*ref*: Rangwala, 2004). The disk's normal displacement relative to thickness is small enough to avoid non-linear effects. The pair of modes is essentially the inverse of each other, with nodal lines of one becoming the anti-nodal lines of the other. The natural frequencies for the pair are separated by 0.5 to 1.5 Hz, indicative of the presence of radial symmetry of the disk's mass and stiffness properties. With perfect symmetry the mode pair degenerates into a single mode shape and corresponding natural frequency.

From equation (3.28), resonance occurs in a rotating disk when excitation frequency $\omega = p_n \pm n\Omega$, so two frequencies are exciting the same mode. Thus, at a specific disk speed the 2*N* mode of vibration, for instance, will be in resonance at (2*N* + 2Ω) and at (2N – 2Ω) frequencies. Figure 3.17 illustrates the situation of the electromagnetic exciter frequencies required to generate resonance for the first 5 modes. The upper curve for each mode is a plot of $p_n + n\Omega$ and the lower plot of $p_n - n\Omega$. As disk speed approaches zero the two branches converge to the corresponding resonant frequency for the stationary disk.

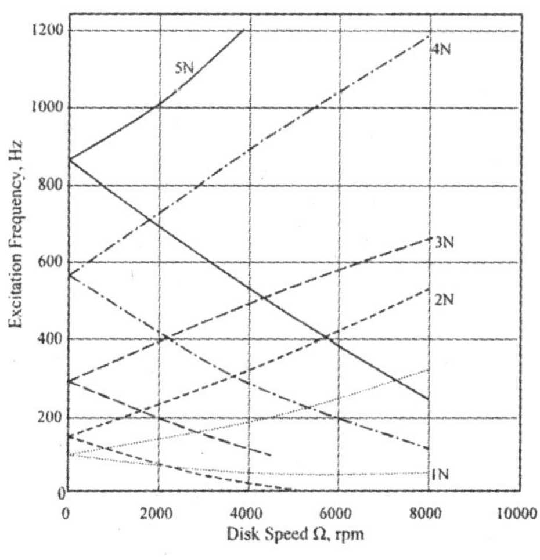

Fig. 3.17. *Variation of Resonant Excitation Frequency with Spin Speed (MacBain, 1978)*

Extrapolated data for the 2N and 3N modes cross the abscissa at 5,600 rpm and 7,600 rpm where resonant modes and dangerous regions of operation can be expected because of second and third engine orders of excitation. Figure 3.18 shows the same disk resonant mode shapes of fig. 3.17 when the disk is not spinning. Each mode in fig. 3.18 corresponds to some point on one of the curves of fig. 3.17. For example, the 5N mode shape is excited at 1193 Hz when the disk is running at 3,910 rpm, lying on the upper branch of the curve. The actual natural frequency for the mode at this operating speed is calculated to be:

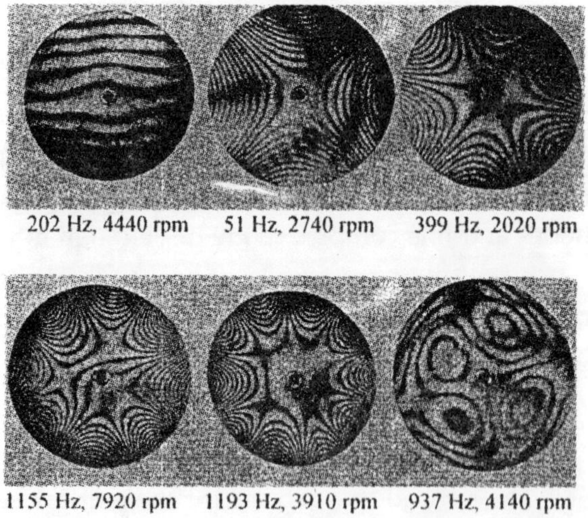

Fig. 3.18. *Resonant Mode Shapes for Rotating Disk (McBain, 1987)*

3.6 RANDOM BLADE MISTUNING

Traditional dynamic analysis of a compressor rotor assumes the blades to be identical. The assumption of cyclic symmetry enables considerable reduction in computation time by modeling a single sector instead of the full blade assembly. However, in actual practice there are small differences in structural characteristics of individual blades arising from manufacturing and material tolerances or in-service degradation, and is referred to as blade mistuning. Mistuning may lead to confinement of vibration energy to a few blades or even a single blade. Thus, one or more blades may experience deflections and stresses that are much larger than predicted by a tuned analysis. Also, since this is a random occurrence, statistical analyses such as Monte Carlo simulation are critical for safely predicting response amplitudes in the design process. But this procedure for a full finite element blade assembly is enormously large and expensive, hence a low order model with efficient response prediction is needed.

A reduced-order model (ROM) procedure (*ref*: Bladh, *et al.*, 2002) assumes disk and blade degrees of freedom are ordered in a manner that yields a block-diagonal form for the assembled stiffness matrix for the full structure:

$$\mathrm{K} = \begin{bmatrix} K_d & 0 \\ 0 & K_b \end{bmatrix} = \begin{bmatrix} I \otimes \tilde{K}_d & 0 \\ 0 & I \otimes \tilde{K}_b \end{bmatrix} \tag{3.29}$$

where I is the identity matrix and $\otimes$ denotes the Kronecker product. The tilde notation indicates the quantity refers to either the blade or the disk. Thus, the uncoupled setup implies that blade-disk interface degrees of freedom appear in both component matrices. A key idea of the technique is to describe motion of the assembly using two particular sets of component modes. The first mode set consists of modes of a single blade that is clamped at the chosen interface location, while the second set has modes of the disk with massless blade attached. Although they are not classical component modes, they include disk induced blade motion as a combination of rigid body and elastic deformation at the interface.

Figure 3.19 depicts the finite element mesh of the second stage of a four-stage compressor rotor with 29 blades. The blades and disk are machined from a single continuous block of metal. Material properties of the metal are: Young's modulus E = 203.4 GPa, density ρ = 7909 kg/m^3, Poisson's ratio ν = 0.305 and structural damping γ = 0.006. MSC/NASTRAN cyclic symmetry routines have been employed in the evaluation. The sector model is clamped at ribs located at outer edges of the disk, approximating the effects of the neighboring sections.

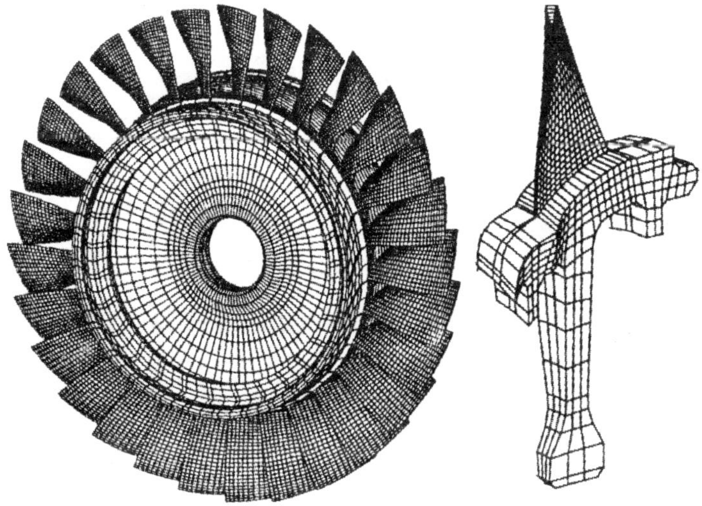

Fig. 3.19. *Finite Element Model (left) Full Assembly, (right) Fundamental Sector (Bladh, 2002)*

Table 3.1. ***Blade Mistuning Pattern***

Blade #	*Mistuning,* δ_n	*Blade #*	*Mistuning,* δ_n	*Blade #*	*Mistuning,* δ_n
1	0.05704	11	−0.03631	21	0.02919
2	0.01207	12	−0.03570	22	−0.00328
3	0.04670	13	−0.03631	23	0.00086
4	−0.01502	14	−0.03631	24	−0.03654
5	0.05969	15	0.00242	25	−0.03631
6	−0.03324	16	0.04934	26	−0.01665
7	−0.00078	17	0.04479	27	0.00783
8	−0.01688	18	0.03030	28	−0.01169
9	0.00242	19	0.00242	29	−0.01332
10	−0.02747	20	0.01734		

The mistuned finite element model consists of the entire assembly. The mesh pattern in the single sector model is duplicated around the periphery. Mistuning is introduced into the assembly by allowing each blade to have a different Young's modulus by using the scheme $E_n = E_0(1 + \delta_n)$, $n = 1, .., N$, where E_0 is the Young's modulus of a tuned blade and δ_n is a dimensionless tuning parameter associated with the nth blade.

A specific mistuning pattern based on experimental natural frequency measurements is shown in table 3.1.

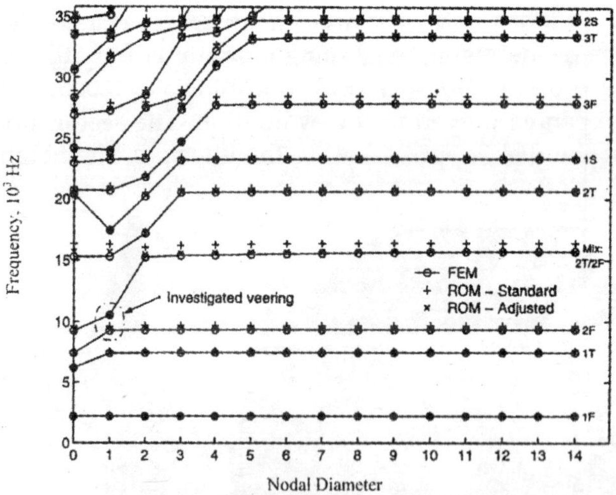

Fig. 3.20. *Free Vibration Modes of Tuned Rotor (Bladh, 2002)*

Mode shapes of a tuned rotor can be conveniently described in terms of nodal diameters (nodal lines across the diameter of disk) and nodal circles (nodal lines in the circumferential direction. The number of nodal diameters for a mode shape corresponds to a certain phase shift between adjacent blades, called the inter-blade phase angle, and is given by $\phi_h = 2\pi h / N$, where h is the number of nodal diameters and N is the number of blades. Characteristics of the free vibration modes are summarized by a plot of natural frequencies against the number of nodal diameters, as shown in fig. 3.20. Two interesting features of the tuned rotor's modal structure may be noted. As the nodal diameter number increases the disk stiffens, hence the sloping portion of the curves indicates some effects from the disk modes. In the absence of interaction between the blades, as may be expected from shrouded blades, the blade dominated modes do not stiffen considerably with the increase in nodal diameters. The other notable feature is the numerous eigen frequency veerings, regions in which blade and disk dominated mode families appear to veer away from each other. Physically the separation of the eigen frequencies indicate the degree of coupling between the families of disk and blade modes. The strength of a veering may be measured by the spacing between the frequencies and the local curvature in the veering region (*ref*: Pierre, 1988).

The external excitation force vector used for forced response analysis consists of point loads applied at the tip and mid-chord of the blades in a direction normal to the surface of the blade, together with the structural damping factor. If more realistic pressure loads are to be simulated, specification of equivalent nodal loads on all blade degrees of freedom is needed. In the tuned system all blades have identical peak displacement norms. But in the mistuned case the blades have substantially different amplitudes, and the maximum displacement norm in the assembly may be selected at each driving frequency to define the maximum response amplitude. Figure # 3.21 provides a comparison of mistuned finite element and reduced order model (ROM) maximum blade responses for blade tip excitation for the fourth engine order.

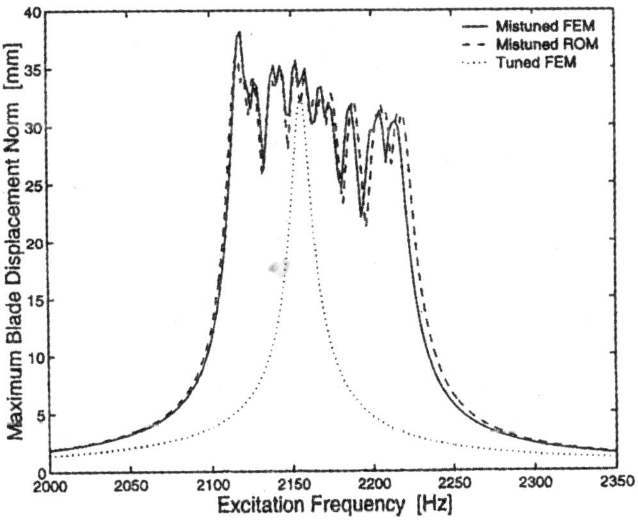

Fig. 3.21. *Mistuned Blade Response Comparison for 4th Engine Order (Bladh, 2002)*

Discrepancies in resonant frequencies and response amplitudes between the two methods are small. Maximum mistuned response is about 17 percent higher than the tuned resonant amplitude. Significant shifts in amplitude may be expected when some inter-blade coupling is available (*ref*: Ottarson *et al.*, 1995, Rangwala, 2004). Most of the modes are dominated by the cantilever blades, so they yield a relatively modest response amplitude variation due to mistuning.

Excitation close to a veering has some significance for the reduced order method, since coupling between the blades must be captured. The first two flexural modes of the cantilever blade are located in the heart of the eigen frequency veering. The reduced order method yields a good representation of the mistuned response pattern, but response amplitudes do not compare as well. Accuracy of the mistuned response of blade dominated modes is of some significance in the capture of any existing veering frequencies. Small residual errors remaining after eigenvalue adjustments among the modes comprising an investigated veering give rise to discrepancies in the mistuned response. But the maximum mistuned response amplitude is still within 6 percent of the full finite element solution with only a fraction of the computational effort. Note also that maximum mistuned response amplitude exceeds tuned response by 57 percent, mostly due to increased disk-blade modal interaction in the veering region.

3.7 CUMULATIVE DAMAGE THEORY IN LIFE PREDICTION

Low cycle fatigue (LCF) has been a primary design consideration for bladed disk assemblies, and this has succeeded in controlling LCF related failures to a considerable extent. But a number of developments in the design of blades and disks have created problems due to high cycle fatigue (HCF). The role of mistuning in HCF failures has been recognized and accepted throughout the industry. Transient response of blades during startup and shutdown of the machines is another significant contributor. Failure due to resonance at lower modes such as the first bending and torsion may include fracture of an entire blade and its attachment to cause substantial secondary damage, and even greater if the fragments are not contained in the engine. At the higher second torsion or second and third bending modes mostly the outer portion of the blade is separated. At still higher modes only a portion of the tip may be released.

A number of cumulative damage theories have been proposed in making estimates of the life of a blade subjected to variable stress amplitude (*ref*: Rao *et al.*, 1999). A commonly used linear method in predicting fatigue life of a blade is based on work done by Palmgren (1924) and Miner (1945). Dynamic stress around a critical speed may be split into a convenient number of steps on either side of peak stress. Operation at a stress level S_i gives a life of N_i cycles. If the blade is subjected to n_i cycles, it suffers a damage fraction of $D_i = n_i / N_i$. Failure is then predicted to occur when $\Sigma(n_i / N_i) \geq 1$. The assertion is that the damage fraction at any stress level is linearly proportional to the ratio of number of cycles of operation to the total number of cycles that would produce failure at that stress level. Since the blade is subjected to a mean stress, the S – N plane is shifted to the location of the applied mean stress level on the fatigue failure surface.

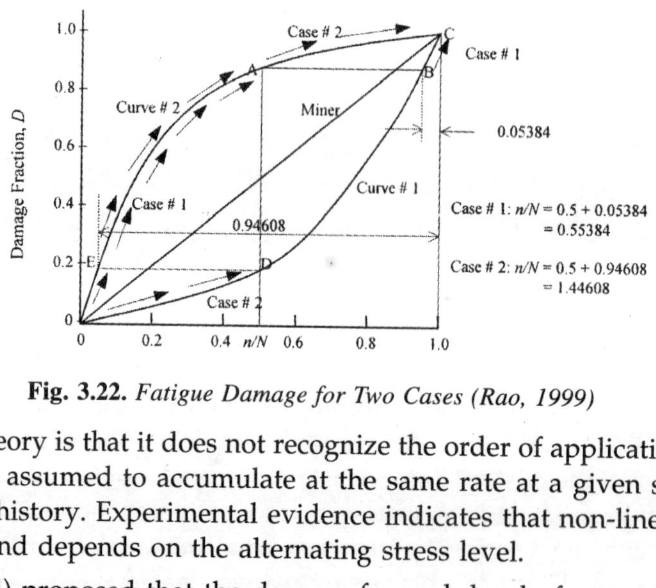

Fig. 3.22. *Fatigue Damage for Two Cases (Rao, 1999)*

A drawback to this theory is that it does not recognize the order of application of various stress levels, and damage is assumed to accumulate at the same rate at a given stress level without consideration of past history. Experimental evidence indicates that non-linear accumulation of the fatigue damage, and depends on the alternating stress level.

Marco *et al.*, (1954) proposed that the damage for each level of reversed sinusoidal stress amplitude is expressed by $D = (n/N)^m$, where the exponent m depends on the amplitude of the alternating stress. A specimen is deemed to have failed when D reaches a value c irrespective of the sequence in which the stresses are applied. Failure occurs when $\Sigma(n/N)$ expressed by:

$$\sum \frac{n}{N} = c\int_0^1 \frac{1+\dfrac{N_1}{N_2}+\dfrac{N_1}{N_3}+\ldots+\dfrac{N_1}{N_i}}{1+\dfrac{N_1}{N_2}r_2 D\left(\dfrac{r_2-1}{r_2}\right)+\dfrac{N_1}{N_3}r_3 D\left(\dfrac{r_3-1}{r_3}\right)+\ldots+\dfrac{N_1}{N_i}r_i D\left(\dfrac{r_i-1}{r_i}\right)}\, dD \tag{3.30}$$

where N_i are cycles of completely reversed stresses S_i to produce failure, with i denoting order of application of stress levels, D is the damage ratio, r_i is the ratio of exponents m_i / m_1 representing stress levels S_i and S_l and m_i is the exponent in the damage equation associated with i^{th} stress level. Aside from the difficulty in evaluating the integral, the primary difficulty is the necessity of generating a lot of experimental data for the exponent m_i. Qualitatively, the procedure is illustrated by fig. 3.22 using available data for two stress levels σ_1 and σ_2, with $\sigma_1 > \sigma_2$. First, the specimen is loaded for $n/N = 0.5$ at stress level σ_1 (line O – A), followed by the lower stress level σ_2 (line B – C), until damage occurs at $D = 1$. Then:

$$\sum \frac{n}{N} = \left(\frac{n}{N}\right)_{\sigma_1} + \left(\frac{n}{N}\right)_{\sigma_2} \tag{3.31}$$

$$= 0.5 + 0.05384 = 0.55384$$

On the other hand, if the lower stress level σ_2 is applied first for $n/N = 0.5$ (*OD* in fig. 8.15) and then with the higher stress level σ_1 (line *EC* in fig. 8.15) until damage occurs at $D = 1$, then:

$$\sum \frac{n}{N} = \left(\frac{n}{N}\right)_{\sigma_2} + \left(\frac{n}{N}\right)_{\sigma_1} = 0.5 + 0.94608 = 1.44608 \tag{3.32}$$

When the higher stress is applied first it takes only a few more lower stress cycles before damage occurs. When the stress application order is reversed, the specimen requires many more higher stress level cycles before damage is encountered. Experiments support such a situation.

When stress levels are increased gradually, Palmgren-Miner's rule provides satisfactory results. When the stress levels are decreasing from a peak value, application of a factor simplifies the situation. The average value of $\Sigma(n_i/N_i)$ in this period may be multiplied by a factor $f = 1.67$ for the sum in the full period. This implies that the damage is higher when stress levels are decreasing.

Corten, *et al.*, (1956) rely on the fracture mechanics theory for postulating initiation of permanent damage. Once initiated, fracture propagation is assumed to occur at stress levels that are well below the minimum stress required to initiate a crack. Damage per nucleus is given by $D = mrN^a$ where m is the total number of fatigue damage nucleii, r is the coefficient of damage propagation rate and a function of stress level and a is the damage propagation exponent, also a function of stress level. From an experimental relation between the stress dependent ratio $R = r_2/r_1$, damage exponent a and a material property d (given by $R^{1/a} = [S_3/S_1]^d$), failure can be expected when:

$$\frac{n_1}{N_1} + \frac{n_2}{N_1}\left(\frac{S_2}{S_1}\right)^d + \frac{n_3}{N_1}\left(\frac{S_3}{S_1}\right)^d + \ldots + \frac{n_i}{N_1}\left(\frac{S_i}{S_1}\right)^d = 1 \tag{3.33}$$

where N_1 is the number of cycles to failure at the highest stress amplitude S_1, and n_i are the number of cycles imposed at each stress S_i. For steels exponent d is found to be between 6.2 and 6.9 and mean value of 6.57. A modification of the $S-N$ diagram calls for defining a line $N = N_e \times (S/S_e)^{-k}$ where N_e is the number of stress cycles endured at the fatigue limit S_e, and k is between 0.8 and 0.9. This yields a stress cycle with slope more than the classical $S-N$ curve but continued into the range of fatigue strength.

Marin (1962) assumed that the equivalent number of cycles at a reference level S_1 produce the same damage as n_i cycles of operation at S_i level of stress can be expressed by $n_{ie} = n_i \times (S_i/S_1)^y$ where y is an exponent to be determined experimentally. The damage ratio corresponding to each operation is $R_i = n_m/N_1$, so the condition of failure is $R_1 + R_2 + \ldots + R_i = 1$. The exponent y is the same as Corten's exponent d, so the $S - N$ relation becomes $S^xN = k$. The failure criterion then becomes:

$$\frac{n_1}{N_1}+\frac{n_2}{N_2}\left(\frac{S_2}{S_1}\right)^q+\frac{n_3}{N_3}\left(\frac{S_3}{S_1}\right)^q+\ldots+\frac{n_i}{N_i}\left(\frac{S_i}{S_1}\right)^q=1 \tag{3.34}$$

where $q = y - x$. Note that when $q = 0$, Marin's theory takes the same form as Palmgren-Miner's expression, and Corten's expression for $q = d$.

Henry (1955) takes into consideration reduction in fatigue limit when a specimen suffers damage, so the $S - N$ curve shifts as a consequence. This curve is an equilateral hyperbola about the stress axis, and a line passing through S_e parallel to N axis as the asymptotes. Damage D and fatigue limit are related by $D = S_e - S_{ed}/S_e$, where S_e is the fatigue limit of the virgin material and S_{ed} is the limit after the damage. By assuming no damage at cyclic stress levels below the fatigue limit, the curve may be expressed by $N_r = N - n = k/S - S_{ed}$, where N_r is the number of remaining cycles to failure at stress level S and k is a material constant. The damage equation then takes the form:

$$D = 1-\frac{S_{ed}}{S_e} = \frac{n/N}{1+\left(\frac{S_e}{S-S_e}\right)\left(1-\frac{n}{N}\right)} \tag{3.35}$$

The expression may be extended to a sequence of different alternating stress amplitudes in the order of application.

The Gatts (1961) damage theory is based on the dependence of fatigue strength and fatigue limit on the number of cycles of stress, and that this change is proportional to a damage function $D(S) = (-1/k_1)(dS_1/dN)$, with S_i as the instantaneous value of strength, n the number of applied stress cycles and k_1 a constant of proportionality. Damage is expressed as a function of stress level $D(S) = (S - S_e)^p$, with p as a material constant. k_1 and p are related with strain energy associated with stresses exceeding the fatigue limit. It is assumed fatigue limit can be expressed by instantaneous strength $S_e = CS_i$ where C is a material constant. By applying the boundary conditions, the $S - N$ curve takes the form:

$$KN = \frac{1}{\gamma-1}-\frac{1}{\gamma(1-C)} \tag{3.36}$$

where $\gamma = S/S_{eo}$, S_{eo} is fatigue limit when $n = 0$, $\beta = n/N$, $\gamma_e = S_e/S_{eo}$ and $K = kS_e$. For most steels C takes the value of 0.5. For $C = 0$ this expression is the same as that by Henry.

Manson *et al.*, (1967) recognize the role of crack initiation and propagation during the damage sustained by a component. Crack initiation period is denoted by N', and crack propagation period is defined by the number of cycles for failure after the crack initiates. Hence, $N_p = PN_f^{\,p}$ and $N' = N_f - PN_f^p$ where N_f is the total number of cycles for failure including crack initiation, P is propagation coefficient and p is propagation exponent. $P = 14$ and $p = 0.6$ from experiments. Except for short life, Miner's rule is adopted for crack initiation and propagation phases separately. Fatigue nucleii of critical size initiate when:

$$\sum_{i=1}^{m} \frac{n_i}{N_i'} = 1 \tag{3.37}$$

Fatigue cracks then propagate to failure when:

$$\sum_{j=1}^{q} \frac{n_i}{N_{Pj}} = 1 \tag{3.38}$$

In both phases n is number of cycles applied at i^{th} or j^{th} stress level. Thus, Manson uses a double linear damage rule.

3.8 ACOUSTIC RESONANCE IN COMBUSTOR

Combustion of the air and fuel mixture is accompanied by noise directly as a consequence of the process and indirectly due to the flow of burned gases through the turbine and exhaust nozzle. Combustion noise can become detrimental when instabilities arising in the burning process couple with acoustic modes inside the chamber. The natural frequencies of the combustor can be excited by resonant pressure waves in the main gas flow along the axial and radial directions, as also by lateral modes in the tangential direction (*ref*: Paxson *et al.*, 1995, Ohtsuka *et al.*, 1998, Rangwala, 2004). Sustained oscillating phenomena due to a higher level of mixing of the fuel and air prior to combustion lead to engine noise and vibration problems.

Premixed combustion in gas turbines helps produce low levels of NO_x emissions, but practical application of this concept is limited by self-excited combustion oscillations. When operation in a lean, premix combustor is close to the flammability limit slight changes in operating conditions can lead to sudden flame extinction or to excessive CO emissions.

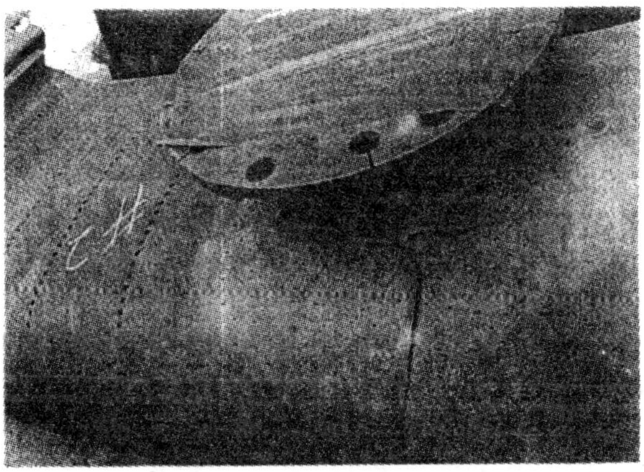

Fig. 3.23. *Acoustic Oscillations Damaged Transition Piece (Lieuwen, 2002)*

In addition to static stability, lean premix combustors must achieve dynamic stability, meaning the combustion must not oscillate. Oscillation must be eliminated in a combustor design because the associated pressure oscillations tend to have life shortening consequences (*ref*: Richards *et al.*, 1997). Figure 3.23 shows cracks experienced in a transition piece due to excessive acoustic oscillations. Operation near the lean limit is especially prone to oscillation problems, where minor variations in fuel/air ratio lead to appreciable variations in combustion reaction rate. When these variations in the reaction rate couple with the acoustic modes, significant pressure oscillations occur, with frequencies ranging from hundreds to a few thousand Hertz.

The task of studying and eliminating combustion oscillations in a gas turbine is complicated by the specific acoustic response of a combustor's design. The combustion process interacts with the acoustic field, leading to instabilities. Rapid changes in air and fuel supply and aerodynamic disturbances may lead to the instability because of a sequence of extinction and re-ignition of the flame in parts of the combustor. If the heat release rate does not take place uniformly and periodic spikes occur, acoustic waves of the same frequency may be expected in the combustion zone. Reflection from the liner causes pressure waves to be returned to the combustion zone after a time delay, and the waves are reinforced when the heat release and pressure wave peaks coincide. As defined by Lord Rayleigh's criterion, oscillations set in when changes in heat release are in phase with acoustic pressure disturbances. Conversely, oscillations are dampened when heat release fluctuations are out of phase with pressure fluctuations. This criterion serves as the cornerstone for the development of combustion oscillation analysis. Variation in heat release result from changes in flame structure produced by acoustic pressure disturbances. Time delay between pressure disturbance and heat release variation determines the phase and, consequently, the stability of the system.

Based on these observations, lean premix combustors can be characterized by a simple time lag approach. Figure 3.24 shows for a specific case a schematic of important processes, where a sinusoidal pressure disturbance produces a sinusoidal variation in airflow 180° out of phase with the pressure. Time lag τ is estimated from the distance between the point of fuel injection and the flame front divided by average axial velocity, or

$$\tau = (L + L')/U_{avg} \tag{3.39}$$

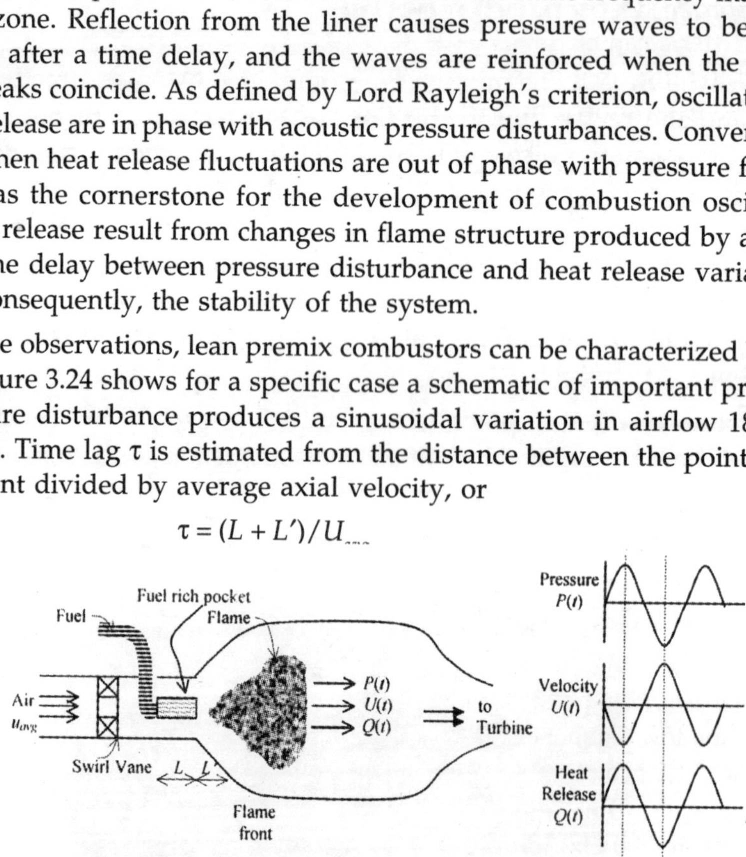

Fig. 3.24. *Flow Characteristics During Acoustic Oscillation (Richards, 1997)*

where L is the distance from fuel injection point to nozzle tip, L' is distance of nozzle tip to flame front and U_{avg} is average velocity of air-fuel mixture in the nozzle.

A positive pressure fluctuation in the combustor produces a momentary decrease in airflow. If the fuel supply is choked, fuel flow rate will not change with pressure variation. Thus, the reduced airflow will receive a proportionally higher amount of fuel, creating a fuel rich pocket. This richer pocket arrives at the flame front with a time lag, indicated by the equation given above. If the additional fuel produces an immediate increase in heat release, oscillations will be most likely when the pressure fluctuation peak is in phase with the increased heat release; that is, when time lag ($t_2 - t_1$) is an integer multiple of acoustic period. This criterion for oscillations may be stated as: (time lag)/(acoustic period) = 1, 2, 3... Since acoustic period is the reciprocal of frequency f, then: (time lag) × (frequency) = 1, 2, 3... or, $f(L + L')/U_{avg}$ = 1, 2, 3... This is a restatement of Rayleigh's criterion. In practice, heat release and pressure do not necessarily need to be exactly in phase to drive oscillations. Heat release fluctuations leading or lagging pressure by as much as ¼ of the acoustic cycle will also cause some oscillations, although driving is greatest for integer values where pressure and heat release are exactly in phase.

The discussion above is specific to the example where positive pressure produces an immediate decrease on airflow, and assumes that the fuel rich pocket produces an immediate increase in reaction rate when arriving at the flame front. Other mechanisms for variable heat release can complicate the criterion for oscillations such that the expression may have values other than 1, 2, 3, ... Similar criteria can be developed to account for fuel system impedance, or to describe oscillations linked to tangential velocity component in the fuel nozzle swirl vane. Geometry of flame front has also been shown to produce a numeric series.

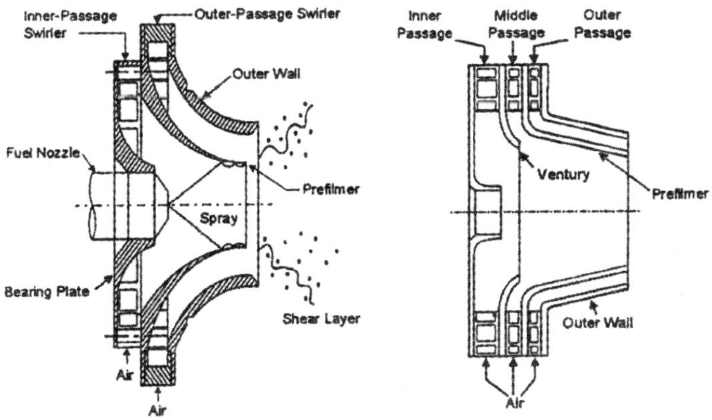

Fig. 3.25. *Baseline* (left) *and New* (right) *Fuel Injector Designs (Arana, 2000)*

Radiated sound may have frequencies ranging from 100 to 2000 Hz. Sound pressure frequencies mostly do not depend on engine power or flame temperature, but radiated noise level tends to vary with these factors. In the presence of combustion instability, a rumbling or growling form of noise is audible in the low frequency 50 to 180 Hz range when the engine may be in the sub-idle operating condition. The growl is objectionable because it increases the time to start an engine, while also reducing the stall margin in the compressor. At higher frequencies corresponding to takeoff condition (200 to 500 Hz) the generated noise takes a more distinct howling or humming pattern. Unstable operation in the compressor tends to play a role, and may even act to trigger the noise. Increase in air temperature to combustor inlet has been noted to decrease the rate of occurrence and intensity of growling noise, while raising combustor pressure has the opposite effect. Fluctuations in fuel pressure may also induce high frequency noise.

Thermo-acoustic response of a gas turbine engine combustor for two different fuel injectors has been investigated in a study conducted by the US Air Force (*ref*: Arana *et al.*, 2000), with the intention of identifying design features that cause increase in the acoustic pressure. A hybrid air blast injector presently in use with inner and outer flow passages is selected as a baseline design. To lower the smoke production level the investigation focused on using higher swirl flow in the proximity of the spray point, while lean blowout and dynamic stability can be obtained with lower swirl in the zone. Simultaneously achieving the apparently conflicting requirements for high and low swirls near the spray point led to the development of a new design concept with variances. Figure 3.25 provides details of the baseline and new fuel injector designs.

The new injector design differs from the baseline in the configuration of the venturi, the counter-rotating swirlers of the venturi and the middle passage and the ratio of vane and discharge areas. The last parameter is 50 percent larger than in the baseline design, suggesting that the new swirler exhibits less resistance to dynamic changes in pressure at higher frequencies. Use of velocity on the downstream side may be expected to maintain a higher level of the transfer function.

Co-rotating and counter-rotating swirlers are characterized for different passages. The injectors are initially tested at atmospheric and high-pressure conditions in an ignition rig, then assembled in the combustor of a development engine demonstrator employing 24 injectors around the circumference of the bulkhead. Air is fed to the combustor through a stepped diffuser.

The radial swirlers are the primary conduits of air between the external combustor shrouds and the internal combustion chamber. If coupling and amplification between the chambers is the root cause of the instability, then swirler response to a forcing function needs to be checked by measuring the impedance of the conduit. Impedance defines the total resistance and reactance opposition exerted by the swirlers to the forced, or pulsed, airflow of a given frequency, and is determined by measuring the transfer function of the swirlers. Pressure is measured as the upstream parameter and velocity on the downstream side for a number of frequencies.

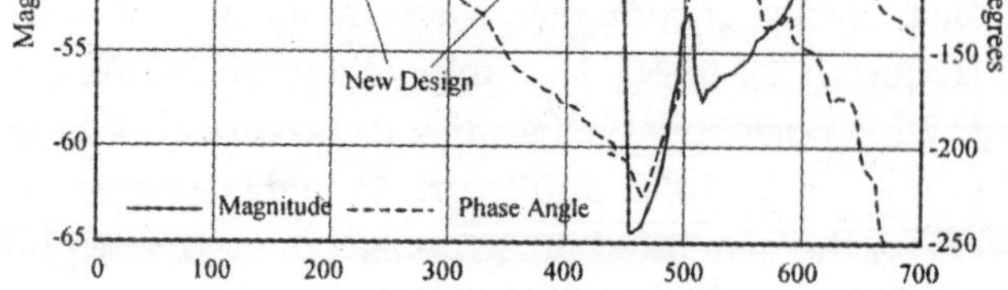

Fig. 3.26. *Transfer Function of Upstream Velocity and Downstream Pressure (Arana, 2000)*

Figure 3.26 shows the measured transfer functions and corresponding phase angles for the two designs. The new design swirlers exhibit a higher value of the transfer function in the 400 to 500 Hz frequency range, where the natural frequency of the annular combustor occurs. The phase angle relation between the pressure and velocity oscillations also points to this aspect, and is indicative of a dynamic response as opposed to a static one. The implication is that if the frequency of the acoustic chamber of less than 400 Hz is obtained, the new fuel injector design provides better attenuation and less acoustic response.

3.9 ACTIVE COMBUSTION INSTABILITY CONTROL

Combustion instabilities are difficult to predict analytically in the design phase for all operating conditions due to the complex geometry of the system. Noticeable humming caused by self-excited vibrations can occur during shop tests in the premixed mode operation of the turbine. Pressure oscillations may exceed unacceptable levels, and a quick and flexible response in the form of control mechanism may be necessary while the combustor design is optimized.

A similar situation has been experienced on a Siemens model V84.3 gas turbine equipped with a new ring (or annular) combustor design (*ref*: Seume *et al.*, 1997). Dynamic pressure and heat release rate are measured at different locations in the combustor, and dominant signals are recognized at 217 Hz and 433 Hz due to oscillations in the combustor. Several cross power density spectra and transfer functions are derived from two dynamic pressure signals in different areas of the combustor. Modal analysis indicated the oscillations excite standing sound waves in the structure. The standing waves consist of alternating regions of high and low sound pressure amplitudes, related to each other by a characteristic difference in phase. Azimuthal modes in the form of waves are distributed along the circumferential coordinate. With a mean diameter of d = 2.5 m and speed of sound c = 844 m/s at a mean temperature of 1500°C, the frequency of vibration $f_n = nc/\pi d$ yields 215 Hz and 430 Hz for the second and fourth harmonics, indicating good agreement with the measurements. Significant amplitudes are not observed for the first and third harmonics of 108 and 326 Hz. Figure 3.27 provide details of the second and fourth modes.

Passive methods rely on making changes in operating parameters (such as equivalence ratio) or geometry of the combustion system to hinder the self-exciting mechanism. The sound pressure amplitude can also be decreased to a tolerable level by dissipative baffles or mufflers (*ref*: Culick, 1988). By contrast, active methods use a feedback control loop. Heat release or pressure in the combustor is processed by a controller and used as an input signal for an actuator to influence the oscillating combustion that counteracts the self-excitation process (*ref*: Candel, 1992, McManus *et al.*, 1993). Fluid stream inside the combustion chamber can be modulated to reduce pressure fluctuations by introducing inversed sound pressure oscillations, as in a loud speaker.

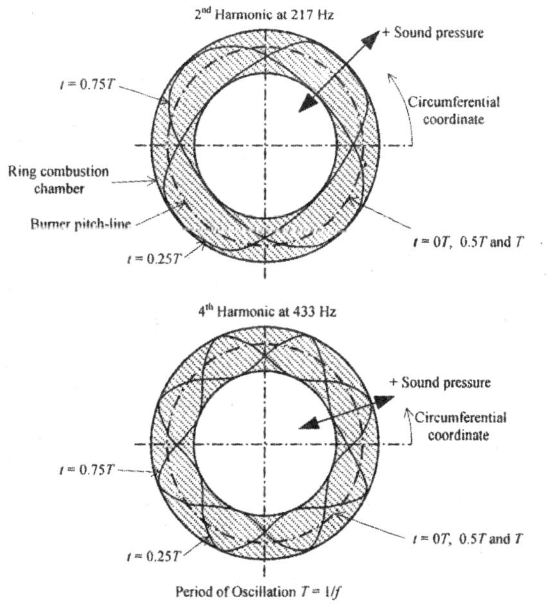

Fig. 3.27. *Excited Modes in Combustion Chamber (Seume, 1997)*

The method is impractical for bigger turbines because of large amounts of air and exhaust gas to be handled. Combustion oscillations can be suppressed if the rate of fuel reaching the flame is anti-cyclical to the oscillations of the heat release rate. In either case, modulation of gases or fuel must take place at the frequency of the self-excited vibrations. Since these can often reach 1000 Hz, suitable actuators must be used to meet the requirement.

The active stability control system for the V84.3 gas turbine uses pressure transducer measurements in the chamber, and the signals are sent to a control unit to derive an input signal for the actuator to modulate the fuel flow rate. A few basic problems needed to be solved for operation on a ring combustor. In the premixed mode and at base load the engine uses 9 kg/s of gas. Thus, the actuator has to handle this mass flow rate at the observed frequencies. Active control is secured through additional diffusion flames that contribute about 10 percent of the total power of the burner to stabilize the main premixed flame. Fuel flow rate to the pilot flame is modulated to influence the heat release in the main flame accordingly. A special high-speed direct-drive valve serves to actuate the pilot gas flow. The success of the control mechanism heavily depends on the pressure amplitudes in the pilot gas pipes, with higher amplitudes increasing the heat release in the flame. Hence, the length of the pipe to the pilot must be acoustically tuned to the frequency to be controlled. When the pilot's gas piping layout is complex and introduces damping, a suitable device may be placed upstream of the actuator to acoustically decouple it. Figure 3.28 shows a schematic of the control mechanism.

The second problem is associated with the azimuthal modes of the instabilities. Since several control systems are placed in different locations along the circumference, the control devices are also situated in different regions of the excited acoustic field, with prevailing oscillating parameters strongly differing in amplitude and phase. The symmetry of the azimuthal modes is marked by a characteristic distribution of nodes and anti-nodes, with regions of high and low amplitudes related to each other by a constant phase shift. Consequently, it is possible to use a signal measured at a certain circumferential location on the ring combustor to calculate not only the actuator signal for the particular location but also for other defined locations.

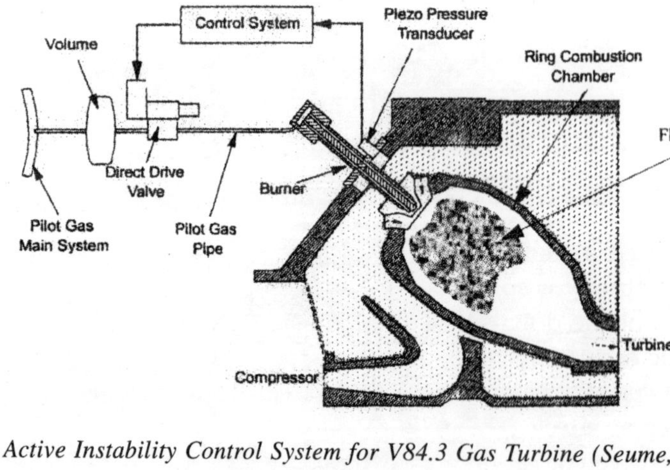

Fig. 3.28. *Active Instability Control System for V84.3 Gas Turbine (Seume, 1997)*

One control unit can then be used for all the actuators in the system. Figure 3.29 depicts this principle for the second harmonic to provide the input signal for four actuators located 90° from each other.

Performance of the active control system can be gauged from shop test measurements of the 170 MW gas turbine's ring combustor (fig. 3.30). The mechanism reduced the oscillations at the dominant frequency of 433 Hz by up to 17 dB. With active control turned off, measured sound pressure amplitudes rose to 210 mbar (corresponding to sound pressure level of 177 dB), falling to about 30 mbar with the control system turned on (*ref*: Rangwala, 2004).

Modulation of fuel flow rate is commonly achieved by using reciprocating flow devices where instability occurrence is at about 200 Hz or when the level of modulation required is small. In instances where instability frequencies are in the 200 to 500 Hz range and attenuation requires modulation of large fractions of engine fuel flow rate of hundreds of pounds per hour, a spinning drum valve has proved more useful (*ref*: Barooah *et al.*, 2002). The spinning valve design is based on a rotary concept to generate maximum frequency response. A rotating drum with a selected number of holes equally spaced around the circumference is used, with the holes aligned in the surrounding enclosure to pass the liquid fuel flow. By minimizing the clearance between the drum and the enclosure, leakage is reduced when the holes in the rotating and stationary components come in line. The holes in the enclosure are radially opposed to balance the pressure and to minimize the traverse loads. Unlike a reciprocating device, the upper frequency limit is not affected by the inertia of the spool or the low power requirement to accelerate it.

Liquid fueled low NO_x combustors can mitigate combustion instability at realistic operating conditions by modifying the fuel nozzle (*ref*: Cohen *et al.*, 1998). The fuel is injected through axial tubes with spray tips protruding from the nozzle centerpiece.

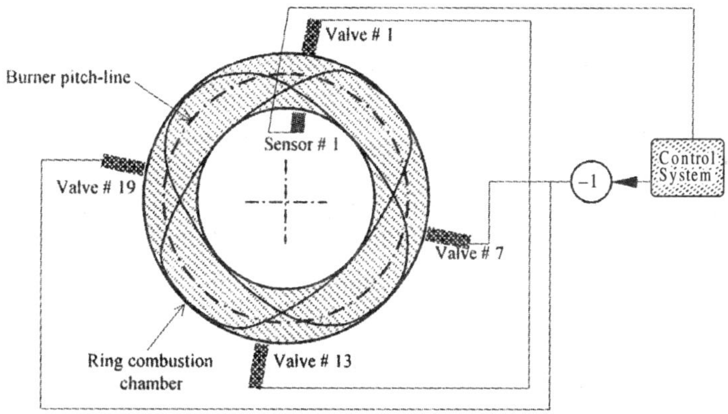

Fig. 3.29. *Sensor and Signal Input Controller for Second Harmonic Mode (Seume,1997)*

A pilot injector is placed a short distance downstream of the fuel ejection plane. After passing through a venturi the airflow is split between the fuel nozzle and a bypass segment. Airflow from the bypass is injected at the downstream end combustor, and represents dilution air. To obtain control of the acoustic oscillations, one tube delivers the fuel to a metering system and a solenoid valve. The tubing between the valve and the injection point is minimized to reduce attenuation and time lag due to capacitive effects. The main fuel flow takes place through the other injection tubes. The solenoid valve is driven at varying frequencies independent of the combustor behavior using a signal generator, with an on/off duty cycle of 50 percent. Combustor pressure and heat release rate are measured. When the control sensor signal crosses a predetermined threshold level, a command sent to the solenoid valve turns it on or off. Time delay between the instant of crossing and valve command is also taken into account. The threshold level and the time delay are manipulated through a user interface to the control algorithm. A proper choice of the two parameters yields 15 dB attenuation of the objectionable oscillating mode. The control system is also effective in holding the NO_x emission relatively constant across the range of equivalence ratios.

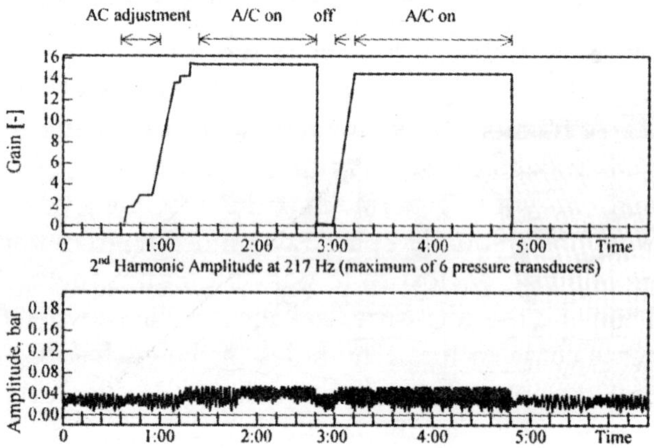

Fig. 3.30. *Control System Test Operation at Base Load (Seume, 1997)*

3.10 STRUCTURAL DESIGN OF COMBUSTOR LINER FOR DYNAMIC PRESSURE

The liner and transition piece of a dry low NO_x combustor contain and direct the high temperature combustion gases. To allow for thermal growth the components are restrained at the fewest locations. The combustion process often creates pressure oscillations at discrete frequencies, especially during premixed burning. This stimulus on the flexibly attached structure can lead to substantial vibrations, develop fatigue cracks and eventual failure.

A tradeoff between adequate flexibility to minimize thermal stress and stiffness to avoid vibration problems is thus required. Once a specific temperature distribution is available, thermal stresses may be computed with relative ease. Vibratory stresses, however, cannot be so readily defined. Complexities in geometry, loading pattern and contact behavior are some reasons. Strain measurements may also be of limited usefulness because critical locations are not generally known, often are inaccessible and have high levels of metal temperature (*ref*: Rangwala, 2004).

Field experience gained from years of operation may permit setting limits on combustion dynamic pressure fluctuations to ensure structural integrity. Direct application of this experience to new applications, however, is questionable. Analytical prediction of component life and development of a method to include effects of dynamic pressure loading may thus be essential. Non-linear transient finite element analytical procedures are useful in predicting the dynamic behavior, and can compare measured strains and accelerations with reasonable accuracy (*ref*: Barnes, 1996). Component stiffness and mass characteristics, as also distributed pressure loading, can then be accurately modeled. Contact behavior, including sliding friction at supports and seals, is required to compute response, since they add considerable damping. Non-linear gap conditions capable of maintaining or breaking physical contact in accordance with relative displacements between components is also essential. Verification of natural frequencies and mode shapes computed from modal analysis may be made through laboratory modal testing, and to determine proximity to resonant conditions.

Figure 3.31 shows the finite element model. Mesh density requires special attention be paid to supports, fillets and areas of known stress concentration. This level of detail captures stress concentration in a forced response analysis, but may not be required for a modal analysis. Linear eight-noded brick elements help to maintain the number of elements within limits in a large model, especially in a computationally intensive solution.

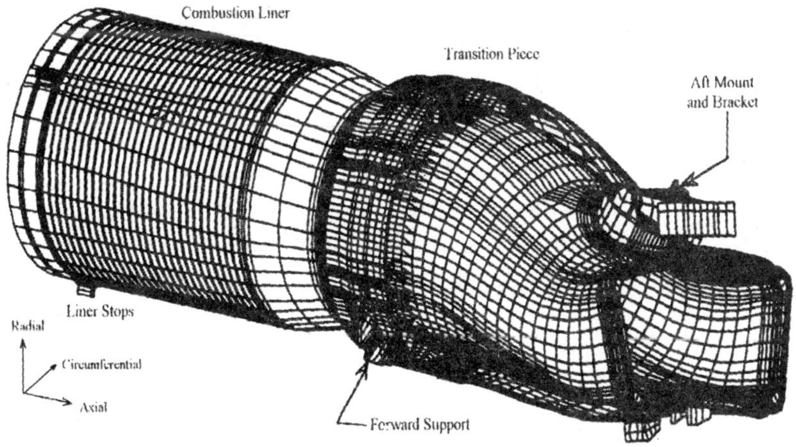

Fig. 3.31. *Combustion Liner and Transition Piece Finite Element Model (Barnes, 1996)*

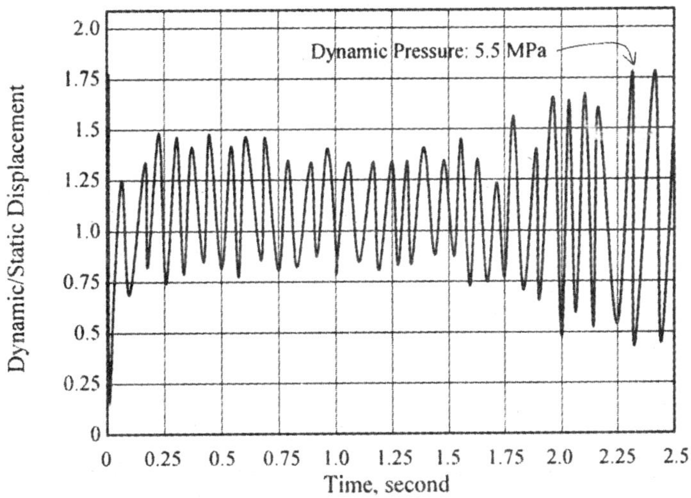

Fig. 3.32. *Dynamic Axial Displacements at Forward End (Barnes, 1996)*

Effects of non-symmetric loading and support features generally do not permit taking advantage of geometric symmetry. Seals on forward and aft ends of the liner may be modeled as a combination of a normal spring and a tangential friction element. Seal stiffness may be computed independently from its own finite element model, while friction coefficient must be measured from seal fretting tests. Contact force at the seals will depend on relative thermal expansion of the components, hence the sequence of calculations may need repetition for various operating conditions.

Crossfire tubes between adjacent combustion liners are flexible and generally will not impact the calculations; hence they may be deleted in the analysis. Responding to pressure variation with unknown phase difference due to the crossfire tubes between adjacent combustors is at best a difficult proposition. Note that liner supports are considerably less rigid than turbine cylinders to which they are attached, hence liner supports may be grounded at the other end.

The pressure loading represents test laboratory measurements simulating the machine's operating conditions. In this example peak-to-peak pressure variation of 21 kPa at 160 Hz defines the dynamics of combustion. Figure 3.32 provides computed oscillating displacements at the forward end of the transition piece. Strain gauges approximately centered in this region measures deformations that correlate very well with the analytical predictions. Radial motion at the aft end of the inner panel is affected by the damping provided by the seals, and the effectiveness of the damper is controlled by wear and by alignment of the mating components.

High cycle fatigue capability of the components depends on vibratory and mean stresses, metal temperature and material characteristics. Critical locations are determined by computing the dynamic strain range, metal temperature and mean stress in the elements. Figure 3.33 provides a plot of equivalent vibratory strain divided by allowable strain at a critical location between the aft mount and the aft frame at the centerline of the transition piece. Measurement of strain at this location poses practical problems.

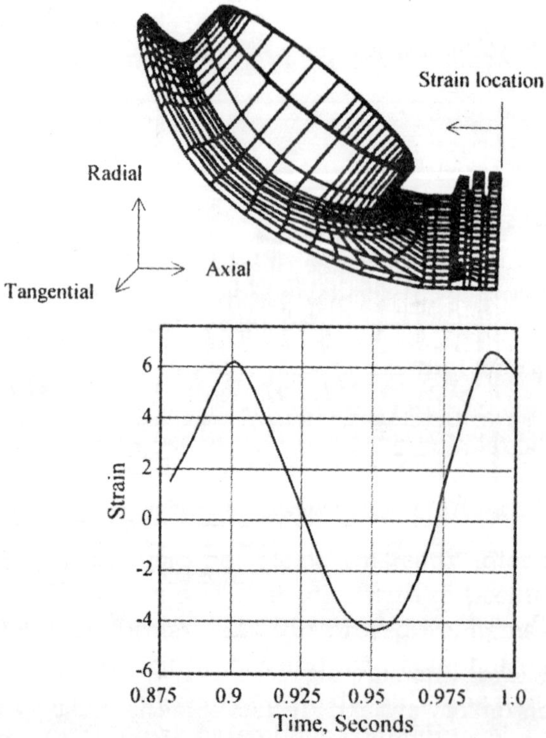

Fig. 3.33. *Equivalent Strain History (Barnes, 1996)*

Determination of response to initial impact of the dynamic pressure requires non-linear transient dynamic analysis, assuming oscillation to initiate at full amplitude. Dynamic amplification peaks at the forward support of the transition piece, mostly because a gap at the location causes it to alternate between contacting and free modes of vibration. This results in the reaction force at impact to multiply by a factor of 12 over the static reacting force value. Thus, the seat for the support must be designed to reflect this situation. To assure structural dynamic response is not significant due to small variations in stimulus frequency, a wider frequency range can be expeditiously checked through modal analyses of individual components in a freely supported condition. Analytical prediction of natural frequencies must agree with measurements, as also the corresponding mode shapes. Variations due to differing mass and stiffness properties from manufacturing tolerances and measurement inaccuracies must be maintained within acceptable limits. Where the natural frequencies are deemed too close to those of the combustion dynamic pressure, the structural response for these modes must be captured with the non-linear dynamic analysis process.

3.10 EXAMPLE PROBLEMS

Example Problem # 3.1: Explain the terms fatigue and limiting fatigue range as applied to materials for turbo-machinery components. How is the limiting fatigue range related to the mean stress during a load cycle?

Solution: Experiments indicate that an alloy may fail at a stress considerably lower than its ultimate strength in a normal tensile test if this stress is repeated a large number of times. The term fatigue is used for the effects of repeated load cycles on the material. If the limits of stress during the cycle are of the same sign, for example both tensile, the stress is said to be fluctuating. If the lower limit is zero, the term repeated stress is sometimes used. Reverse, or alternating, stress implies limits which are numerically equal but opposite in sign.

As the range of stress during the cycle decreases, the number of applications of the load required to initiate failure is increased. In the case of steels, it is found that for a given mean stress there is a limiting range within which failure does not occur, however many cycles are applied. This is called the limiting fatigue range, and experiments reveal that it is approximately equal to the range which the material can withstand for 10 million cycles. For some non-ferrous materials such a limiting range may not exist, and failures have been reported after a hundred million cycles.

The following parabolic relation has been suggested by Gerber on the basis of experiments:

$$s = s_0 \quad m(f_{average})^2$$

where s_0 = limiting fatigue range for zero mean stress, or alternating stress

s = limiting fatigue range for mean stress $f_{average}$

m = an experimental material constant.

Example Problem # 3.2: A certain alloy has an ultimate strength of 122.0 kpsi. The limiting range for alternating stress is ±19.5 kpsi. Estimate the probable safe maximum stress for an unlimited number of cycles if the minimum stress is 17.2 kpsi.

Solution: The limiting fatigue range $s_0 = 2 \times 19.5 = 39.0$ kpsi. When $f_{average}$ reaches the ultimate tensile stress the range must be zero. Thus, $s = 0$ when $f_{average} = 122.0$, and substituting in Gerber's expression:

$$0 = 39.0 - m \times 122.0^2$$

or $m = 0.00262$. If f_{max} and f_{min} are the upper and lower limits of stress during the load cycle, then $f_{min} = 17.2$ and $s = f_{max} - f_{min} = f_{max} - 17.2$. Also, $f_{average} = (f_{max} + f_{min})/2 = (f_{max} + 17.2)/2$. Use these results in the Gerber expression:

$$f_{max} - 17.2 = 39.0 - 0.00262 \times \{(f_{max} + 17.2)/2\}^2$$

or: $0.000655(f_{max})^2 + 1.045(f_{max}) - 56.394 = 0$

Taking the positive root of the quadratic equation provides the probable safe maximum stress is:

$$f_{max} = 52.25 \text{ kpsi.}$$

Example Problem # 3.3: Discuss the merits of various theories of elastic failure. A certain steel has a proportionality limit of 40 kpsi in simple tension. In a two-dimensional stress system the principal stresses are 15 kpsi tensile and 5 kpsi compressive. Determine the factor of safety from the theories.

Solution: The greatest principal stress theory may be applied to most brittle materials such as cast iron. The greatest principal strain theory, on the other hand, holds little significance. The maximum shear stress theory is widely used for ductile materials, especially for a rotating shaft experiencing a combination of bending and torsion.

Experimental results on ductile materials tend to support the total strain energy theory, but are more in agreement with the Mises-Hencky criterion. The latter finds extensive usage for the design of mechanical components.

The stresses are $\sigma_1 = 15$, $\sigma_2 = 0$ and $\sigma_3 = -5$. By the maximum shear stress theory the equivalent single tensile stress is $s = \sigma_1 - \sigma_3 = 15 - (-5) = 20$ kpsi, so the factor of safety is 40/20 = 2.0. The Mises-Hencky theory for combined stresses is:

$$\begin{aligned} 2s^2 &= (\sigma_1 - \sigma_2)^2 + (\sigma_2 - \sigma_3)^2 + (\sigma_3 - \sigma_1)^2 \\ &= (15 + 0)^2 + (0 + 5)^2 + (-5 - 15)^2 \\ &= 650 \end{aligned}$$

Hence: $s = 18.03$ kpsi

The factor of safety $= 40/18.03 = 2.22$

Example Problem # 3.4: Describe the various theories put forward to obtain the failure criterion when a component is subject to a state of complex stress. Illustrate the situation for a thin walled component subjected to perpendicular stresses of 12 kpsi and 5 kpsi, both tensile, assuming a Poisson's ratio of $\nu = 0.3$.

Solution: Failure refers to elastic breakdown and onset of permanent strain. The stress s at which this occurs in simple tension may be assumed to be the limit of elastic proportionality. Consider a three-dimensional complex stress system, where the principal stresses are σ_1, σ_2 and σ_3 in descending order, tensile being positive.

The greatest principal stress theory postulated by Rankine states that failure occurs when this stress reaches the critical value s. Hence, in this case, $s = \sigma_1$.

The greatest principal strain theory of St. Venant considers the greatest strain as the relevant quantity. In this case the value is (using E for Young's modulus):

$$\varepsilon_1 = \sigma_1/E - \sigma_2\nu/E - \sigma_3\nu/E$$

In simple tension, the strain is s/E. By equating the strains, the expression for stress is:

$$s = \sigma_1 - \sigma_2\nu - \sigma_3\nu$$

Coulomb's maximum shear stress theory is based on the maximum shear stress on an interface being half the difference of the corresponding principal stresses, or $(\sigma_1 - \sigma_3)/2$ for the complex stress system and $s/2$ when in simple tension. Hence, using this theory, $\sigma_1 - \sigma_3 = s$.

Beltrami's total strain energy theory may be explained as follows. Strain energy per unit volume due to a single direct stress is half the product of stress and strain, or $\sigma\varepsilon/2 = \sigma^2/2E$. In a complex system the principal stresses and strains must be used. Since strain in the direction σ_1 of in two dimensions is $(\sigma_1/E - \sigma_2\nu/E)$, due to σ_1 the strain energy per unit volume is:

$$[\sigma_1 \times (\sigma_1/E - \sigma_2\nu/E)]/2 \quad = \quad (\sigma_1^2/2E - \sigma_1\sigma_2\nu/2E)$$

By extending this reasoning to three-dimensional conditions, the total strain energy is:

$$\{\sigma_1^2 + \sigma_2^2 + \sigma_3^2 - 2\nu(\sigma_1\sigma_2 + \sigma_2\sigma_3 + \sigma_3\sigma_1)\}/2E$$

In simple tension the strain energy is $s^2/2E$, and this leads to the relationship:

$$\sigma_1^2 + \sigma_2^2 + \sigma_3^2 - 2\nu(\sigma_1\sigma_2 + \sigma_2\sigma_3 + \sigma_3\sigma_1) = s^2$$

The Mises-Hencky theory is based on the quantity $(\sigma_1 - \sigma_2)^2 + (\sigma_2 - \sigma_3)^2 + (\sigma_3 - \sigma_1)^2$, and the expression represents the shear strain energy. In simple tension the principal stresses are s, 0 and 0, so the corresponding expression is $2s^2$. The criterion then takes the form:

$$(\sigma_1 - \sigma_2)^2 + (\sigma_2 - \sigma_3)^2 + (\sigma_3 - \sigma_1)^2 = 2s^2$$

In the numerical example, $\sigma_1 = 12.0$ kpsi, $\sigma_2 = 5.0$ kpsi and $\sigma_3 = 0$. Using St. Venant's principal strain theory, the equivalent stress in simple tension is:

$$s = \sigma_1 - \nu(\sigma_2 + \sigma_3) = 12.0 - .3 \times 5 = 10.5 \text{ kpsi}$$

The total strain energy (Beltrami) theory gives:

$$s^2 = \sigma_1^2 + \sigma_2^2 + \sigma_3^2 - 2\nu(\sigma_1\sigma_2 + \sigma_2\sigma_3 + \sigma_3\sigma_1)$$

$$= 12^2 + 5^2 + 0 - 2 \times 0.3 \times (12 \times 5) = 133.0$$

Hence $\quad s = 11.53$ kpsi.

■■■

REFERENCES & BIBLIOGRAPHY

Arana, C.A., Sekar, B., Mawid, M.A., Graves, C.B., "Determination of Thermo-Acoustic Response in a Demonstrator Gas Turbine Engine", ASME paper # 00-GT-091, New York, 2000.

Barnes, J.E., "Structural Integrity of a Gas Turbine Combustion System Subjected to Increased Dynamic Pressure", ASME paper # 96-GT-473, New York, 1996.

Barooah, P. Anderson, T.J., Cohen, J.M., "Active Combustion Instability Control with Spinning Valve Actuator", Proceedings, ASME Turbo Expo, paper # GT-2002-30042, Amsterdam, June, 2002.

Bladh, R., Pierre, C., Castanier, M.P., Kruse, M.J., "Dynamic Response Predictions for a Mistuned Industrial Turbo-Machinery Rotor Using Reduced Order Modeling", ASME Journal of Engineering for Gas Turbines and Power, Volume 124, Pages 311–324, New York, 2002.

Candel, S.M., "Combustion Instability Coupled by Pressure Waves and Their Active Control", 25th International Symposium on Combustion, Sydney, Australia, 1992.

Cohen, J.M., Rey, N.M., Jacobson, C.A., Anderson, T.J., "Active Control of Combustion Instability in a Liquid Fueled Low NO_x Combustor", ASME paper # 98-GT-267, New York, 1998.

Culick, F.E.C., "Combustion Instabilities in Liquid Fueled Propulsion Systems – An Overview", AGARD Conference on Combustion Instabilities in Liquid Fueled Propulsion Systems, Bath, AGARD-CP-450, Page 1-1 – 1-73, 1988.

Chen, J.J., Menq, C.H., "Periodic Response of Blades Having Three-Dimensional Non-Linear Shroud Constraints", ASME paper # 99-GT-289, New York, 1999.

Corten, H.T., Dolan, T.J., "Cumulative Fatigue Damage", Proceedings, International Conference on Fatigue of Metals, ASME and IME, page 235, New York, 1956.

Ewins, D.J., "Bladed Disk Vibration – A Review of Techniques and Characteristics", Recent Advances in Structural Dynamics: Proceedings of the International Conference, Volume I, Southampton, England, 1968.

Filsinger, D., Szwedowicz, J., Schafer, O., "Approach to Uni-directional Coupled CFD-FEM Analysis of Axial Turbocharger Turbine Blades", ASME paper # 2001-GT-288, New York, 2001.

Filsinger, D., Szwedowicz, J., Schafer, O., Dickman, H.P., "Pulse Charged Axial Turbocharger Turbines – A Challenge for Numerical Design Methods", Proceedings, CIMAC World Congress on Combustion Engine Technology, Volume 2, Pages 712 – 722, 2001*a*.

Gatts, R.R., "Application of a Cumulative Damage Concept to Fatigue", Transactions, ASME, Volume 83, Series D, Page 529, 1961.

Hall, R.M., Armstrong, E.K., "The Vibration Characteristics of an Assembly of Interlock Shrouded Turbine Blades", Structural Dynamics Aspects of Bladed Disk Assemblies, Srinivasan, A.V., editor, ASME, New York, 1976.

Harris, C.M., Crede, C.E., "Shock and Vibration Handbook", Mc-Graw Hill Book Co., 4th edition, New York, 1995.

Henry, D.L., "Theory of Fatigue Damage, Accumulation in Steel", Transactions, ASME, Volume 77, Page 913, New York, 1955.

Jay, R.L., MacBain, J.C., Burns, D.W., "Structural Response Due to Blade-Vane Interaction", ASME paper # 83-GT-133, New York, 1983.

MacBain, J.C., Horner, J.E., Stange, W.A., Ogg, J.S., "Vibration Analysis of a Spinning Disk Using Image De-Rotated Holographic Interferometry", Experimental Mechanics, SESA, Pages 17–22, 1978.

McManus, K.R., Poinsot, T., Candel, S.M., "A Review of Active Control of Combustion Instabilities", Prog. Energy Combustion Science, Volume 19, Pages 1–29, 1993.

Manson, S.S., Frecke J.C., Ensign, C.R., "Application of a Double Linear Damage Rule to Cumulative Fatigue", Fatigue Crack Propagation, STP-415, ASTM, Philadelphia, Page 384, 1967.

Marco, S.M., Starkey, W.L., "A Concept of Fatigue Damage", Transactions ASME, Volume 76, Page 627, New York, 1954.

Marin, J., "Mechanical Behavior of Materials", Prentice-Hall, Englewood Cliffs, NJ, 1962.

Menq, C.H., Yang, B.D., "Non-Linear Spring Resistance and Friction Damping of Frictional Constraint Having Two-Dimensional Motion", Journal of Sound and Vibrations, No. 1, Pages 127 – 143, 1998.

Miner, M.A., "Cumulative Damage in Fatigue", ASME Transactions, Journal of Applied Mechanics, Volume 67, Page A159, 1945.

Ohtsuka, M., Yoshida, S., Inage, S., Kobayashi, N., "Combustion Oscillation Analysis of Premixed Flames at Elevated Pressures", ASME paper # 98-GT-581, New York, 1998.

Ottarsson, G.S., Pierre, C., "On the Effects of Inter-Blade Coupling on the Statistics of Maximum Forced Response Amplitudes in Mistuned Bladed Disks", Proceedings, 36th AIAA/ASME/ASCE/AHS Structures, Structural Dynamics and Materials Conference, Volume 5, AIAA, Pages 3070–3078, New York, 1995.

Palmgren, A. "Die Lebensdauer von Kugellagern", ZV'DI, Volume 68, Page 339, 1924.

Paxson, D.E., "A Comparison Between Numerically Modeled and Experimentally Measured Loss Mechanisms in Wave Rotors", AIAA Journal of Propulsion Power, Volume 11, Number 5, Pages 908–914, 1995 (also NASA TM 106279).

Pierre, C., "Mode Localization and Eigen-value Loci Veering Phenomenon in Disordered Structures", Journal of Sound Vibration, 126, Pages 485–502, 1988.

Rangwala, A.S., "Turbo-machinery Dynamics: Design and Operation", McGraw-Hill, New York, 2004.

Rao, J.S., Pathak, A., Chawla, A., "Blade Life: A Comparison by Cumulative Damage Theory", ASME paper # 99-GT-287, New York, 1999.

Richards, G.A., Janus, M.C., "Characterization of Oscillations During Premix Gas Turbine Combustion", ASME paper # 97-GT-244, New York, 1997.

Seume, J.R., Vortmeyer, N., Krause, W., Hermann, J., Hantschk, C.C., Zangl, P., Vortmeyer, D., Orthmann A., "Application of Active Combustion Instability Control to a Heavy Duty Gas Turbine", ASME paper # 97-GT-119, New York, 1997.

Stange, W.A., MacBain, J.C., "An Investigation of Dual Mode Phenomena in a Mistuned Bladed Disk", ASME paper # 81-DET-133, New York, 1981.

Stetson, K.A., Elkins, J.N., "Optical System for Dynamic Analysis of Rotating Structures", Air Force Aero Propulsion Lab Contract F33615-75-C-2013, AFAPL-TR-77-51 (October, 1977).

Szwedowicz, J., "Harmonic Forced Vibration Analyses of Blade Assemblies Modeled by Cyclic Systems, Part I – Theory and Vibration", ABB Technical Reports HZX-ST 5849, Baden, Switzerland, 1996.

Thomson, W.T., "Theory of Vibration with Applications", Prentice-Hall, Englewood Cliffs, NJ, 1988.

Wadell, P., "Strain Pattern Experimentation", Engineering and Materials Design, 17 (3), 1973.

CHAPTER 4

Bearing and Seal Characteristics

4.1 INTRODUCTION

As a direct consequence of the transmission of radial load between the rotating assembly and the reacting structure, the bearings play a key role not only to characterize the steady state performance but also the dynamic behavior of the rotor-bearing system in the form of dynamic elements. The bearings are represented by stiffness and damping entities in the overall mathematical model, where the rotor is described by a distribution of mass and elastic elements. The lateral degree of freedom involves motion of translation of points on the axi-symmetric rotor in planes perpendicular to the axis of rotation, accompanied by rotation of the axis of rotation relative to its nominal position. Stability is another important benchmark for journal bearings. Aside from other external stimuli, bearing hydrodynamic forces can induce unstable operation in the system. The degree of stability of a given bearing is characterized by a whirl inducing frequency, and if it exceeds the natural frequency of the rotating shaft unstable operation ensues. The value of the threshold whirl inducing frequency is governed by the spring and damping coefficients of the bearing. These bearing dynamic coefficients also yield a critical load W_{cr}, representing an upper load limit for stable operation. Thus, an additional criterion for stable bearing operation requires applied load to be less than the critical load.

The differences between an aircraft power plant and a power generation turbine are probably best highlighted by the type of bearings used to support the rotor. In the former, overall weight of the rotating system, speed range and length of the bearing make rolling element type of bearings most suitable, while load carrying capacity and durability considerations in the latter make it imperative to use hydrodynamic journal bearings. Power generation steam and gas turbines operate at two most typical speeds, 3,000 rpm for 50 Hz units and 3,600 rpm for 60 Hz units. However, gas turbines not tied directly to the generator can have a higher operating speed. Siemens V84.3 gas turbine, for example, runs at 5,400 rpm, with a speed reducer placed between the turbine and the generator. Turbine speed for naval and merchant marine ships must be reduced to low propeller rpm. Since rotor weight may run into several thousand pounds for larger power generation units, rolling element bearings cannot provide the required capacity to carry the load. Journal bearing diameter may reach 27.5 in. and length of 25 in.

Rolling element bearings are designed to operate at high speeds, and provide a large power density to reduce the overall size of the package and to control the weight of the engine. Aircraft engines rely exclusively on ball and roller element bearings to provide a rotating support for the shaft. Limiting static and dynamic axial and radial excursions of the rotor within a safe range is important in avoiding dynamic instability, rubs on the casing and fatigue failures. Deceptively simple in geometry, their operational characteristics are complex. Substantial progress has been made to enhance the understanding of the principles of operation, directly aiding in overcoming problems related to lubrication, surface finish and operational life. Ball bearings are employed to absorb the thrust and to carry some radial load, while roller bearings are primarily for radial components. Bearing geometry, material and lubrication characteristics play a major role on the functional reliability and life of the bearing within a given environment.

Selection of a rolling element for a particular situation is generally based on design fatigue life considerations. Even because of minor differences in the bearing's makeup, apparently identical rolling element bearings subjected to identical load, speed, lubrication and environmental conditions will not possess the same fatigue life. A common standard for fatigue life calculations uses a rating life of L_{10}, referring to the life of an identical group of bearings 90 percent of which will equal or exceed the rated life. The L_{10} parameter is a function of the dynamic load capacity of a bearing, the equivalent radial load acting on it and the shaft speed. Compensation may be made to account for variations in material and lubricant characteristics and effectiveness of lubrication to complete the life predictions. The presence or lack of a sufficiently thick oil film at the contact points between the balls and races may enhance or deplete the bearing fatigue life. The life adjustment factor is dependent on the surface finish of the mating parts, lubricant viscosity, speed and, to a limited extent, load. Hardness is known to seriously affect the fatigue life, and the drop in hardness when the bearing operates above 400° F can be accounted for by applying a correction factor. Permanent deformations to the extent of 0.0001 times ball diameter do not substantially affect the operation of the bearing, and this aspect provides the concept of static load rating. The static load corresponds to a total permanent deformation of the ball and the raceway at the most heavily stressed contact of the mentioned size.

The fluid film separating the moving and stationary surfaces in a hydrodynamic journal bearing is of varying thickness. As the lubricant travels the annular space from the inlet to the outlet end, velocity, pressure and temperature of the fluid undergo changes, which bear directly on the bearing performance. The load that the bearing can safely support at a given minimum film thickness h_{min} depends on the likelihood of physical contact between the mating surfaces (leading to eventual failure), the intensity of peak pressures and temperatures and the reserve available in the bearing to accommodate unexpected excursions in the imposed shaft load. The significance of the maximum temperature T_{max} is often on a par with that of h_{min}. Excessive temperatures cause failure either by softening or melting the bearing surface, even when the film thickness is ample. In a well-aligned journal bearing T_{max} usually occurs on the axial centerline, and in thrust bearings it is found at the outer radius where the linear speed and the circumferential path of the lubricant are at a peak. Increase in the bearing temperature difference ΔT between the oil discharge and supply points is often deemed to be a suitable criterion for bearing performance, however, T_{max} is a more reliable indicator since lubricant breakdown and the consequent bearing failure are due to unacceptably high temperature. Furthermore, ΔT is controlled by frictional power loss and oil flow rate, and increasing oil inlet pressure or the flow rate will alter the rise in oil temperature but it does replace the role of T_{max}.

4.2 FLUID FILM BEARING

Journal bearings have been used for a long time in all types of rotating machines.

High levels of damping in the fluid film of the bearing have made possible successful operation of flexible rotors at high speed. Dynamic characteristics of the lubricant film in the form of its stiffness and damping are in a sense a newer development.

Turbulence in the fluid film and temperature are two aspects that play a major role in bearing operation. The Reynolds number determines whether the bearing operates in the laminar, transition or turbulent regime, and is given by the expression:

$$\text{Re} = \frac{\pi DNc}{\mu/\rho} \tag{4.1}$$

where ρ = lubricant density, μ/ρ = lubricant kinematic viscosity, D = journal diameter, c = radial clearance and N = shaft speed. Laminar flow prevails when the Reynolds number is under 750, the turbulent regime is marked by Reynolds number in excess of 1500 and the transition occurs between these two values. Aside from various operating conditions, within the bearing domains of both laminar and turbulent operation may coexist because the values of μ and film thickness h vary throughout the film. Generally, the turbulence raises the load capacity by increasing the film thickness but it also boosts the power losses, and the consequent temperature gain lowers the operating viscosity. This in turn lowers the load capacity. It may be surmised that the effects of turbulence on stability are deleterious because of the increased film thickness.

Inclusion of variable viscosity in the bearing film stems from three different considerations. The variable viscosity affects the performance characteristics, mapping of turbulence in the film and calculation of the temperature field T. Use of average viscosity may offer an approximate determination of the load capacity and power loss, but it does not hold true of T_{max}. A rigorous analysis of thermal effects in bearings may be based on the energy equation, heat transfer relationships between the film, runner and bearing shell and the governing equation for the flow in the annular region. The basic differential equation for bearing fluid films, referred to as generalized Reynolds equation, is given by the expression:

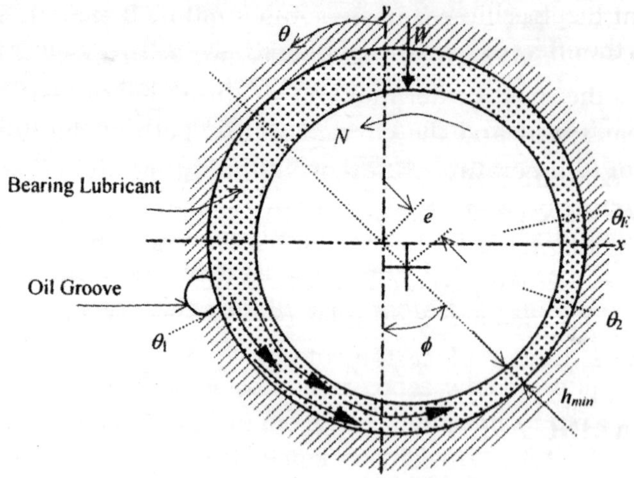

Fig. 4.1. *Fluid Film Journal Bearing*

$$\frac{\partial}{\partial\theta}\left[\frac{G_\theta h^3}{\mu}\left(\frac{\partial p}{\partial\theta}\right)\right]+R^2\frac{\partial}{\partial z}\left[\frac{G_z h^3}{\mu}\left(\frac{\partial p}{\partial z}\right)\right] = 6R^2\omega\left(\frac{\partial h}{\partial\theta}\right)+12R^2\left(\frac{\partial h}{\partial t}\right) \tag{4.2}$$

where θ and z are circumferential and axial coordinates, G_θ and G_z are turbulence coefficients in the two directions and are functions of the Reynolds number, h is film thickness and is a function of θ and z, R is journal radius, μ is lubricant viscosity and p is pressure. In fig. 4.1 the arc from the oil groove to θ_E defines the extent of the pad. Temperature variation in the film may be approximated by the one-dimensional convective energy equation:

$$\frac{wbu}{R}\frac{dT}{d\theta} = \frac{\mu}{G_\tau}\left(\frac{u}{h}\right)^2 \tag{4.3}$$

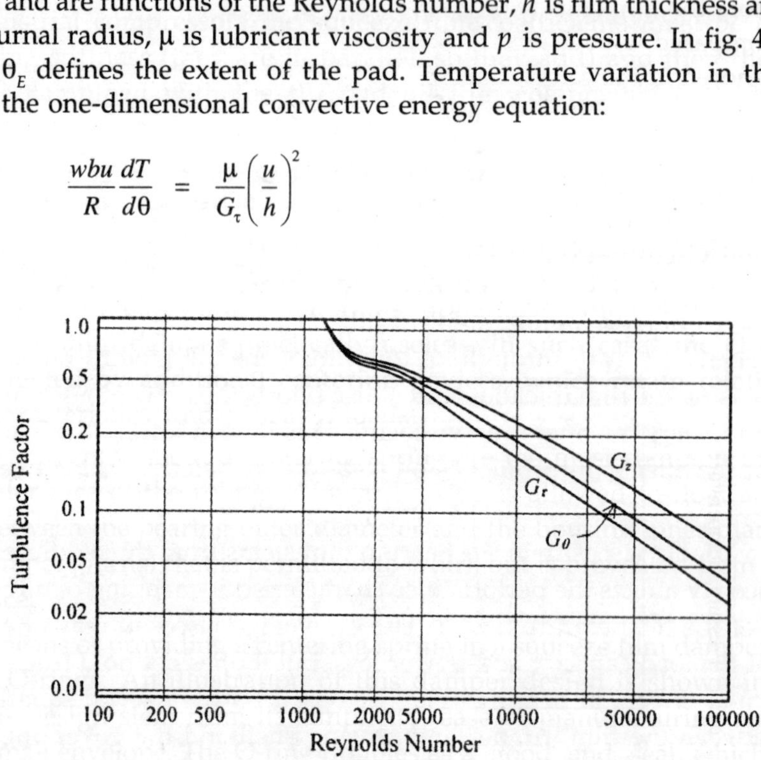

Fig. 4.2. *Turbulence Coefficient Values*

where w and b represent the specific weight and heat of the lubricant, u is journal peripheral speed and $G\tau$ represents the effect of turbulence on viscous shear. The relation between viscosity and temperature is

$$\mu = \mu_1 e^{-\alpha(T-T_1)} \tag{4.4}$$

where α is viscosity temperature coefficient of the lubricant and the subscript refers to values of the parameters at a reference point. A combination of the last two expressions provides two relevant equations for the thermal aspects of the problem:

$$\Delta T = -\alpha(T-T_1) = \ln\left\{1 + E\int_{\theta_1}^{\theta} \frac{d\theta}{G_\tau h^2}\right\} \tag{4.5}$$

$$\frac{\mu_1}{\mu} = 1 + E\int_{\theta_1}^{\theta} \frac{d\theta}{G_\tau h^2} \tag{4.6}$$

where E is an adiabatic parameter. Typical values for the turbulence coefficients are provided in fig. 4.2.

Hydrodynamic pressures start developing at the downstream edge of the oil feed groove, and progress to the other end of the pad. Further away from h_{min} a region of cavitation may occur in the diverging area θ_2 to θ_E of fig. 4.1, where the oil flows in streamlets mixed with air and foam. The minimum film thickness occurs before the end of the pressure profile. The temperature T_1 of the lubricant admitted at the groove rises along the pad, and reaches a maximum just before the film ends at θ_2. Oil pressure at the inlet, usually of the order of 10 – 30 lbs/in², is small compared with as high as 600 lbs/in² hydrodynamic pressure developed in the film and can be ignored in the calculations. ω represents angular velocity. Figure 4.3 shows the temperature and pressure profiles developed in a typical journal bearing.

The attitude angle ϕ in fig. 4.1 defines the line between the centers of the bearing and the journal from the direction of the load. When the direction of the load is vertical ϕ also denotes the angle between h_{min} and the load vector. The adiabatic constant E making its appearance in the temperature and viscosity expressions for journal bearings is given by:

$$E = \frac{2\mu_1 \alpha\omega}{bw}\left(\frac{R}{c}\right)^2 \tag{4.7}$$

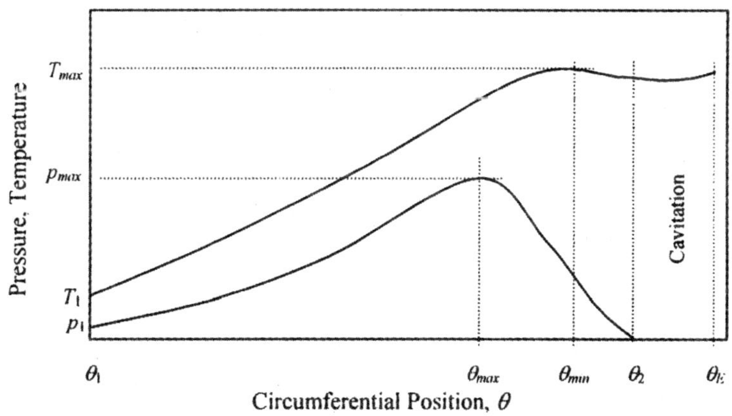

Fig. 4.3. *Temperature and Pressure Profiles in Bearing Lubricant Film*

E is thus a function of lubricant properties and conditions of bearing operation. For a well-aligned journal the film thickness for eccentricity ratio $\varepsilon = e/c$ is defined by:

$$h/c = 1 + \varepsilon \times \cos(\theta - \phi) \tag{4.8}$$

Minimum film thickness is obtained when $\theta - 180°$ coincides with ϕ, hence h_{min} equals $c(1-\varepsilon)$. An important bearing parameter is the Sommerfeld number, given by:

$$S = \frac{\mu N}{P}\left(\frac{R}{c}\right)^2 \tag{4.9}$$

where static pressure $P = W/LD$ and L is the bearing length. The friction coefficient, the ratio between frictional (F_τ) and bearing loads (W) is:

$$f = \frac{F_\tau}{W} \tag{4.10}$$

The general shape of the coefficient of friction in terms of the Sommerfeld number is given in fig. 4.4. For low values of S the friction factor tends to be high because of boundary condition effects, when partial contact between the bearing and journal surfaces may occur. Power loss H due to friction is given by Petroff's equation:

$$H = F_\tau \times R \times \omega \tag{4.11}$$

Hydrodynamic flows induced by shearing action and pressure gradients in the fluid film cause the lubricant flow rate Q_1 to enter the bearing at the leading edge, Q_2 flows out at the trailing end and an amount Q_z leaks out from the two sides. Q_z is the minimum amount to be delivered to maintain a full fluid film with all its potentialities. For isothermal conditions a bulk temperature rise can be obtained from the friction power loss and side leakage:

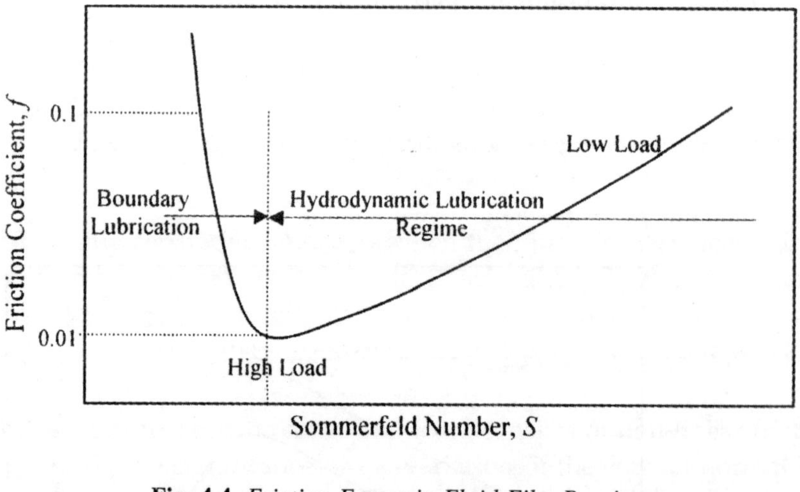

Fig. 4.4. *Friction Factor in Fluid Film Bearings*

$$\Delta T = (T_{average} - T_1) = \frac{H}{bwQ_z} \tag{4.12}$$

When thermal effects are included the variable temperatures may be combined by the factor correlating temperature with viscosity. Evaluation of the bearing performance parameters for specific operating conditions can then be obtained for the desired bearing geometry.

4.3 JOURNAL BEARING TYPES

Many different types of bearings are used to support a rotating shaft. Some representative types are shown in fig. 4.5. The partial arc and the grooved bearings are adaptations of the plain cylindrical bearings. With a fixed geometry a cylindrical bearing tends to be an unstable form of support. The elliptical, three-lobed and offset cylindrical bearings are capable of providing some pre-load, which is a function of the ratio of diameter D and the clearance c. A value of 0.5 for the pre-load is widely used. With the pre-load, bearings tend to operate with a greater minimum film thickness. When the direction of the steady load on the bearing is varying, the three-lobed configuration is preferred over the offset and the elliptical types.

The effect of a decreasing viscosity is to lower load carrying capacity, but the trend is partially mitigated by the fact that adiabatic solutions yield a higher attitude angle and thus a more extensive pressure profile. Friction will decrease with a rise in the adiabatic constant E, but here too partial offset is obtained by a larger extent of the fluid film. Except for certain combinations of high E and eccentricity ratio ε, the friction factor is mostly lower in isothermal bearings.

Bearings with fixed geometry under some conditions exhibit unstable operation. Thermal effects tend to increase ϕ which diminishes stability, but it also increases ε, which tends to increase stability, so the overall trend is not readily predictable. Lower viscosity mostly

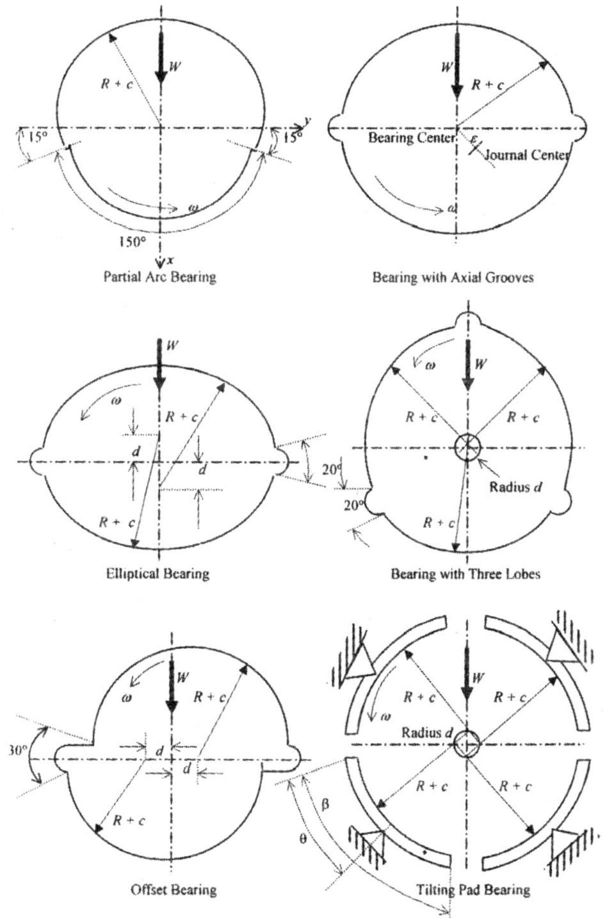

Fig. 4.5. *Journal Bearing Types*

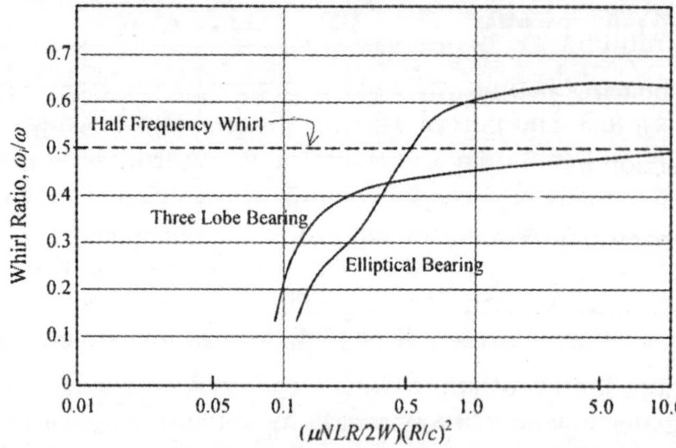

Fig. 4.6. *Threshold Frequency for Unstable Operation (Allaire, 1980)*

tends to have a small reduction in the critical load. Operation at differing levels of turbulence and E implies that a comparison based on the same ε and other bearing parameters such as c, N, R and μ to calculate W and Q cannot provide an adequate indication, and serves only to provide a trend.

Elliptical bearings are designed to provide an enhanced capacity to suppress instability. In addition to the two axial grooves, the two halves are put together such that their centers of curvature do not coincide. Each lobe is displaced inward by a fraction of the machined radial clearance, referred to as ellipticity or preload. Thus, even if the bearing is operating at the center of the bearing, the shaft center is placed at some eccentricity in relation to the two halves, thus providing some measure of stability. Commonly used in industrial turbo-machines because of ease of manufacture, the drawbacks of this form are higher friction loss and lubricant flow rate, while load capacity diminishes at low eccentricity. In the evaluation of bearing performance characteristics the journal position is referred to the geometric center of the bearing, but the eccentricity ε and attitude angle ϕ are measured with respect to the lobe centers. Thus, for a bearing with an ellipticity d and eccentricity ratios ε_1 and ε_2, $h_{min} = c \times (1 - \varepsilon_2)$ for the lower lobe # 2.

The three lobe bearing design further accentuates the features of the elliptical design, and has even better stability characteristics. Three pads ranging from 80° to 120° angular extent, with the lower lobe placed symmetrically below the applied load, are used in this design. But the configuration entails a lower average clearance, and the lower lobe has a smaller arc length, resulting in higher losses and lower load capacity. The lobes play a similar role as in the elliptical bearing, hence the remarks pertaining to eccentricity ratio, attitude angle and other operating parameters are applicable to this bearing style also. The shaft does not rise above the horizontal centerline, hence the minimum film thickness stays in the bottom lobe. To understand stability aspects of the lobe design as opposed to the elliptical version, consider the Sommerfeld number S. As speed increases, the corresponding increase in S causes the operation to shift towards the unstable regime, with bearing diameter and lobe clearance playing a more dominant role than length and viscosity. This has the effect of the three-lobe design being more stable at light loads and small clearance than the elliptical version. At heavier loads, however, S moves towards the unstable region at a faster pace than the elliptical bearing style. Oftentimes the low load mode of operation is more troublesome when considering bearing stability. Whirl ratio approaches zero when operation is in the fully stable mode. Figure 4.6 provides a comparison in the variation of this ratio for the two types of bearings (*ref*: Ehrich, 1999).

The tilting pad bearing has much better stability characteristics, and may be selected if there is a possibility for instability. Its primary distinguishing trait is that the pads are supported at a pivot. Load direction may be towards the center of a pad or between two pads, and can be designed to provide a proper level of preload. Rather than h_{min} serving as the means for determining load capacity, film thickness over the pad, h_p, acts as a more suitable criterion. The center of curvature of the pad does not stay fixed because the pad can swivel in either direction by $\pm\gamma$. If the preload is not sufficient the pads on the upper half may lose the load and fluid film forces may cause the pad to scrape the journal, leading to fluttering type of motion in the pads. To avoid this condition, shaft position in the form of ε and preload need to be carefully selected. Preload is defined by $m = 1 - (c_m/c)$, where c_m is the smallest clearance for $\varepsilon = 0$. Figure 4.7 provides the range of values for $\varepsilon_m = e/c_m$ and m when unloading occurs. Typically the number of pads varies between 3 and 8, with the angular extent of the pads β varying accordingly. The

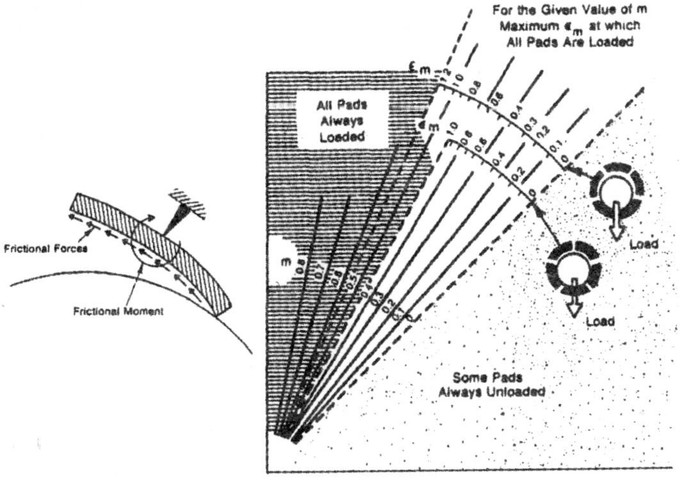

Fig. 4.7. *Regime of Unloading in Five-Pad Tilting Pad Bearing (Allaire, 1980)*

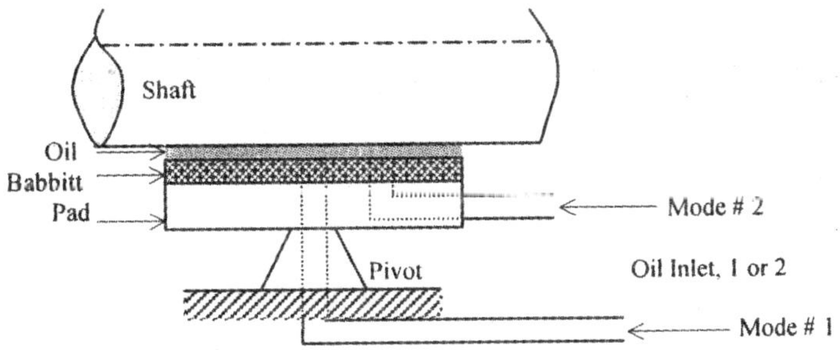

Fig. 4.8. *Lubrication Methods for Individual Pad of Tilting Pad Bearing*

option of varying β among the pads is also available. Pivot location may be symmetrical at the center of the pad or be asymmetrical in one direction. Inertia of the pad must be taken into consideration when evaluating its ability to track the journal. Because of the extent and number of variables, a general solution for the characteristics of tilting pad bearings cannot be obtained (Allaire, 1980).

A striking feature of tilting pad bearings is that for symmetrical vertical loading the locus of the shaft center is restricted to a vertical straight line. Shaft eccentricity tends to reduce when loading occurs over a pivot. Vertical stiffness and damping improve for loading over the pivot, but deteriorate in the lateral direction.

In the matter of number of pads, in a comparison between 3 pad and 5 pad designs with central pivot, zero preload and $L/D = 1$, the bearing with fewer pads has been found to carry more load when the load is in line with the pivot. However, when the load direction is between the pads the 5-pad bearing has a higher load carrying capacity and experiences lesser friction losses than the 3-pad bearing. Central pivoting of the pads in a tilting pad bearing permits rotation of the journal in either direction, and is easier to assemble. A higher preload of $m > 0.5$ offers advantages from a number of points, such as avoiding unloading of the upper pads, improved stiffness and damping and smaller amplitudes of motion due to enhanced stability characteristics. But the higher preload results in reduced film thickness over the pivot, even when eccentricity is not large. Power loss and temperature also tend to increase with m (Sawyer, 1982).

Oil supplied to a tilting pad bearing is mostly accomplished by flood lubrication, with the pads submerged in a pool, and the method is responsible for increased friction losses because of churning of the lubricant in the space between the pads. If the freedom of movement of the pads to tilt is not compromised, the lubricant may be introduced through connections inside the pad pivot or from a side of the pad as shown in fig. 4.8. Flexible hoses may be used to accommodate relative movement between the oil delivery line and the pad.

4.4 DYNAMIC CHARACTERISTICS

Experimental determination of the stiffness and damping characteristics of some bearing types was first attempted by Hagg *et al.*, (1956). Applying selected values of steady and rotating loads, and measuring the elliptical orbit of the journal, stiffness (K_1 and K_2) and damping constants (B_1 and B_2) were calculated relative to the major and minor axes of the ellipse. Figure 4.9 shows these constants plotted in a dimensionless form as a function of eccentricity ratio ε for the plain cylindrical bearing (Lund *et al.*, 1978).

The experimental approach has its drawbacks: it does not give information about the cross coupling terms, which are often a significant portion of the stiffness and dynamic characteristics of a bearing. By definition, cross stiffness and damping terms of the lubricant film relate shaft journal motion in a direction perpendicular to the direction of the force. Later evaluations were based on a numerical solution of the lubricant film's generalized equation (4.2). Rotor amplitude is assumed to be sufficiently small, so fluid forces may be replaced by their gradients around the steady state operating eccentricity. The forces are then proportional to the vibratory displacement and velocity, where the stiffness and damping coefficients provide the constant terms. For displacements X and Y in the x and y coordinates as shown in fig. 4.1, the fluid film forces may be expressed as:

$$\begin{aligned} F_x &= -K_{xx}X - B_{xx}\frac{dX}{dt} - K_{xy}Y - B_{xy}\frac{dY}{dt} \\ F_y &= -K_{yx}X - B_{yx}\frac{dX}{dt} - K_{yy}Y - B_{yy}\frac{dY}{dt} \end{aligned} \tag{4.13}$$

Lund *et al.*, (1978) give tabulated values of all the coefficients in the above equation as a function of Sommerfeld's number or eccentricity ratio. Four of the bearing types illustrated in fig. 4.5 are considered, with L/D ratios of 0.5 and 1.0. Figures 4.9 to 4.11 depict these coefficients as dimensionless numbers as a function of Sommerfeld's number for three geometric bearing types and $L/D = 1$. Note that some coefficients assume negative values under certain circumstances. Use of this data permits calculations for determining the effects of journal bearing characteristics on the stability of the rotor system, and also the steady state response to unbalance. The cross coupling coefficients are responsible for destabilizing effects in the fixed bearing geometry.

If the bearings are known to operate without instability in a speed range, response to unbalance can be determined more conveniently by using an axi-symmetric model for the bearings. The information thus available is oftentimes useful for many design tasks. An average value of the stiffness and damping coefficients from the set of eight parameter values simplifies the analytical work. The assumption in this case will be that the shaft trajectory will be circular instead of elliptical, with the circle radius equal to the average of the semi-major and minor radii of the ellipse. An alternate method calls for performing two sets of calculations, one using maximum values of the stiffness and damping coefficients from the full set, the other using minimum values. The results will then provide upper and lower bounds for natural frequencies and response to unbalance. Average bearing oil film forces in the radial and tangential directions for angular velocity ω and circular whirl radius r_a are given by the following expressions:

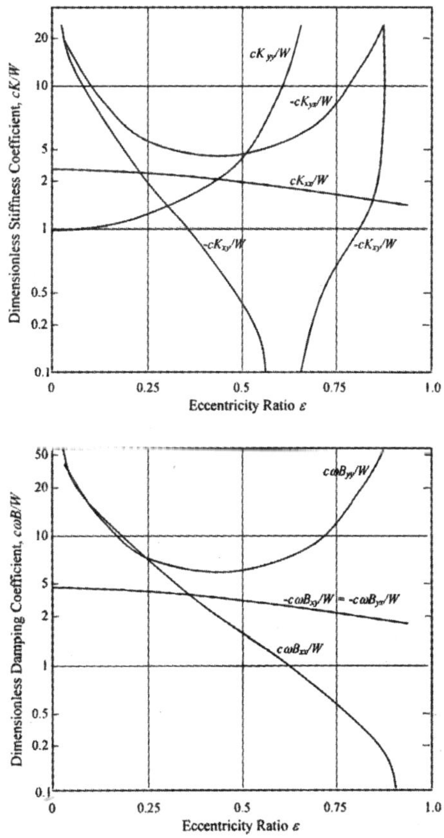

Fig. 4.11. *Plain Cylindrical Bearing Stiffness and Damping Coefficients,* L/D = *l (Lund, 1978)*

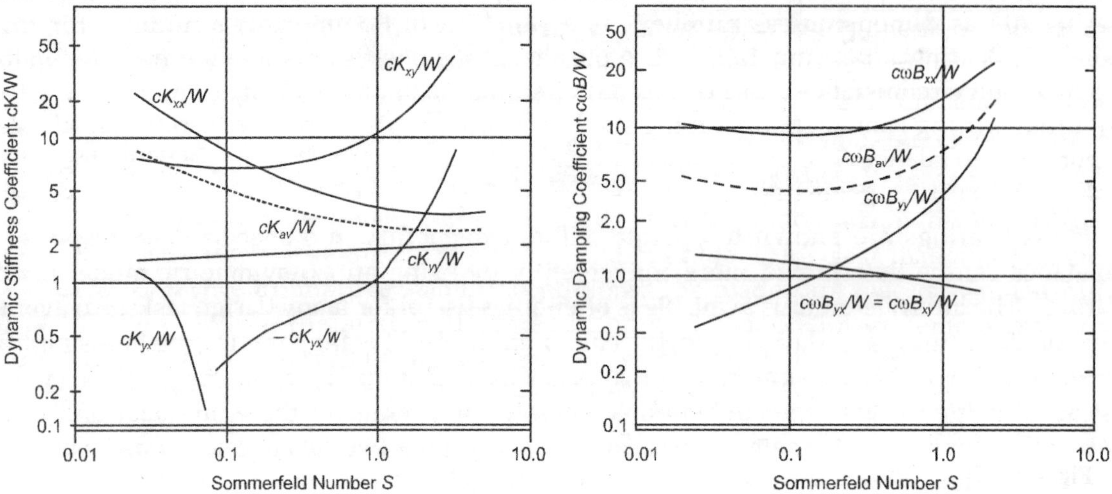

Fig. 4.10. *Two Axial Groove Bearing Stiffness and Damping Coefficients,* L/D = *1 Lund. 1998*

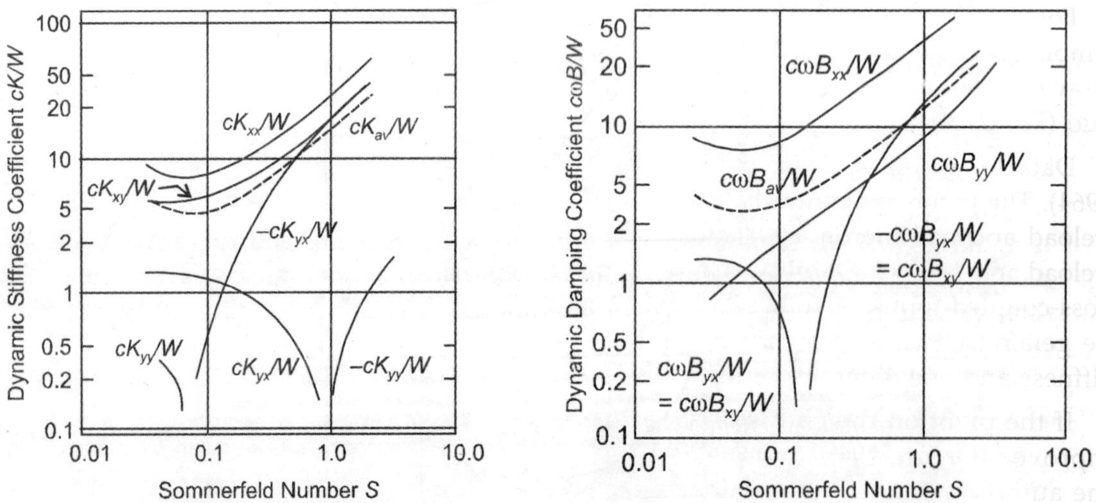

Fig. 4.11. *Elliptical Bearing Stiffness and Damping Coefficients,* L/D = *1 (Lund, 1978)*

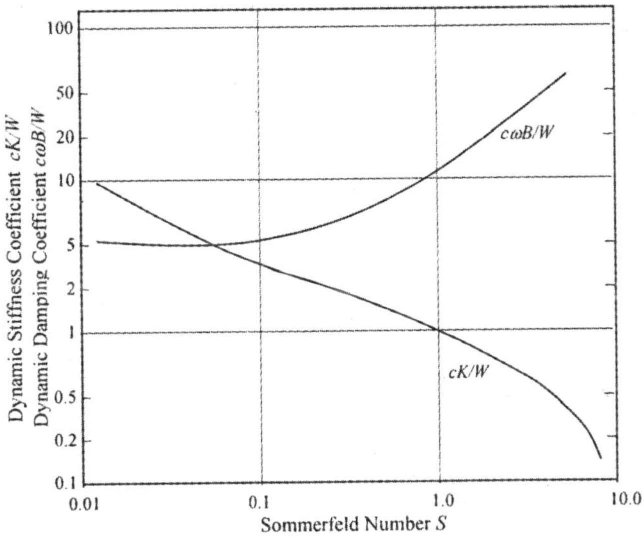

Fig. 4.12. *Tilting Pad Bearing Stiffness and Damping Coefficients, Four Pads Central Pivot,* L/D = 1. *(Lund, 1964)*

$$\frac{F_r}{r_a} = K_{av} = \frac{K_{xx} + K_{yy}}{2} + \frac{\omega(B_{xy} - B_{yx})}{2} \tag{4.14}$$

$$\frac{F_t}{r_a} = \omega B_{av} = \frac{-K_{xy} + K_{yx}}{2} + \frac{\omega(B_{xx} + B_{yy})}{2} \tag{4.15}$$

For an assumed small shaft center motion about an equilibrium position, the stiffness and damping coefficients relate the load and displacement. The dynamic characteristics for a bearing are available from bearing manufacturers, and tend to be displayed as a function of the eccentricity ratio (Ehrich, 1999).

Data for tilting pad journal bearings with 4, 5, 6 and 12 pads is obtained from Lund (1964). The pivots are centrally located on the pads, with several values of L/D. The effects of preload and pad inertia are also included. Figure 4.12 shows data for four pads, with zero preload and $L/D = 1$. Stability in a tilting pad is excellent, which is borne out by the absence of cross-coupled terms. A four pad bearing will be symmetric in the x and y directions, hence the generally two direct stiffness and two direct damping coefficients are reduced to one stiffness and one damping terms.

If the pivot on the pad is offset by a small amount, the hydrodynamic performance is improved (Orcutt, 1967), but the bearing is limited to operation in one direction of rotation. The author mentions that laminar flow assumption for fluid film bearings is appropriate when using hydrocarbon oils with high kinematic viscosity. Turbulence may set in if lubricant viscosity is low (water, liquid metals) and if rotor speed is high, resulting in Reynolds number to exceed 1,500. For bearing films the expression for Reynolds number is given in eqn. 4.1.

4.5 ROLLING ELEMENT BEARING

High operating speeds coupled with the need to reduce axial length make rolling element bearings the preferred choice for supporting the main rotors in aircraft power plants. Careful selection of the many different variables, among them load, speed, materials, lubrication method, alignment and fit-up will determine the degree of success attained in the operation of a bearing. Figure 4.13 shows examples of ball and roller bearings. Grease packing is common on some class of machines such as small compressors. Bearing performance improves substantially if spray lubrication in the form of a mist is used. This serves also to reject any heat developed. Most of the disadvantages of rolling element bearings arise from rubbing or sliding contact between the rolling elements, races and cages, with life limiting consequences.

In the preliminary selection of a bearing for a given application various criteria are employed that place particular emphasis on operating speed. The *DN* value takes into account the bore *D* (mm) and shaft speed *N* (rpm) to estimate the high-speed limitations. A relatively coarse indicator, the value gauges the acceptability of the bearing, since load characteristics tend to increase in complexity at higher speeds to adversely generate effects arising from cooling, excessive tolerance variation and flaws in the material. Another factor suggested by Bailey *et al.*, (1981) is the TAC factor τ to address centrifugal forces generated by an epicyclic ball or roller motion:

$$\tau = d_m N^3 D_W{}^3/\cos^3\alpha \tag{4.16}$$

where d_m is pitch diameter, mm, N = inner race speed, rps, D_W = rolling element diameter, mm and α = nominal contact angle, degrees. The upper limit on this factor is 31×10^8, but higher values have been successfully attained in bearings. Acceptable lubrication characteristics may be established from the minimum film thickness given by the equation:

$$h = 9 \times 10^{-4} \times D_o \times [(LP) \times N_d]^{0.74} \tag{4.17}$$

where h = minimum film thickness, μ-inch, D_o = outer bearing diameter, mm, LP = a lubrication parameter and N_d = speed difference between the inner and outer raceways, rpm. Metal to metal contact can be avoided by maintaining h at a minimum of 12 μ-inch, and LP varies between 100 and 1000 at moderate temperatures. For thin oils $D_o \times N_d > 4000$, and > 400 for thick oils to avoid boundary lubrication related problems. Guidelines have been established by the bearing manufacturing industry in an effort to control the quality of the bearings, for interchangeability and for parts replacement, with emphasis on component dimensions and tolerances. The tolerance range diminishes to enhance precision as the class level increases.

The choice of a bearing for a given task is closely associated with its fatigue life predictions. Not so significant differences in the bearing's configuration may cause identical bearings subject to the same load, speed and lubrication to have differing fatigue characteristics. Bearing manufacturers recommend the use of L_{10} rating life, since it is representative of 90 percent operating reliability. The operating life in hours is determined from the relationship:

$$L_{10} = (16667/N) \times (C/W)\gamma \tag{4.18}$$

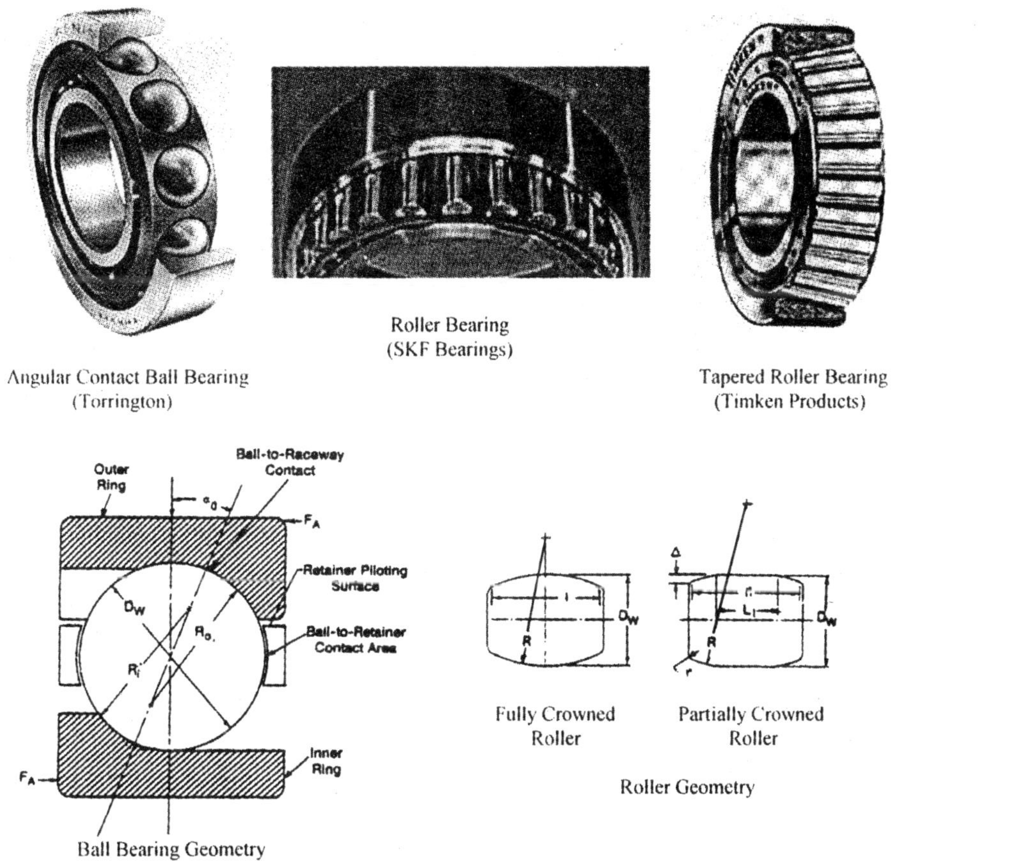

Fig. 4.13. *Ball and Roller Bearings (Courtesy: SKF Bearings, Timken, Torrington)*

where C = dynamic load capacity of the bearing, W = equivalent radial load on the bearing, N = shaft speed, rpm and $\gamma = 3$ for ball bearings and 3.3 – 4.0 for roller bearings. Dynamic load capacity of a bearing of bore diameter D defines the endurance load of a bearing for a fatigue life of 1×10^6 cycles, and is calculated by the equations:

For ball diameter less than 1 in.: $C = f_c \times (i\cos\alpha)^{0.7} \times Z^{2/3} \times D_W^{1.8}$ (4.19)

For ball diameter more than 1 in.: $C = f_c \times (i\cos\alpha)^{0.7} \times Z^{2/3} \times D_W^{1.4}$ (4.20)

For roller length less than 2.5 D_W: $C = f_c \times (i_{eff}\cos\alpha)^{7/9} \times Z^{3/4} \times D_W^{1.074}$ (4.21)

where Z = number of rolling elements, D_W = element diameter, inches, α = contact angle, degrees, and i = number of rows. Equivalent load W for ball bearings is based on a proportional linear combination of axial and radial loads acting on the bearing. Factor f_c varies between 3500 and 7500 depending on rolling element size.

The fatigue life calculation procedure must be corrected for material properties, lubrication effectiveness, reliability and hardness at elevated temperature. An array of data based on experiments of many different bearing materials is available from which a correction factor may be derived to account for differences in material characteristics. A fall in the material Rockwell hardness below 58 can compromise the fatigue life of a bearing operating above 400° F. In the normal hardness range of R_c = 58 – 62, the correction factor for fatigue life is zero.

Continuously applied large static loads beyond the basic capacity can also cause permanent deformation in the ball elements and the races. To account for sudden overloads, or an overload of short duration, a correction factor must be applied to the predicted fatigue life.

Boundary friction will determine behavior at the contact in the event an adequate lubricant film is not present at the mating surfaces. For a heavily loaded contact a full film may separate the surfaces, but the elasticity of the parts will result in surface deflections, causing the film to be altered. Coupling among the elastic deformation equations and the hydrodynamic Reynolds equation is then essential for realistic simulation of the contact region. Figure 4.14 represents the lubricant pressure profile using the full film concept and including the effects of elastic deformation of the contact surfaces. The region may be split into an area where the oil is pressurized, a full film zone where Hertzian deflections occur to cause the extent of separation and a zone where the pressure drops sharply to the atmospheric level.

Thermal effects pertaining to the lubricant film behavior must also be included in the evaluation. Film thickness may first be calculated based on isothermal conditions and then modified by a thermal reduction ratio. Note that line contact occurring in a roller bearing will be different from an essentially point contact between a spherical ball and a cylindrical race, and the consequent reduction in film size will affect temperature increase and side leakage.

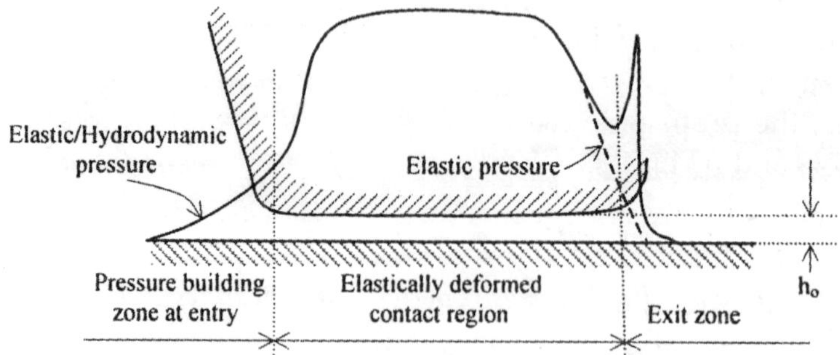

Fig. 4.14. *Lubricant Film Pressure Profile in Rolling Element Bearing*

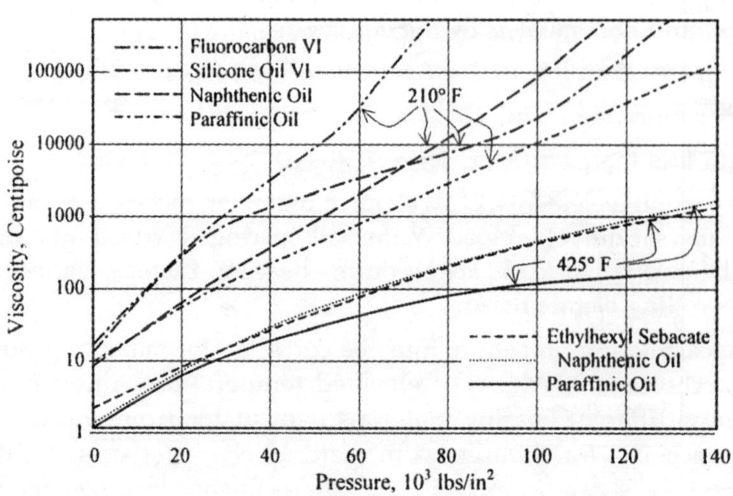

Fig. 4.15. *Pressure/Viscosity Curves for Lubricant Oils (ASME, 1954)*

Lubricant viscosity plays a considerable role, not merely from engineering consideration but from the relationship between pressure and viscosity. At a nominal Hertzian stress level of 150,000 lbs/in^2 the viscosity of a paraffin-based lubricant may be 100,000 cps, as opposed to 10 cps at atmospheric pressure, and this increase is responsible for developing the oil film in ball bearings. Figure 4.15 provides pressure/viscosity data for various oils at a number of temperatures. Viscosity as a function of temperature of several petroleum oils commonly used for turbo-machinery bearings is shown in fig. 4.16. With a sufficient film thickness between the contacting elements the fatigue life of the bearing experiences substantial enhancement when the operating temperature is lower.

Surface finish of the contacting components also affects the formation of the lubricant film due to the protrusion of asperities from both surfaces. In superior bearings falling within the ABEC Class 5 or higher designations, surface finish of 4 rms is available on the races and 2 rms on the balls. Corresponding finishes on lower grade bearings run at 8 and 4 rms. Rougher surfaces may be used on roller bearings, with RBEC Class 1 bearings provided with surface finish in the 8 to 16 rms range.

AISI M-50 is commonly used for rolling element bearings for turbo-machines, as also AISI 52100. The vacuum re-melting and degassing processes used for these steels improve the fatigue characteristics of the metal by reducing the level of impurities around which the material nucleates and where fatigue cracks initiate. But the materials may not have appropriate resistance to corrosion and fracture failure at high operating speed and load. Failures are experienced by fatigue spalls and subsurface cracks, often in the inner ring of the bearing. If a crack reaches critical proportions, the propagation rate increases rapidly to cause failure. Other materials such as the AMS 5749 and AMS 5900 have been determined to have greater corrosion resistance.

Rolling elements also have inherently low damping features, and if the rotor dynamic aspects of the machine's system require it, some damping elements must be built into the system.

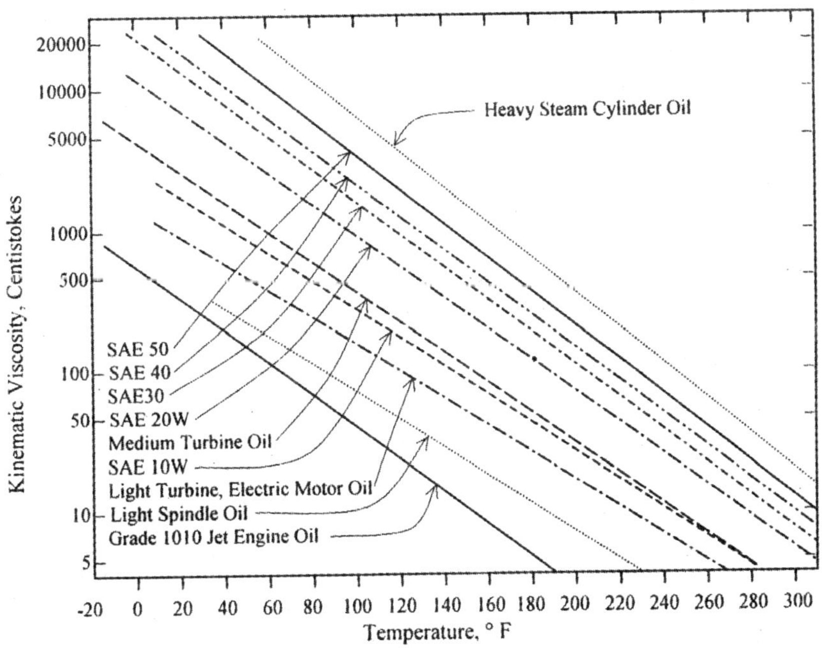

Fig. 4.16. *Effect on Temperature of Viscosity of Various Lubricant Oils (Wilcox, 1957)*

The interaction between rolling element bearings and the rotor is of special interest when looking at the dynamics of the system (Rangwala, 2004). Significant factors that come into play are:

- Due to lack of damping, systems operating through a critical speed will require attenuation of critical speed response amplitudes. To accommodate this need, squeeze film dampers or dampers made of a resilient material are typically used.
- Rolling element bearings are free of destabilizing forces common in hydrodynamic bearings, such as oil-induced whip.
- Unlike journal bearings, rotating elements of the bearing are fixed to the rotor. The rollers and the cage rotate at approximately half the rotor speed as an assembly. As a result, problems may occur during balancing and in calculating non-synchronous response due to slight variations in component dimensions, but for the most part cause minor problems.

Differential thermal expansion among the moving and non-moving components of the bearing may result in compression and premature fatigue. To avoid this the bearings must be designed with radial clearance between the elements and the races. But if radial load is reduced during certain operating cycles, bearing life is compromised due to skidding of the elements. Effective softening of the bearing support stiffness may also develop due to larger than normal clearances, with a consequent reduction in natural frequencies and critical speeds.

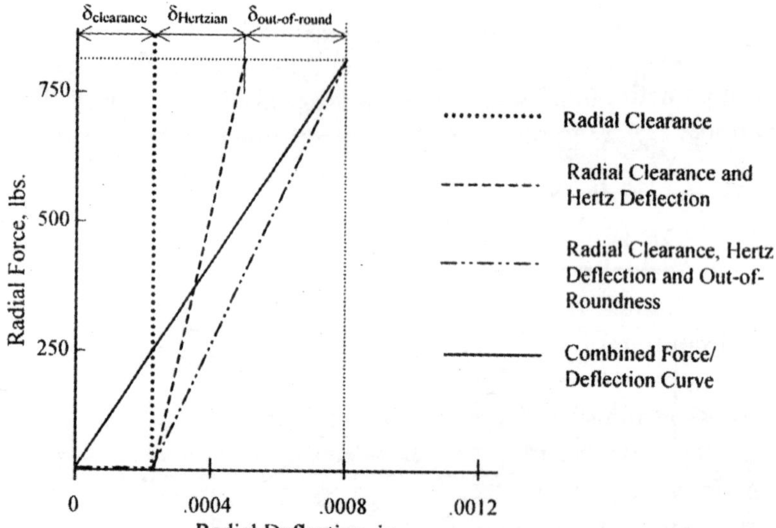

Fig. 4.17. *Rolling Element Bearing Stiffness Calculations (Ehrich, 1999)*

Other disadvantages as a result cf larger clearances (and bi-linear stiffness of the bearing) are reported to cause anomalies in vibration response to unbalance, such as peaks at whole number multiples of the critical speed and hysteresis (Ehrich *et al.*, 1967). One possibility to overcome the problem is to machine lobes in one of the races, so at least some parts of the bearing circumference provide roller contact with no clearance. At the same time the race in these close contact areas elastically deforms to permit the rolling elements to move along without applying a large compressive load.

As in hydrodynamic sleeve bearings, the stiffness of the rolling element bearing is required for rotor dynamic calculations. Stiffness level plays a major role when dealing with a stiff rotor.

Due to its geometric complexity and the number of parts involved, its stiffness is not easily calculated. Deformation is due to three factors: deflection through the radial clearance, elastic compression in the rollers and deformation of bearing races from a circular to an oval shape. Compression of rollers at the points of contact with the races may be determined from equations developed by Hertz (see example problem # 4.1). The change in the profile produces a change in load distribution used for roller deformation, so an iterative procedure is required for the two combined effects. Because of the presence of the radial clearance, the overall stiffness curve is not linear. Figure 4.17 shows a linear approximation of the overall characteristic, with slope of the force/deflection being 1.0×10^6 lbs/in. If in the design process it is determined that correlation between calculated and measured rotor critical speeds is not obtained, an adjustment to this stiffness value may be required.

Magnetic bearings are used in high-speed rotors to avoid stability problems of journal bearings and life limitation problems of rolling element bearings (see section 4.6). The bearing levitates the rotor through magnetic forces set up by electromagnets that are opposed. Magnetic bearings use displacement measurements between the rotor and the magnet to actively control forces acting on the rotor. Forces proportional to relative displacement yield effective bearing stiffness, and forces proportional to relative velocity yield damping. Thus, both stiffness and damping are adjustable. There are no cross-coupled destabilizing stiffness terms in magnetic bearings, so the bearing is stable. But the bearing can become unstable if looked at as a classical linear sampled data feedback control system. Magnetic bearings have a lower stiffness value than journal bearings with oil film, and its use markedly increases rotor length and diameter at the bearing location.

4.6 TIP CLEARANCE ACTUATION WITH MAGNETIC BEARINGS

Stable operation of axial flow compressors as they are employed in modern jet engines and gas turbines is often limited by two flow breakdown processes, known as surge and rotating stall. Surge is a circumferentially uniform pulsation of the mass flow through the machine, while rotating stall appears as a reduced flow region in part of the circumference, which travels around the compressor annulus at a fraction of the rotor speed. Theoretical and experimental investigations for the active control of rotating stall have been conducted at the NASA Glenn Research Center on a single stage transonic core compressor inlet stage (Spakovszky *et al.*, 2000). Active stabilization of rotating stall and surge using unsteady air injection was first presented by Weigl *et al.*, (1988) in the NASA high-speed stage. The experiments showed a significant benefit in the stable operating range. Blade tip clearance in axial flow compressors is known to have a strong impact on compressor performance and stability; it also plays a major role in the interaction between the rotor dynamic shaft deflections and the aerodynamic behavior of the compressor (Rangwala, 2004).

Magnetic bearings are widely used as active suspension devices in rotating machinery, mainly for active vibration control purposes. The concept of active tip-clearance control suggests a new application of magnetic bearings as servo-actuators to stabilize rotating stall in axial compressors. The magnetic bearing servo-actuator is used to actively whirl the shaft, inducing an unsteady variation of the rotor blade tip-clearance distribution as shown in fig. 4.18. The general sequence of steps required in the analysis and design of the magnetic bearing system is shown in fig. 4.19.

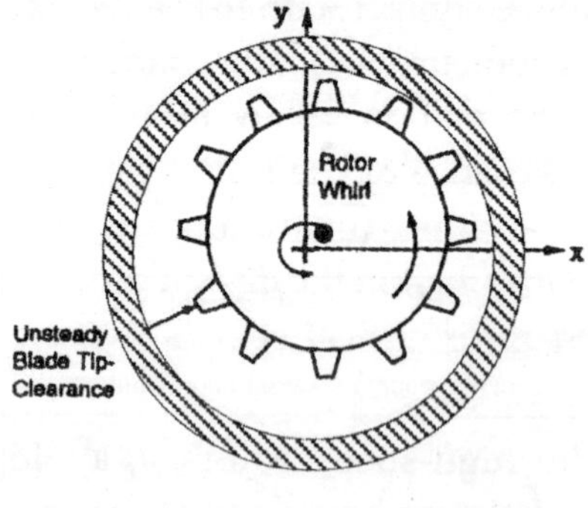

Fig. 4.18. *Active Tip Clearance Control Concept (Spakovszky, 2000)*

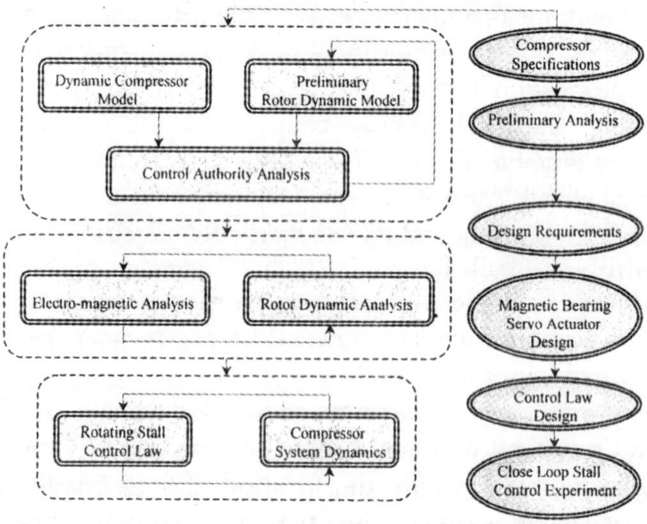

Fig. 4.19. *Active Tip Clearance Control Concept (Spakovszky, 2000)*

Starting with the compressor flow specifications, an unsteady compressor tip clearance and a rotor dynamic model of the compressor are implemented to determine the control authority and the detail design of the magnetic bearing's actuation system.

The NASA Stage 37 test compressor, originally designed as an inlet stage of an eight-stage 20:1 pressure ratio core compressor, has a total pressure ratio of 2.05, mass flow of 20.2 kg/s, rotor tip speed of 454 m/s, and rotation frequency of 286 Hz at design conditions. The rotor consists of 36 blades with an aspect ratio of 1.19, hub-to-tip radius ratio of 0.7, and blade tip diameter of approximately 50 cm. The mean-line rotor chord length is 56 mm. Atmospheric air is drawn into the test facility through an orifice plate and a plenum chamber upstream of the test section. Downstream of the compressor the flow is regulated with a sleeve type throttle valve. The compressor shaft is coupled through a drive train to a 2.2 MW DC drive motor. The shaft setup of the test compressor is an overhung rotor with radial fluid film bearings at the front (near the rotor disk) and at the back of the compressor (near the motor drive coupling), as well as a fluid film thrust bearing on the motor coupling side. A schematic of the test section and the compressor shaft is shown in fig. 4.20.

The effect of tip-clearance asymmetries due to shaft deflections on compressor performance and stability is addressed next. The objective of the preliminary analysis is to determine the magnetic bearing force bandwidth and the stall control authority required to conduct the rotating stall control with tip-clearance actuation.

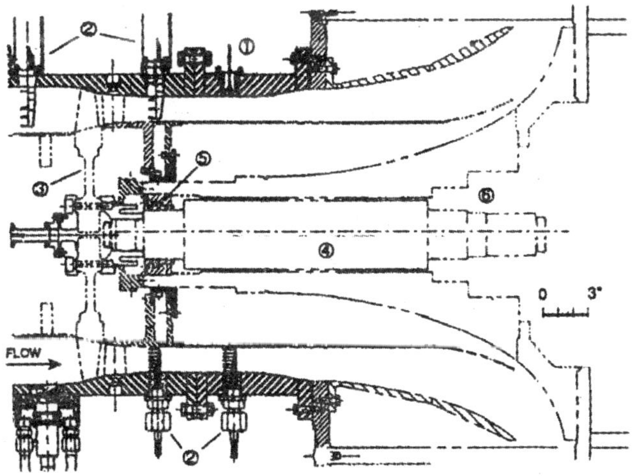

1. High Frequency Static Pressure Sensors 2. Steady/Unsteady Survey Probes 3. Rotor Disk 4. Shaft 5. Front Fluid Journal Bearing 6. Rear Fluid Journal Bearing

Fig. 4.20. *NASA High Speed Single Stage Compressor Rig (Spakovszky, 2000)*

The specific question is: how much shaft motion and magnetic bearing force is required to stabilize a rotating stall? To answer this question a rotor dynamic design analysis and a unique stochastic estimation and control analysis are conducted. A design of the rotor with a magnetic bearing rotor is shown in fig. 4.21. The solid shaft in fig. 4.20 is replaced by a hollow shaft, and includes the magnetic bearing rotor laminations. The shaft is pinned at the rear journal and thrust fluid film bearings, and coupled to the motor drive train. Typical catcher bearing designs do not contact the shaft during magnetic bearing suspension. However, for the proposed stall experiments a fail-safe suspension system is mandatory. In particular, the compressor blades must be protected from possible rubs at the blade tips; destructive impacts must also be avoided in the case of a loss of magnetic levitation. A possible failsafe solution is to use a spring-loaded catcher bearing that is always in contact with the shaft. This allows for the shaft deflections, but still yields a hard stop in case of an emergency. To ensure safe transient operation without large vibrations when critical frequencies are crossed during an emergency shut down, the damping must be added in parallel to the soft spring-loaded support. One possible compact solution is an integral squeeze film damper (ISFD) setup as reported by Santiago *et al.*, (1998). The ISFD's are comprised of accurate squeeze film pads rendering visco-elastic support and wire-electrical discharge machined webs acting like a squirrel cage. The open loop whirl speeds, natural frequencies and mode shapes are obtained from an eigen-value problem resulting from the equations of motion. Assuming that the rotor is spinning at design speed (286 Hz) the first four eigen-values are plotted in fig. 4.22 and the mode shapes are reconstructed from the corresponding eigen-vectors (fig. 4.23). In order to determine the effective shaft motion (i.e., blade tip deflection) for the control of a rotating stall the closed loop system is employed.

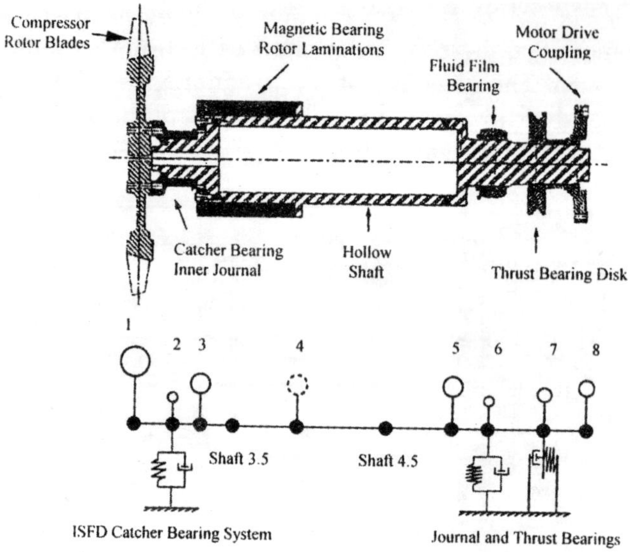

Fig. 4.21. *Rotor Dynamic Model of Compressor with Magnetic Bearing (Spakovszky, 2000)*

The compressor pre-stall dynamics are denoted by the transfer function $G(s)$. The outputs of $G(s)$ are the pre-stall pressure perturbations sensed upstream of the rotor $\delta p(i)$, which are fed back to the rotating stall feedback controller $K(s)$.

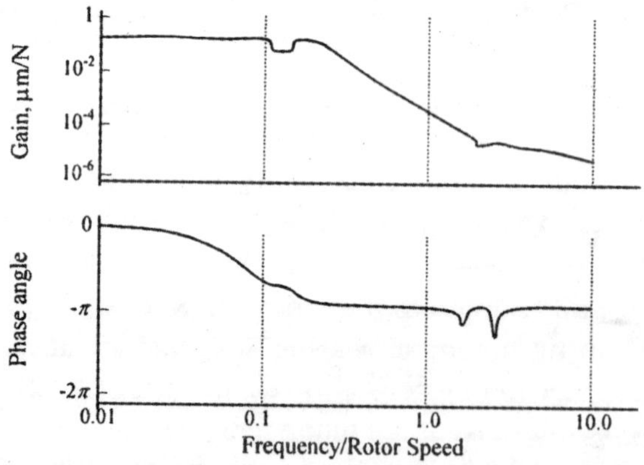

Fig. 4.22. *Transfer Function Between Bearing Force and Tip Deflection (Spakovszky, 2000)*

The controller outputs are the six actuator commands $\delta c(t)$, which are modified by the magnetic bearing servo-actuator dynamics to yield the actual shaft position and the corresponding tip clearance distribution $\delta \varepsilon(t)$.

The open loop stable magnetic bearing servo-actuator dynamics parameters consist of the shaft rotor dynamics and the magnetic bearing servo control loop. The inputs to the compressor pre-stall transfer function $G(s)$ are the tip-clearance distribution $\delta \varepsilon(t)$ and background noise modeled by unsteady velocity fluctuations $\delta w(t)$. pole configuration and a new soft magnetic material with relatively small magnetic losses and very high saturation flux densities are considered.

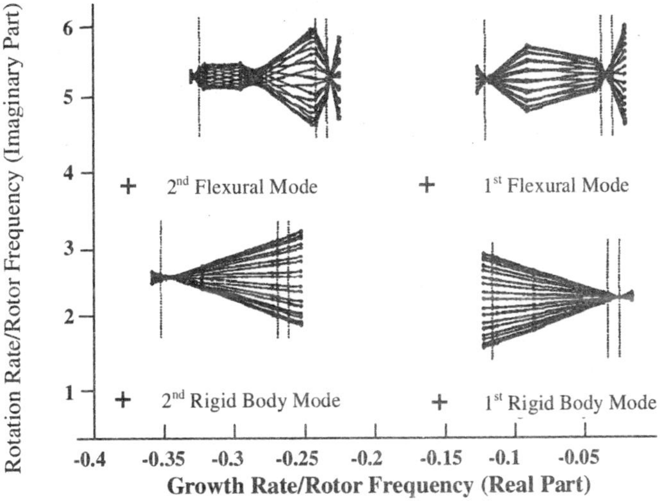

Fig. 4.23. *Preliminary Rotor Model Eigen-values and Mode Shapes (Spakovszky, 2000)*

Apart from the magnetic bearing actuator itself, the above issues will also strongly influence the actuator electronics design. A 16-pole N-S-N-S configuration with rotor laminations, illustrated in fig. 4.24, is selected to satisfy the requirements.

Predicted results of the overall dynamic system are shown in the Campbell diagram of fig. 4.25. The two rigid body forward and backward whirling modes are crossed by the one-per-rev line (1E) at about 30, 180, and 200 Hz. Note that none of the flexural modes are crossed despite the strong gyroscopic effects, and the rigid body mode frequencies compare well to the preliminary results obtained from the simple lumped parameter model. The frequencies of the two rigid body modes mostly depend on the stiffness of the fluid film and the catcher bearings, the mass properties of the compressor shaft and the actuator feedback control. In order to limit the vibration level when the rigid body critical frequencies are crossed, either the rotor must be well balanced or the bearings must dissipate a substantial amount of energy, thus calling for sufficient external damping.

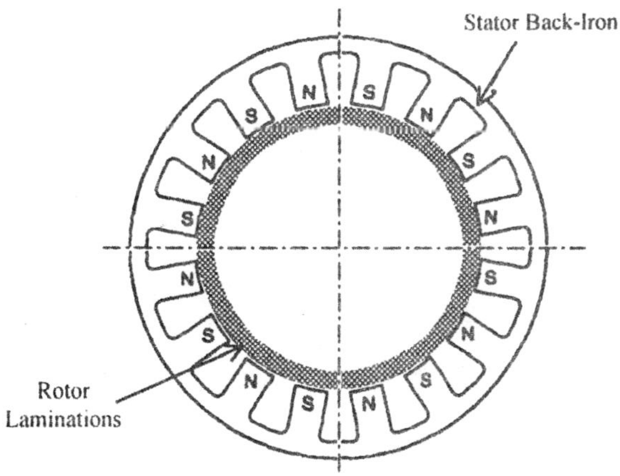

Fig. 4.24. *Magnetic Bearing Servo-Actuator (Spakovszky, 2000)*

Non-linear effects are included in the voltage and current saturation characteristics of the power amplifier because of the influence of the actuator inductance and resistance and the back electro-motive force term in the rotor. The resulting higher order system of the non-linear differential equations is solved by an iterative procedure. The results are shown in fig. 4.26 for the shaft spinning at the design speed of 286 Hz. The plot depicts a family of nonlinear current-force characteristics of the magnetic bearing servo-actuator and the locus of operating points with the excitation frequency ranging from 0 to 300 Hz at maximum available power (thick solid line).

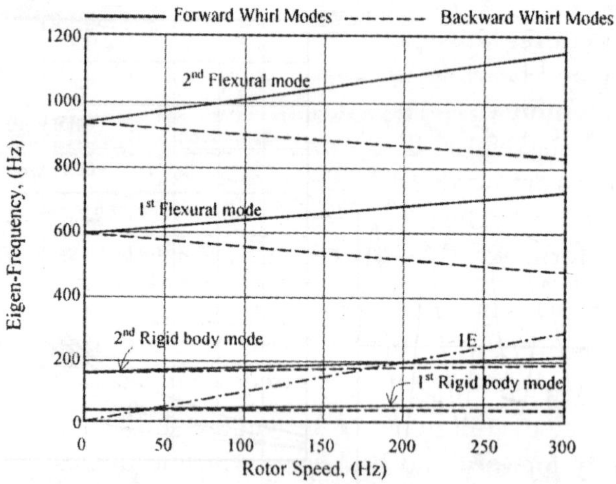

Fig. 4.25. *Campbell Diagram (Spakovszky, 2000)*

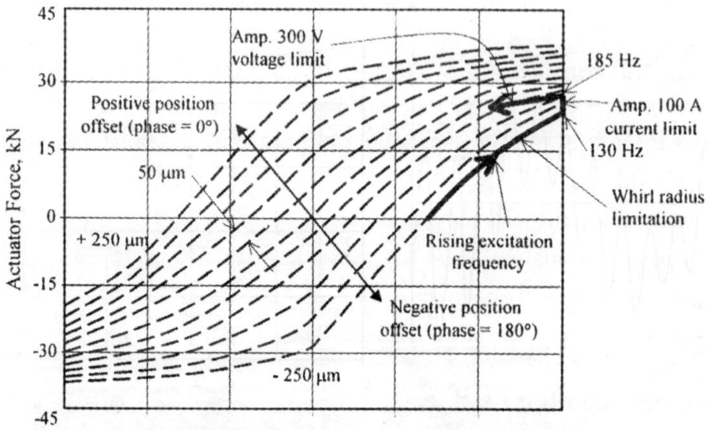

Fig. 4.26. *Current-Force Characteristic of Magnetic Bearing Actuator (Spakovszky, 2000)*

The effect of a varying air gap is included in the calculation. The achievable whirl radii at several shaft locations are plotted on the right hand side within the operational constraints of the amplifier and the whirl radius limitation at the magnetic bearing location (250 μm). Thus, the results show that the stated design requirements of 250 μm compressor blade tip deflection up to 185 Hz are fulfilled.

4.7 IMPACT OF FLEXIBLE SUPPORT

The support structure for a bearing oftentimes plays a large role on the behavior of a rotating machine (Rangwala, 2004). The dynamic characteristics of the support combine with the bearing's stiffness and damping coefficients to modify the impedance observed by the rotor at the bearing locations. API standard 617 for compressors recommends including the effects of the support structure in unbalance response analysis when the ratio of the support to bearing stiffness is less than 3.5, and specifies the use of calculated critical frequency dependent support stiffness and damping values or values derived from modal testing.

The calculation procedure, however, is sometimes difficult to accomplish because of the complexity of the casing and bearing supports. Experimental testing to measure the frequency response function (FRF) is more commonly taken advantage of. The supports may be excited by impact (or by mechanical shakers) at the bearing locations and measuring the response in the vicinity. The measurements are needed in the horizontal and vertical directions at the supports and also between the supports. The measured functions are then used to generate equivalent physical models in the form of mass supports (Redmond, 1996) or with many degrees of freedom (Stephenson *et al.*, 1992).

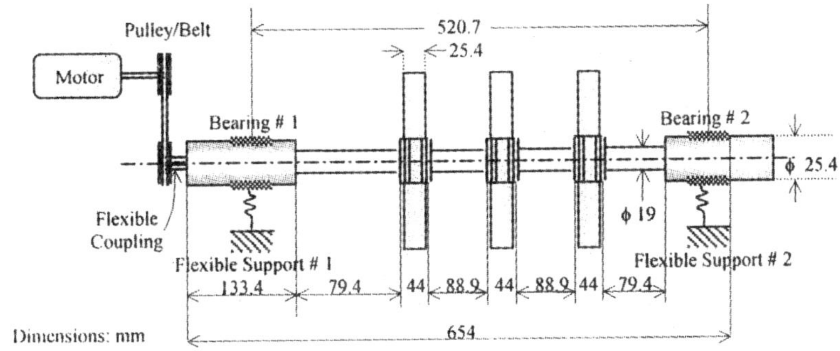

Fig. 4.27. *Test Rotor (Vazquez, 2000)*

Modal parameters may also be calculated by using the unbalance response of a rotating system, then create the equivalent foundation model.

To establish the effects of the support structure on the dynamics of the rotating system, a research project was initiated at the University of Virginia (Vazquez *et al.*, 2000). A flexible rotor supported by three-lobe bearings on flexible supports is tested for unbalance response and stability. The numerical predictions are extended to study the effects of cross coupling in the support structure between the vertical and horizontal directions and of cross talk between the pedestals. The test apparatus consists of the rotor with three equally spaced disks and supported by two identical three-lobe fluid film bearings (fig. 4.27). The bearings are supported by anisotropic flexible supports, with their other ends clamped to a concrete block. Table 4.1 lists the physical traits of the three lobe bearings, and fig. 4.28 shows their calculated dynamic coefficients as a function of speed. The housing for the bearings are rigid, and mounted on flexible structural aluminum members with stiffening plates at both ends, as shown in fig. 4.29.

Horizontal direction stiffness of the support can be varied by using plates of differing thickness, although the vertical direction stiffness remains unaffected. The support's dynamic

Table 4.1. *Lobe Bearing Data*

Journal diameter, mm	25.4
Lobe radial clearance, mm	0.833
Bearing radial clearance, mm	0.033
Lobe length, mm	12.7
Bearing preload factor	0.604
Lobe arc length, degrees	93
Static load, N	62.2
Oil inlet pressure, kPa	20.7
Oil inlet temperature, °C	48.3
Oil viscosity - 48 °C, Pa-s	0.021

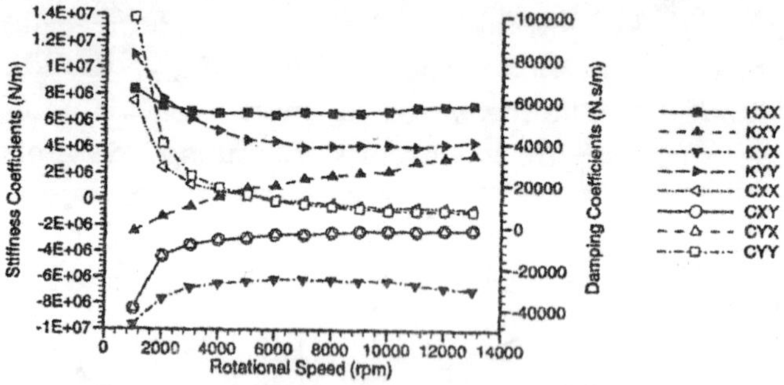

Fig. 4.28. *Bearing Dynamic Coefficients (Vazquez,2000)*

characteristics are determined by attaching electromechanical shakers to the bearing housing through piezoelectric force transducers. A sine sweep excitation is applied to each support in a single direction, and the response at the bearing housings measured in the vertical and horizontal directions. A total of 16 acceleration responses are then obtained to assemble the support matrix:

$$\begin{Bmatrix} x_1 \\ y_1 \\ x_2 \\ y_2 \end{Bmatrix} = [DC(\omega)] \begin{Bmatrix} F_{x1} \\ F_{y1} \\ F_{x2} \\ F_{y2} \end{Bmatrix} \tag{4.22}$$

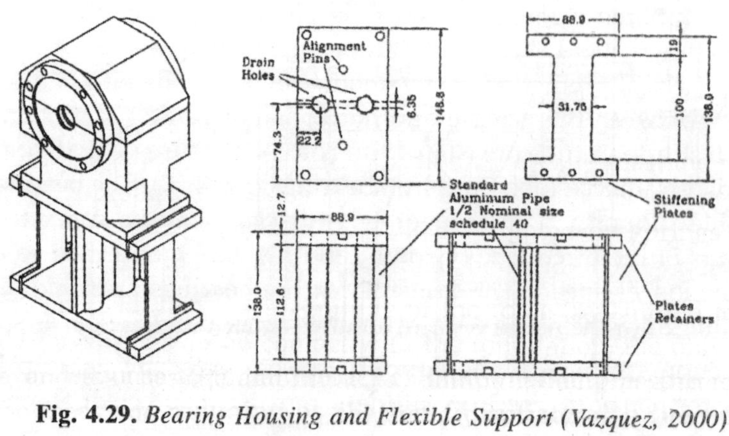

Fig. 4.29. *Bearing Housing and Flexible Support (Vazquez, 2000)*

where $[DC(\omega)]$ is the dynamic compliance matrix of the support structure, where the matrix terms are integrated twice for convenience to perform the calculations in displacements and rotations. The matrix is defined for each of the frequencies used for the excitation of the supports. The matrices may then be used directly for unbalance response calculations. For stability analysis polynomial transfer functions may be obtained for each of the matrix terms using the synthesis (Santhanan *et al.*, 1963) or the rational fraction polynomial (Friswell *et al.*, 1993) methods. Equation 4.34 then takes the form

$$\begin{Bmatrix} x_1 \\ y_1 \\ x_2 \\ y_2 \end{Bmatrix} = [G(s)] \begin{Bmatrix} F_{x1} \\ F_{y1} \\ F_{x2} \\ F_{y2} \end{Bmatrix} \tag{4.23}$$

where $[G(s)]$ defines the transfer function matrix of the support structure, with its terms defined in terms of the complex variable s valid in the whole complex plane. For unbalance response $s = i\omega_s$, and for stability analysis $s = p \pm i\omega_d$. Flexible support equations require stiffness and damping coefficients that may be combined to form complex stiffness coefficients. These coefficients are obtained by inverting the transfer function matrix for each complex frequency s_i. The support stiffness and damping coefficients are then used with a standard technique to include support effects in rotor dynamics analysis. Note that variation of the coefficients with the complex frequency must be available in the analytical tool. Since the support structure is passive, the dynamic compliance matrix $[DC(\omega)]$ must be symmetric. If one of the two support structures is more active at a particular resonance, the magnitude of the cross talk dynamic compliance is reflected in the direct dynamic compliance, indicating that the cross talk between the supports has a large influence in the response of the rotor.

The unbalance response of the rotor is measured during a slow run-up from 1,000 to 6,000 rpm. An unbalance weight of 9 gm-mm, designed to excite the first critical mode, is located on the middle disk. In a second run the unbalance is shifted 180° away. The unbalance response is then obtained by subtracting the response of the second run from the first run and dividing the difference by 2. The intention is to eliminate the effects of mechanical and electrical run-out, shaft bow and residual unbalance by using this scheme. Figure 4.30 shows the unbalance response near the middle of the disk, where three main responses are identified. Structural resonance is

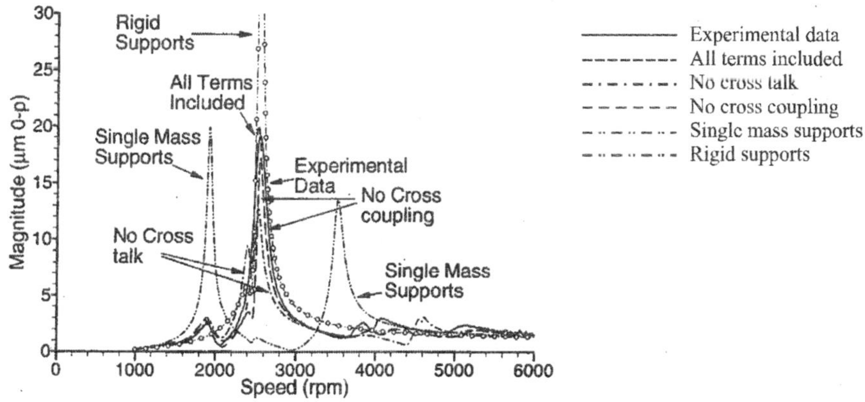

Fig. 4.30. *Unbalance Response at Middle Disk (Vazquez, 2000)*

detected at 1,900 rpm, the peak at 2,550 rpm corresponds to the first shaft critical speed and the response at 4,000 rpm is attributed to coupling with the second critical speed of the rotor. Predicted response using different models of the support structure is also included with the measured data in the figure. The single mass support representation does not agree with the measured data, showing two peaks at 1,900 and 3,500 rpm. The analysis using rigid supports predicts the first critical speed at 2,530 rpm, but the peak amplitude is considerably larger than the measured data.

Results from the transfer function representation show better agreement with the experimental data. The first critical speed is predicted at 2,520 rpm, and the deflection is just 6 percent larger than the measured value. Neglecting the cross talk between the supports also gives good results. Ignoring the cross coupling between the horizontal and vertical directions also provides comparable data.

Establishing the threshold of stable operation requires the rotor to be accelerated until it becomes unstable. The sub-synchronous vibrations are self-sustained and grow with time as the stability limit is crossed. In a lightly damped system this definition is significant, because sub-synchronous vibrations may be present but the overall vibrations do not grow with time. The spectral map indicates instability at 2,640 rpm, synchronous and two times the running speed. Figure 4.31 shows the stability map calculated for the rotor system using different support models. The logarithmic decrement is plotted against the rotor speed, with the bearing cross coupling term being the only instability mechanism in the system. With all terms in the transfer function matrix included, the predicted stability limit is slightly below the measured threshold speed of 9,350 rpm.

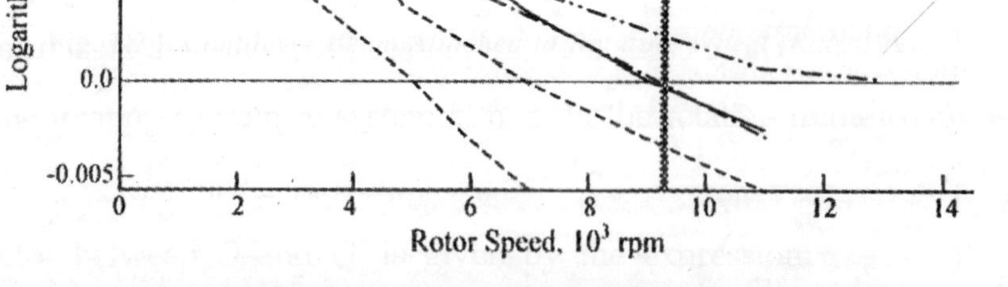

Fig. 4.31. *Calculated Stability Map (Vazquez, 2000)*

4.8 LABYRINTH AND HONEYCOMB SEAL EVALUATION

Labyrinth seals are extensively used in centrifugal compressors operating at high pressure and speed to control leakage. Field experience and tests indicate the labyrinth seals can generate unstable operation of the rotating system. In comparison, honeycomb seals of up to 50 mm in length have displayed better stability characteristics, and are also insensitive to pre-swirl at the inlet (Childs *et al.*, 1992).

Shunt injection is one approach to improve the performance of labyrinth seals. High-pressure gas, usually from the compressor's discharge, is injected into an intermediate labyrinth cavity at several circumferential locations in the radial direction or against the shaft rotation. The negative swirl flow injection is sufficient to reverse the flow direction in the seal, and the consequent reduction of circumferential velocity leads to a reduction in the cross-coupled stiffness k. The performance penalty arising from the diverted flow from the compressor discharge is the disadvantage of the technique.

An experimental study to compare the sub-synchronous vibrations and instability characteristics of a conventional labyrinth seal, labyrinth seal with shunt injection and a honeycomb seal has been conducted in a joint NASA-USAF funded program (Soto *et al.*, 1998). The test seal, 130 mm in diameter and 63 mm in length, is provided with feed holes for shunt injection, and choking of the holes is avoided by controlling the exit Mach number to less than 0.33. The seal has 20 teeth with a pitch of 3.2 mm and height of 3.17 mm. Shunt injection is facilitated by deleting the fourth tooth, and the flow occurs in both directions of the seal (fig. 4.32). The rig is designed to measure axial pressure distribution, airflow rate, temperature, vibration amplitudes and seal reaction forces. Rotor dynamic coefficients are then calculated from transient measured reaction force and motion data. The test apparatus in the form of a rotor coupled to an electric motor has provision for controlling the eccentricity in the seal. The rotor is suspended in the style of a pendulum from another rigidly mounted shaft at the top, and permits control of the motion of the rotor in the horizontal plane. A cam within the pivot shaft controls the system's vertical position. The equation of motion for the seal stator housing is:

$$\begin{Bmatrix} f_{sx} - M_s d^2X_s/dt^2 \\ f_{sy} - M_s d^2Y_s/dt^2 \end{Bmatrix} = \begin{bmatrix} K & k \\ -k & K \end{bmatrix} \begin{Bmatrix} X_s \\ Y_s \end{Bmatrix} + \begin{bmatrix} C & c \\ -c & C \end{bmatrix} \begin{Bmatrix} dX_s/dt \\ dY_s/dt \end{Bmatrix} \tag{4.24}$$

where M_s is mass of stator housing, f_{sx} and f_{sy} are measured forces on the housing, K and k are the direct and cross-coupled stiffness and C and c are the direct and cross-coupled damping. Acceleration components of the hmousing, d^2X_s/dt^2 and d^2Y_s/dt^2 are measured by accelerometers. The coefficients are described in the frequency domain by using a swept si ne wave in the 40 to 70 Hz range.

Rotor dynamic performance and seal leakage flow comparisons are carried out for the labyrinth seal with pre-swirl and with shunt injection and for the honeycomb seal. Shunt injection pressure ratios are set at 0.85, 0.90 and 0.95. In normal applications this pressure varies between 0.4 and 0.6. Rotor surface speeds corresponding to speed points between 30 and 110 m/s, against a normal operating range of 100 to 170 m/s.

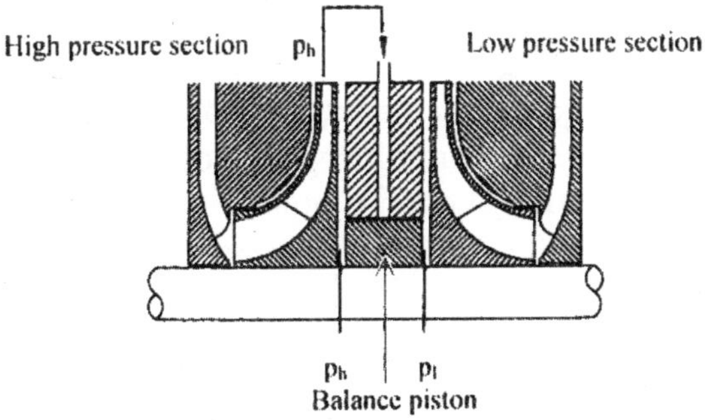

Fig. 4.32. *Shunt Injection at Balance Piston Between High and Low Pressure Sections*

Table 4.2. *Cross-Couple Stiffness and Direct Damping Values*

Shunt Injection Pressure Ratio	*Direction of Injection*	*Cross-Couple Stiffness k, kN/m*	*Direct Damping C, kN-s/m*
0.85	with rotation	–175	0.82
0.85	against rotation	–2025	0.83
0.90	with rotation	–	0.76
0.90	against rotation	–1810	0.77
0.95	with rotation	–	0.69
0.95	against rotation	–1475	0.72

Effective damping, defined by $C_{eff} = C \times (1 - k/C\omega)$, is used to evaluate overall seal stability characteristics. The parameter is useful in comparing annular seals since it combines the effects of direct damping and cross-coupled stiffness, both of which are associated with rotor stability. Since k is responsible for instability, it must have a low or negative value. Measured values for shunt injection at the maximum speed of and constant pressure ratio across the seal are shown in table 4.2. Injection against rotation provides the largest negative values of k, and increasing injection pressure also tends to increase k. Cross-coupled stiffness values due to radial shunt injection, on the other hand, remain practically the same for all cases.

A positive and large magnitude for direct damping in the annular seal improves rotor dynamic response, since it counteracts the destabilizing effects of cross-coupled stiffness in the rotor. Table 4.2 shows the coefficient magnitudes at the maximum speed of 16,500 rpm and constant pressure ratio across the seal of 0.45. The direct damping value increases with injection pressure, and the direction of injection has minimal impact.

The tangential force on a synchronous precessing (or forward whirl) seal is given by $f_\theta = (k - C\omega)A$, where A is the radius of precession. The whirl frequency ratio is expressed by $f_\omega = (k / C\omega)$. For the shunt injection seals with radial injection the whirl frequency is calculated to be negative and small, implying the seal could destabilize the seal into a backward whirl mode. However, backward whirl normally associated with rubbing is rarely experienced in the operation of turbo-machines, as opposed to the prevalent forward whirl instability condition.

Table 4.3. *Comparison of Seal Performance*

Seal/Configuration	*Effective Damping kN-s/m*	*Cross-Coupled Stiffness, kN/m*	*Whirl Frequency Ratio*	*Mass Flow Rate g/s*
Labyrinth/pre-swirl	0.3	20	0.15	110
Labyrinth/radial injection	0.8	–200	–0.20	165
Labyrinth/injection against rotation	1.5	–1500	–1.20	170
Honeycomb	1.6	1700	0.35	80

Table 4.3 shows the comparison for all seals. The honeycomb seal has better effective damping than the labyrinth seal with radial injection, but shunt injection nearly doubles this value without injection. The highest positive cross-coupled stiffness (destabilizing) is obtained from the honeycomb seal, but the labyrinth seal has superior numbers with or without the injection. Whirl frequency ratio is negative for labyrinth seal with shunt injection, and is caused by the

negative cross-coupled stiffness. In terms of leakage, the honeycomb seal performance is superior than that of the labyrinth with or without shunt injection.

4.9 DAMPING SEAL DYNAMIC CHARACTERISTICS

Vibration problems in the turbo-machinery of the space shuttle's main engine have required closer evaluation of rotor dynamic stability margins. Annular seals are used in the machines for leakage control, but they also play a major role in controlling vibration response levels. Prior to entering the seal the fluid has a significant tangential velocity component, in contrast with the axial velocity down the length of the seal. A good example of the pre-swirl condition is the discharge from an impeller entering a seal, and is a significant source of destabilizing forces in the machine.

In an experiment conducted at NASA's Marshall Space Flight Center, test measurements focused on the capacity of annular damping seals to minimize the whirl frequency ratio in the presence of a highly pre-rotated fluid at the inlet (Darden *et al.*, 1999). Dynamic coefficients for smooth seals tend to have a strong dependence on eccentricity over a wide range of speed, pressure and eccentricity ratios (Marquette *et al.*, 1997). Results from the smooth seal are used as a baseline for the investigation of damping seals. Details of the test rig are shown in fig. 4.33. Designed to support testing of seal pressure differential of 2,000 lbs/in^2, the rig is powered by a 250 horsepower steam turbine and has a maximum speed of 20,000 rpm. De-ionized water between 70° and 90° F is used in the experiment. The test bearing is located in the seal carrier assembly at the mid-span of the shaft, and is supported axially by the upper and lower orifice compensated thrust bearings. This allows the test bearing to translate in response to an external load without recourse to additional load paths, while ensuring the seal assembly remains parallel to the shaft during the test. Pre-rotation of the fluid medium is

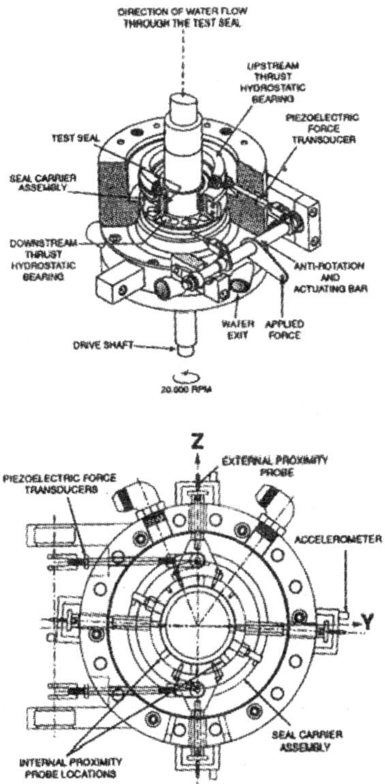

Fig. 4.33. *Seal Test Rig: (upper) Cut-away View, (lower) Top View; (Darden, 1999)*

achieved by using a shaft with 10 equally spaced radial fins placed just upstream of the entrance to the seal. Blade passing frequencies are well outside the frequency range of interest to avoid degradation of the frequency response functions.

The test seals are characterized at a combination of shaft speeds, seal pressure differentials and inlet pre-swirl conditions. The applied force is a band limited random excitation between 20 – 200 Hz. High-frequency data comprising of displacements, forces and pressures is recorded. Duration of each test is limited to 30s. Relative displacement of the fluid film between the test seal and the shaft are measured by 4 inductive probes spaced equally around the periphery of the seal assembly. These measurements are fed back for inertial correction of the excitation force applied to the carrier assembly. Displacements are differentiated in the frequency domain to obtain accelerations. Acceleration levels of rig housing, measured by accelerometers at same locations as proximity probes, must also be taken into account. Seal reaction forces are:

$$-\begin{Bmatrix} f_y \\ f_z \end{Bmatrix} = \begin{bmatrix} K & k \\ -k & K \end{bmatrix}\begin{Bmatrix} \delta y \\ \delta z \end{Bmatrix} + \begin{bmatrix} C & c \\ -c & C \end{bmatrix}\begin{Bmatrix} d\delta y/dt \\ d\delta z/dt \end{Bmatrix} + \begin{bmatrix} M & m \\ -m & M \end{bmatrix}\begin{Bmatrix} d^2\delta y/dt^2 \\ d^2\delta z/dt^2 \end{Bmatrix} \tag{4.25}$$

where M and m are the direct and cross-coupled mass of the seal carrier assembly, f_y and f_z are measured fluid film reaction forces, K and k are the direct and cross-coupled stiffness and C and c are the direct and cross-coupled damping. Based on the symmetry of the centered seal state and flow conditions, all of the coefficients are assumed to be skew-symmetric. The measured parameters in the individual time history are transformed to the frequency domain, and the frequency response functions are curve fitted to yield the rotor dynamic coefficients.

The test matrix is considerably influenced by the high pre-swirl condition, and the stability of the seal is predicted to degrade with increasing shaft speed. Theoretical predictions are obtained from the code developed by Padavala *et al.*, (1993) using Hir's friction factor and inlet pre-swirl values of 0.19 and 1.1. The results of the experimental investigation, with the predicted data, are shown in table 4.4. Non-dimensional values are obtained by multiplying the stiffness coefficients by $[c/(\pi d l \Delta P)]$ and the damping coefficients by $[c\omega_s/(\pi d l \Delta P)]$, where ω_s = shaft speed, c, d and l are seal radial clearance, length and diameter and ΔP = seal pressure differential.

The direct stiffness term is over predicted by the theory by 50 percent for both the high and low pre-swirl conditions, but there is agreement in the overall upward trend with increasing seal pressure differential. The cross-coupled stiffness term is over predicted by the theory by 15 percent for both the high and low pre-swirl conditions, and there is agreement in the upward trend with increasing shaft speed. Static tests were also conducted concurrently by applying a series of loads and recording the resulting displacements. Agreement between the static and dynamic tests provided verification of the lower than expected stiffness values in the operational tests. The damping coefficients under-predict the calculated values by 15 to 25 percent for the low and high pre-swirl cases (Rangwala, 2004).

Table 4.4. *Damping Seal Rotor Dynamic Coefficients*

Test Condition			Stiffness		Damping		Inertia		Leakage
Pre-swirl	Pressure kPa	Speed rpm	K N/m	k N/m	C N-s/m	c N-s/m	M Kg	m Kg	Rate ml/s
0.19	10300	5850	6.82E6	1.81E6	2.01E4	7.55E2	2.19	–0.60	14.8
0.19	10300	10500	7.00E6	3.19E6	2.09E4	8.90E2	2.87	–0.76	14.7
0.19	10300	15100	7.07E6	5.33E6	2.06E4	2.11E3	1.94	–0.32	14.7
0.19	13400	5780	8.85E6	1.98E6	2.31E4	1.02E3	2.61	–0.81	17.2
0.19	13300	10100	9.58E6	3.28E6	2.31E4	1.23E3	4.03	–0.91	17.1
0.19	12900	15800	9.18E6	6.04E6	2.37E4	1.72E3	2.38	–0.60	16.8
1.10	13100	5130	8.32E6	9.23E6	2.39E4	5.41E3	7.44	–3.84	16.7
1.10	12700	6140	8.37E6	1.04E7	2.44E4	6.54E3	7.85	–4.17	16.6
1.10	12300	4020	7.11E6	7.48E6	2.40E4	2.07E3	4.82	–2.24	16.3
1.10	16000	5280	1.01E7	1.10E7	2.73E4	3.57E3	5.62	–3.40	17.4
1.10	15700	6350	1.07E7	1.26E7	2.53E4	7.02E3	8.32	–3.08	17.1
1.10	14400	3980	8.89E6	8.15E6	2.52E4	2.17E3	5.20	–1.87	16.7

The whirl frequency ratio is given by $k/(C\omega_s)$, and is over predicted by 20 percent for the low pre-swirl condition. The average whirl frequency ratio for the high pre-swirl case is 0.73, representing a large increase in the damping seal's stabilizing capacity.

4.10 SQUEEZE FILM DAMPER

Squeeze film dampers are more complex to design and analyze than other machinery elements. They have been sometimes been used in the process industry as a last resort to reduce vibration or to improve rotor dynamic characteristics of a machine. This stigma has often tainted their use as an emergency solution and resulted in rejection even when properly applied, with the exception of the aircraft gas turbine industry where their use is widespread. The influence of cavitation in squeeze film damper bearings that is distinctly different than cavitation in journal bearings has only been recently understood. Problems associated with cavitation, however, are still not amenable to a rigorous quantitative analysis, thus making it essential to rely on experience and knowledge of the application at hand.

In a squeeze film damper the bearing is a large cylinder attached to a fixed frame. The inside diameter has a very wide central oil supply groove. The squeeze film is formed by two lands on each side of the supply groove.

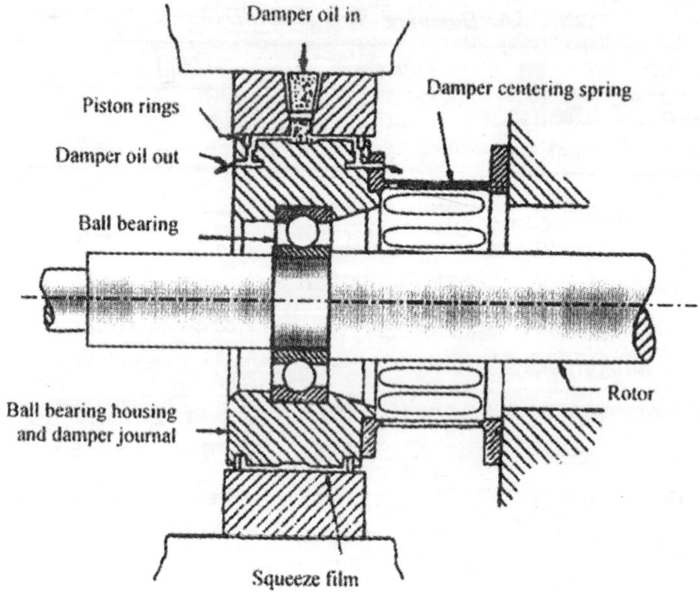

Fig. 4.34. *Squeeze Film Damper in Parallel with Squirrel Cage Spring (Gunter, 1975)*

The journal is the outside diameter of a sleeve that is fitted to the outer race of a ball bearing and supported by coil springs. This configuration has the squeeze film in parallel with the spring supports so that they are subject to the same displacements and share the dynamic load unequally. Most squeeze film dampers in aircraft engines have a similar arrangement with the 'journal' (and roller or ball bearing) arc spring-supported, and with the bearing attached to the engine frame as shown in fig. 4.34 (Gunter *et al.*, 1975). The squeeze film and the spring support are subject to the same deflection, and the dynamic loads are divided unequally between them.

The most commonly recurring problems in rotor dynamics are excessive steady state synchronous vibration levels and sub-synchronous rotor instabilities (Rangwala, 2004). The first problem may be reduced by improved balancing or by introducing modification into the rotor bearing system to move the system critical speeds out of the operating range, or by introducing external damping to limit peak amplitudes at traversed critical speeds. Sub-synchronous rotor instabilities may be avoided by eliminating the instability mechanism or by raising the natural frequency of the rotor-bearing system as high as possible or by introducing damping to raise the onset speed of instability above the operating speed range.

The performance of squeeze film dampers is not only determined by the geometry (length, diameter and radial clearance) and the viscosity of the lubricant used, but affected greatly by a number of specific design and operating conditions. The level of supply pressure, the feeding and discharge flow mechanisms, the type of end seals, fluid inertia and dynamic cavitation are but a few of these important factors.

In the dynamic analysis of rotor bearing systems, squeeze film dampers are regarded as highly nonlinear mechanical elements, providing forces obtained from relationships based on the instantaneous journal center eccentricity. Current analysis of rotor-disk assemblies supported on a squeeze film damper are based on overly simplified analytical expressions for fluid film forces as derived from the short journal bearing model with the so called π-film cavitation assumption. Oftentimes a jump is encountered in calculating the response of the system. The jump phenomenon finds its roots in the stiffness theory of nonlinear oscillations. In particular, systems with a cubic non-linearity exhibit a response that is characterized by multiple values. It is the multi-values of the response curve that leads to the jump for a hardening spring. As the frequency of excitation is varied, the harmonic response will increase. A jump down in amplitude occurs when this characteristic takes place in a squeeze film damper due to non-linearity of the cross-coupled damping (stiffness term). This nonlinear response (jump down while accelerating through resonance) may not be experienced in field installations. However, this hardening effect is not due to the squeeze film stiffness but due to journal lockup that produces a moment stiffness, similar to what happens with gear coupling when they lockup (Zeidan *et al.*, 1998).

Squeeze film dampers operating with or without a centering spring and operating with a high unbalance may experience a bilinear spring effect. The higher stiffness results from the damper journal bottoming out and contacting the higher stiffness housing. This bilinear spring characteristic may lead to nonlinear response often characterized by excitation of the first critical speed or resonance of the structure. Sub-harmonic response at an exact fractional speed may also be exhibited under these conditions. At certain speeds the damper journal will impact the housing and excite the first critical speed, while at other instances, depending on the ratio of the running speed to the first critical, may result in sub-harmonic vibrations at 1/2, 1/3, 1/4 or 1/5, 2/5, 3/5 or 4/5 of shaft speed. In addition to the sub-harmonic activity, super-harmonic vibrations at exact fractions are also evident due to this bilinear spring effect.

Aircraft engine designers around 1970 had a strong predilection to specify the radial clearance as 3.0 mil regardless of the rotor size, speed and mass. Some designers also believed that the squeeze film clearance would subtract from the available tip clearance around compressor blades. Experiments and analysis have since shown that the radial squeeze film clearance must be at least 2.3 times the local imbalance in order for the damper to be effective, and that the clearance is a strong factor in determining the damping coefficient, which has an optimum value for each different application. Also, the damper often makes more clearance available for compressor blade tips, especially at critical speeds. It is true, however, that most squeeze film damper designs are severely constrained in their size and shape by the available space around the bearing, so most designs will continue to be strongly influenced by past experience with similar machines. Some general guidelines for the design are listed below:

- A specific rotor dynamic phenomenon to be suppressed by the squeeze film damper must be kept in mind, such as the response to unbalance at critical speeds or sub-synchronous instability. Conduct rotor dynamic computations to determine the optimum damping coefficients required to suppress the selected phenomena. At this point, it will sometimes be discovered that no amount of damping at the bearings has a significant effect on the selected phenomena because the bearings are located near the nodes of the relevant mode shapes. For aircraft engine applications, such a discovery makes the remaining procedure irrelevant unless the mode shapes can be changed by rotor modifications or by softening the squirrel cage. For industrial compressor applications the squeeze film and O-ring (or centering spring) flexibility will be in series with the tilt pad bearing, so both the resultant stiffness and damping will be less than the coefficients of the tilt pad bearing alone. The lowered resultant stiffness can move the nodes away from the bearing and make the available damping effective.

- Decide which geometry, lubricant feed mechanism and end seals best fit the available space and configuration of the machine. An annular feed groove provides the option of making it wider at a later date if the damping (or the effective stiffness) turns out to be too large. This is especially important in industrial compressor applications where the damper is in series with the oil film bearing. Also, piston ring end seals usually leak more than may be expected. O-ring seals generally have low leakage.
- Obtain access to equations or computer codes which are applicable to the configuration, and use them to calculate the force coefficients of the squeeze film at the whirl frequencies of interest. The calculated coefficients will probably have a large error unless the damper has no end seals, narrow land (short bearing), and high supply pressure of the same order of magnitude as the peak pressure in the squeeze film. The coefficients for this case can be calculated from the full film version. For an aircraft engine application these are the coefficients to use after adding the squirrel cage stiffness in the rotor dynamic computer simulation to predict critical speed response or instability. For circular centered orbits, the coefficients are equal in the two directions.
- For industrial compressors the squeeze film coefficients must be combined with the O-ring stiffness (they are in parallel, so they add) and then these coefficients must be combined with the tilt pad oil film coefficients (in series). The most important factor for success in these cases is that the O-ring stiffness should not be too high. The resultant damping will always be less than the damping which is already in the tilt pad bearing, so the resultant stiffness must be reduced to allow the available damping to work. Elastomeric materials have a dynamic stiffness that varies with frequency. The frequency dependence information can usually be obtained from the material supplier. Very high O-ring stiffness renders the squeeze film damper practically useless. In such cases the O-ring should be supplemented by additional centering devices, which retain relatively low stiffness characteristics under heavy loads.

Squeeze film dampers without a centering spring are the most common in use for turbomachines. The outer race of the rolling element bearing is allowed to float and whirl in the clearance space between the bearing outer diameter and the housing inner diameter. The race than forms the damper journal that is permitted to whirl, but is prevented from spinning by an anti-rotation device.

The simplest means of providing a centering spring in a squeeze film damper is through the use of elastomer O-rings. An illustration of this damper design is shown in fig. 4.35. The advantages of this design stem from its simplicity, ease of manufacturing and its ability to incorporate in a small envelope. The O-ring doubles as a good end seal, which helps increase the effectiveness of the damper by reducing side leakage. Some of the disadvantages with

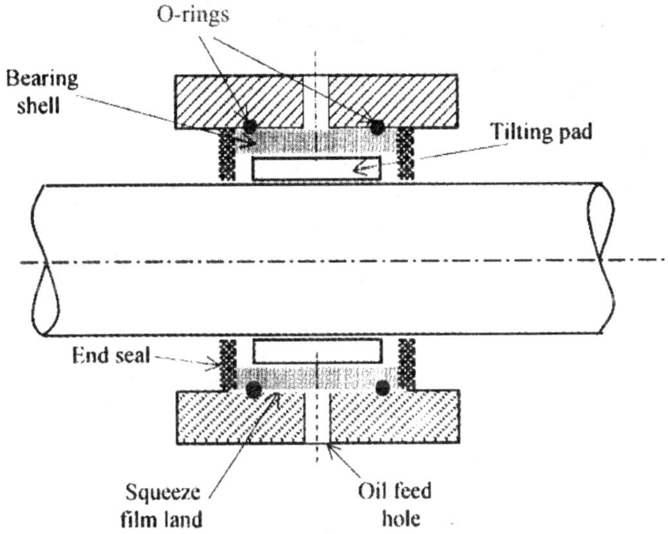

Fig. 4.35. *Schematic Diagram of O-Ring Supported Damper*

this design are attributed to the limited range of stiffness that can be achieved with elastomers. Predicting the stiffness with a good degree of certainty is difficult in elastomeric materials due to material variance and the influence of temperature, frequency and time on its properties. The O-ring design is also susceptible to creep, causing the damper to bottom out and lead to a bilinear spring behavior. O-ring dampers are not capable of taking thrust loads, and cannot be easily manipulated for centering of the damper journal within the damper clearance space. One means of achieving some centering capability is by making the O-ring groove eccentric. This limitation makes them only suitable for lightweight rotors.

The squeeze film damper design is the most commonly used, particularly in aircraft engines where its use is widespread. Most large aircraft gas turbine engines employ at least one, and in many instances, two or three of these dampers in one engine. A schematic of this damper is shown in fig. 4.36. A distinctive feature necessary with such a design (and apparent from the schematic) is the relatively large axial space required in comparison to the damper length. This is one of the major drawbacks of this damper design. The squirrel cage forming the centering spring for the damper quite often requires three to four times as much axial space as the damper itself. Assembling the squirrel cage spring and centering the journal within the clearance space requires special tools and skills. The squirrel cage spring also complicates the damper end seal design and assembly. It is also difficult to offset the spring assembly in order to account for the gravity load due to the shaft weight. Maintaining parallelism between the damper journal and housing is another factor that adds uncertainty and complications to this design.

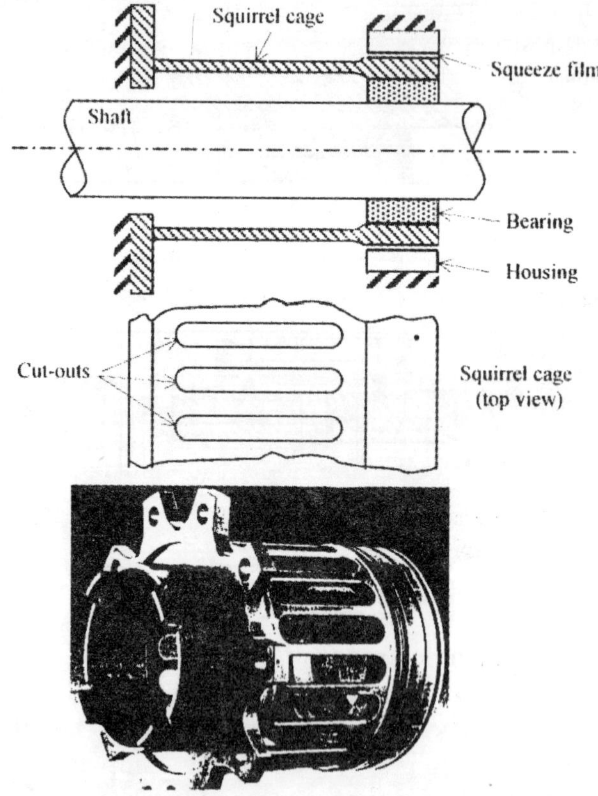

Fig. 4.36. *Squirrel Cage Supported Damper Details*

An integral centering spring squeeze film damper has cantilevered support ribs, and along with the sector, they are supported at both ends to form a centering spring element. The small gap between the sector and the outer ring forms the squeeze film damper clearance space. The complete assembly may contain any number of sectors depending on the load and required stiffness and damping for the particular application. The sector in the lower half of the damper carrying the rotor weight can also be machined with an offset to counter the deflections caused by the rotor weight. More than one stiffness range can be achieved by using an additional pivot located at a specified distance from the support rib. This can provide a more gradual non-linearity that is desirable in case of excessive unbalance or side force. Such a configuration will help in absorbing impact loads, high side loads and high vibration excursions when traversing the critical speeds or in the event of a blade loss. The circumferential location of the stop pivot and the gap between the pivot end and the support rib are two additional design variables that provide additional flexibility.

4.11 EXAMPLE PROBLEMS

Example Problem # 4.1: A state of tri-axial stress occurs when two bodies with curved surfaces are compressed against each other. This causes a point or a line contact to change to an area contact, causing the state of stress in both bodies to change to a three-dimensional one.

Such situations arise in the contact of mating gears, a wheel against a rail and in rolling element bearings. Computing the stress levels resulting from loading one body against another may prevent surface failures.

Consider two solid spheres of diameters d_1 and d_2 with a contacting force F (fig. 4.37). A circular contact area of radius a is obtained. If the elastic constants (Young's modulus and Poisson's ratio) of the two spheres are E_1, ν_1, E_2, and ν_2, the contact circle radius a is given by the following expression:

$$a = \sqrt[3]{\frac{3F[(1-\nu_1^2)/E_1]+[(1-\nu_2^2)/E_2]}{8[(1/d_1)+(1/d_2)]}} \qquad (4.26)$$

According to Lubkin (1962), stresses in the x, y and z directions are principal stresses, and are defined by the maximum pressure at the center of the contact area as defined by the expression:

$$p_{\max} = 3F/(2\pi a^2) \qquad (4.27)$$

For an element on the z-axis the stresses are:

$$\sigma_x = \sigma_y = p_{\max}\left[(1+\nu)\left(\frac{z}{a}\mathrm{Tan}^{-1}\frac{a}{z}-1\right)+\frac{1}{2}\frac{a^2}{a^2+z^2}\right] \qquad (4.28)$$

$$\sigma_z = -p_{\max}\left(\frac{a^2}{a^2+z^2}\right) \qquad (4.29)$$

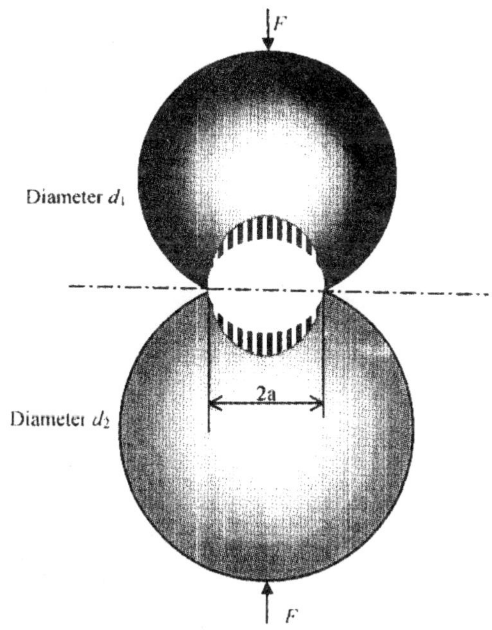

Fig. 4.37. *Pressure Distribution Between Two Spheres*

The equations given above also apply to the contact of a sphere and a plane surface or to a sphere and an internal spherical surface. For a plane surface use $d = \infty$. For an interna' surface the diameter is expressed as a negative quantity.

A ball bearing's inner raceway may be considered to be the segment of a sphere. The rolling elements have a mating surface. Hence, the two contacting surfaces may be idealized as spheres loaded against each other. If the raceway diameter = 2.25 in., ball diameter = 0.775 in. and contact force F = 15 lbs, determine the contact area and stress levels. Use Young's modulus $E_1 = E_2 = 30 \times 10^6$ lb/in² and Poisson's ratio $\nu_1 = \nu_2 = 0.3$ for both bodies.

Solution: The pressure within each sphere has a semi-elliptical distribution, as shown in fig. 4.37. Maximum pressure occurs at the center of the contact area. For an internal surface the diameter is negative. Then the contact area and maximum pressure are calculated to be:

$$a = \{(3 \times 15/8) \times [(1-0.3^2)/30 \times 10^6 + (1-0.3^2)/30 \times 10^6]/(-1/2.25 + 1/0.775)\}^{1/3}$$

$$= 0.007425 \text{ in.}$$

$$p_{max} = 3 \times 15/(2 \times \pi \times 0.00739^2) = 129895 \text{ lbs/in}^2.$$

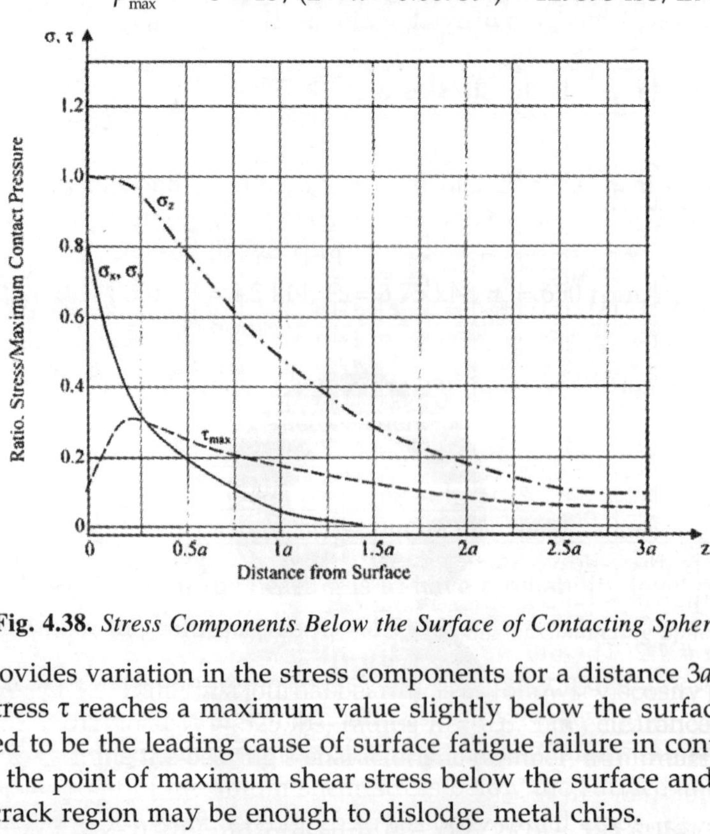

Fig. 4.38. *Stress Components Below the Surface of Contacting Spheres*

Figure 4.38 provides variation in the stress components for a distance $3a$ below the surface. Note that shear stress τ reaches a maximum value slightly below the surface. Maximum shear stress is considered to be the leading cause of surface fatigue failure in contacting elements. A crack initiating at the point of maximum shear stress below the surface and lubricant pressure flowing into the crack region may be enough to dislodge metal chips.

When the contacting surfaces are cylindrical the contact area is a narrow rectangle of half width b, given by the equation

$$b = \sqrt{\frac{2F[1-\nu_1^2/E_1]+[(1-\nu_2^2)/E_2]}{\pi l[(1/d)+(1/d_2)]}} \tag{4.30}$$

where l is the length of the contact area. The pressure distribution across the width $2b$ is elliptical, and maximum pressure is given by $p_{max} = 2F/\pi bl$.

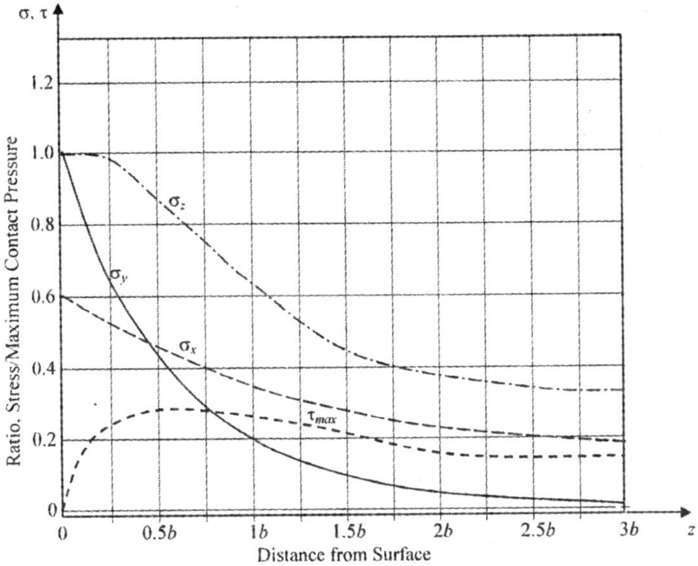

Fig. 4.39. *Stress Components Below the Surface of Contacting Cylinders*

When applied to a cylinder and a plane surface, for the plane use $d = \infty$. For an internal cylinder d is negative. To evaluate stresses, select the origin of a reference system at the center of the contact area with x parallel to the axes of the cylinders, y perpendicular to the plane formed by the two cylinder axes and z in the plane of the contact force. For elements on the z-axis principal stresses σ_x, σ_y and σ_z exist. Figure 4.39 shows a plot of the stresses for depths up to $3b$ below the contact surface. For the contacting spheres three different shear stresses are created, given by:

$$\tau_{xz} = \tau_{yz} = \frac{\sigma_x - \sigma_x -}{2} = \frac{\sigma_y - \sigma_z}{2} \tag{4.31}$$

This shear stress is also shown in fig. 4.49, labeled τ_{max}. It reaches a peak value slightly below the surface, similar to what is seen in the case of the contacting spheres.

Example Problem # 4.2: The life of a bearing is defined as the total number of revolutions, or the number of hours at a fixed speed, of bearing operation required for the failure criteria to develop. Under ideal conditions the first evidence of fatigue failure will consist of a spalling of the load carrying surfaces. The Timken Company goes by the failure criterion of pitting or spalling of an area of 0.01 in^2. In testing a group of bearings the objective is to establish the median and the L_{10}, or rated, life. The L_{10} implies that 90 percent of identical bearings operating at a constant speed and load will complete or exceed the test before the failure criterion develops. When a number of batches of bearings are under test, the median life is usually between four and five times the L_{10} life.

The concept of probable survival of a batch of bearings also needs examination. If a machine uses N bearings with each bearing having the same reliability R, then the reliability of all the bearings is $(R)^N$. The distribution of bearing failures can be approximated by the Weibull procedure. By making adjustments to the Weibull parameters (Mischke, 1965), the distribution of bearing failures takes the form:

$$R = exp\left[-\left(\frac{L}{6.84L_{10}}\right)^{1.17}\right] \tag{4.32}$$

If a certain application requires a reliability of 98 percent for the bearing to last for 2500 hours, determine the rated life.

Solution: Using $L = 2500$ and $R = 0.98$ in the above equation

$$0.98 = \exp\left[-\left(\frac{2500}{6.84L_{10}}\right)^{1.17}\right]$$

Then the rated bearing life L_{10} is calculated to be 10262 hours.

Example Problem # 4.3: Experiments indicate that identical bearings acting under different radial load F_1 and F_2 and operating at speeds n_1 and n_2 have lives L_1 and L_2 according to the relation

$$F_2 = F_1\left[\frac{n_1L_1}{n_2L_2}\right]^{1/a} \tag{4.33}$$

where $a = 3$ for ball bearings and 10/3 for roller bearings. A roller bearing can safely accept a load of 4.5 kN at 650 rpm for an L_{10} life of 1400 hours. Determine its life if the load is reduced to 3.75 kN and the speed is increased to 725 rpm.

Solution: In equation (4.33), F_1 = 4.5 kN, F_2 = 3.75 kN, n_1 = 650 rpm, n_2 = 725 rpm, L_1 = 1400 hours and a = 10/3. Then L_2 = 2305 hours.

Example Problem # 4.4: Equations (4.32) and (4.33) may be combined to obtain a specified levels of load, speed, life and reliability factor. Then:

$$F_2 = F_1\left[\frac{n_1L_1}{n_2L_2(6.84)}\right]^{1/a}\frac{1}{[\ln(1/R)]^{1/1.17a}} \tag{4.34}$$

Determine the load rating F_2 if the bearing is to have a reliability level of 95 percent.

Solution: Substituting the values gives F_2 = 4.51 kN.

Example Problem # 4.5: Data for a journal bearing is as follows: viscosity $\mu = 3.95 \times 10^{-6}$ reyn, speed N = 1800 rpm, radial load W = 525 lbs, radius R = 0.875 in., clearance c = 0.0014 in. and length l = 1.625 in. Determine the bearing's characteristic number, minimum film thickness and its angular location.

Solution: Bearing l/d = 0.929 and pressure $P = (W/2Rl)$ = 184.6 lbs/in^2. Equation (4.9) provides the bearing Sommerfeld, or characteristic, number: $S = (.875/.0014)^2 \times \{3.95 \times 10^{-6} \times 1800/(184.6 \times 60)\} = 0.251$. From charts for minimum film thickness and eccentricity ratio (Raimondi *et al.*, 1958), the ratio of minimum film thickness and clearance h_o/c = 0.54 and eccentricity ratio ε = 0.46, and since clearance c = 0.0014, then minimum film thickness h_o = 0.00076 in. and eccentricity e = 0.00064 in. For no load, eccentricity e = 0.0 and film thickness h = 0.0014. As the load is increased the journal is forced downward. Figure 4.40 shows the distribution of hydrodynamic pressure in the lubricant film.

***Example Problem # 4.6*:** In example problem # 4.5 determine the coefficient of friction, lubricant flow, film pressure and temperature rise.

***Solution*:** From the chart for coefficient of friction (Raimondi *et al.*, 1958), the value is $(R/c)f = 5.2$. Hence, the coefficient $f = 0.00832$. The torque required to overcome this frictional loss is $T = fWR = 3.82$ in-lbs, which is equivalent to HP = $TN/63000 = 0.1092$ hp. From the chart for lubricant flow, $60Q/RcNl = 4.08$ and $Q = 0.244$ in^3/s. Leakage at the two ends of the bearing is obtained from the charts, where $Q_s/Q = 0.54$, so $Q_s = 0.132$ in^3/s.

Heat generated by friction is dissipated by conduction, convection and radiation, and also carried away by the oil flow. A conservative assumption calls for all the heat to be extracted by the oil. The temperature rise in degrees Fahrenheit is given by the expression:

$$\begin{aligned}\Delta T_F &= \{0.103 \times P \times (Rf/c)\}/\{[1 - ½(Q_s/Q)][60Q/RcNl]\} \\ &= 0.103 \times 184.6 \times 5.2/[(1 - 0.54/2) \times 4.08] \\ &= 33.2° \text{ F.}\end{aligned}$$

The maximum pressure developed in the film is obtained from the charts in the form of pressure ratio $P/P_{max} = 0.47$. Since $P = 184.6$, then $P_{max} = 392.8$ lbs/in^2.

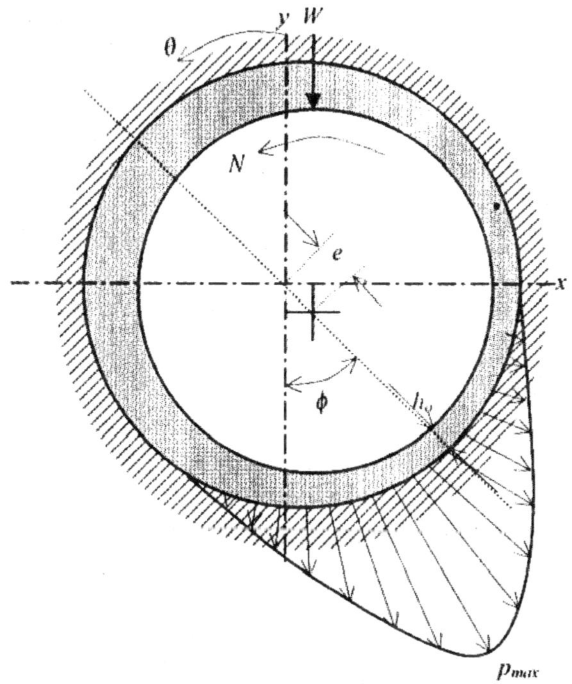

Fig. 4.40. *Hydrodynamic Pressure Distribution in Fluid Film Journal Bearing (Raimondi, 1958)*

REFERENCES & BIBLIOGRAPHY

"AFBMA Standards for Ball and Roller Bearings", Revision # 4, June, 1972.

Allaire, P.E., Li, D.F., Choy, K.C., "Transient Unbalance Response of Four Multi-Lobe Journal Bearings", Journal of Lubrication Technology, ASME, New York, 1980.

Allison Gas Turbine Division, TM # 55-2840-231-23, 1981.

American Petroleum Institute, "Centrifugal Compressors for Petroleum, Chemicals and Gas Service Industries", API Standard, 617, Sixth Edition, 1995.

ASME Report – "Pressure/Viscosity in Rolling Element Bearings", Volume II, New York, 1954.

Bently, D.E., "Monitoring Rolling Element Bearings", Orbit 3, No. 3, Pages 2–15, 1982.

Boto, P.A., "Detection of Bearing Damage by Shock Pulse Measurement", Ball Bearing Journal, SKF, 1971.

Childs, D., Kleynhans, G., "Experimental Rotor Dynamic and Leakage Results for Short (L/D = 1/6) Honeycomb and Smooth Annular Pressure Seals", Proceedings, Institute of Mechanical Engineers, 5th International Conference On Vibrations in Rotating Machinery, London, 1992.

Darden, J.M., Earhart, E.M., Flowers, G.T., "Comparison of Dynamic Characteristics of Smooth Annular Seals and Damping Seals", ASME paper # 99-GT-177, New York, 1999.

Ehrich, F.F., "Handbook of Rotor Dynamics", Krieger Publishing Co., Malabar, FL, 1999.

Ehrich, F.F., "The Influence of Trapped Fluids on High Speed Rotor Vibrations", Journal of Engineering for Industry, Volume 89(4), Pages 806–812, 1967.

Forster, N.H., "High Temperature Lubrication of Rolling Contacts with Lubricants Delivered from the Vapor Phase and as Oil Mists", Ph. D. Thesis, University of Dayton, 1996.

Friswell, M.I., Penny, J.E.T., "The Choice of Orthogonal Polynomials in the Rational Fraction Polynomial Method", International Journal of Analytical Experimental Modal Analysis, Volume 8, Number 3, Pages 257–262, 1993.

Gunter, E.J., Barrett, L.E., Allaire, P.E., "Design and Application of Squeeze Film Dampers for Turbo-Machinery Stabilization", Proceedings of the 4th Turbo-Machinery Symposium, Texas A & M University, College Station, TX, Pages 127–141, 1975.

Hagg, A.C., Sankey, G.O., "Some Dynamic Properties of Oil Film Journal Bearings with Reference to Unbalance Vibration of Rotors", Journal of Applied Mechanics, Volume 23(2), Pages 302 – 306, New York, 1956.

Harris, T.A., "Rolling Bearing Analysis", John Wiley & Sons, New York, 1984.

Hertz, H., "The Contact of Elastic Solids", J. Reine Angew, Math. 92, Pages 156–171, 1881.

Jones, A.B. "Analysis of Stress and Deflections", New Departure Engineering Data, Bristol, CT, 1946.

Holmes, R., Box, S., "On the Use of Squeeze Film Dampers in Rotor Support Structures", Machine Vibration, Volume 1, Pages 71–92, 1992.

Kirk, R.G., "Oil Seal Dynamics: Considerations for Analysis of Centrifugal Compressors", Proceedings, 15th Turbo-machinery Symposium, Texas A & M University, College Station, Texas, 1986.

Kirk, R.G., "A Method for Calculating Labyrinth Seal Inlet Swirl Velocity", Rotating Machinery Dynamics, Volume 2, Pages 345 – 350, ASME, New York, 1987.

Lees, A.W., Friswell, M.I., Smart, M.G., Prells, U., "The Identification of Foundation Vibration Parameters from Running Machine Data", Proceedings, ISROMAC-7, 7th International Symposium on Transport Phenomena and Dynamics of Rotating Machinery, Honolulu, Pages 715–724, SME, Bethel, CT, 1998.

Liao, N.T., Lin J.F., "A New Method for the Analysis of Deformation and Load in a Ball Bearing with Variable Contact Angle", ASME Journal of Mechanical Design, New York, 1999.

Lubkin, J.L., "Contact Problems" in Handbook of Engineering Mechanics, W. Flugge (editor), Section 42-1, McGraw Hill, New York, 1962.

Lund, J.W., "Spring and Damping Coefficients for the Tilting Pad Journal Bearing", Transactions, American Society of Lubrication Engineers, Volume 7(4), Pages 342–352, 1964.

Lund, J.W., Thomsen, K.K., "A Calculation Method and Data for the Dynamic Coefficients of Oil Lubricated Journal Bearings", Topics in Fluid Film Bearing and Rotor Bearing System Design and Optimization, ASME, Pages 1–28, New York, 1978.

Marquette, O.R., Childs, D.W., San Andres, L., "Eccentricity Effects on Rotor Dynamic Coefficients of Plain Annular Seals: Theory versus Experiment", ASME, Journal of Tribology, Volume 119, Pages 443–448, New York, 1997.

Mathew, J., Alfredson, R.J., "The Condition Monitoring of Rolling Element Bearings Using Vibration Analysis", ASME Journal of Vibration and Acoustics, Pages 447–453, New York, 1984.

Mischke, C., "Bearing Reliability and Capacity", Machine Design, Volume 37, No. 22, Pages 139 – 140, September, 1965.

Orcutt, F.K., "The Steady State and Dynamic Characteristics of the Tilting Pad Journal Bearing in Laminar and Turbulent Flow Regimes", Journal of Lubrication Technology, Volume 89(3), Pages 392–404, 1967.

Padavala, S., Palazzolo, A.B., Vallely, D.P., Ryan, S.G., "Application of an Improved Nelson-Nguyen Analysis to Eccentric Arbitrary Profile Liquid Annular Seals", Workshop on Rotor Dynamic Instability Problems in High Performance Turbo-machinery, Texas A & M University, Pages 113–115, 1993.

Pinkus, O., Sternlicht, B., "Theory of Hydrodynamic Lubrication", Mc-Graw Hill, New York, 1961.

Raimondi, A.A., Boyd, J., "A Solution for the Finite Journal Bearing and Its Application to Design and Analysis, Parts I, II and III", Transactions of ASLE, Volume 1, Number 1, Pages 159 – 209 in Lubrication Science and Technology, Pergamon, New York, 1958.

Rangwala, A.S., "Turbo-machinery Dynamics: Design and Operation", McGraw-Hill, New York, 2004.

Redmond, I., "Rotor Dynamic Modeling Utilizing Dynamic Support Data Obtained from Field Impact Tests", Proceedings, 6th International Conference on Vibrations in Rotating Machinery, Paper # C500/055/96, Oxford, 1996.

Santhanan, C.K., Koerner, J., "Transfer Function Synthesis as a Ratio of Two Complex Polynomials", IEEE, Transactions Automatic Control, SME, Pages 56–68, Bethel, CT, 1963.

Santiago, O., San Andres, L., Oliver, J., "Imbalance Response of a Rotor Supported on Open End Integral Squeeze Film Dampers", ASME paper # 98-GT-6, New York, 1998.

Sawyer, T., "Gas Turbines – Volumes I, II and III", International Gas Turbine Institute, Atlanta, ASME, 1982.

Sedy, J., "Improved Performance of Film-Riding Gas Seals Through Enhancement of Hydrodynamic Effects", ASLE Transactions, Volume 23 (1), Pages 35–44, 1979.

Shigley, J. E., Mitchell, L. D., "Mechanical Engineering Design", 4th edition, Mc-Graw Hill, New York, 1983.

Soto, E.A., Childs, D.W., "Experimental Rotor Dynamic Coefficient Results for (a) a Labyrinth Seal With and Without Shunt Injection and (b) a Honeycomb Seal", ASME paper # 98-GT-8, New York, 1998.

Spakovszky, Z.S., Paduano, J.D., Larsonneur, R., Traxler, A., Bright, M. M., "Tip Clearance Actuation with Magnetic Bearings for High-Speed Compressor Stall Control", ASME paper # 2000-GT-528, New York, 2000.

Stephenson, R.W., Rouch, K.E., "Generating Matrices of the Foundation Structure of a Rotor System from Test Data", Journal of Sound and Vibrations, 154, Number 3, Pages 467–484, 1992.

Stribeck, R., "Ball Bearing for Various Loads", ASME, 29, Pages 420–463, New York, 1907.

Vazquez, J.A., Barrett, L.E., Flack, R.D., "Flexible Bearing Supports Using Experimental Data", ASME paper # 00-GT-404, New York, 2000.

Wilcox, D.F., Booser, E.R., "Bearing Design and Application", Mc-Graw-Hill, New York, 1957.

Yu, J.J., Bently, D.E., Goldman, P., Dayton, K.P., Van Slyke, B.G., "Rolling Element Bearing Defect Detection and Diagnostics Using Displacement Transducers", ASME paper # 01-GT-028, New York, 2001.

Zeidan, F.Y., San Andres, L., Vance, J.M., "Design and Application of Squeeze Film Dampers in Rotating Machinery", Proceedings of the 25th Turbo-machinery Symposium, Texas A & M University, College Station, TX, 1998.

Part II
Structural and Operational Analysis of Engine System

CHAPTER 5

Shaft Torsional Vibration

5.1 INTRODUCTION

Torsional vibration is an oscillatory angular motion causing twisting in the shaft of a rotating system; the oscillatory motion is superimposed on the steady rotational motion of the shaft. Even though the vibration cannot be detected without special measuring equipment, its amplitude can be destructive. Motion by itself is rarely a concern with torsional vibration unless it affects the operating function of the system. It is stresses that affect the structural integrity and life of the components, and thus determine the allowable magnitude of the torsional vibration. Torsional vibratory motions can produce stress reversals that cause metal fatigue. Components tolerate less stress reversals than steady stress. In addition, stress concentration factors associated with machine components decrease the effectiveness of load bearing materials.

A classic example of this situation is an electric generator coupled to a power generation turbine, with the generator electrically connected to the regional power grid. The turbine and generator set operates at 3,600 rpm for 60 Hz electrical power (or 3,000 rpm for 50 Hz electricity). If the torsional natural frequency of vibration of the turbine-generator rotor train lies in the vicinity of the electrical power frequency, substantial disturbances are created in the combined electrical power for the regional grid due to the oscillatory torsional motion in the rotating system. Thus, even though the elastic periodic motion in the shaft is purely mechanical in nature, it electrically upsets the power distribution system. In addition, the second harmonic 120 Hz (or 100 Hz for a unit operating at 3,000 rpm) has been recognized as a major source of disruption. Note that the power generating turbine and the generator also will suffer consequences in the form of considerable oscillating torsion loads at the critical frequency as the regional power grid tends to impose its own electrical power characteristics on the generator.

In modeling the system for torsional analysis, customary practice is to calculate the natural frequencies of vibration of the rotor as a body free in space, not connected to the supporting structure. For rotating machinery this assumption is a very good approximation. While it makes for simple and accurate analysis, it implies that such systems have a level of damping that is extremely low and essentially limited to internal material damping of the shaft. The torsional damping constant of the bearings – the proportionality constant between bearing torque and rotational speed – is negligibly small for both rolling element and journal bearings. Thus, unlike lateral vibrations of a rotor system where significant damping is derived from the bearings and support structure, the simple torsional system is one with very little inherent damping.

Large amplitudes of torsional vibration can be present in the rotor of a machine with little or no vibration induced in the stationary structure. If a harmonic exciting torque is present in the system, either in the driving element or in the reactive torque of the driven element, then the possibility of a destructive failure of the shafting elements of the rotating system due to torsional fatigue exists. Systems with gearing tooth contact force between mating gear elements impose lateral forces on the bearings and moment reactions on the gear case. If the system is in a state of torsional vibration, then cyclic force reactions are superimposed on these steady forces. Some vibration can be induced in the stationary structure by the cyclic forces, but in general, it is masked by the background machinery noise. However, if torsional vibration amplitude is large enough to unload the gear teeth, then clattering noise ensues as a result of the backlash in the gearing, providing a warning of the presence of torsional vibrations. This large amplitude vibration, if allowed to continue, can damage tooth contact surfaces. Under such circumstances the cyclic component of the load on the bearings provides some damping as a result of squeeze film action.

When the frequency of any harmonic component of the exciting torque is equal to one of the natural torsional frequencies of the rotor, then a condition of resonance exists. The speed at which this coincidence of frequencies occurs is called a 'critical speed'. For each natural frequency, a series of critical speeds correspond to the various harmonics of the exciting torque. It is important that a dynamic analysis of the system be made and that the forced response of the system be evaluated for these critical speeds in or near the operating range of speed. High levels of response in the operating range of speed should be avoided if at all possible; otherwise special damping devices are required to limit the resonant response.

Gear sets used to alter speeds of power transmission systems transmit the vibration to the casing. Because of their construction features reciprocating machines are subjected to unbalance arising from the reciprocating components and from the rotational motion of the crankshaft. Thus, slider crank mechanisms in engines, compressors and pumps convert torque to radial force that is discernible to human perception but is not measurable due to the insensitivity of the test equipment and background noise. If gearboxes or reciprocating machines are part of a drive train, excess noise and vibration can indicate trouble. Standards and measurement methods dealing with acceptable magnitudes are required to ensure trouble free operation of such a system.

Initially considering a simple two-inertia system connected by a shaft with torsional flexibility will highlight the basic concepts in torsional analysis. Traditional calculation procedures as they relate to reciprocating machinery will next be reviewed, each system characterized by a different source of torsional excitation.

5.2 SIMPLE TWO-INERTIA SYSTEM

The simple two-inertia system is a convenient staring point for making estimates, particularly where the driving and driven elements of a machinery system may each be approximately represented by a lumped inertia. Figure 5.1 shows two disks with polar mass moments of inertia J_1 and J_2 connected by a mass-less shaft with torsional stiffness K_{12}. Angular rotations of the disks are denoted by angles θ_1 and θ_2. A sinusoidal exciting torque equal to $T_1 \text{Sin}(\omega t)$ is applied to disk 1. The exact solution will be based on the governing differential equations of vibration (Ehrich, 1992). Three different conditions of damping may be considered

1. A viscous damping torque proportional to the rate of change of the angle of twist in the shaft connecting the disks.
2. A viscous damping torque proportional to the angular velocity of disk 1.
3. A viscous damping torque proportional to the angular velocity of disk 2.

With the damping torque equal to the viscous damping coefficient B_{12} times the difference in angular velocities of the two disks ($d\psi/dt = d\theta_1/dt - d\theta_2/dt$), a steady state solution yields a relationship for the amplification factor of the angle of twist. In terms of undamped natural frequency

$$\omega_n = \sqrt{\frac{K_{12}}{J_1 J_2 /(J_1 + J_2)}} \tag{5.1}$$

and the damping ratio $$\xi = \frac{B_{12}}{2\omega_n J_1 J_2 /(J_1 + J_2)} \tag{5.2}$$

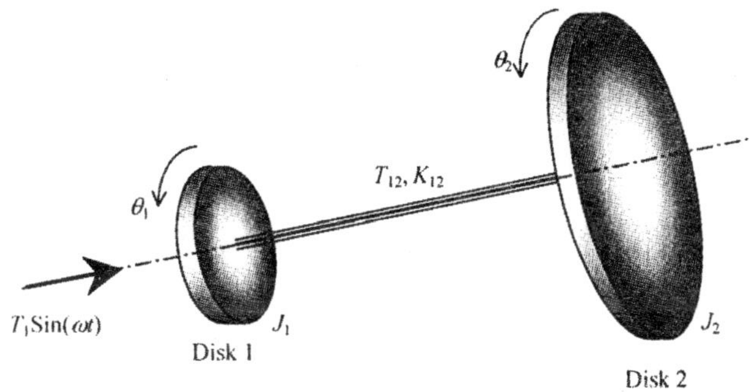

Fig. 5.1. *A Simple Two-Inertia System in Torsion*

The amplitude of twist angle is given by:

$$\frac{\psi}{\psi_{st}} = \frac{1}{\sqrt{\{[1-(\omega/\omega_n)^2]^2 + [2\xi(\omega/\omega_n)]^2\}}} \tag{5.3}$$

where ω is the exciting frequency and ψ_{st} is the shaft static angle of twist due to an exciting torque T_1 applied statically to disk 1. The amplitude of torque T_{12} in the connecting shaft at resonance is:

$$|T_{12}| = \frac{T_1[J_2/(J_1+J_2)]\pi}{\delta} \tag{5.4}$$

where the damping ratio is a single function of δ (the logarithmic decrement) and $\xi = \delta/2\pi$. A representative value for the logarithmic decrement of a rotor steel might be 0.01, giving a value of the damping ratio equal to 0.0016 and an extremely high amplification factor of 314 at resonance. This damping is too low to be of practical significance. It may be necessary to add special damping provisions in the machinery if a condition of torsional resonance exists.

For the case in which the damper acts on the same disk where the sinusoidal exciting torque is applied, the viscous damping torque on disk 1 is equal to $B_1(d\theta_1/dt)$. Here, the simplicity of the solution for damping condition 1 is not possible. The general solution for angular displacement at each disk must be derived separately from the solution of the differential equations. A more manageable perspective can be obtained by examination of the response at resonance ($\omega/\omega_n = 1$), which will yield

$$\theta_1 = -\frac{T_1}{B_1\omega_n}\cos(\omega_n t) \tag{5.5}$$

$$\theta_2 = \frac{T_1}{B_1\omega_n}\frac{J_1}{J_2}\cos(\omega_n t) \tag{5.6}$$

shaft torque is then:

$$|T_{12}| = \frac{T_1}{B_1}\left[\frac{K_{12}(J_1+J_2)J_1}{J_2}\right]^{1/2} \tag{5.7}$$

For the case in which the damper acts on the disk remote from where the sinusoidal exciting torque is applied, the viscous damping torque on disk 2 is equal to $B_2(d\theta_2/dt)$. The angular amplitudes at resonant condition ($\omega/\omega_n = 1$) is obtained from the expression:

$$\theta_1 = \frac{-T_1}{K_{12}}\frac{J_2}{J_1}\left\{\sin(\omega_n t) + \frac{J_2}{J_1}\frac{K_{12}}{B_2\omega_n}\cos(\omega_n t)\right\} \tag{5.8}$$

$$\theta_2 = \frac{T_1}{K_{12}}\frac{J_2}{J_1}\frac{K_{12}}{B_2\omega_n}\cos(\omega_n t) \tag{5.9}$$

and the amplitude of shaft torque is found to be:

$$|T_{12}| = T_1\frac{J_2}{J_1}\left[1+\frac{K_{12}}{B_2^2}(J_2+J_1)\left(\frac{J_2}{J_1}\right)\right]^{1/2} \tag{5.10}$$

5.3 ENERGY BALANCE METHOD

The energy balance method is often used to calculate the response of a system to a harmonic exciting torque when the frequency of the exciting torque is equal to an undamped natural frequency of the system. The energy input from the exciting torque and the energy dissipated by the damping are evaluated on the basis of the undamped mode shape of the system. Thus, the method is particularly useful in the case of systems where there is little damping due to shaft twisting.

For the system of Fig. 5.1 the natural frequency is given by (5.1), and the mode shape is given by the expression:

$$\theta_2 = -\theta_1(J_1/J_2) \tag{5.11}$$

For a harmonic exciting torque of amplitude T_1 and frequency ω_n acting on disk 1, energy input per cycle E_i for the component of torque in phase with the harmonic angular velocity of amplitude $\omega_n\theta_1$ is given by:

$$E_i = \pi T_1\theta_1 \tag{5.12}$$

With the damping torque equal to the viscous damping coefficient B_{12} times the difference in angular velocities of the two disks ($d\psi/dt = d\theta_1/dt - d\theta_2/dt$) (damping condition #1), the amplitude of viscous damping torque acting on disk 1 is equal to $\omega_n B_{12}(\theta_1 - \theta_2)$, which is equal and opposite to the torque acting on disk 1, and is also in phase with it. The energy dissipated per cycle is:

$$E_d = \pi\omega_n B_{12}(\theta_2 - \theta_1)^2 \tag{5.13}$$

Equating the input energy with the dissipating energy, and expressing the amplitude θ_2 in terms of amplitude θ_1 gives:

$$E_i = \pi T_1\theta_1 = E_d = \pi\omega_n\theta_1^2 B_{12}[(J_1 + J_2)/J_2]^2 \tag{5.14}$$

From this the angular amplitude of each inertia is derived:

$$\theta_1 = \left[\frac{T_1}{\omega_n B_{12}}\right]\left[\frac{J_2}{J_1 + J_2}\right]^2 \tag{5.15}$$

$$\theta_2 = -\left[\frac{T_1}{\omega_n B_{12}}\right]\frac{J_1 J_2}{(J_1 + J_2)^2} \tag{5.16}$$

and the amplitude of shaft torque is:

$$|T_{12}| = \frac{T_1[J_2/(J_1 + J_2)]\pi}{\delta} \tag{5.17}$$

which is identical to (5.4).

If hysteretic damping is assumed instead of viscous damping, the energy dissipated per cycle E_d, for low values of damping, may be taken as twice the logarithmic decrement times the total vibrational energy of the system. This concept is useful for systems of greater complexity. The total vibrational energy of the system can be expressed as a summation of kinetic energies for the inertial elements in the system at zero amplitude, when the kinetic energy is a maximum and the potential energy is zero.

$$E_d = \delta\omega_n^2[J_1\theta_1^2 + J_2\theta_2^2] \tag{5.18}$$

and the amplitude of vibratory shaft torque is:

$$T_{12} = T_1(\pi/\delta)[J_2/(J_1 + J_2)] \tag{5.19}$$

Due to the relation between δ and ξ, this result is effectively identical to (5.17) and also to (5.4).

With a damper located at disk 1 the viscous damping torque on disk 1 is equal to $\omega_n B_1\theta_1$. The energy balance then becomes:

$$E_i = \pi T_1\theta_1 = E_d = \pi\omega_n B_1\theta_1^2 \tag{5.20}$$

From this energy balance, together with (5.11), the angular amplitudes of the inertias are the same as (5.5) and (5.6), and the amplitude of the shaft torque is the same as (5.7).

With the damper located at the remote inertia end from torque application location, the amplitude of viscous damping torque on disk 2 is equal to $\omega B_2\theta_2$, and the energy balance equation becomes:

$$E_i = \pi T_1\theta_1 = E_d = \pi\omega_n B_2\theta_2^2 \tag{5.21}$$

From this energy balance together with (5.11), the angular amplitudes of the inertia are given by the expressions:

$$\theta_1 = [T_1/(\omega_n B_2)](J_2/J_1)^2 \tag{5.22}$$

$$\theta_2 = -[T_1/(\omega_n B_2)](J_2/J_1) \tag{5.23}$$

and the amplitude of shaft torque is:

$$|T_{12}| = \frac{T_1 J_2}{B_2 J_1}\left[K_{12}(J_2 + J_1)(J_2 / J_1)\right]^{1/2} \tag{5.24}$$

The amplitudes given by (5.22), (5.23) and (5.24) are not in precise agreement with the amplitudes given by (5.8), (5.9) and (5.10) of the exact solution. Instead, they represent limiting values of the exact solution amplitudes as the damping B_2 approaches zero. This is consistent with the previously mentioned limitation of the energy balance method, and is applicable only to systems with low damping.

5.4 MATRIX METHOD

The common approach for analyzing the dynamics of rotating systems is to use the state vector matrix. The method is particularly well suited for 'chain-like' structures such as turbo-machines with its many shafts, disks and couplings at a number of locations. The first use of this method is attributed to Holzer (1921) in the area of torsional vibrations. Other refinements have made this a reliable and widely used analytical procedure for rotating machinery manufacturers and rotor dynamics experts throughout the world.

The primary advantage of this method is that it does not require the storage and manipulation of large system arrays. Matrix algebra is used for the analysis of torsional vibration problems. The procedure is also applicable for calculation of natural frequencies and critical speeds of other eigen-value problems.

Two basic calculation procedures are available, the fundamental difference between them lying in the assumed independent variable. In any eigen-value problem a unique mode shape of the system is associated with each natural frequency. In the matrix iteration method the mode shape is the independent variable. A mode shape is assumed and improved by successive iterations until the desired accuracy is obtained; its associated natural frequency is then calculated.

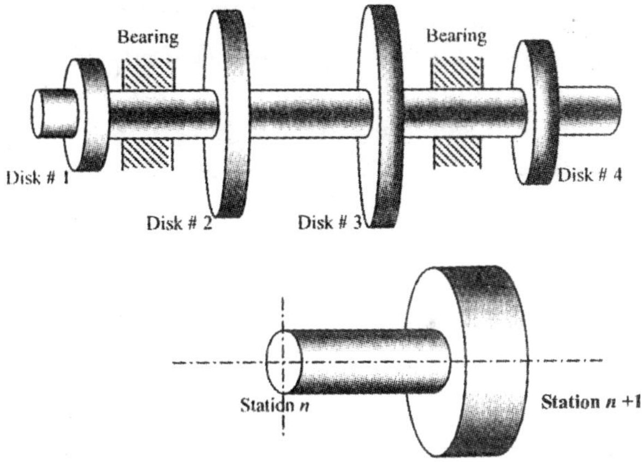

Fig. 5.2. *Typical Torsional Vibration Model*

On the other hand in Holzer's transfer matrix method a frequency is assumed, and the mode shape of the system is calculated. If the mode shape fits the boundary conditions, the assumed frequency is a natural frequency, and a critical speed is thus obtained. Determining the correct natural frequencies amounts to a controlled trial and error process. Some of the essential boundary conditions arising from the geometry and natural boundary conditions from force considerations are assumed. For example, in the torsional system of Fig. 5.2 the natural boundary conditions would be zero torque at both ends. The torque at the left end is made zero, and the torsional vibration amplitude is set at unity. The right end torque is calculated as a function of ω and plotted to find the natural frequencies. The calculations proceed according to the system transfer matrices, or functions, the elements of which reflect the characteristics of the system.

No accuracy is lost with the transfer matrix method because of coupling of the mode shapes. Accuracy is lost with the matrix iteration method, however, because each frequency calculation is independent of the others. A disadvantage of the transfer matrix method is the large number of points that must be calculated to obtain the T_{right} versus ω curve, but with digital computers the problem is readily overcome.

A typical station at location n is shown in Fig. 5.2. This general station and the transfer matrix equation shown below are used in the execution of the Holzer type of calculations for a given angular velocity ω across the full model.

$$\begin{bmatrix} \theta \\ T \end{bmatrix}_n = \begin{bmatrix} 1 & 1/k \\ -\omega^2 J & -(\omega^2 J/k)+1 \end{bmatrix}_n \begin{bmatrix} \theta \\ T \end{bmatrix}_{n-1} \tag{5.25}$$

where θ = torsional motion, radians

T = torque, in-lbs.

J = station polar mass moment of inertia, in-lbs-second^2

k = torsional stiffness of shaft section between stations, in-lbs/radian.

Because of the complex geometric configuration of a shaft, stiffness k deserves special mention. In the case of a hollow circular shaft with outer diameter D, inner diameter d and length l, the stiffness k is given by:

$$k = GI_p/l = G \times (\pi/32) \times (D^4 - d^4)/l \tag{5.26}$$

where G is the shaft material shear modulus of elasticity and I_p is the polar section area moment of inertia. Expressions for torsional stiffness coefficients for typical shaft sections are given in table 5.1. The stiffness and polar moment of inertia of each station are entered into the matrix equation (5.25) to determine the transfer effect of each element of the model. Thus, the calculations begin with station number 1 at the left end to include its stiffness and inertia, which results in the output torque T_1 and rotation θ_1 for the given input values. At station number 1 these values generally are set at 0 and 1. The equations will next be applied at station number 2, yielding T_2 and θ_2. The process is repeated to obtain values T_n and θ_n at the other end of the rotor. The calculation sequence is carried out for the assumed frequency ω. If the assumed value is a natural frequency, the value of torque T_n at the right end reaches zero.

5.5 FINITE ELEMENT METHOD

The finite element method is a numerical procedure to calculate the natural frequencies of vibration, mode shapes and forced response of a discretely modeled rotor system. Finite element computer programs based on this procedure perform three useful functions:

- They incorporate modeling objects that transform a conceptual model such as the shaft into a set of matrix equations of the form expressed by (5.25).
- They incorporate a rich set of rules for performing a wide range of analyses, including transient and frequency response.
- They contain a rich set of data processing operations, including random response and response spectrum analyses.

The rotating system is composed of an assemblage of discrete smaller finite elements that are continuous structural members. The displacements and rotations are forced to be compatible, and force and torque balance is required at the joints, commonly referred to as nodes (*ref:* Harris, 1995).

Figure 5.3 shows a uniform torsional element in local coordinates. The x-axis is taken along the axis of centroid. Physical properties of the element are density (ρ), area (A), shear modulus of elasticity (G), length (l) and polar area moment of inertia (I). $T(t)$ are the torsional forcing functions at the ends of the element.

The torsional displacement within the element can be expressed in terms of rotations $\theta_1(t)$ and $\theta_2(t)$ at the joints as:

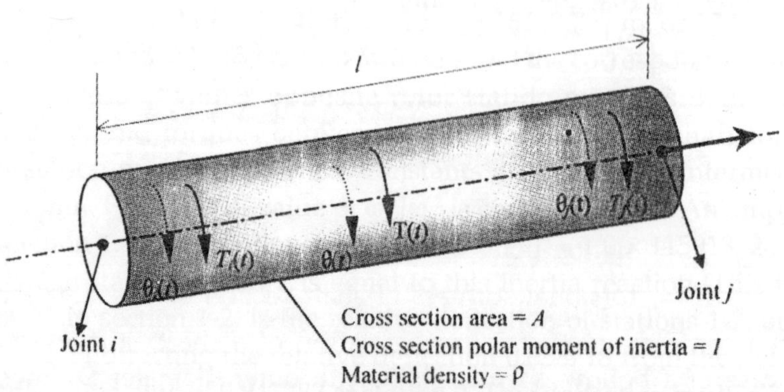

Fig. 5.3. *Finite Element Model for Torsional Vibration*

$$\theta\,(x,t) = U_1(x)\theta_1(t) + U_2(x)\theta_2(t) \tag{5.27}$$

where $U_1(x)$ and $U_2(x)$ are called 'shape functions'. Since $\theta\,(0, t) = \theta_1(t)$ and $\theta\,(l,t) = \theta_2(t)$, the shape functions must satisfy the following boundary conditions:

$$U_1(0) = 1 \quad U_1(l) = 0$$
$$U_2(0) = 0 \quad U_2(l) = 1$$

The shape function for the torsional element is assumed to be a polynomial with two constants of the form:

$$U_i(x) = a_i + b_i x$$

where $i = 1, 2$ (5.28)

Selection of the shape function is left to the engineering analyst's judgment, and is a part of the art and science required for performing an accurate modeling of the finite element model of the rotating configuration at hand.

Thus, with four known boundary conditions the values of a_i and b_i can be determined from (5.28):

$$U_1(x) = 1 - x/l \qquad U_2(x) = x/l$$

Then from (5.27):

$$\theta(x,t) = \left(1-\frac{x}{l}\right)\theta_1(t) + \frac{x}{l}\theta_2(t) \tag{5.29}$$

Kinetic energy, strain energy and virtual work are used to formulate the finite element mass and stiffness matrices and force vectors. The quantities are then used to form the equations of motion. The derived matrices are:

$$\{J\} = \frac{\rho Il}{6}\begin{bmatrix} 2 & 1 \\ 1 & 2 \end{bmatrix}$$

$$\{K\} = \frac{GI}{l}\begin{bmatrix} 1 & -1 \\ -1 & 1 \end{bmatrix}$$

$$\overline{T} = \begin{Bmatrix} T_1(t) \\ T_2(t) \end{Bmatrix} = \begin{Bmatrix} \int_0^l T(x,t)\left(1-\frac{x}{l}\right)dx \\ \int_0^l T(x,t)\left(\frac{x}{l}\right)dx \end{Bmatrix} \tag{5.30}$$

where
$\{J\}$ = mass matrix
$\{K\}$ = stiffness matrix
$\overline{T}$ = torque vector
ρ = shaft material density
I = polar area moment of inertia
G = shaft material shear modulus
l = length of element

For modeling convenience the elements thus defined are in local coordinates. Since the system as a whole must be analyzed as a complete unit, the elements must be transformed into one global coordinate system. Figure 5.4 shows the local element within a global coordinate system. The mass and stiffness matrices and the joint force vector of each element must be expressed in the global coordinate system to find the vibration response of the complete system.

Using transformation matrices, mass and stiffness matrices and force vectors are used to set up the system equation of motion for a single element in global coordinates, given by the expression:

$$[J]_e\{d^2\Theta/dt^2\,(t)\} + [K]_e\{\Theta(t)\} \;=\; \{T_e(t)\} \tag{5.31}$$

The complete system is an assemblage of a number of finite elements required to adequately model and analyze. The joint displacements of the elements in the global coordinate system are labeled as $\Theta_1(t)$, $\Theta_2(t)$, . . . $\Theta_m(t)$, or can be expressed as a column vector:

$$\{\Theta(t)\} \;=\; \begin{Bmatrix} \Theta_1(t) \\ \Theta_2(t) \\ \cdot \\ \cdot \\ \Theta_m(t) \end{Bmatrix}$$

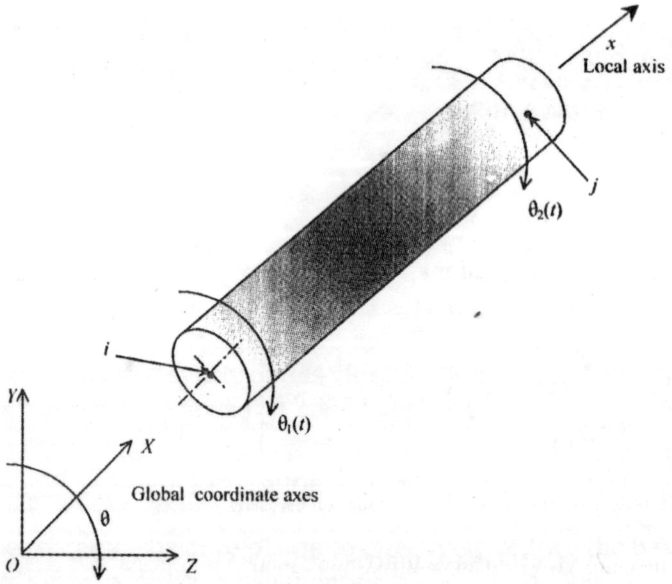

Fig. 5.4. *Local and Global Joint Displacements of an Element*

Using global joint displacements, mass and stiffness matrices and force vectors, the equations of motion are developed:

$$[J]_{n\times n}\{d^2\Theta/dt^2\}_{n\times 1} + [K]_{n\times n}\{\Theta\}_{n\times 1} = \{T\}_{n\times 1} \tag{5.32}$$

where n denotes the number of joint displacements in the system. In the final steps prior to solution appropriate boundary conditions and constraints are introduced into the global model.

Table 5.1. *Torsional Stiffness Equations of Common Rotating Elements*

Series Spring Connection	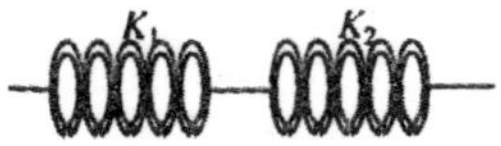	$K = 1/[1/K_1 + 1/K_2]$
Parallel Spring Connection	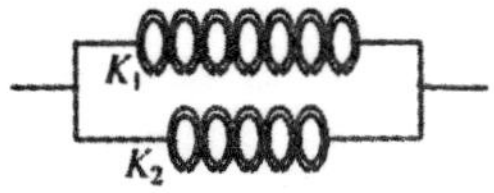	$K = K_1 + K_2$
Uniform Hollow Shaft	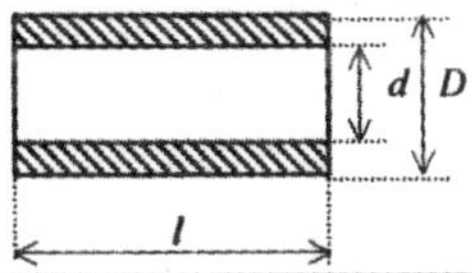	$K = \pi G(D^4 - d^4)/(32l)$
Circular Tapered Shaft	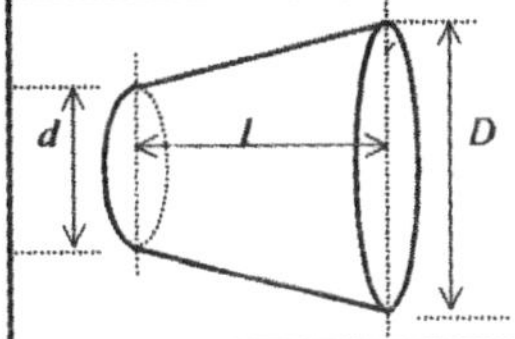	$K = \dfrac{3\pi}{32} \dfrac{d^4}{l(n+n^2+n^3)}$ $n = d/D$
Pair of Geared Shafts (With reference to shaft # 1)	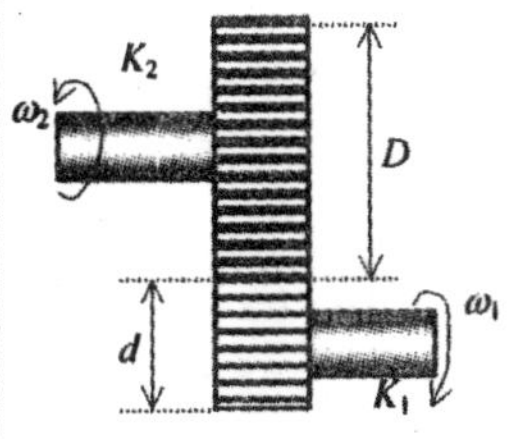	$K = [K_1K_2]/[n^2K_1 + K_2]$ $n = D/d = \omega_1/\omega_2$
$K = T/\theta = GI_p/L$ = Torsional Stiffness, G = Material Shear Modulus ω = Shaft Speed		

The equations of motion for free vibration are solved for the eigen-values (natural frequencies) using the matrix iteration method. Modal analysis is used to solve the forced torsional response. The finite element method is available in commercially available computer programs for the personal computer. The analyst is required to select the information for the joints (nodal location, materials, shape functions, geometry, torque and constraints) to model the system for computation of natural frequencies, mode shapes and torsional response. Similar to any other analytical effort, experience in the form of capabilities of the computer program and the rotating system enable the engineer to obtain reasonably accurate results.

Besides crank webs in reciprocating compressors, pumps and engines, crankshaft designs include elements such as changes in cross section, collars, shrunk and keyed armatures and other design features. An assessment of the stiffness can only come from experience on similar other design of the elements. For a fillet radius equal to 10% of the smaller diameter, the stiffness can be estimated by assuming that the smaller shaft is lengthened, and the larger shaft shortened, by a length λ obtained from the curve of Fig. 5.5. This may also be applied to flanges where D is the bolt diameter. Stiffening effects from collars may be ignored.

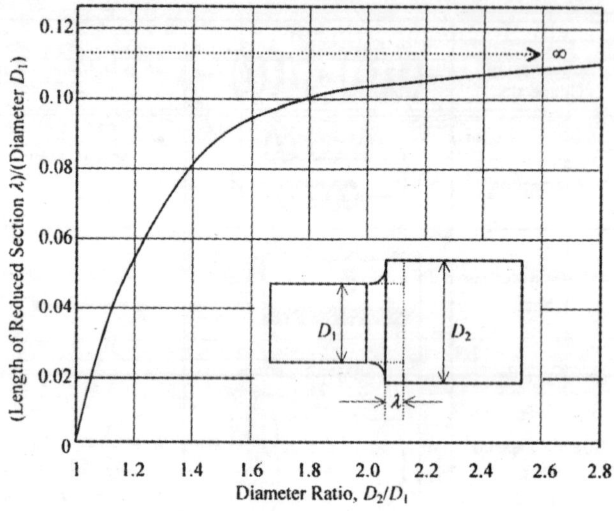

Fig. 5.5. *Reduction in Stiffness due to Change in Shaft Diameter* (ref. *Porter, 1943*)

Stiffness of shrunk and keyed parts is difficult to estimate because the stiffening effect depends on how tight the shrink fit remains during the machine's operating condition. The interference fit present at the mating surfaces of the two components during assembly will change due to inertia effects, distortion due to operating loads and thermal conditions if existing at the location. The preferred approach is to ignore the stiffening effects of the interference between, say, an armature and a shaft. In that case only the inertia effects from the mass of the armature may be included at the proper location on the shaft. Some armature designs call for spiders and flywheels with significant stiffness in the arms; the treatment of these is discussed at length in example problem # 10.8.

A similar procedure is used to determine the polar inertia of propellers and other irregular parts. For propellers, hydraulic couplings and impellers an allowance must be made for the inertia of the fluid rotating with the component. For marine propellers this is generally 26% of the propeller inertia.

For complex rotating elements such as couplings and small flywheels, it may be easier to determine the polar moment of inertia experimentally than to calculate it. In one procedure the component is suspended from three equally spaced vertical wires, as shown in Fig. 5.6. The component is then set into torsional motion and the period of vibration measured. The polar moment of inertia J is then obtained from the following expression:

$$J = Wr^2\tau^2/[(2\pi)^2 l] \text{ in-lbs-sec}^2 \tag{5.33}$$

where τ = period of vibration, second/cycle

W = weight of element, lbs.

l = length of cables, in.

r = radius of cables, in.

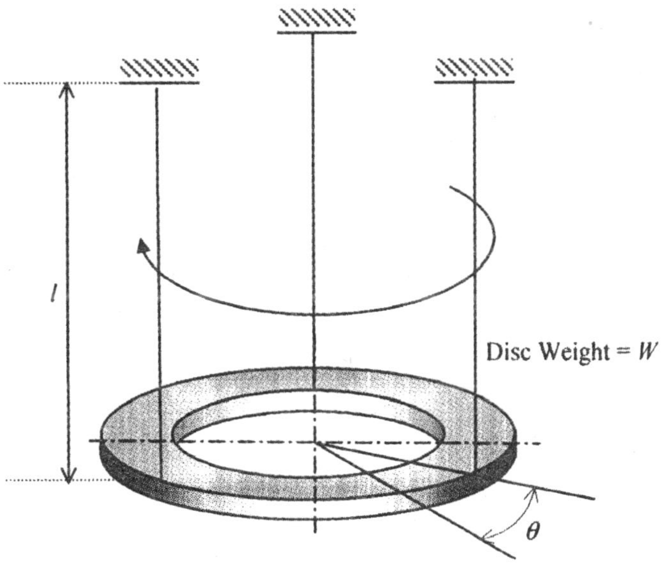

Fig. 5.6. *Measurement of Polar Moment of Inertia from Torsional Oscillations*

5.6 GEARED AND BRANCHED SYSTEMS

Natural frequencies of vibration of a geared system may be calculated by replacing it with a system in which the speed of the driver unit is a multiple of n, where n is the ratio of the speeds of the driver unit and the driven unit. The inertia and elastic constants on the driven side of the unit are multiplied by the factor $1/n^2$, then calculate the system natural frequencies as if the gears do not exist. If the system calls for a damping device on the driven side, the damper constant of the device must also be multiplied by $1/n^2$. Torques and deflections thus obtained on the driven side of the substitute system, when multiplied by n and $1/n$ respectively, are equal to those in the actual geared system. Alternatively, the driven side may be used as the reference, in which case the driver constants are multiplied by the factor n^2.

When two or more drivers are geared to a common load, hydraulic or electrical type of couplings may be placed between the drivers and the gear set. The couplings serve a dual-purpose role, one of isolating the gears from driver induced vibrations, the other as a clutch for connection and disconnection of load between the components. In many such coupling designs the vibration isolation may be complete enough to permit termination of the driver end of the unit at the coupling. The damping effect upon the vibration in the driver end of the system is small and may be disregarded in the amplitude calculations.

In many applications calling for two identical driver units hydraulic or electrical couplings are not employed. Then the modes of vibration are of two types:

- Opposed phase modes, in which the drivers vibrate against each other, with a node at the gear. The modes are calculated for a single branch in the usual manner, with the model terminating at the gear. The condition for a natural frequency is that shaft twist amplitude be set to zero at the gear.

- In phase modes, in which both drivers vibrate in the same direction against the driven machinery. To calculate these frequencies the inertia and stiffness constants of the driver side of one branch are doubled; the calculations are then performed for only one driver. The condition for a natural frequency is zero residual torque at the end.

The two drivers can be phased such that certain like-phase modes can be eliminated. If the drivers are connected with clutches, these phasing possibilities cannot be utilized. In the general case of non-identical branches the calculation scheme proceeds as follows. Reduce the system to a 1:1 gear ratio. Call the two branches a and b. Perform the sequence calculation for the a branch, with initial twist angle equal to unity, and for the b branch the initial amplitude set to an unknown value x. At the junction equate the amplitudes and find x. With this numerical value of x substituted, the torques in the two branches and the torque of the gear are added; the sequence calculation is then continued up to the last mass.

The branch may consist of a single member elastically connected to the system. Examples of such a branch are a flywheel with flexible spokes or an armature with a flexible spider. Assume J to be the moment of inertia of the flywheel's rim and k the elastic constant of the connection. Then the flexibly mounted flywheel is equivalent to a rigid flywheel of an equivalent moment of inertia given by the expression:

$$J' = \frac{J}{1 - J\omega^2/k} \tag{5.34}$$

5.7 CRITICAL SPEEDS AND MODES

The rotating components and the shafting of a turbo-machinery attached to it comprise a torsional elastic system. Such a system has several modes of free torsional oscillation. Each mode is characterized by a natural frequency and by a pattern of relative amplitudes of all parts of the system when it is oscillating at its natural frequency. The harmonic components of the driving torque provide the excitation for vibration of the modes. If the frequency of any harmonic component of the torque is equal to, or close to, the frequency of any mode of vibration, a condition of resonance prevails, and the machine is said to be running at a critical speed. Operation of the machinery system at such critical speeds can be dangerous, and in an extreme condition result in fracture of the shafting.

The number of complete oscillations of the elastic system per unit revolution of the shaft is called the order of a critical speed. An order of a critical speed that corresponds to a harmonic component of the torque from, for example, a reciprocating engine is called a major order. A critical speed can also be excited by the harmonic component of the torque curve of a single cylinder. The fundamental period of torque from a single cylinder in a four-cycle internal combustion engine is 720°; critical speeds in this engine will then be of ½, 1, 1½, 2, 2½, etc., order. In a two-cycle engine only critical speeds of 1, 2, 3, etc. order will prevail. All critical speeds except those of the major orders are called minor critical speeds. Note, however, that this term does not necessarily mean they are not important.

When a reciprocating machine's crankshaft and other machinery's attached rotors have many masses, the Holzer method can be used to calculate the natural frequencies. A frequency is assumed; then starting at one end of the system a balance of torques and rotations is obtained in a step-by-step procedure. The final external torque required to achieve balance is called the

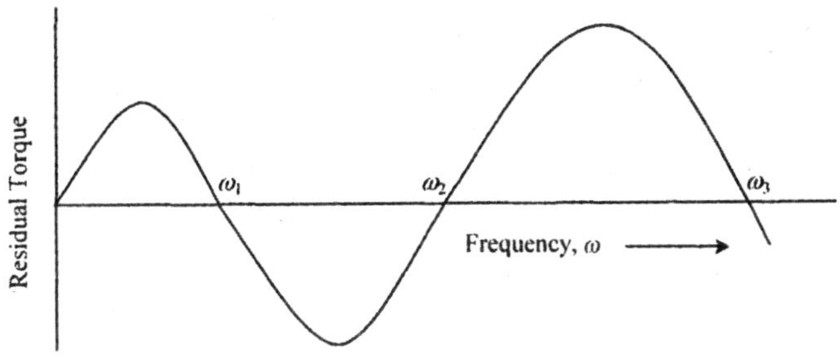

Fig. 5.7. *Typical Curve of Residual Torque as a Function of Assumed Frequency* ω

residual torque. If the residual torque is zero, the assumed frequency is one of the natural frequencies of the system. The mode associated with this natural frequency is identified by the number of changes of the sign of the amplitudes at various mass locations. Figure 5.7 gives a plot of residual torque versus frequency, from which the natural frequencies of the system within the frequency range of interest can be obtained.

In many multi-cylinder reciprocating machines the polar inertia representing the cranks and pistons is equal, as are also the stiffness constants of the crankshaft between these locations. Advantage of this similarity in the stiffness and inertia properties can be taken to reduce the amount of computational effort. Assume J to be one of the equal polar moments of inertia and k one of the equal spring constants. Divide all the polar moments of inertia by J and all shaft stiffness values by k. The modified system, dynamically similar to the original system, is referred to as the unity system. Then if ω_n is the angular natural frequency of the actual system and ω_s that for the unity system,

$$\omega_n^2 = \omega_s^2 \frac{k}{J} \tag{5.35}$$

Before starting the sequence calculation it is necessary to know the approximate frequency of the first mode. For a reciprocating machine with small inertias in the presence of large ones the following procedure is useful, giving generally satisfactory results. Add all the polar inertias within the reciprocating machinery portion of the system. Assume 40% of this amount is placed at the crank farthest from the flywheel or attached machine's rotor (number 1 crank). Find the stiffness with reference to the flywheel, or driven or driving machine's rotor, to the number 1 crank. If a large mass such as a flywheel is not attached, the system reduces to a two-mass system, and equation (5.1) gives the approximate lowest natural frequency of the system. For complicated systems with many masses and stiffness coefficients of differing values, general rules of this type are not available.

5.8 RESONANT RESPONSE

The torsional vibration response of a turbo-machine's rotating system is determined by the magnitude, points of application and phase angle relations of the exciting torques produced by gas pressure and inertia, and by the magnitude and point of application of the damping torques. Damping may be due to a number of sources, including pumping action in the bearings, hysteresis in the shafting and between fitted parts, energy absorbed by the machine's frame and by the foundation. In some cases, such as a marine propeller or a pump's impeller, damping from the fluid on the propeller or the impeller predominates. When a machine is fitted with a damper, the effects of damping will be substantial.

Analytical procedures are available for calculating the torsional vibration amplitudes in the transfer matrix method and in the modal analysis, which can be implemented in the finite element or the lumped parameter models. The Holzer table calculation sequence can be readily extended for this purpose. The initial amplitude is treated as an algebraic unknown θ. At each cylinder where an exciting torque may be present, the torque is added. Initially, assume that there is no torque from damping. Then the residual torque at the last rotor station is of the form $a\theta + b$, where a and b are numerical constants. Since torque at the last station is zero, $\theta = -b/a$. The amplitude and torque at any point of the system are found by substituting this numerical value of θ at the proper point in the calculation sequence. At frequencies removed from resonance damping has little effect and can be neglected.

Damping may be introduced into the system in the form of an exciting torque equal to the imaginary quantity $-jc\omega\theta$, where c is the damping constant and θ the amplitude at the point of application of the damper. Relative damping between two inertia points can be treated as a spring of stiffness constant equal to the imaginary quantity $+ jc\omega$.

For major critical speeds the exciting torques are in phase, and are real numbers. For minor critical speeds the exciting torques are out of phase; these are entered as complex numbers of amplitude and phase as determined from the phase diagram for the critical speed of the order under consideration. With damping and out of phase exciting torques introduced, a and b in the equation $a + b\theta = 0$ are complex numbers, and θ must be entered as a complex number in the tabulation in order to determine the angle and torque at any point. These angles and torques are then of the form $r + js$, where r and s are numerical constants and the amplitudes are $\sqrt{(r^2 + s^2)}$.

In the forced vibration calculations the amplitude of the exciting torque is also assumed to be unity. It is advisable to start the sequence calculation at the flywheel end of the system, particularly when the intent of the analysis is to design the damper. This way introduction of the exciting and damping torques is deferred further down the calculation sequence.

Classical modal analysis of vibrating systems can be used to obtain forced response of multi-station rotor systems in torsional motion. Natural frequencies and mode shapes of the system are found using the transfer matrix method. The response of the rotor to periodic phenomena (not necessarily a harmonic or a shaft frequency) is determined as a linear weighted combination of the mode shapes of the system. Damping is to be entered in modal form; damping forces are thus a function of the different modal velocities. Formation of equivalent viscous damping constants that are some percentage of critical damping is required; the critical damping factor is obtained from the system modal inertia.

The modal analysis technique can be used for a torsional distributed mass model of a reciprocating machine using modal damping. Non-synchronous speed excitations are permitted. Shaft sections of the modeled rotor have distributed mass properties and lumped end masses (including rotary inertia). A transfer matrix analysis is performed to obtain a finite number of natural frequencies. The number required depends on the range of forcing frequencies used in the problem. The natural frequencies are substituted back into the transfer matrices to obtain the mode shapes. A function consisting of a weighted average of the mode shapes is formed to determine the torsional response.

Example problem # 5.4 provides the procedure for determining the torsional response using Holzer's sequence of calculations. The problem presents a widely accepted and used method to calculate the torsional natural frequencies and mode shapes. For a given excitation maximum torque, vibration amplitude and shaft stress can then be obtained due to damping introduced at a location in the system.

5.9 DYNAMIC VIBRATION ABSORBERS

In order to improve upon a machine or its component experiencing obnoxious vibrations when subjected to a steady alternating force of constant frequency, especially when the frequency is close to resonance, dynamic vibration absorbers have been applied quite successfully. The technique was invented by Frahm in 1909, and does not include any damping device. In Fig. 5.8 let the combination K, M be the schematic representation of a machine, with force $P_0\text{Sin}(\omega t)$ acting on it. The vibration absorber consists of a comparatively small vibratory system k, m attached to the main mass M. The natural frequency $\sqrt{(k/m)}$ of the attached absorber is chosen to be equal to the frequency ω of the disturbing force. It can be mathematically proved that the main mass M does not vibrate at all, and that the smaller system k, m vibrates in such a way that its spring force at all instants is equal and opposite to $P_0\text{Sin}(\omega t)$. Because of this no force acts on mass M, and so it will not vibrate at all (*ref*: den Hartog, 1985).

To prove this concept, start with the equations of motion:

$$Md^2x_1/dt^2 + (K+k)x_1 - kx_2 = P_o \sin(\omega t)$$

$$md^2x_2/dt^2 + k(x_2 - x_1) = 0 \tag{5.36}$$

The forced vibration of the system will be of the form:

$$x_1 = a_1 \sin(\omega t) \tag{5.37}$$

$$x_2 = a_2 \sin(\omega t)$$

After differentiating twice the sine function is still a sine function. Substituting into equations (5.36) will produce the algebraic entities, which are easier for calculating the solution.

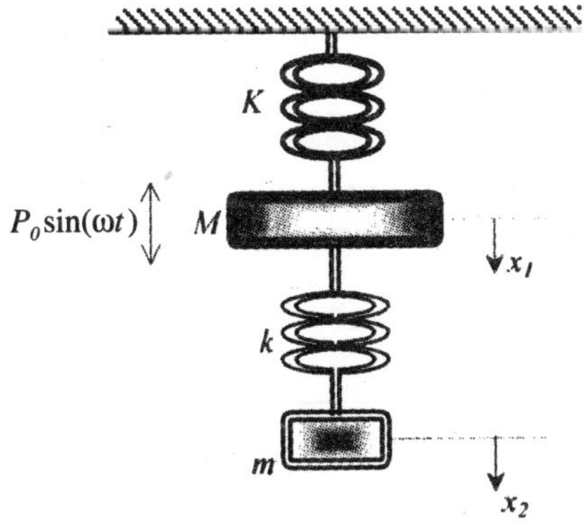

Fig. 5.8. *Dynamic Vibration Absorber*

$$
\begin{aligned}
a_1(-M\omega^2 + K + k) - ka_2 &= P_0 \\
-ka_1 + a_2(-m\omega^2 + k) &= 0
\end{aligned}
\tag{5.38}
$$

For simplification, the dimensionless form may be introduced using the following symbols:

$$
\begin{aligned}
x_{st} &= P_0/K = \text{static deflection of main spring} \\
\omega_a^2 &= k/m = \text{natural frequency of absorber} \\
\Omega_n^2 &= K/M = \text{natural frequency of main system} \\
\mu &= m/M = \text{mass ratio} = \text{absorber mass/main mass}
\end{aligned}
$$

Then the amplitudes of vibration will be:

$$
\frac{a_1}{x_{st}} = \frac{1 - \dfrac{\omega^2}{\omega_a^2}}{\left(1 - \dfrac{\omega^2}{\omega_a^2}\right)\left(1 + \dfrac{k}{K} - \dfrac{\omega^2}{\Omega_n^2}\right) - \dfrac{k}{K}} \tag{5.39}
$$

$$
\frac{a_2}{x_{st}} = \frac{1}{\left(1 - \dfrac{\omega^2}{\omega_a^2}\right)\left(1 + \dfrac{k}{K} - \dfrac{\omega^2}{\Omega_n^2}\right) - \dfrac{k}{K}} \tag{5.39a}
$$

The first equation reveals the contention that the amplitude of the main mass is zero if $\omega = \omega_a$, and this occurs when the frequency of the forcing function is the same as the natural frequency of the absorber. In the second equation, if $\omega = \omega_a$, the equation reduces to:

$$
a_2 = -\frac{K}{k} x_{st} = -\frac{P_0}{k} \tag{5.40}
$$

With the main mass standing still the damper mass has the motion $[-P_0/k] \times \sin(\omega t)$, and the force in the damper varies as $-P_0 \sin(\omega t)$, which is equal and opposite to the external force.

These relations are good for any value of ω/Ω_n. However, adding the absorber is useful only if the original system is in resonance, or near it. Hence, consider the case where:

$$
\omega_a = \Omega_n \quad \text{or} \quad k/m = K/M \quad \text{or} \quad k/K = m/M
$$

The ratio $\mu = m/M$ then defines the size of the damper when compared with the size of the main machine. For this special case equations (5.39, *a*) can be expressed as follows:

$$
\frac{x_1}{x_{st}} = \frac{1 - \dfrac{\omega^2}{\omega_a^2}}{\left(1 - \dfrac{\omega^2}{\omega_a^2}\right)\left(1 + \mu - \dfrac{\omega^2}{\omega_a^2}\right) - \mu} \sin(\omega t) \tag{5.41}
$$

$$
\frac{x_2}{x_{st}} = \frac{1}{\left(1 - \dfrac{\omega^2}{\omega_a^2}\right)\left(1 + \mu - \dfrac{\omega^2}{\omega_a^2}\right) - \mu} \sin(\omega t)
$$

The denominators in the two equations are equal, which is due to a physical reason. When multiplied out and equated to zero, the denominator becomes a quadratic in $(\omega/\omega_a)^2$ having two roots. Thus, for two values of the external frequency ω both denominators become zero, and consequently x_1 and x_2 become infinitely large. These two frequencies are the natural frequencies of the system. As a result the compression and extension of the damper spring, as well as the force in it, are appropriately balanced.

For a torsional system such as the crankshaft of a reciprocating machine the Frahm dynamic vibration absorber takes the shape of a mass inside the flywheel A rotating freely on the shaft on bearings B and held to it by mechanical springs k only, as shown in Fig. 5.9*a*. Since the torsional impulses, for instance, in an internal combustion engine, are harmonics of the firing frequency (which is dependent on the engine speed), the absorber will work for one engine speed only. Note that there will also be two neighboring speeds at which the shaft will resonate. In order to overcome this situation the system is modified by replacing the mechanical springs by the centrifugal spring of Fig. 5.9*b*. The pendulum in the centrifugal field acts in the same manner as an ordinary gravity pendulum, where the field g is replaced by the centrifugal field $\omega^2 r$.

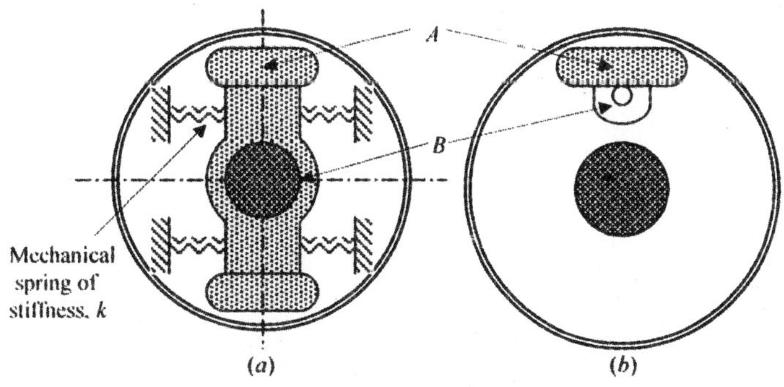

Fig. 5.9. *Torsional Dynamic Vibration Absorber For Reciprocating Machines (a) With Mechanical Springs, (b) With Centrifugal springs*

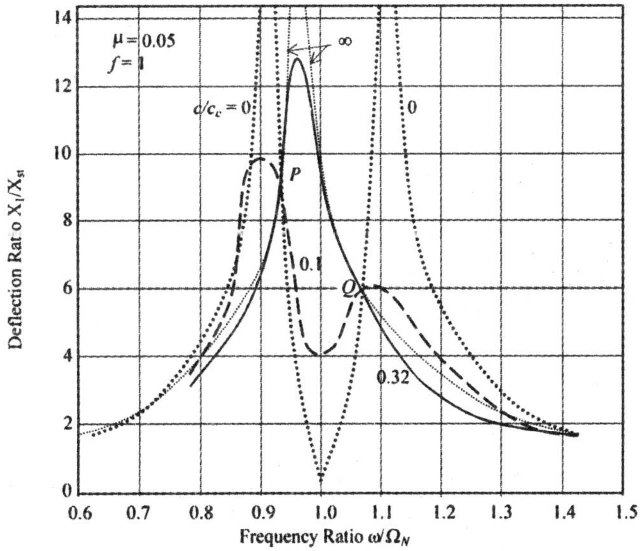

Fig. 5.10. *Amplitudes of Main Mass of Fig. 5.8 as Function of Absorber Damping*

Since the frequency of a gravity pendulum is $\sqrt{(g/l)}$, the frequency of a centrifugal pendulum becomes $\omega\sqrt{(r/l)}$, hence it is proportional to the machine's shaft speed. Thus, a centrifugal pendulum will act as a Frahm dynamic vibration absorber at all speeds.

It is worth noting that several reciprocating machines operate at a fixed speed. For example, a diesel engine coupled to a power generator needs to generate alternating current electricity with a reasonably constant frequency, which calls for the diesel prime mover to operate at a constant shaft speed. The dynamic absorber with mechanical springs should be able to take care of the vibrations arising in the system.

In the case of damped vibration absorbers, a dashpot c is arranged parallel to the damper spring k, between the masses m and M. Following a similar scheme of mathematical arguments, it is instructive to observe and analyze the expression for the amplitude of motion of the main mass:

$$\frac{x_1^2}{P_o^2} = \frac{(k - m\omega^2)^2 + \omega^2 c^2}{[(-M\omega^2 + K)(-m\omega^2 + k) - m\omega^2 k]^2 + \omega^2 c^2 (-M\omega^2 + K - m\omega^2)^2} \tag{5.42}$$

For particular cases this equation reduces to known results as previously obtained. Figure 5.10 shows a plot of x_1/x_{st} as a function of frequency ratio $g = \omega/\Omega_n$, $f = \omega_a/\Omega_n = 1$, $\mu = 1/20$, for various values of c/c_c. The plots describe the behavior of a system where the main mass is 20 times the dynamic absorber mass and the frequency of the damper is equal to the frequency of

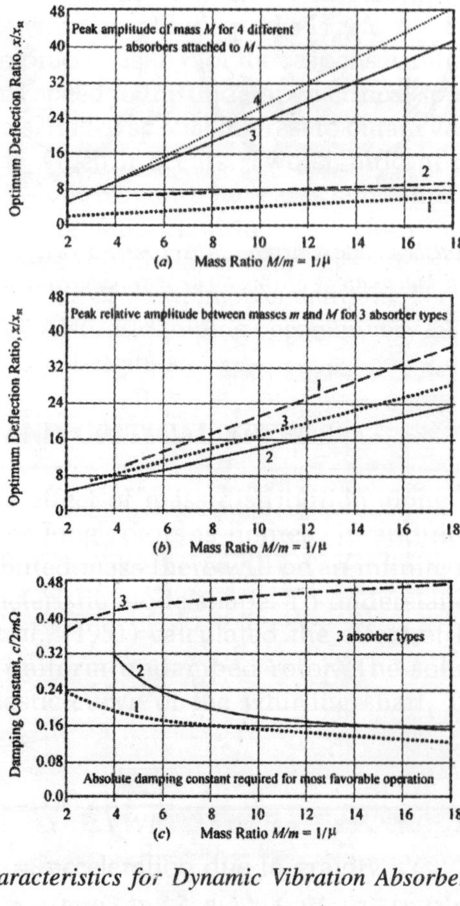

Fig. 5.11. *Vibration Characteristics for Dynamic Vibration Absorbers (Den Hartog, 1985)*

the main system. All curves pass through points P and Q. Some other cases commonly encountered in actual machines are shown in Fig. 5.11. Figure 5.11*a* provides peak amplitudes of the main mass as a function of μ for various absorbers attached to it. Figure 5.11*b* gives peak relative amplitudes between masses m and M, while Fig. 5.11*c* determines damping constants required for the most favorable operation of the absorber. In all three plots curve 1 represents the most favorably tuned and damped absorber; curve 2 represents the most favorably damped absorber tuned to the frequency of the main system; curve 3 considers the most favorably damped viscous Lanchester damper (discussed at length in the next section); and curve 4 for the most favorably damped Coulomb Lanchester damper (also discussed in the next section).

5.10 TORSIONAL DAMPERS

The Lanchester damper provides an effective method of mitigating torsional vibrations in rotating machines. It consists of two disks a (Fig. 5.12*a*), which rotate freely on the shaft bearing b. Between the two parts is a hub disk h solidly keyed to the machine's shaft. Hub h carries a brake lining c on its faces against which disks a can be pressed by screwing down the springs s (Den Hartog, 1985).

If the machine hub h is in uniform rotation, the friction carries the disks a with the shaft, so the disks then merely increase the inertia of the machine by some percentage. If, however, the hub executes a torsional vibration, the motion of the disks depends upon the amount of friction between them and the hub. If the frictional torque is extremely small the disks rotate uniformly, and there is a relative slip between the hub and the disks that is equal to the amplitude of the hub's motion. Since the friction torque is nearly zero, little work is converted into heat. But if

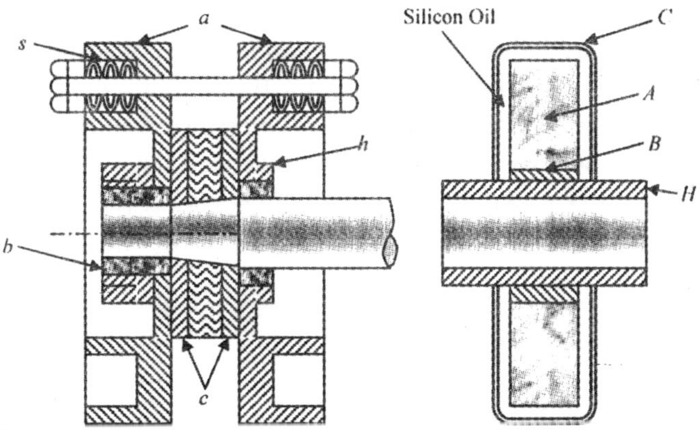

(*a*) (*b*)

Fig. 5.12. *(a) Lanchester Damper with Friction Element, (b) Houde Type of Viscous Damper*

the friction torque is very large, the disks lock on the hub and follow its motion. There is then no relative slip and hence no energy dissipation. Between these two extremes there is both slip and a friction torque, so that energy is destroyed. There is an optimum value of the friction torque at which the energy dissipated is a maximum, as shown in Fig. 5.13.

The earlier version of Lanchester damper depends on Coulomb, or dry, friction between the steel and asbestos-type of brake lining. The coefficient of friction of such an assembly is dependent on the degree of contamination with oil, and since the damper is located at the end of the crankshaft, this is difficult to avoid. In order to be more certain of the coefficient of friction, the dry friction can be replaced by viscous friction. Figure 5.12*b* shows the Houde type of viscous damper. *H* is a hub, splined inside to facilitate assembly of the damper on a shaft. *A* represents a flywheel, which, through a bronze ring *B*, can rotate freely about the hub. C is the assembly housing, welded to the hub. The clearance between the flywheel and the housing is small and accurately held to a dimension, and completely filled with a fluid permanently sealed in the space. The fluid is a type of silicone oil selected because of its advantage that its viscosity does not change substantially due to temperature change. A rotating motion of *A* relative to C and *H* is associated with a viscous-friction torque, so the damper substantially acts like the dry friction type.

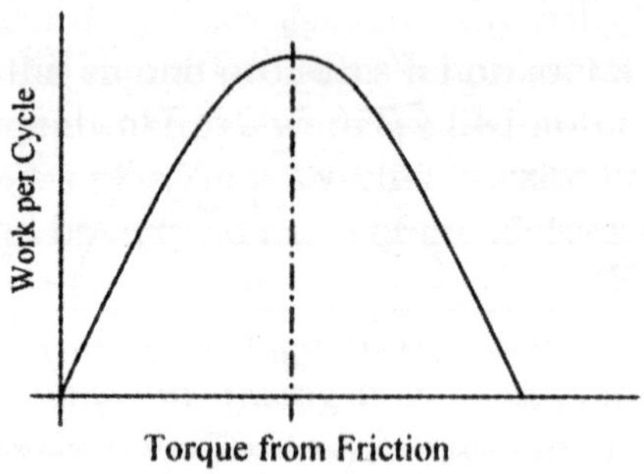

Fig. 5.13. *Energy Dissipation in Torsional Damper*

5.11 PERMISSIBLE AMPLITUDES

Failure caused by torsional vibrations in the shafts of rotating machinery invariably initiates in the form of fatigue cracks that start at points of stress concentration. Typical locations are at fillets where there is a change in shaft diameters, at oil holes in a crankshaft and at the ends of keyway slots. At shaft oil holes the crack begins on a line at 45° to the shaft axis, then grows in a spiral pattern until failure occurs. The stress at the edge of the oil hole is theoretically 4 times the mean shear stress in the shaft, and failure may be expected if this concentrated stress exceeds the fatigue limit of the material. The problem of estimating the stress required to result in failure is further complicated by the presence of steady stress from the mean driving torque and the variable bending stresses. In practice, the severity of a critical speed is judged by the maximum nominal torsional stress from the equation:

$$\tau = 16T_m/(\pi d^3) \tag{5.43}$$

where T_m is the maximum torque amplitude and d is the shaft diameter. The calculated nominal stress must be modified to include the effects of stress increase due to manufacturing procedures and operational and environmental considerations, and then compared with the fatigue strength of the material.

Military standards issued by the U.S. Department of Navy state that the limit of acceptable nominal torsional stress within operating range is:

$$\tau = (\text{Ultimate strength})/25 \text{ for steel}$$

$$\tau = (\text{Ultimate strength})/6 \text{ for cast iron}$$

If the full-scale shaft has been fatigue tested (which is a difficult test), then:

$$\tau = \text{(Torsional fatigue limit)}/2 \text{ for both materials.}$$

For critical speeds inside the operating range, passed through during machine start up and shut down, the nominal torsional stress must not exceed 1.75 times the above values. Crankshaft steels having ultimate tensile strength between 75,000 and 115,000 lbs/in^2 usually have a torsional stress limit between 3,000 and 4,600 lbs/in^2.

For gear drives the vibratory torque across the gears at any operating speed must not be greater than 75% of the driving torque at the same speed, or 25% of the full load torque, whichever value is smaller.

The American Petroleum Institute considers the following sources of torsional excitation significant: problems arising from gears, including unbalance, pitch line run-out and eccentricity; start up conditions resulting from inertial impedances; and torsional transients from synchronous and induction electric motors. Torsional natural frequencies of the machine train must be at least 10 percent above or below any possible excitation frequency within the specified operating speed range. Torsional critical speeds at integer multiples of operating speed (for example, pump vane passing frequency) should be avoided, or should be shown to have no adverse effect if excitation frequencies exist. Non-synchronous torsional excitation frequencies must be considered in the evaluation. Identification of torsional excitations is the mutual responsibility of the purchaser and the vendor.

When torsional resonance is calculated to fall within the ± 10 percent margin, and the purchaser and vendor have agreed that all efforts to remove the natural frequency from the limiting frequency range have been exhausted, a stress analysis should be performed to demonstrate lack of adverse effect on any portion of the machine system.

In the case of synchronous motor driven units, the vendor is required to perform a transient torsional vibration analysis, with acceptance criteria previously agreed upon by the purchaser and the vendor.

5.12 SHIFTING OF CRITICAL SPEEDS

Among other procedures, shifting critical speeds by making appropriate changes in mass and elasticity has been successfully implemented in new machine designs. If the stiffness of all the shafting in a system is increased by ratio a, then all the frequencies will increase by the same ratio a, provided that there in no increase in the inertia accompanying it. In crankshafts the diameters are made as large as practical to ensure bearing pressures are maintained within limits, hence it is rarely possible to stiffen the main and crank pin journals. If bearing diameters are increased the increase in the critical speeds will be much smaller than indicated by the ratio a, because the corresponding increase in inertia will also be substantial. Changes in the stiffness of a system made near a nodal point will have the maximum effect. Also, changes in inertia near a loop will have the greatest effect; those near a node will have little impact.

Elastic couplings have been used very effectively to place certain critical speeds below the operating speed, where they are passed through only in starting and stopping the machine. This leaves a clear range above the critical speed. The procedure must be used with caution. For example, in an eight-cylinder four-cycle internal combustion engine, fourth order torsional vibrations are so violent that it may be dangerous to pass through them. If the acceleration through the critical speed is rapid, some reduction in the amplitude may be attained, but with a controlled rate of engine acceleration the reduction in the amplitudes may not be large enough. The rate of deceleration when the machine is stopped is sometimes more important from this consideration. Mechanical clutches are sometimes used to disconnect the driven machinery until the driver has reached a speed above the dangerous critical speeds.

Elastic couplings take many forms. Helical coils arranged tangentially are the examples; flat leaf springs arranged longitudinally or radially are other examples. Smaller diameter shaft sections of high tensile strength and shafts with grooves machined in them serve a similar purpose. Other configurations call for use of disks made of some resilient material between flanged coupling halves.

In marine installations damping arising from the hydraulic action on the propeller can be used to reduce torque amplitude in the shaft below a desired level by making the flywheel moment of inertia J of sufficient magnitude. The ratio of the propeller amplitude to the engine amplitude increases as the flywheel becomes larger, hence increasing the effectiveness of the propeller as a damper.

5.13 EXAMPLE PROBLEMS

Example Problem # 5.1: Explain the practical significance of torsional stiffness of a rotating shaft.

Solution: An adequate shaft design calls for it to transmit power uniformly with a steady motion. If a shaft permits excessive angular motion, it contributes to torsional as also lateral vibrations. This can affect performance of gears, couplings and bearings. As an example, torsional deflection in camshafts for reciprocating machines should be limited to 0.5° over its full length, regardless of its length. Stiffness of a rotating shaft element is determined from $T/\theta = GI_p/l$, where G is the material shear modulus, I_p is the polar moment of inertia of the shaft section and l is the length of the shaft segment. For a round hollow shaft I_p is given by the expression: $I_p = [\pi/32][d_o^4 - d_i^4]$. For shafts having other than round cross sections table 5.2 is presented as a designing aid.

When the geometric configuration of a rotor bearing system gets more complex, finite element representation is probably desirable. Oftentimes it is required to find the impact on the dynamic behavior of the rotating system due to relatively small changes. For example, in the bearing sump area the material properties of a component or an assembly are affected by temperature related variations in a characteristic. In such a case parametric evaluation of the problem provides quantitative results, which may be used as the basis for making design decisions. Experimental results to confirm analytical predictions may also be necessary for the purpose. Accelerometer readings at bearing locations can be obtained in most cases with relative ease.

Since the torsional oscillatory motion is superimposed on the regular rotating shaft motion, measurement of the elastic twisting action is a difficult proposition at best. Dynamic strain gages provide useful data, but are more difficult to install. The strain gages necessitate use of a slip ring system for picking up the signals. As the rotating system increases in length and complexity, the number of locations where measurement is required increases. Relative closure data between the stator and the rotor at selected locations around the periphery using telemetry may also be required.

Example Problem # 5.2: Figure # 5.8 shows a spring-mass combination K and M to represent a machine with force $P_o\text{Sin}(\omega t)$ acting on it, resulting in unacceptable vibrations in the machine. To remedy the situation a dynamic absorber in the form of a comparatively smaller vibratory system k, m is attached to the main mass. If $Mg = 10$ lbs, $mg = 1$ lbs, $P_o = 10$ lbs, $K = 102$ lbs/in and $k = 10.2$ lbs/in, calculate the following parameters:

(*a*) The best damping coefficient across the absorber.

(*b*) The amplitude of the main mass.

(*c*) The maximum stress in the absorber spring.

(*d*) If the requirement $k/K = m/M$ is dropped, for what value of k is the best overall effect obtained?

Solution: The answers can be found in Fig. 5.11 *a, b, c*.

(*a*) From Fig. 5.11*c*: $c/2m\Omega_n = 0.152$, so $c = (2 \times 0.152 \times 2\pi \times 10)/386 = 0.0495$ lbs-sec/in, where $\Omega_n = \sqrt{(K/M)} = \sqrt{(102 \times 386/10)} = 62.80 = 2\pi \times 10$ cps.

(*b*) Figure 5.11*a* gives $x/x_{st} = 4.3$, then $x = 4.3 \times [1/102] = 0.0042$ in.

(*c*) Figure 5.11*b* gives $x_{rel}/x_{st} = 11.4$, so $x_{rel} = 11.4/102 = 0.112$ in. The force kx_{rel} is $10.2 \times .112 = 1.14$ lbs.

Table 5.2. ***Torsional Stiffness and Shear Stress in Rotating Beams***

Section	Stiffness	Maximum Shear Stress
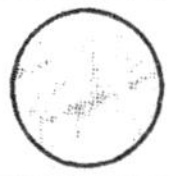	$I_P = \frac{\pi d^4}{32}$, $K = T/\theta = GI_p/L$ d = solid shaft diameter, L = length of shaft	$\tau = 0.5 \times d \times T/I_P$
	d = outer diameter, d_i = inner diameter, L = length of shaft $I_P = \frac{\pi[d^4 - d_i^4]}{32}$, $K = T/\theta = GI_p/L$	$\tau = 0.5 \times d \times T/I_P$
	r = mean radius, t = thickness, L = length of shaft $I_P = \frac{2\pi r t^3}{3}$, $K = T/\theta = GI_p/L$	τ • $0.5 \times d \times T/I_P$
	a = length of major axis, b = length of minor axis, L = length of shaft $I_P = \pi ab(a^2 + b^2)/64$, $A = \pi ab/4$ $K = GA^4/(4\pi^2 LI_P)$	$\tau = 0.5 \times a \times T/I_P$
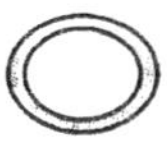	Outer Ellipse: a = length of major axis, b = length of minor axis Inner Ellipse: a_1 = length of major axis, b_1 = length of minor axis $I_P = \{\pi[ab(a^2 + b^2) - a_1b_1(a_1^2 + b_1^2)]\}/64$ $A = \pi(ab - a_1b_1)/4$, $K = GA^4/(4\pi^2 LI_P)$	$\tau = 0.5 \times a \times T/I_P$
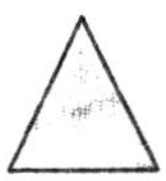	$I_P = \frac{b^4\sqrt{3}}{80}$, $K = T/\theta = GI_p/L$ b = length of equilateral triangle side	$\tau = 0.7454 \times b \times T/I_P$
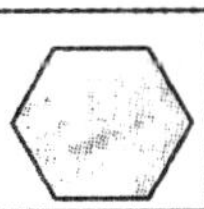	$I_P = 1.082b^4$, $A = 2.598b^2$ $K = GA^4/(4\pi^2 LI_P)$ b = length of uniform hexagon side	$\tau = 0.7454 \times b \times T/I_P$
	$I_P = [ab(a^2 + b^2) - a_1b_1(a_1^2 + b_1^2)]/12$ $A = ab - a_1b_1$ $K = GA^4/(4\pi^2 LI_P)$ a, a_1 = length of outer, inner larger side b, b_1 = length of outer, inner shorter side	$\tau = 0.5 \times (a^2 + b^2)^{1/2} \times T/I_P$
	$I_P = b^4/12$, $A = b^2$ $K = GA^4/(4\pi^2 LI_P)$ b = length of side of square.	$\tau = 0.7071 \times b \times T/I_P$
	K = torsional stiffness, I_P = section polar moment of inertia, A = section area	

(*d*) The most favorable tuning follows from $\omega_a/\Omega_n = 1/[1 + m/M] = 10/11$, so $(\omega_a/\Omega_n)^2 = 100/121$. Since *m*, *M*, and *K* are defined, $(\omega_a/\Omega_n)^2$ is proportional to *k*, and the new absorber spring stiffness will be:

$$k = 10.2 \times (100/121) = 8.43 \text{ lbs/in.}$$

Example Problem # 5.3: Explain the terms 'torsional rigidity' and 'polar modulus of section'. A solid steel shaft of 11.0 in. diameter is to be replaced by a hollow alloy steel shaft. If the polar modulus has the same value for both and the ratio of the diameters is 1.5, find the diameters. If alternatively the torsional rigidity is the same for the two shafts, calculate the ratio of the polar moduli.

Solution: The angle of twist is obtained from the torsion equation $\theta = TL/GI_p$. For a given torque and shaft length, the angle is determined by the product GI_p, and is called the torsional rigidity of the shaft. Maximum shear stress in the shaft is obtained from the expression $\tau_{max} = T/(I_p/R_{max})$. The ratio I_p/R_{max} is called the polar modulus of the section, analogous to section modulus in the beam bending theory. For a solid shaft of diameter *d* its value is $\pi d^3/16$.

The solid steel shaft of 11.0 in. diameter has $I_p = \pi d^4/32 = 1437.38$ in^4 and $I_p/R_{max} = 261.34$ in^3. If the polar modulus has the same value for both shafts, $261.34 = (2/1.5d) \times \pi/32 \times \{(1.5d)^4 - d^4\}$ and $d = 7.89$ in. So the inner diameter is 7.89 in. and the outer diameter is 11.84 in.

Example Problem # 5.4: Use the Holzer method to determine the torsional natural frequencies of vibration and their associated mode shapes for the rotor geometry defined in table 5.3. Shaft diameters are provided at station locations, and length defines axial offsets between the stations.

Solution: Expressions for torsional stiffness of the shaft segment are given in table 5.2. The values are provided in table 5.4. Since the beams connecting the mass elements do not have mass, calculated mass inertia of the beams are divided equally between the connected points. Mass moment of inertia of shaft segment is given by WR^2, where *R* is the radius of gyration and r_o and r_i are outer and inner shaft radii

$$WR^2 = \pi \times (r_o^2 - r_i^2) \times \rho \times l \times (r_o^2 + r_i^2)/4$$

The Holzer sequence of calculation is started with $\omega = 16469.7$ rpm, or 1724.7 radians/s. The calculations are shown in table 5.5. Rotor station numbers and the corresponding inertia constants are placed on alternate lines. β values opposite rotor station numbers define the first-mode relative amplitude curve, and torques between station numbers are shaft torques between corresponding rotor station numbers. Stiffness constants are placed on intermediate lines 1-2, 2-3, etc., in the last column. For the trial value of ω, $J\omega_s^2$ is computed next. An amplitude value of 1.0 is taken for β at station 1. In the torque column on line 1, $T = 1.0 \times 115913226$. The torque in the shaft section between stations 1 and 2 is equal to this inertia reaction. This torque divided by stiffness, 1974083933 in section 1-2, is the relative deflection of stations 1-2, and is placed in the β column of line 1-2. Subtracting the relative deflection 0.058718 from the deflection at rotor station 1 will give deflection value of 0.941283 at station 2. This deflection multiplied by 30618014 ($J\omega^2/g$) for station 2) gives inertia reaction 28820201 of station 2. The inertia torque of station 2

Table 5.3. *Shaft Geometry Definition*

Stn	Shaft Diameter, in.		Length in.	Material Properties		Added WR2
#	Outer	Inner	in.	G, lb/in2	r, lb/in3	lb-in2
1	14.775	12.85		1.12E+07	0.281	13452.25
			11.357			
2	16.325	13.25		1.10E+07	0.281	
			8.577			
3	17.535	14.125		1.09E+07	0.281	
			14.127			
4	19.289	14.25		1.08E+07	0.281	21247.3
			12.358			
5	22.875	14.25		1.06E+07	0.281	22574.75
			9.25			
6	23.5	14.25		1.06E+07	0.281	23874.12
			8.477			
7	24.445	14.25		1.08E+07	0.281	21475.55
			11.433			
8	22.975	14.25		1.09E+07	0.281	18957.43
			14.847			
9	21.772	14.25		1.10E+07	0.281	14587.23
			9.378			
10	21.175	14		1.11E+07	0.281	8754.44
			14.897			
11	18.125	14		1.14E+07	0.281	3425.45
			11.324			
12	16.347	14		1.17E+07		
			0.02			
13	0.25	0		5.00E+01	0.28	

is added to the torque in section 1-2 to give the torque in section 2-3. The sequence calculation is executed up to the last station number (13). The final Fig. in torque column gives residual torque value of (–2.15793E-6). Note that all additions and subtractions are done algebraically.

The residual torque has the following physical significance: If a simple harmonic torque of amplitude T_r and angular frequency ω is applied at station 10, then all rotors will oscillate with the amplitudes as calculated in the β column, and torque values will be as given in the T column. If the Holzer calculations are made over a sufficiently large range of ω to cover all the natural frequencies, T_r will follow the general shape of the curve shown in Fig. 5.13. The number of intersections the curve makes with the x-axis, excluding the origin, is equal to the number of rotor stations minus 1. The location of the trial values of ω^2 on this diagram can be determined from the sign of T_r and the number of changes in the sign of the relative amplitudes β of the rotor stations in the sequence table. In this example there is no sign change in β (except at station 13, which is nearly zero), and T_r is also small. Hence, the assumed value of ω^2 lies below the first mode.

Table 5.4. ***Shaft Inertia, Stiffness Characteristics***

Station #	J lb-in²	K in-lbs./radian
1	15049.3	
		1974083933
2	3975.2	
		5061931268
3	7711.0	
		4146128082
4	34864.4	
		8300784950
5	45696.0	
		26115735187
6	54130.7	
		32439009751
7	61799.9	
		29155262462
8	68169.8	
		17125838277
9	50761.2	
		21068762273
10	37329.0	
		11923385376
11	25562.6	
		6839453702
12	5432.9	
		1.89486E+12
13	4.551	

The first mode shape is defined by the deflection β values at the different stations, and is provided in table 5.6. A plot of the relative deflections along the length of the shaft is shown in Fig. 5.14. A review of the mode shape indicates that considerably greater level of rotation is taking place between axial locations from 80 ins. to the right hand end, with the remainder of the shaft showing lesser participation in this mode. Thus, if it is desired to increase this frequency, stiffening of this portion or reduction of the added masses will achieve this result.

The mode shape plot of Fig. 5.14 indicates that the zero axis is intersected once, hence the calculated mode corresponds to the primary natural frequency. The procedure for calculating the second and higher natural frequencies is identical to the one illustrated here, with the trial frequency used in the calculations higher than the preceding calculated natural frequency. The mode associated with a given natural frequency is identified by the number of times

Table 5.5. ***Holzer Calculation Sequence (Trial Frequency ω = 16469.7 rpm)***

Stn. #	J	Jw^2/g	Deflection, b	Torque, T	Stiffness, k
1	15049.3	115913227	1	115913226	
			0.058718	115913226	1974083933
2	3975.2	30618014	0.941283	28820201	
			0.028593	144733427	5061931268
3	7711.1	59392193	0.91269	54206660	
			0.047982	198940088	4146128082
4	34864.4	268533842	0.864708	232203322	
			0.05194	431143410	8300784950
5	45696.1	351961034	0.81277	286062587	
			0.027462	717205997	26115735187
6	54130.7	416927285	0.812768	327415155	
			0.032203	1044621152	32439009751
7	61799.9	475993855	0.753103	358472188	
			0.048125	1403093341	29155262462
8	68169.8	525059667	0.704978	370155346	
			0.103542	1773248688	17125838277
9	50761.2	390974012	0.601435	235145603	
			0.095326	2008394292	21068762273
10	37329	287516586	0.50611	145514930	
			0.180646	2153909222	11923385376
11	25562.6	196888715	0.325464	64080171	
			0.324293	2217989394	6839453702
12	5432.9	41845776	0.001171	48982	
			0.001171	2218038377	1.89E+12
13	4.551	35051	–6.16E-11	–2.16E-06	

Table 5.6. ***Mode Shape Corresponding to* ω = *16469.7 rpm***

Station #	Relative Rotation, Radians
1	1.000
2	0.9413
3	0.9127
4	0.8647
5	0.8128
6	0.7853
7	0.7531
8	0.7050
9	0.6014
10	0.5061
11	0.3255
12	0.0012
13	–6.16E-11

the mode shape intersects the zero deflection line. Thus, the second mode will see two sign changes in its mode shape, the third mode shape experiences three sign changes and so on. In this regard there is a close analogy to the natural frequencies and their mode shapes in the beam bending theory.

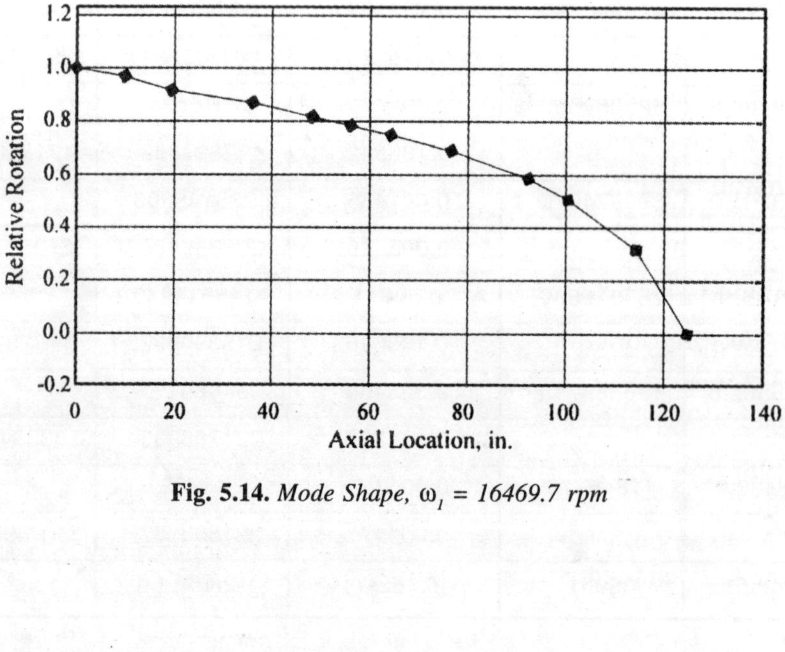

Fig. 5.14. *Mode Shape,* ω_1 = *16469.7 rpm*

REFERENCES & BIBLIOGRAPHY

American Petroleum Institute, "Centrifugal Compressors for General Refinery Service", API STD 617, 5th edition, Washington, D.C., 1988.

Anwar, I., "Computerized Time Transient Torsional Analysis of Power Trains", ASME paper No. 89-DET-74, New York, 1989.

Cataford, G.F., Lancee, R.P., "Oil Free Compression on a Natural Gas Pipeline", ASME paper # 86-GT-293, ASME, New York, June, 1986.

Constant, H., British. Aerospace Research Committee R. and M, No. 1201, 1928.

Crede, C.E., Ruzcicka, J.E., "Theory of Vibration Isolation", Shock & Vibration Handbook, McGraw-Hill, New York, 1988.

Den Hartog, J.P., "Mechanical Vibrations", Dover Publications, Inc., New York, 1985.

Ehrich, F.E., "Handbook of Rotor-dynamics", Krieger Publishing Co., Malabar, FL, 1999.

Eshelman, R.L., "Torsional Response of Internal Combustion Engines", Transactions, ASME, 96(2):441, 1984.

Eshelman, R.L., "Torsional Vibrations in Machine Systems", Vibration, 3(2):3, 1987.

Eshelman, R.L., Eubanks, R.A., "Effects of Axial Torque on Rotor Response: An Experimental Investigation", ASME paper # 80-WA/DE-14, New York, December, 1980.

Harris, C.M., "Shock and Vibration Handbook", Mc-Graw Hill, 4th edition, New York, 1995.

Holzer, "Die Berechnung der Dreschwingungen", Springer-Verlag, Berlin, 1921.

Jery, B., Acosta, A., Brennen, C., Caughey, T., "Hydrodynamic Impeller Stiffness, Damping and Inertia in the Rotor Dynamics of Centrifugal Pumps", CP2338, NASA, Pages 137-160, May, 1984.

Lewis, F.M., Transactions, ASME, 78:APM 377, 1955.

Myklestad, N.O., "Fundamentals of Vibration Analysis", McGraw-Hill, New York, 1956.

Nestorides, E.J., "A Handbook of Torsional Vibration", Cambridge University Press, 1958.

Porter, F., "Evaluation of Effects of Torsional Vibration", Society of Automotive Engineers, War Engineering Board; also Transactions, ASME, 65:A-33, 1943.

Porter, F., "Harmonic Coefficients of Engine Torque Curves", Journal of Applied Mechanics, 10(1):A33-A48, March, 1983.

Rangwala, A.S., "Reciprocating Machinery Dynamics – Design and Analysis", Marcel Decker, Inc., New York, 2001.

Rao, S.S., "Mechanical Vibration", Addison-Wesley Publishing Co., Reading, Mass., 1990.

Roark, R.J., Young, W.C., "Formulas for Stress and Strain", 5th edition, McGraw-Hill, New York, 1985.

Shemeld, D.E., "A History of Development in Rotor Dynamics, A Manufacturer's Perspective", CP2443, NASA, Washington, D.C., 1986.

Sohre, J.S., "Transient Torsional Critical Speeds of Synchronous Motor-Driven, High Speed Compressor Units", ASME paper No. 86-FE-22, June, 1985.

Taylor, C.F., "The Internal Combustion Engine in Theory and Practice – volume 2: Combustion, Fuels, Materials, Design", The MIT Press, Cambridge, Massachusetts, 1985.

Timoshenko, S., Young, D.H., Weaver, W., "Vibration Problems in Engineering", John Wiley & Sons, New York, 1984.

U.S. Navy Department: "Military Standard Mechanical Vibrations of Mechanical Equipment", MIL-STD-167 (SHIPS).

Wehrli, V.C., "Uber Kritische Drezahlen unter Pulsierender Torsion", Ingenieur Arch, 33:73-84, 1993.

Zdanowich and Wilson, T.S., Proceedings, Institute of Mechanical Engineers, London, 1943:182, 1940.

CHAPTER 6

Flexural Rotor Dynamics

6.1 INTRODUCTION

Consider a disk of mass m mounted on a shaft of length l between two bearings, and spinning at a fixed speed ω. Assume the center of gravity of the disk to be located at a radial distance e away from the centerline of the shaft. Then a rotating centrifugal force $m\omega^2 e$ acts on the disk. The rotating force vector may be divided into horizontal and vertical components, causing the disk to execute vibratory motions in the two directions. And if the spin speed ω coincides with the elastic shaft's natural frequency of vibration in the transverse direction, a state of resonance exists, and the shaft and disk assembly will experience considerable vibrations. Neglecting the effects of gravity, there are two forces acting on the disk, the elastic pull of the shaft tending to restore it to its original position and the centrifugal force on the center of gravity of the disk, causing it to travel in a circle. The elastic force depends on the bending stiffness of the shaft, and is proportional to its deflection. This observation is also true of a shaft with many disks running in two or more bearings.

The sag, or deflection, of the rotating shaft bent by various concentrated (such as the centrifugal force of the eccentric disk) or distributed (such as the weight of the rotor) loads is calculated by integrating a differential form that expresses the local deformation of a small element of the beam. By considering the local deformations, bending moment M is given by:

$$M = EI\frac{d\varphi}{dx} \tag{6.1}$$

where E is the Young's modulus of beam material, I is the area moment of inertia, or the second moment of the cross-section and $d\varphi$ is the change in the angle of slope of the deflection curve of a beam element of length dx. If the beam bends through only slightly, then approximately $\tan\varphi = \varphi = dy/dx$, and:

$$M = EI\frac{d^2y}{dx^2} \tag{6.2}$$

This differential equation of flexure of a beam is written to express the varying bending moment M along the length of the beam in terms of the deflection y and the location coordinate x. For small slopes the term d^2y/dx^2 is the curvature of the beam, and the quantity EI is referred to as the bending or flexural stiffness of the beam. The corresponding bending stress in the beam is determined from $\sigma_b = (My/I)$.

The objective in a rotor dynamic study is to determine the deflections y due to dynamic forces. Failures in turbo-machinery shafts result from fatigue due to stress as it fluctuates between extreme values. In addition to torsion, fibers on the surface of a rotating shaft are subjected to bending loads, undergoing both tension and compression for each revolution of the shaft. If the shaft of a centrifugal compressor is coupled to the shaft of an electric motor operating at 1,740 rpm, the shaft is stressed in tension and in compression 1,740 times. If the shaft is also axially loaded (caused, for example, by a helical or worm gear), the shaft has an axial stress superimposed upon the bending component. These and other loads occurring in the shaft produce stresses often referred to as repeated, alternating or fluctuating stresses.

Bending in the shaft may arise from a number of operational factors, some of which arise internally in the machine and others from outside the machine. Dynamic balance of the machine's rotating assembly is an internal machine factor that causes bending in the shaft. As a result of manufacturing procedures and limitations, tolerances have to be provided on machined components. Thus, a minimum diameter shaft journal may be paired with a maximum diameter bearing, causing a larger than normal radial clearance in the bearing. In contrast with torsional vibrations, bearing and support stiffness plays a significant role in the shaft's lateral vibrations. Since the hydrodynamic stiffness of the bearing's lubricant film is reduced as a consequence of the larger bearing clearance, the shaft's natural frequency of vibration also is reduced, causing the machine to operate at or close to the natural frequency. Rotor unbalance arises from many sources attributable to manufacturing and field operation. With the appropriate amount and location of unbalance in the rotor, substantial bending moments are caused when operation takes place near a critical speed, resulting in flexural shaft vibrations along its lateral axes.

Shaft misalignment is another factor responsible for flexural vibrations. At the coupling between an internal combustion engine and an electric power generator thermal conditions may throw out the rotor from an aligned position during cold condition. Radial and angular misalignment of the rotor can radically alter the flexural vibration signature of a rotating assembly, with consequent bending moments in the shaft. This also has the adverse effect of putting unusual wear pattern in the bearings.

For flexural vibrations the most commonly encountered stimulus is rotor unbalance and shaft bow, resulting in whirling or whipping of the shaft. Unstable operation of the machine in the form of chatter (or stick-slip) and large deflections and loads are other undesirable consequences. To deal with such vibrations appropriate design strategies are required. For instance, placing the system's critical speeds out of the steady state operating range will minimize the degree of super-critical whirling and whipping phenomena. If for practical reasons it is not possible to raise the critical speed, or operation above the fundamental mode of vibration cannot be avoided, introduction of damping to reduce critical response peaks and to cancel the destabilizing forces in the operating range will be required. Minimizing the excitation through better balancing and aligning procedures will go a long way towards the stated objective of smoother operation of the machine.

An easy to use and accurate computational tool for prediction of the vibration characteristics of any equipment design is an indispensable requirement in the design process. The design process is too complex to permit automation in a direct sequence from specification to definition of the finished design. In spite of this drawback, a direct sequence can be approximated by a speedy iterative process that starts with posing design alternatives, then analytically evaluating them, followed by selecting and quantitatively refining the design that satisfies design requirements. Such analytical tools are made available in the form of computer codes, and are discussed in this and other chapters of this treatise. Analytical methods are useful for predicting the rotor dynamic behavior of the preliminary design configuration, to assess its impact and to provide quantitative data for design refinement. The mathematical model developed provides valuable insight during testing of a new machine; when unexpected data from accelerometers during operations indicates trouble, analytical models are studied to provide understanding and guidance in remedial design, fault detection and correction. The analytical model may range from an idealized version of a single mode to refined and detailed finite element model of the total system to provide precise information in the form of natural frequencies, mode shapes, deflections and loads throughout the assembly.

Mathematical models for determining the vibration characteristics of a rotating machine simulate the kinematics of important degrees of freedom. They permit the determination of the response of the system due to the interaction of the geometric configuration of the rotor, bearings and support structure, with the dynamic loads acting on the system. The analytical model is used to analyze free vibrations, forced vibrations and self excited vibrations.

For each cycle or revolution of the shaft the most careful analysis may reveal that the maximum dynamic stresses are below the ultimate strength of the material, and quite frequently even below the yield strength; even then a fatigue failure is encountered. The primary distinguishing characteristic of fatigue failures is that the stresses have been repeated a very large number of times, and hence the failure is called a fatigue failure. A fatigue failure begins with a small crack, which may be small enough to prevent detection by the naked eye or even to locate in a Magnaflux or x-ray inspection. The crack will develop at a point of discontinuity in the material such as a keyway, a change in the cross section or a hole. Less obvious points at which fatigue failures are likely to start are inspection or stamp marks, internal cracks or irregularities caused by machining. Once a crack has developed, the stress concentration effect is increased and the crack propagation progresses at a faster pace. As the stressed area decreases in size, the stress increases in magnitude until the remaining area fails abruptly. The fatigue failure is characterized by two distinct areas of failure, the first due to progressive development of the crack and the second due to sudden fracture. This form of failure is in contrast to machine parts failing from a static loading; in this case the parts develop large deflection because the stress has exceeded the yield strength. Thus, many static failures are visible ones and give advance warning. Fatigue failures give no warning; they are sudden, total and hence dangerous. It is relatively easier to design against a static failure since load conditions are better defined and understood. Fatigue is a more complicated phenomenon, and requires intensive design engineering effort to come up with a successful product.

6.2 DYNAMICS OF SHAFT MOTION

Transverse, or lateral direction, vibrations of a shaft are analytically different from torsional and axial vibrations in several respects:

- The deformation of the rotor is due to a combination of bending and shearing, and is inherently more complex than simple stretching or twisting.
- Shaft bending is not confined to a single plane, and motion must be looked at in two orthogonal planes. Within each plane it is necessary to characterize the motion of each rotor station by two independent parameters, displacement and angular rotation.
- Gyroscopic and other asymmetric effects cause the motion of the shaft in the horizontal and vertical directions to be coupled.
- The location and dynamic characteristics of the bearings play a substantial role in transverse direction vibrations. This calls for development of detailed models of the bearings and their support structure.

Despite these complications, it is possible to develop equations of motion for transverse vibration of the rotor that are analogous to torsional vibrations. To establish some important concepts, consider a simplified model of a high-speed rotor that retains many of the essential characteristics of the complex system in its response to unbalance.

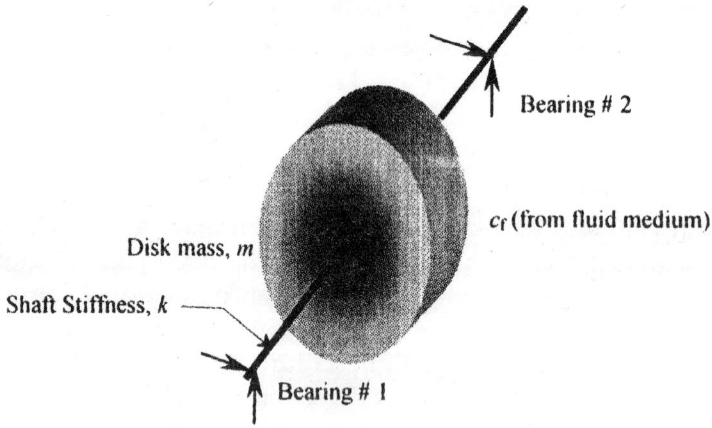

Fig. 6.1. *Shaft Supported in Two Bearings*

A disk is carried at the center of a flexible but massless shaft supported in two bearings (Fig. 6.1). Two types of bearing supports will be considered, rigid and flexible. In the model with rigid bearings the only damping c_f will come from the fluid medium surrounding the disk. Two types of unbalance must be considered separately, one due to eccentricity in the disk where the center of the disk is offset from the elastic axis of the shaft by a radial distance q, the other due to shaft bow, s.

In the eccentric disk case the whirl radius at disk r_{wd} is calculated from the steady state solution of the differential equations for response to rotating unbalance (Myklestad, 1956):

$$r_{wd1} = \frac{q(\omega/\omega_{ns})^2}{\{[1-(\omega/\omega_{ns})^2]^2 + (2\xi_f\omega/\omega_{ns})^2\}^{1/2}} \tag{6.3}$$

where ω is the rotational speed, $\omega_{ns} = \sqrt{(k/m)}$ is the undamped natural frequency, m is the disk mass, k is the rotor stiffness and $\xi_f = c_f/(2m\omega_{ns})$ is the damping ratio. To understand the nature of critical speed consider a situation where the rotor has no unbalance, or $q = 0$, and no damping. The inertia force $m\omega^2 r_{wd}$ will equal shaft elastic spring force kr_{wd}. At the critical speed this force balance is satisfied for all values of whirl radius. There exists a state of indefinite unbalance, which will be easily upset by the slightest amount of unbalance or damping. Thus, a critical speed is defined as one where the inertia forces and elastic restoring forces are in balance for an undamped system.

For the case of the bowed shaft, assume mass eccentricity q to be zero. The rotor would be in perfect balance except for a permanent bow in the elastic axis of the shaft, which could be due to thermal distortion during operation, manufacturing deviation or physical damage. Thus, at standstill the elastic center is assumed to be displaced from the rotation axis by a radial distance s, the amount of shaft bow. A steady state solution of the differential equations yields:

$$r_{wd2} = \frac{s}{\{[1-(\omega/\omega_{ns})^2]^2 + (2\xi_f\omega/\omega_{ns})^2\}^{1/2}} \tag{6.4}$$

Here, too, an analogy with planar beam vibration exists. If a radial sinusoidal force is applied to the stationary disk the resulting planar beam vibration is governed by the same equations that apply to the whirling motion of the bowed rotor. However, the analogy is limited by factors such as gyroscopic stiffening associated with the mass moment of inertia of the rotor, internal damping in the rotor and certain system parameters (as in fluid film bearings) which are dependent on shaft rotation.

When the shaft is supported on flexible bearing supports both the stator and the rotor are treated as flexible elements. Damping in the system, c_b, will be associated with the bearing support only. If the rotor is infinitely stiff, the undamped critical speed is given by $\omega_{nb} = \sqrt{(K_s/m)}$, where K_s is support stiffness and the new damping ratio $\xi_b = c_b/(2m\omega_{nb})$. If ω_{nb} is substituted for ω_{ns} and ξ_b for ξ_f, then equations (6.5) and (6.6) for the flexible rotor on rigid bearing supports

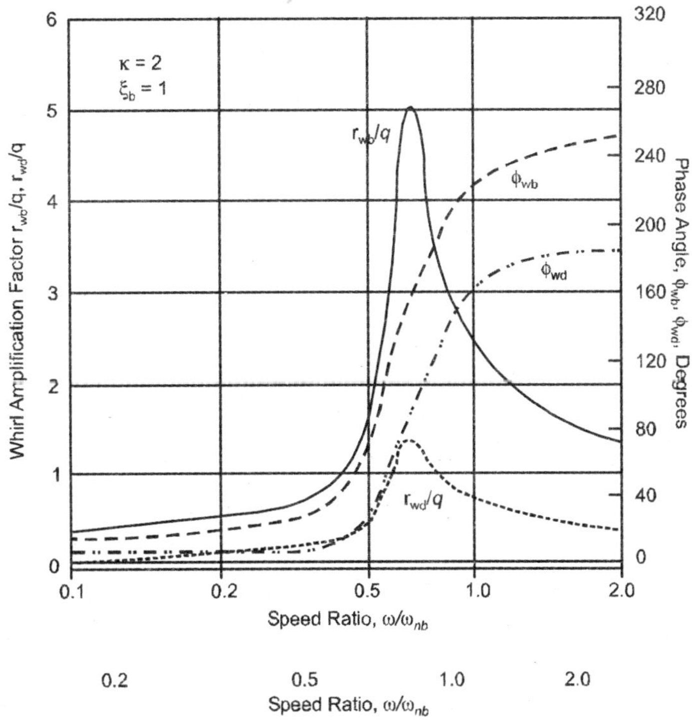

Fig. 6.2. *Shaft Supported in Flexible Bearings – Response to Unbalance*

may be used to represent the response to the unbalance of the rigid rotor on flexible supports. A steady state solution of the differential equations of motion for the full system with a flexible rotor and flexible supports yields the following response at the bearing and at the disk to unbalance:

$$r_{wb} = \frac{q(\omega/\omega_{nb})^2}{[c_1^{\,2} + (c_3 c_2)^2]^{1/2}} \tag{6.5}$$

$$\phi_{wb} = \arctan\left(\frac{-c_3 c_2}{c_1}\right) \tag{6.6}$$

$$r_{wd} = r_{wb}[(1+\kappa)^2 + (c_3\kappa)^2]^{1/2} \tag{6.7}$$

$$\phi_{wd} = \arctan\left[\frac{-c_3}{(1+\kappa)c_1 + c_3^{\,2}\kappa c_2}\right] \tag{6.8}$$

where

$$c_1 = 1-(1+\kappa)(\omega/\omega_{nb})^2$$

$$c_2 = 1-\kappa(\omega/\omega_{nb})^2$$

$$c_3 = 2(\omega/\omega_{nb})\xi_b$$

κ = stiffness ratio of support and rotor = K_s/k

The phase angle is the relative angle between the displacement vector and the unbalance vector q. The speed at which the whirl amplitude peaks is referred to as the damped, or actual, critical speed. A common design practice is to plot the response at the disk and at the bearing as functions of the ratio of the actual speed to the undamped critical speed. Figure 6.2 shows a plot for a given set of parameter values. It is of special interest to obtain values of maximum response and the corresponding speed at which it occurs. Two significant observations can be made here. For a given damping level shaft response increases substantially with an increase in the stiffness ratio κ. Thus, as the bearing support stiffness increases (or shaft stiffness decreases), the rotor becomes more responsive to unbalance. For a given stiffness ratio, shaft response minimizes at some intermediate value of damping. As damping is reduced to negligible levels, the shaft response reaches infinity. Also, as damping becomes very large, shaft response becomes very large because of rigidity in the bearings.

6.3 SUPPORT FLEXIBILITY AND CRITICAL SPEEDS

In the discussion so far the effect of mass distribution along the length of the rotor is absent, and so its usefulness for long rotors is limited. In addition to the fundamental, or first, critical speed, with distributed mass there will be an infinite number of critical speeds. Each critical speed has a characteristic mode shape. To understand this, consider a rotor of uniform cross section. Linn *et al.*, (1951) calculated the effect of bearing flexibility for the first four critical speeds of a uniform undamped rotor. The solution is derived from the differential equation of the elastic curve of the whirling shaft. The first critical speed on rigid bearings is given by:

$$\omega_1 = \pi^2 \times \sqrt{\frac{gEI}{WL^3}} \tag{6.9}$$

where

g = acceleration due to gravity

E = shaft material modulus of elasticity

I = moment of inertia of shaft cross section

W = weight of rotor

L = length of rotor.

Figure 6.3 shows the impact of bearing flexibility on the first four critical speeds of a uniform rotor. A parameter representing support flexibility is plotted along the abscissa and the ratio of critical speed to first critical speed on rigid bearings along the ordinate. For a given value of rigid bearing, the first and second critical speeds continuously decrease with increasing bearing flexibility. This is accompanied by a gradual transformation of the mode shapes for these critical speeds. The numerical value between the first two critical speeds is equal to $\sqrt{3}$ for large values of bearing flexibility, the ratio obtained from theoretical considerations for rigid body modes. When bearing flexibility approaches infinity, these rigid body critical speeds approach zero. The third and fourth critical speed ratios initially decline rapidly with increasing bearing flexibility, then level out asymptotically to definite values. The third and fourth modes are referred to as free-free modes. This arises from the fact that conditions in the bearings at both ends have little impact on them.

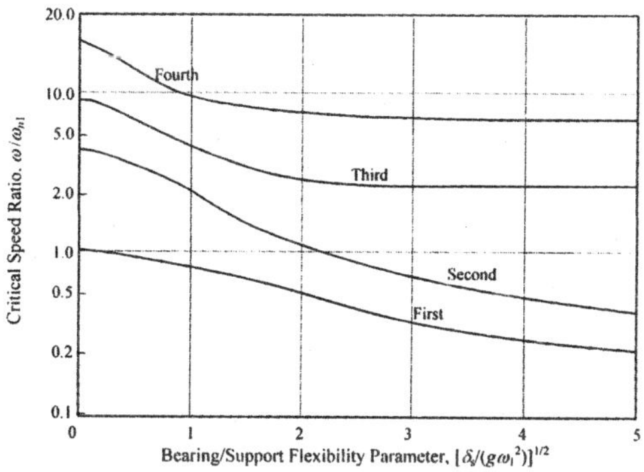

Fig. 6.3. *Critical Speed as Function of Bearing Flexibility – Uniform Rotor*

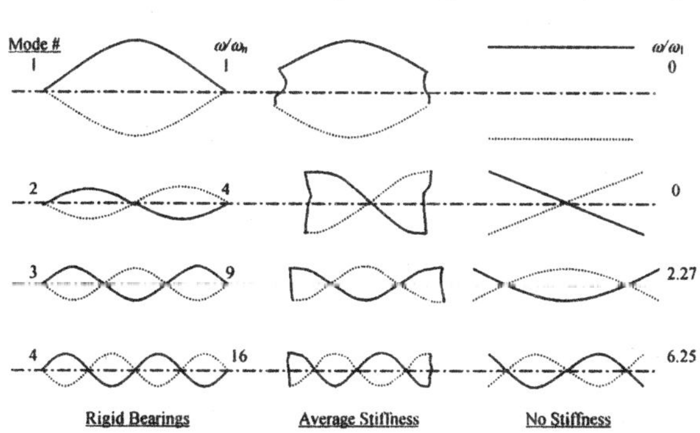

Fig. 6.4. *Mode Shapes of Uniform Rotor as Function of Bearing Stiffness*

The location of the bearings along the length of a rotor will have a major effect on shaft response to unbalance. For instance, if the bearings were placed closer to each other and away from the ends to the nodal points of the first mode of the free-free rotor, as shown in Fig. 6.4, supporting structure damping will not be effective. Excessive response to unbalance will be experienced when operating near the first critical speed.

6.4 PROCEDURE FOR CRITICAL SPEEDS

The direct stiffness method or the transfer matrix method may be used for calculating free and forced response of a rotating system. Both procedures are developed using force-displacement and force-velocity equations, together with momentum principles. The transfer matrix method, an extension of the Holzer method for torsional vibrations, is well suited for analyzing rotating systems. Improvements to the system by Myklestad (1956) and Lund (1964) have made it suitable for computations of lateral vibrations of rotating beam structures. It does not require storage of large system arrays, and is discussed next in detail.

The method utilizes a sequential procedure, starting with boundary conditions at one end and proceeding along fixed locations in the structure to the other end. Elastic and inertia properties of the system are picked up as the calculation sequence proceeds for any shaft speed. Satisfaction of the boundary conditions at the other end for a specified speed value yields the natural frequency of vibration.

Figure 6.5 illustrates a rotor model for these calculations. Several points along its length are selected to represent the geometric configuration of the system in terms of stiffness and inertia of the rotor and its supports, and are labeled from 1 to *n* from one end of the system to the other. The state vector at a point defines displacements and forces in the element, and is expressed as:

$$\{z\} = [u\ v\ \theta\ \phi\ M_y\ M_x\ V_y\ V_x]^T \tag{6.10}$$

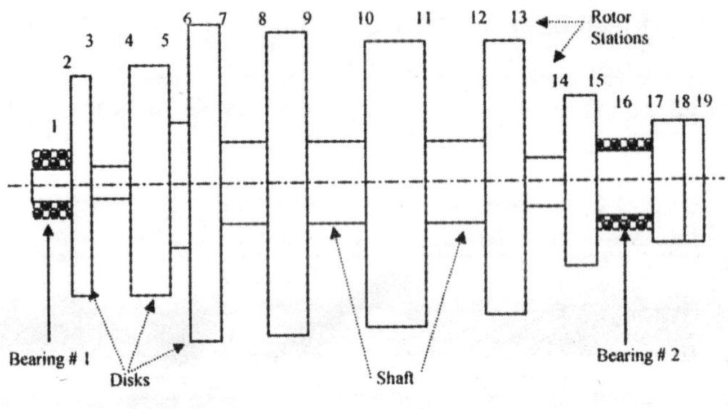

Fig. 6.5. *Rotor Model for Lateral Vibrations*

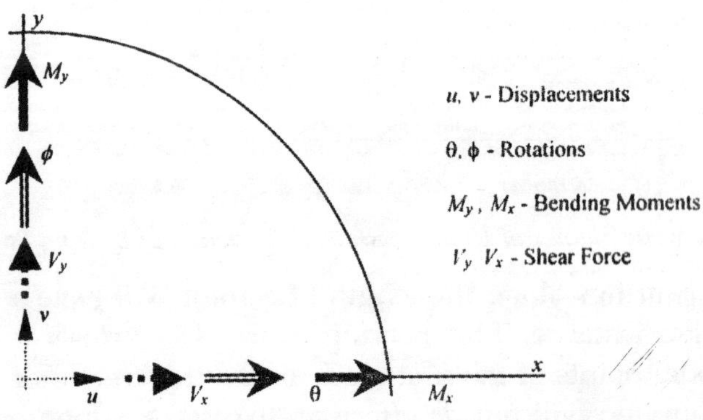

Fig. 6.6. *Forces and Displacements on a Beam's Cross-Section*

The forces and displacements are as shown in Fig. 6.6. Positive internal forces act in positive coordinate directions on beams whose cross sections have a positive outward normal. Displacement u is along the horizontal x direction and v along the vertical y direction. Corresponding rotations are θ and ϕ. M_x and M_y represent bending moments and V_x and V_y give shear forces in the beam. The 'state' on the left hand side of a rotor station will be different from that at the right side, primarily due to the presence of inertia mass or support stiffness at the station. Superscripts R and L denote the right and left sides of the station i. The equations of motion for the rigid disk with a bearing support is:

$$\begin{bmatrix} m & 0 & 0 & 0 \\ 0 & m & 0 & 0 \\ 0 & 0 & I_d & 0 \\ 0 & 0 & 0 & I_d \end{bmatrix}_i \begin{Bmatrix} d^2u_i/dt^2 \\ d^2v_i/dt^2 \\ d^2\theta_i/dt^2 \\ d^2\phi_i/dt^2 \end{Bmatrix} + \begin{bmatrix} c_{xx} & c_{xy} & 0 & 0 \\ c_{yx} & c_{yy} & 0 & 0 \\ 0 & 0 & 0 & \Omega I_a \\ 0 & 0 & -\Omega I_a & 0 \end{bmatrix}_i \begin{Bmatrix} du_i/dt \\ dv_i/dt \\ d\theta_i/dt \\ d\phi_i/dt \end{Bmatrix}$$

$$+ \begin{bmatrix} k_{xx} & k_{xy} & 0 & 0 \\ k_{yx} & k_{yy} & 0 & 0 \\ 0 & 0 & 0 & 0 \\ 0 & 0 & 0 & 0 \end{bmatrix}_i \begin{Bmatrix} u_i \\ v_i \\ \theta_i \\ \phi_i \end{Bmatrix} = \begin{Bmatrix} V_x^R - V_x^L \\ V_y^R - V_y^L \\ M_x^R - M_x^L \\ M_y^R - M_y^L \end{Bmatrix}_i + [m\Omega^2]\mathrm{Re}(e^{i\Omega t}) \begin{Bmatrix} a_{x'} + a_{y'} \\ -ia_{x'} + a_{y'} \\ 0 \\ 0 \end{Bmatrix} \quad (6.11)$$

The rotor stations are connected by elastic beam sections, and are modeled to exclude their mass. Shear deformation of the beam is included, but its internal damping is ignored. Two different transfer matrices are used. A point transfer matrix provides transfer of the state vector at a station from its left side to its right side. A typical rotor point may include a rigid disk and a bearing support, so the point transfer matrix will include the disk's inertia properties and the bearing's dynamic stiffness and damping coefficients. I_d represents the mass moment of inertia about its lateral axis and I_a about the polar axis. $a_{x'}$ and $a_{y'}$ give the mass center component in the rotating reference frame. Ω represents the shaft spin velocity. The disk and the bearing do not have an axial length, so displacements on the right and left hand sides of the point will be identical. Hence:

$$[u_i^L \ v_i^L \ \theta_i^L \ \phi_i^L] = [u_i^R \ v_i^R \ \theta_i^R \ \phi_i^R] \quad (6.12)$$

For free vibration the solution of (6.11) will be of the form $\{z_i\} = \{\bar{Z}_i\}e^{\lambda t}$, and may be expressed as:

$$\begin{Bmatrix} u \\ v \\ \theta \\ \phi \end{Bmatrix}_i = \begin{Bmatrix} \bar{u} \\ \bar{v} \\ \bar{\theta} \\ \bar{\phi} \end{Bmatrix}_i e^{\lambda t} \quad \text{with} \quad \begin{Bmatrix} V_x \\ V_y \\ M_x \\ M_y \end{Bmatrix}_i = \begin{Bmatrix} \bar{V}_x \\ \bar{V}_y \\ \bar{M}_x \\ \bar{M}_y \end{Bmatrix}_i e^{\lambda t} \quad (6.13)$$

where $\lambda = \sigma + i\omega$ represents an eigen-value, ω being the damped natural frequency and σ the damping coefficient. Substitution of (6.13) into (6.11) and rearrangement of terms provides the expression:

$$\{\bar{Z}_i\}^R = [P_i]\{\bar{Z}_i\}^L \quad (6.14)$$

where $$[P_i] = \begin{bmatrix} [I_i] & [O_i] \\ [B_i] & [I_i] \end{bmatrix} \tag{6.15}$$

Here $[I]$ is a 4 x 4 unity matrix, $[O]$ is a null matrix and $[B]$ is defined by:

$$[B] = \begin{bmatrix} 0 & 0 & -\Omega I_a\lambda & I_d\lambda^2 \\ 0 & 0 & I_d\lambda^2 & \Omega I_a\lambda \\ b_{yx} & b_{yy}+m\lambda^2 & 0 & 0 \\ b_{xx}+m\lambda^2 & b_{xy} & 0 & 0 \end{bmatrix}$$

The terms $b_{yx} = c_{yx}\lambda + k_{yx}$ and others similar to it provide point transfer matrix for a rigid disk or bearing point.

For a beam with uniform cross section the field transfer matrix connects two rotor stations i and $i + 1$. Elements of the transfer matrix will include shaft material Young's modulus E, shear modulus G, cross section area A and moment of inertia I and length of shaft segment L between the stations. Since the shaft's mass inertia is included with the disk, the equilibrium requirements in the horizontal plane are:

$$(V_x^R)_i - (V_x^L)_{i+1} = 0$$

and

$$(M_y^L)_{i+1} - (M_y^R)_i + (V_x^L)_{i+1}L_i = 0 \tag{6.16}$$

From Timoshenko's beam theory, expression for deflections in the horizontal plane are given by:

$$u_{i+1} = u_i + \phi_i L_i + (M_y^L)_{i+1}\frac{L_i^2}{2EI_i} + (V_x^L)_{i+1}\frac{L_i^3}{3EI_i}(1+3\varepsilon_i) \tag{6.17}$$

$$\phi_{i+1} = \phi_i + (M_y^L)_{i+1}\frac{L_i}{EI_i} + (V_x^L)_{i+1}\frac{L_i^2}{2EI_i} \tag{6.18}$$

where $\varepsilon = EI/(\kappa GAL^2)$ and κ defines modal stiffness shape factor. Similar relations may be written for the vertical plane motion, with some changes to accommodate the sign requirements. Rearranging the terms in (6.17) and (6.18) and including requirements from (6.16) produces the expression:

$$\{\bar{Z}_{i+1}\}^L = [F_i]\{\bar{Z}_i\}^R \tag{6.19}$$

where: $$[F_i] = \begin{bmatrix} [S] & [T] \\ [O] & [W] \end{bmatrix} \tag{6.20}$$

$$[S] = \begin{bmatrix} 1 & 0 & 0 & L \\ 0 & 1 & -L & 0 \\ 0 & 0 & 1 & 0 \\ 0 & 0 & 0 & 1 \end{bmatrix}$$

$$[T] = \begin{bmatrix} L^2/2EI & 0 & 0 & -L^3(1-6\varepsilon)/6EI \\ 0 & -L^2/2EI & -L^3(1-6\varepsilon)/6EI & 0 \\ 0 & L/EI & L^2/2EI & 0 \\ L/EI & 0 & 0 & -L^2/2EI \end{bmatrix}$$

$$[W] = \begin{bmatrix} 1 & 0 & 0 & -L \\ 0 & 1 & L & 0 \\ 0 & 0 & 1 & 0 \\ 0 & 0 & 0 & 1 \end{bmatrix}$$

and $[O]$ is a null matrix.

Equation (6.20) provides the field matrix for an elastic shaft segment. Pre-multiplication of the rotor's state vector at the left end of the segment just to the right of station i by this matrix determines the rotor's state at the right end of the segment, just to the left of station $i+1$.

For a single shaft machine the system transfer matrix is obtained with the help of the point and field transfer matrices, which sequentially transfer the rotor's state from the left end at station 1 to the right end at station n. Sequential application of the state and field transfer matrices gives the following expressions:

$$\{\bar{Z}_1\}^R = [P_1]\{\bar{Z}_1\}^L$$

$$\{\bar{Z}_1\}^L = [F_1]\{\bar{Z}_1\}^R$$

or

$$\{\bar{Z}_1\}^R = [M]\{\bar{Z}_1\}^L \qquad (6.21)$$

where

$$[M] = [P_n]\prod_{i=n-1}^{1}[F_i][P_i] \qquad (6.22)$$

provides the overall system transfer matrix.

The procedure may be readily generalized for multiple shafts. In the two shaft system of Fig. 6.7 the shafts are denoted by A and B. The system matrix proceeds as for a single shaft, as also the sequential calculations, until it reaches the interconnecting member at the gear and pinion set between the two shafts. A transfer across the gear set requires an inter-connecting transfer matrix.

For a specified shaft spin speed the system transfer matrix is generated. Boundary conditions at the left and right ends of the system are then imposed. Typical end conditions for the horizontal planes may be as follows: Free condition: $V_x = 0$, $M_y = 0$; pinned condition: $u = 0$, $M_y = 0$; fixed condition: $u = 0$, $\phi = 0$. Similar conditions are set up for the vertical plane. As a convenience in calculations, the system matrix may be rearranged and partitioned after the introduction of the boundary conditions so the zeros appear as a subset, as shown below:

$$\begin{Bmatrix} \{z^R\} \\ \{0\} \end{Bmatrix} = \begin{bmatrix} [M_{11}] & [M_{12}] \\ [M_{21}] & [M_{22}] \end{bmatrix} \begin{Bmatrix} \{z^L\} \\ \{0\} \end{Bmatrix} \qquad (6.23)$$

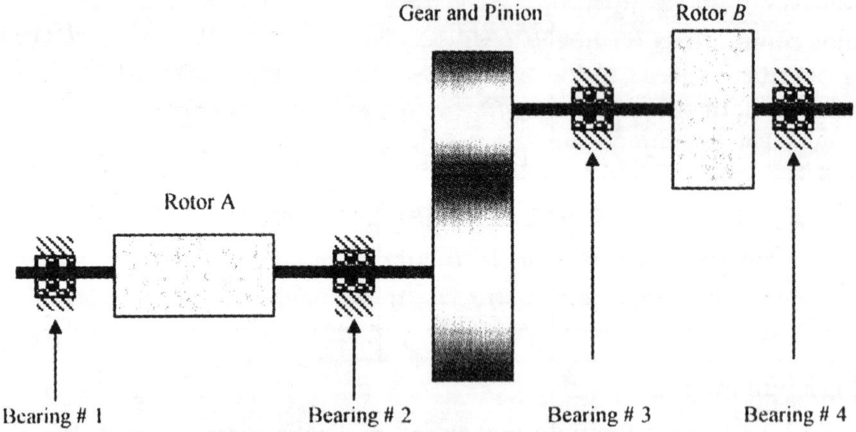

Fig. 6.7. *Two-Shaft Rotating System*

Then from the lower portion of (6.23) a homogeneous expression is obtained:

$$[M_{21}(\lambda)]\{z^L\} = \{0\} \tag{6.24}$$

and a non-trivial solution for $\{z^L\}$ exists only if the coefficient determinant vanishes. Corresponding values for λ provide the system eigen-values.

For an undamped rotor with isotropic supports the orbits of the natural modes of whirl are circular relative to the fixed reference frame. Because of the axial symmetry the order of the system equations may be reduced to half. The point and the field transfer matrices may then be reduced from 8×8 to 4×4 matrices.

6.5 CRITICAL SPEED POSITIONING

Positioning of critical speeds in the operating speed range of a machine plays a major role in the design of rotors. A simple check of the undamped natural frequencies of vibration using proven support stiffness assumptions suffices in many cases. New machines under development and new applications call for a more detailed evaluation using appropriate computational methods. Evaluation of damped response to rotor unbalance is of special importance. Rotor instability as a source of excitation should also be explored.

The design of the rotor and bearing system must be carefully examined for forced synchronous vibrations resulting from residual unbalance. Residual unbalance may be due to a lack of coincidence of axis of the mass centers with the elastic axis of the shaft. It could also be the result of a bow in the elastic axis, or due to a combination of the two factors. Rotor bow often results from a temporary lack of symmetry in the temperature distribution in the rotor. Steam turbine rotors used in power generating plants have been known to end up with a permanent bow at startup or shutdown times if the machine is not permitted to soak the temperatures at the proper speed level and time period. The soak cycle is required since the thickness of the shaft and length between bearing centerlines is large. A similar effect is encountered if splines or couplings are not properly fabricated and assembled. Although differences exist in the response curves away from critical speeds for the two types of unbalance, the importance of positioning of the critical speed in the operating speed range of the machine cannot be overstated.

The discussion will be illustrated for a single shaft two bearing gas turbine shown in Fig. 6.8. All major components of the system will be assumed to be axially symmetric. For proper positioning of critical speeds in the operating speed range of the machine, stiffness and inertia characteristics of both the rotor and the bearings and its supports at the right axial location must be known. The machine stator is generally more rigid than the rotor in large industrial gas turbines, and is not included.

The modes of vibration of the system may be divided into two parts, one where the rotor remains mostly rigid, the other where the rotor undergoes flexure and whirls about the unbent axis.

There are two rigid body modes of the machine. In the first or fundamental mode the rotor supports move in phase with one another, and the shaft axis describes a cylindrical motion. In the second rigid body mode at a higher frequency the supports move out of phase with one another, with a node near the midpoint of the rotor.

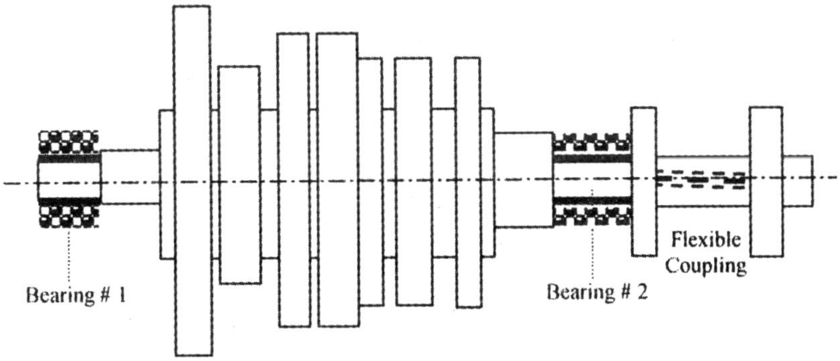

Fig. 6.8. *Gas Turbine Rotor*

The rotor centerline describes two cones with their apex at the node. Usually these frequencies occur well below the frequency of the shaft flexural mode. If the stator mass is very large compared to the rotor mass, low rigid body mode frequencies will prevail, even with stiff mounts. When the rotor is operated at a speed at which these rigid body modes are excited, the rotor and stator will move in phase with little relative motion between them, assuming that the bearings are adequately stiff. With limited motion between rotor and stator, internal structural integrity of the machine will not be affected. There may, however, be undesirable transmission of vibrations to the surroundings.

From design considerations the rotor's flexural modes are of greater interest, and sometimes of concern. At each flexural mode the rotor whirls with a characteristic deflection shape, and the rotor and stator can vibrate out of phase. Thus, the potential for rotor rubbing against the stator exists when machine operation occurs at, or while passing through, a rotor flexural mode. Rubbing of seals and other closely controlled clearance areas may take place, with a resultant loss in efficiency and functionality. In extreme cases major damage to the inside parts of the machine may occur, and in some cases to the supporting structure as well.

In order to define typical options available for the positioning of the rotor's flexural frequencies that are applicable to many different machine types, some examples will be discussed.

First, consider a single shaft, two bearing gas turbine where the operating speed range includes the first flexural rotor mode of vibration, of the type shown in Fig. 6.8. The rotor span between the bearings when evaluated in rigid bearings has its lowest critical speed at 4,474 rpm, and the unit is required to operate up to a maximum speed of 6,200 rpm. The rotor is supported on journal bearings attached to a stiff and massive stator. The ratio of stator mass to rotor mass is 8. Machine stator is supported on a structural steel framework that is determined to have a natural frequency of 1,200 cycles/min. Damping in the steel framework is assumed to result in peak amplification factor (1/2ξ) of 10, implying a value of 5 percent of critical damping. The rotor may be considered detached from driven machinery as far as lateral vibrations are concerned.

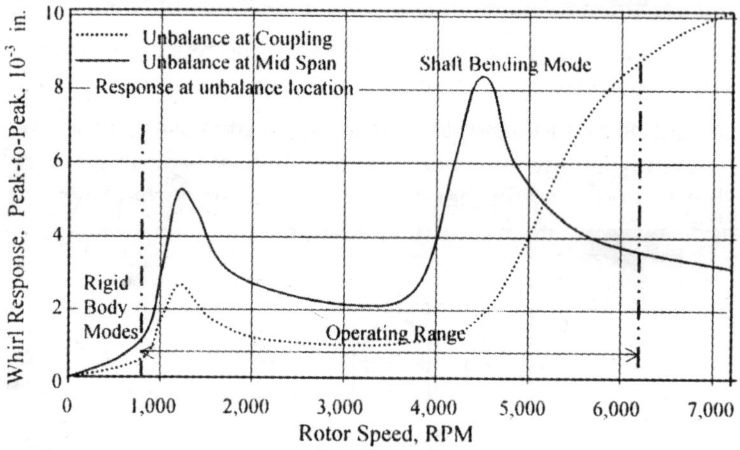

Fig. 6.9. *Calculated Rotor Response due to Unbalance*

The first step in the design process is to calculate the undamped critical speeds, and is determined to be 3,890 rpm for the first mode and 7,580 rpm for the second mode. Of special significance is the response to rotor unbalance, especially since machine operation takes place through the first critical speed. The shape of the modes dictates that unbalance at mid span portion of the rotor will be of interest for the first mode. Unbalance located at the ¼ and ¾ span length from either bearing will result in high deflections in the rotor when operation of the machine is near the second mode of vibration. Unbalance or misalignment at the flexible coupling will also be sensitive in the second mode. Coupling unbalance, however, will have little effect in the first mode. If operation were required through the second critical speed on this machine, then special attention would be needed to balance the coupling. Unbalance here could cause excessive whirl at the overhung coupling, together with high dynamic loads in the bearing at the higher operating speed. Rotor response at the location of unbalance is shown in Fig. 6.9.

Next, consider the case of a power generation steam turbine rotor with two spans and a total of four bearings, connected by a spacer type of coupling (Fig. 6.10).

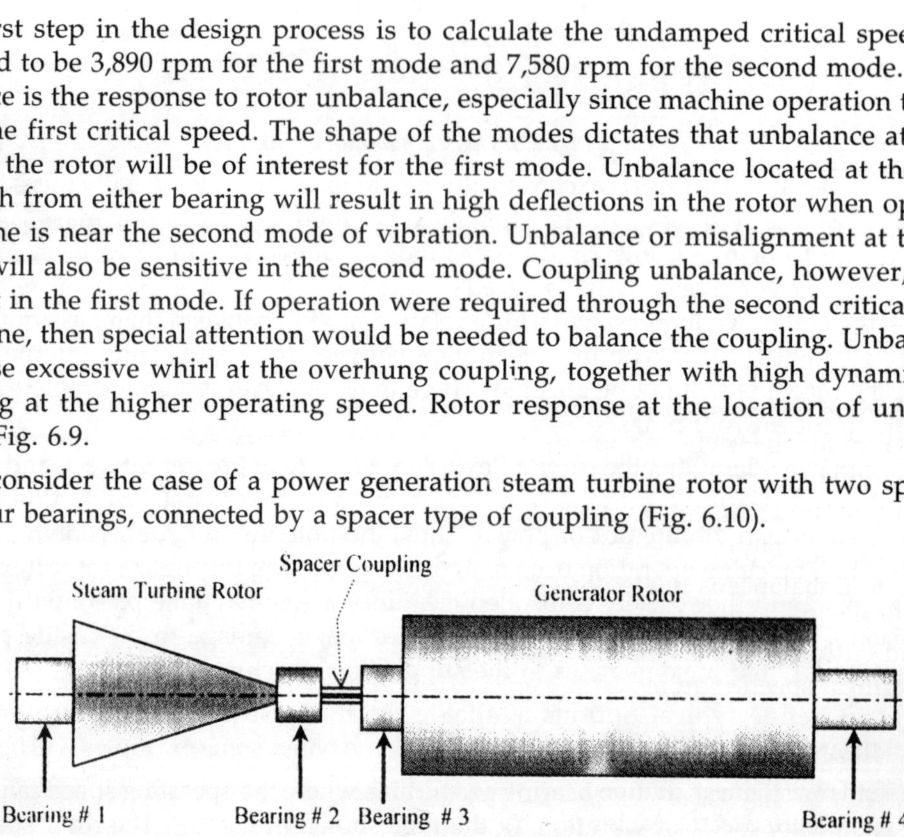

Fig. 6.10. *Steam Turbine/Generator*

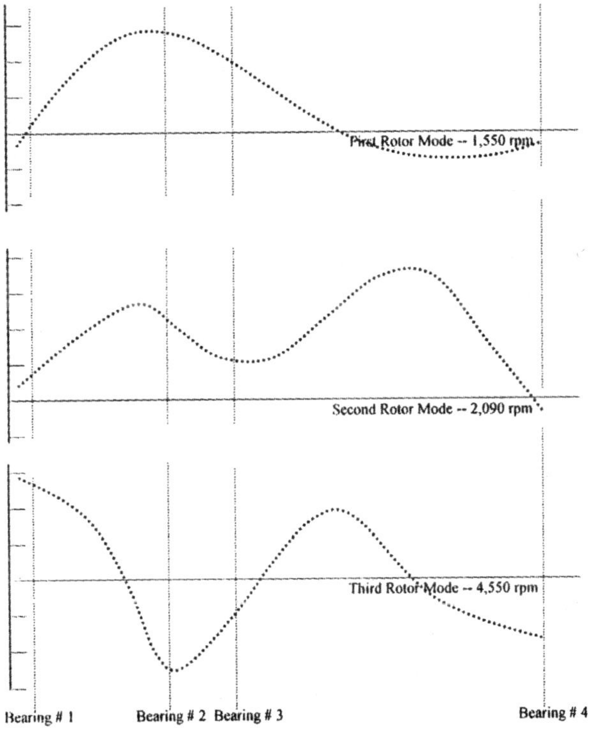

Fig. 6.11. *Mode Shapes for Steam Turbine Rotor*

Individual spans will have their own stiffness and inertia properties, giving rise to a number of critical speeds, each critical speed characterized by a dominant amplitude in the span of the next higher stiffness. The journal bearings are supported in sturdy pedestals anchored to a massive concrete foundation. Due to the substantial characteristics of the mount, the effective spring and damping constants will be due to the journal bearing's lubricant film, explored in detail in section 12.4.

The first three undamped critical speeds with bearing constant $K = 2.5 \times 10^6$ lbs/in are 1,550, 2,090 and 4,550 rpm. The spacing between the second and third critical speeds is large enough to accommodate the constant running speed of 3,000 rpm needed for 50 Hz power generation, or of 3,600 rpm for 60 Hz electric power. Note that most electric power equipment manufacturers prefer a margin of 10 percent above and below the operating speed for vibration free operation. Corresponding mode shapes are plotted in Fig. 6.11.

Rotor response due to unbalance will depend on the location and amount of unbalance. Sensitivity to unbalance is related to the fact that the generator rotor has low stiffness relative to the support stiffness provided by the journal bearings and the associated supporting structure. The second critical speed is responsive to unbalance in the turbine span. To significantly excite the third critical speed requires a moment unbalance in the generator.

6.6 SYNCHRONOUS RESPONSE

For systems with linear damped bearings the steady state unbalance response is determined by using the extended state vector and its associated transfer matrix. The extended state may be defined as the state of equation 6.10 with a unit element adjoined to form:

$$\{\bar{Z}\}_{9\times 1} = [\bar{U}\ \bar{V}\ \bar{\theta}\ \bar{\phi}\ \bar{M}_y\ \bar{M}_x\ \bar{V}_y\ \bar{V}_x\ 1]^T \tag{6.25}$$

The terms in this expression are complex amplitudes of the response, with $u = \mathrm{Re}(Ue^{i\Omega t})$, etc. The field and point transfer matrices are appropriately modified to be 9×9 matrices to include the unit element. Additionally, the point transfer matrix will also include elements for the unbalance at the proper stations. Thus, the expression for the point matrix is:

$$[P_i]_{9x9} = \begin{bmatrix} [P_i]_{8x8} & \{G\} \\ [0...0] & 1 \end{bmatrix} \tag{6.26}$$

where

$$\{G\} = \begin{Bmatrix} 0 \\ . \\ 0 \\ 0 \\ im_i\Omega^2(a_{x'} + ia_{y'}) \\ -m_i\Omega^2(a_{x'} + ia_{y'}) \end{Bmatrix}$$

and for the field matrix is: $[F_i]_{9x9} = \begin{bmatrix} [F_i]_{8x8} & \{0\} \\ \{0\}^T & 1 \end{bmatrix}$. Matrices $[P_i]_{8x8}$ and $[F_i]_{8x8}$ are defined in equations (6.15) and (6.20).

For multi-shaft systems the system unbalance response is determined by successively utilizing the unbalance excitation from each shaft. Total response is then the superposition of separate shaft responses. For a specified spin speed of the shaft the system transfer matrix is determined as before, but the extended state relations are used.

The inclusion of structure support dynamics adds an extra element, which is not a straightforward use of the transfer matrix method, unless the coupling through the support structure between support points is negligible. A single pedestal mass with damping and stiffness to the ground may then be included at each support point. These masses may be treated as additional single point shafts, and the analysis may proceed as if it were a multi-shaft system. If pedestal stiffness is known from experimental or analytical methods, it may be replaced by a dynamic stiffness element. More discussion on supports for rotating shafts may be obtained from Wang *et al.*, (1984) and Li *et al.*, (1981).

Synchronous response of a rotor at critical points that cannot be measured directly can be accurately inferred from measurements taken elsewhere in the machine. Accurate estimates of actual levels of vibration at close clearance points on the shaft such as seals assists machine operators in troubleshooting and in protecting the equipment from expensive damage when the disturbance forces acting upon the rotor are predominantly synchronous. A few sensors can be used to identify a relatively large number of well- characterized synchronous forces, such as from mass unbalance if the measurements are taken at many different speeds (Vance, 1988). Uncertainty in the rotor model can substantially degrade this capability (Sortore, 1999). But the utility of the estimation process is also dependent on a reliable bound on the error in the estimate, and can be obtained if the size of the model errors and forces acting on the rotor are available.

The response estimate is constructed using measurable response in conjunction with an estimator gain matrix derived from a model of the transmissibility of the rotor system. There are two main components influencing the estimation, the nature of the dominant unknown exciting forces and the transmissibility between the forces and the measurable and non-measurable response points. Measured response y_m will typically be at discrete points determined by a proximity probe or an accelerometer, while non-measurable response y_u may be at selected points where, for example, material strain is required because of fatigue concerns. The frequency dependent transmissibility, or transfer, functions between the responses y_m and y_u and the disturbing input forces P_m and P_u can be expressed as:

$$\begin{Bmatrix} y_m(\omega) \\ y_u(\omega) \end{Bmatrix} = \begin{bmatrix} P_m(\omega) \\ P_u(\omega) \end{bmatrix} f(\omega) \tag{6.27}$$

The transmissibility coefficients P_m and P_u must be identified through a combination of modeling and system identification. The non-measurable displacements may then be computed as:

$$y_u(\omega) = P_u(\omega)[P_m(\omega)]^{-1} y_m(\omega) \tag{6.28}$$

assuming $P_m(\omega)^{-1}$ exists. The unknown exciting force can also be identified as an intermediary step:

$$f(\omega) = [P_m(\omega)]^{-1} y_m(\omega) \tag{6.29}$$

In general, at a given speed only a sub-set of f can be estimated, because the dimension of f is greater than that of y_m, or, only those forces lying in the domain of $P_m(\omega)$ produce the measurable y_m. Forces that cannot be estimated through y_m can still produce a part of the response y_u, which portion of the response fundamentally cannot be estimated. By assuming the unknown forces to be negligible, however, a limit may then imposed on the forces acting on the rotor, and a meaningful bound may also be placed on the y_u.

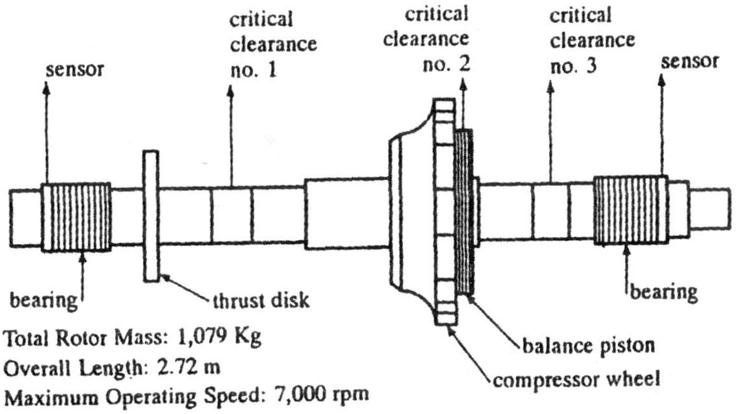

Fig. 6.12. *Compressor Rotor (Maslen, 2000)*

Missing in this formulation is the relative scales of the components of f and the relative importance of errors in y_u. As a dimensioned measure the bound is hard to interpret, hence a better way is to scale the vectors f and e to encapsulate physical limits on the elements of f and engineering requirements on residual error vector e. For instance, rotor balancing specifications typically limit the magnitude of allowed gram-millimeters of unbalance at each balance plane, and this may be converted to unbalance force at a given speed. In a similar manner, target limits on error in estimation will reflect how precisely the measurements need to be known to be useful. So if maximum rotor motion at a seal needs to be of the order of 25 μm, then the corresponding bound on the error might be set at 10 μm.

To illustrate the procedure, consider a large turbo-machine rotor used in a single stage centrifugal pipeline compressor shown in Fig. 6.12, indicating locations for vibration sensors and potentially close clearance points (Maslen, 2000). The free-free natural frequencies of the rotor (at zero speed) and modal parameter uncertainties arising from bearing stiffness and damping coefficients are shown in table 6.1. It is postulated that a strong localized model uncertainty and/or a large localized excitation needs to be estimated at the balance piston seal, which may impose considerable destabilizing forces on the rotor. If unaccounted, this aspect can introduce a substantial difference between the actual and modeled dynamics of the rotor.

Table 6.1. ***Rotor Frequencies and Modal Uncertainties***

Mode	*Natural Frequency, Hz*	*Error, percent*		
		Frequency	*Damping*	*Mode Shape*
1	0	2	3	0.04
2	0	2	4	0.06
3	105.0	3	5	0.08
4	224.8	3	7	0.10
5	442.6	4	9	0.12
6	649.0	4	11	0.14

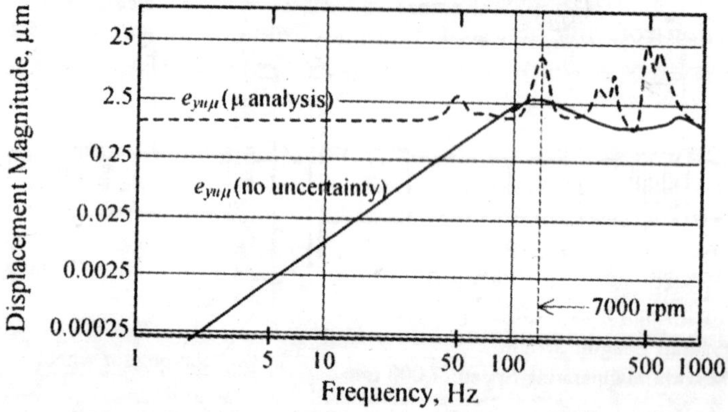

Fig. 6.13. *Maximum Performance and Estimation Errors (Maslen, 2000)*

The results of the analysis of the rotor with a cross-coupled seal acting at the critical location are shown in Fig. 6.13. The maximum performance (e_{yu}) error for clearance # 2 is provided, but actual errors are likely to be smaller. The upper bound in the error estimation ($e_{yu}\mu$), calculated using μ analysis if the model is not perfect, is also shown. The magnitude of the cross-coupled stiffness coefficient is increased linearly with frequency from 0 at standstill to 1.751×10^7 N/m at 7,000 rpm. The peaks in the plot arise from the uncertainty in the lightly damped higher frequency modes. Of particular interest is the visible split peaks in the error bound associated with the higher flexible modes. The peaks occur because of the presence of flexible modes on either side of the nominal value. Despite the combination of large potential unbalance and substantial model uncertainty, the estimator error bound is acceptable.

6.7 COMPONENT MODE SYNTHESIS OF LARGE SYSTEMS

When the total number of degrees of freedom of a dynamic system is large, analysts often resort to the modal method where the system is represented by its free vibration modes. The advantage of using the modal representation consisting of a few low frequency modes accrues from the reduction in the size of the associated problem, and is popular in problems such as bowed shaft analysis (Choy, 1978) and transient behavior of dual rotor aircraft engines (Dennis *et al.*, 1975). In an improved form of this method the dynamic structure is partitioned into a number of smaller sub-structures, with modal characteristics of each unit derived analytically or from vibration testing (Klosterman, 1971). The total structure is then constructed from a reduced number of component modes from each sub-structure.

Component mode synthesis is used extensively in the aerospace industry to calculate undamped natural frequencies of large airframe structures. The sub-structures are required to share common degrees of freedom at the common interconnecting points to obtain a set of constraint equations for the reduced modal equations of motion. In rotating systems the interface between sub-structures consists of bearings and seals, and the supports between the engine casing and the attachment points are also elastic. If a common degree of freedom is not shared by the sub-structures, flexible connections may be introduced in the form of generalized damping, stiffness and force into the modal equations. Together with the rotor's free-free modes the undamped modes of the rotor and casing system are then obtained using the procedure. The system modes aid in developing the system modal equations with damping added.

Linearly elastic non-rotating structures yield a symmetric stiffness matrix, but the rotating subsystem stiffness matrix is asymmetric because of internal friction and asymmetric bearing stiffness coefficients. The symmetric and asymmetric parts may be separated, with only the symmetric portion used in the evaluation of undamped component modes. The asymmetric portion is added to the modal equations at a later stage. This approach provides a choice of evaluating only the free-free rotor modes containing the shaft stiffness or the shaft and bearing modes containing the shaft and principal bearing stiffness values. The shaft's flexural stiffness is then implicit in the mode shape and conveniently expressed by the eigen-values.

The partitioned matrix equation has diagonal and square sub-matrices, with the dimension equal to the number of normal modes employed in the corresponding subsystem. The off-diagonal matrices represent the coupling of the subsystems from the interconnecting elements. Internal friction is represented by the dry friction loss angle and the viscous friction coefficient. The linearly elastic elements omitted in the undamped mode calculation contribute to the coupling stiffness and damping matrices, and consist of bearing stiffness and damping, gyroscopic moments and rotor acceleration. The linearly elastic elements may be generalized into four distinct linking categories, or their combination, with each link containing stiffness and damping coefficients

- Direct links representing a linear element between the degrees of freedom in the subsystem and the stationary ground.
- Cross-coupled links between the coupling force at a degree of freedom in one subsystem and the motion at a degree of freedom in another subsystem.

- Direct/intermediate links consisting of an elastic element between the degrees of freedom in one subsystem and the degrees of freedom in another subsystem.
- Cross-coupled/intermediate links developing equal and opposite forces at various degrees of freedom between two subsystems due to the coupling effect of the relative motion. An example of this is the cross-coupled dynamic coefficient of a journal bearing installed intermediately between two concentric rotating shafts.

The total effect from the linear linking of the elements is the sum of the individual contributions. Thus, a fluid film bearing mounted in a rigid housing may be represented by two direct links and two cross-coupled links between the shaft and the ground. Gyroscopic forces associated with a rotor represent additional modal contributions, and may be dealt with in a similar manner using modal contributions. Modal forces are related to actual forces from unbalance, disk skew or nonlinear forces acting at nodes in accordance with the mode shapes.

Thus, the governing equations of a complex rotor system may be expressed in the generalized coordinates by adding up the contributions of all the flexible links to the basic uncoupled modal equation. After establishing this, critical speeds, stability, forced or transient response may be solved in the modal coordinates, and the final results then transformed back into real coordinates. In non-linear transient analysis it is necessary to transform back and forth between the real and modal coordinates during computation of the nonlinear forces.

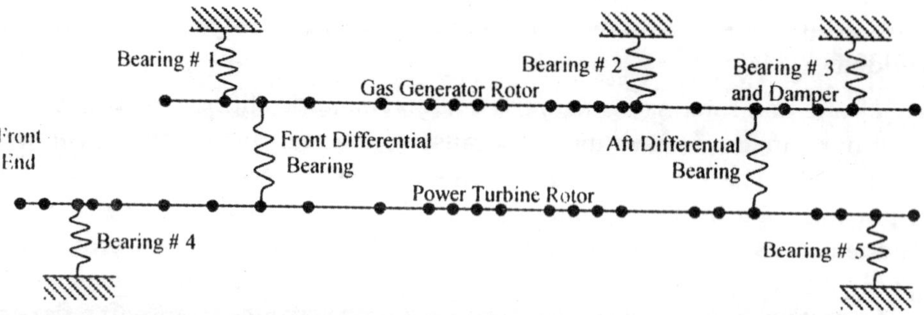

Fig. 6.14. *Dynamic Representation of Two-Spool Aircraft Engine (Li, 1981)*

This scheme of computations is illustrated for a two-spool aircraft gas turbine engine (Li, 1981), shown in Fig. 6.14. The engine consists of an inner, or power turbine, rotor supported by two main bearings placed close to the end extremities. Two differential bearings connect the gas generator and the power turbine rotors. In addition, the gas generator rotor has three primary bearings, with a squeeze film damper incorporated at the third bearing to reduce vibration amplitudes and bearing forces. Component modes of the rotors are obtained independently of each other by omitting the gyroscopic effects of the disk, squeeze film damper and differential bearings. A total of 12 undamped component modes containing two gas generator modes and four power turbine modes in the horizontal and vertical lateral planes below 30,000 rpm are used.

The first portion of the computation calls for transient response of the system due to a suddenly applied unbalance, as may occur in the event of loss of part of the blades. Both rotors are assumed to be initially balanced when the unbalance distribution is applied, with the gas generator operating at 15,000 rpm and the power turbine at 17,000 rpm.

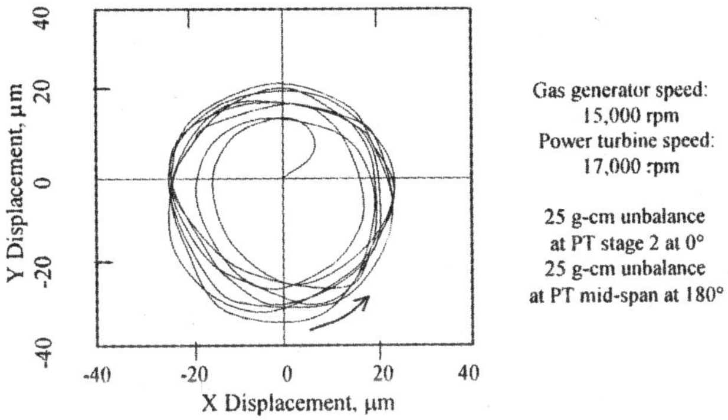

Fig. 6.15. *Transient Orbit of Gas Generator Disk (Li, 1981)*

The unbalance comprises of 25 g-cm in the 2nd stage turbine and an equal amount at the rotor mid-span 180° away. With the assumption of linear system behavior, Fig. 6.15 provides the orbits at the 2nd stage gas turbine for the first 10 cycles of simulation. The steady state orbit is not fully developed even after 40 cycles, but the magnitude and phase angle of the vibration vector can be distinguished. A comparison with an independent steady state response calculation indicates close correlation. The maximum response amplitude at the start of the transient motion is about double the size of the steady state orbit.

The squeeze film damper employed at bearing # 3 is deficient in terms of design requirements, and a non-linear jump is observed during vibration tests of the engine. The damper is 11 mm long, 129 mm in diameter and has 0.1 mm radial clearance. A retainer spring of 2.15×10^7 N/mm is used to center the outer raceway of the rolling element bearing. The diaphragm supporting the squeeze film damper and bearing is assumed to be rigid. Bearing force is calculated using the short bearing theory in the simulation. The second portion of the computation has the power turbine rotor operating at 14,000 rpm with an unbalance of 50 g-cm at the turbine end. The gas generator rotor has the same amount of unbalance at the 2nd stage turbine. The speed of the gas generator rotor is increased from 9,000 rpm in steps of 2,000 rpm until the jump phenomenon occurs. At the start of the transient analysis the motion is set at zero displacement and velocity. Non-linear response is calculated for each speed for 10 gas generator rotor revolutions. Starting conditions at rotor speeds other than the initial speed are taken to be equal to the final displacement, and velocity is computed at the end of the simulation cycles for the previous gas generator speed. The transient response of the system is computed due to the suddenly applied unbalance in the rotor under steady speed operation.

Figure 6.16 provides the transient orbits of the gas generator rotor at the damper location as the speed is increased until the jump phenomenon is encountered. The initial transient motion at 9,000 rpm is shown in (*a*), with the orbits calculated for about 9 turns after release from the center of the bearing. The timing mark is synchronized with the unbalance vector, and is shown as an asterisk at the end of each revolution. Because the two rotors are operating at different speeds, the unbalance forces have two frequency components. Consequently, a response pattern repeating itself for every revolution is not observed even when the system reaches steady state operation. In (*b*) the non-linear motion at the damper location occurs at 13,000 rpm. Limiting orbits of large amplitude are formed at a rapid rate, in less than one revolution of the gas generator rotor. The damper is now operating at eccentricity ratios in excess of 0.9 of the radial clearance, indicating overloading of the damper bearing. The phase angle lag fluctuates around 45° throughout the simulation at this speed point. Finally, in (*c*) the nonlinear jump occurs at the gas generator speed of 17,000 rpm. The orbit of the rotor is reduced to about a quarter of its initial size within 5 revolutions. The shift in the amplitude is accompanied by a shift of the timing mark, and approximately aligns itself with the negative x-axis. The latter change is significant since it indicates that the gas generator rotor has rapidly passed through the critical speed of the gas generator rotor mode due to the sudden reduction in the effective damper stiffness and the consequent smaller orbit size.

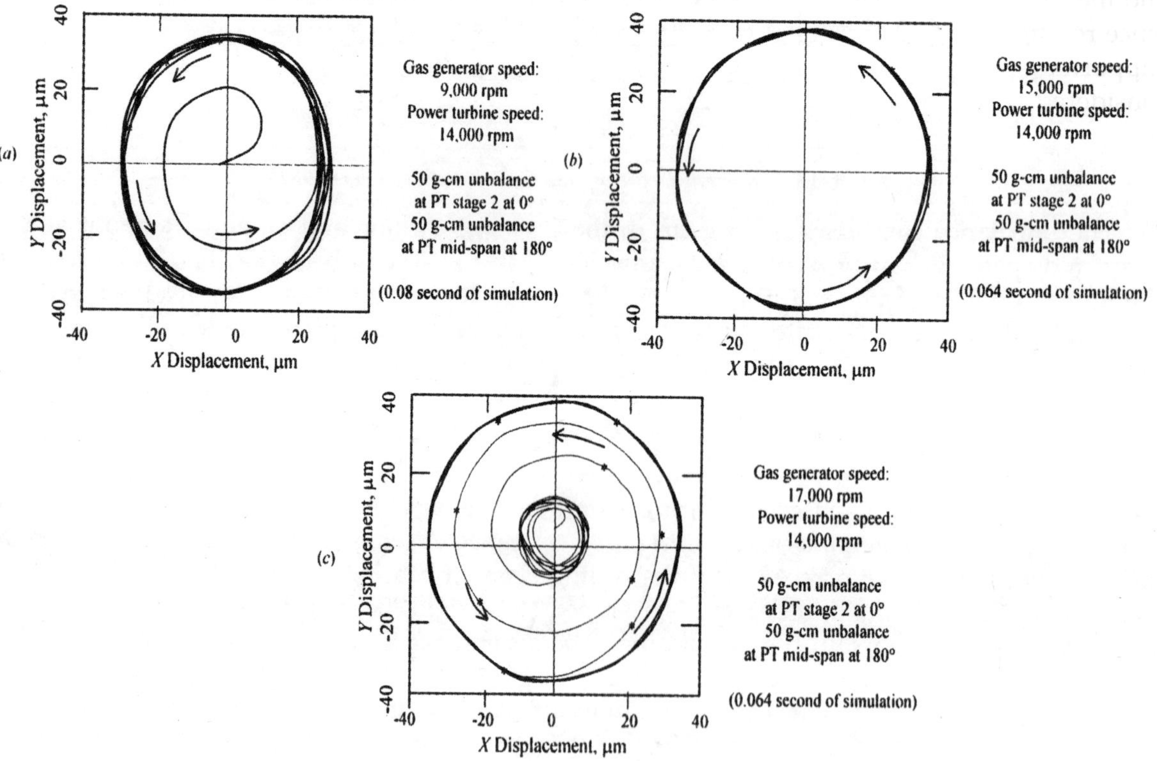

Fig. 6.16. *Non-linear Rotor Orbits at Squeeze Film Damper Bearing (Li, 1981)*

6.8 ROTOR SUPPORTED IN FLEXIBLE CASING

In many turbo-machines the rotor operates above one or more critical speeds, hence the unbalance response at resonance is of major significance. In the early phase of the design, guidelines from parametric studies prove useful. The unbalance chart shown in Fig. 6.17 provides some useful results in this respect (*ref*: Balda, 1975), illustrating the interaction between critical speed ratio ω/ω_0 (ω_0 representing the frequency of a simply supported rotor), response factor a_d/e_d (amplitude/eccentricity) and bearing stiffness and damping ratios K/C and H/K. However, aircraft engine rotors are enclosed and supported in light and flexible casings, consequently the vibration pattern of the rotating system is affected by the stiffness and mass characteristics of the case and its supports. In analyzing the dynamic aspects of the power plant it is necessary to include the casing and support structures in addition to the rotating assembly and the bearings. The first two critical speeds and response factors due to unbalance generally will vary widely with changes in casing mass and bearing and support stiffness, and the damper plays a useful role in reducing the vibratory response in the modes. But the third critical speed and response factor are little affected by the damper (*ref*: Miyachi, 1983).

To understand the impact of the flexible casing and supports, consider a uniform shaft with a disk of mass m_d and mass moment of inertia I_d mounted at the center of the span, two rolling element type bearings at the ends, a casing and its supports. The ratio of the mass of the casing and the rotor is M_c, bearing to shaft stiffness ratio is K_b and support to shaft stiffness ratio is K_s. Since rolling element bearings offer poor vibration damping capacity in comparison to oil film supported sliding bearings, a squeeze film damper type of damping in the right bearing is considered in this evaluation.

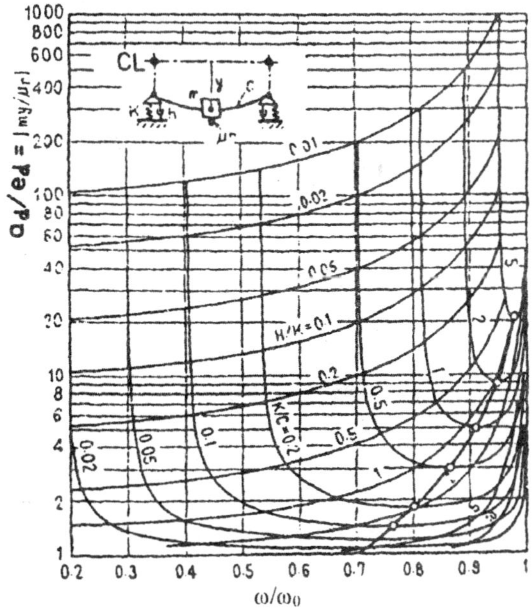

a_d = Disk vibration amplitude
e_d = Disk eccentricity
K = Bearing stiffness
H = Bearing damping
C = Shaft stiffness

Fig. 6.17. *Unbalance Response for Single Degree of Freedom Rotor (Balda, 1975)*

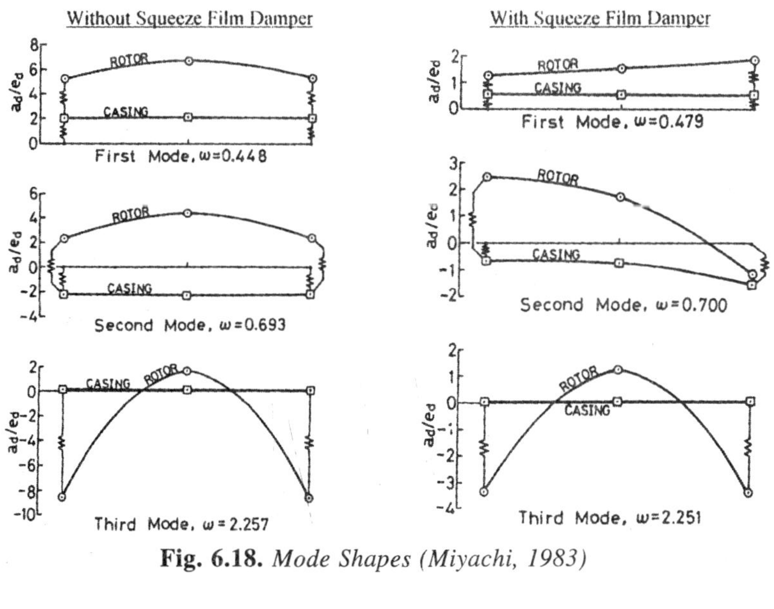

Fig. 6.18. *Mode Shapes (Miyachi, 1983)*

The eccentricity of the disk e_d will be used to obtain the exciting force. The shaft is represented by m_j and I_j concentrated at its ends, m_j being one tenth of m_d, and I_j being one hundredth of I_d. Bending stiffness of the rotor EI_r is constant over the span. The bearings are supported in the casing by struts, while the casing is mounted to the supporting structure, connected to fixed locating points. The mass of the casing m_c, as also its mass moment of inertia I_c, are distributed half at the center and quarter at each of the two ends. Bending stiffness of the casing EI_c, assumed to be ten times that of the rotor, is constant over its length. To simulate the effects of light and heavy engines the mass ratio of casing to the rotor is varied from 0.5 to 20. Rejection of secondary effects of mass and bending stiffness of the casing is achieved by letting the bearings and casing mounts to be positioned at the same axial coordinates. Detailed description of properties such as axial distribution of the casing and rotor mass may affect the vibratory characteristics at higher speeds, but this evaluation is intended to focus on the lower modes.

The transfer matrix method for the forced vibration of the rotating beam is used, with the unbalance force arising from the eccentricity of the disk. Figure 6.18 provides the mode shapes for the model with and without a damper. Bearing stiffness ratio K_b is 0.5, casing support stiffness ratio is 2.0 and casing mass ratio M_c is 5.0. Without the damper the casing sees little participation in the first two modes. When the squeeze film damper is included in the right bearing the mode shapes tend to become asymmetric. Viscous damping coefficient of the squeeze film damper c_q is assumed to be optimum for ω_o, or equal to k_b/ω_o. With the damper, in the first mode the rotor and the casing move in phase, but in the second mode the two components move in reversed phase. In the third mode rotor extremities vibrate in inverse phase with the disk, and the amplitude of the rotor end and relative displacement between the rotating and stationary parts at the bearings is much larger than the amplitudes of the center disk and the case. The dynamic load in the bearing may be expected to maximize in this mode, in spite of the low activity at the disk location. The effective stiffness of the bearing appears to be reduced by the damper. The phase angle of vibration cannot be clearly split into positive and negative, but the two dimensional display defines the intermediate phase angle simply as above or below the reference line. Thus, the relative displacement between the rotor and casing at the right bearing in the second mode appears small, but the difference in the phase angle at this location is 70°, making the actual relative displacement considerably larger.

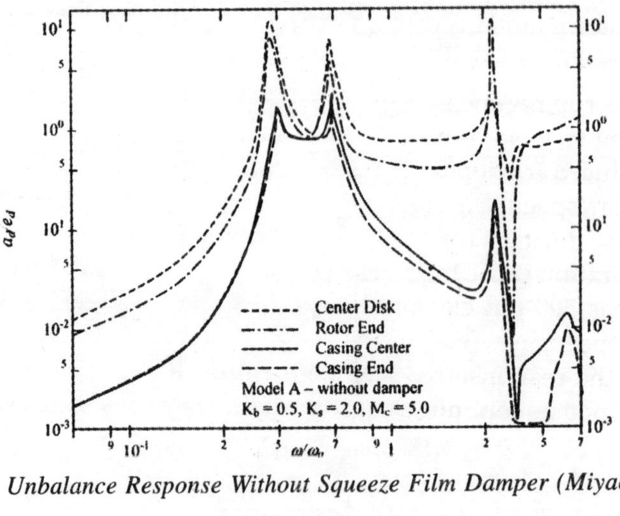

Fig. 6.19. *Unbalance Response Without Squeeze Film Damper (Miyachi, 1983)*

Figure 6.19 provides the unbalance response without the damper, with speed ranging to 5 times the fundamental natural frequency. The disk experiences three peaks of vibratory motion, the higher first two being about equal in magnitude. In the absence of the damper the dynamic energy is dissipated by the structural damping of the casing, the supports and the shaft. Peak vibratory motions do not coincide at the same shaft speed for all the structural members.

The squeeze film damper is effective in reducing the unbalance response at the critical speeds, especially in the region where response is high. Maximum amplitude is reduced by 50 to 80 percent of those experienced without the damper.

Table 6.2. ***Peak Bearing Loads***

Mode #	Without Damper		Squeeze Film Damper	
	Frequency Ratio	Bearing Load	Frequency Ratio	Bearing Load
1	0.448	17.3	0.479	5.1
2	0.693	8.5	0.700	4.7
3	2.257	31.2	2.251	9.8

For another indication of the unbalance response, consider the dynamic load on the bearing, normalized by half the centrifugal force due to the eccentricity of the disk for all three modes. Table 6.2 provides a comparison of the bearing loads.

6.9 DRY FRICTION RUBBING BETWEEN ROTOR AND SEAL

Rubbing between the rotating and stationary parts of a rotating machine is a serious malfunction, holding the prospects of a catastrophic failure. The rub represents a secondary effect, caused by excessive lateral vibrations of the rotor or high displacement of the centerline of the rotor due to misalignment or radial forces arising from the flow of the working fluid. The trend to increase operating efficiency, and also to improve stall margin in the case of a compressor, leads to minimization of the clearances at the blade tip and inner seals, which in turn increases the hazard of rubbing.

There are two types of rub. In the light and partial rub the contact forces are minimal, with the rotor occasionally touching the stationary elements during a fraction of its precessional period. A more serious phenomenon in its destructive abilities is a full annular rub that may occur in seals and bearings. The rotor maintains contact with the seal almost continuously. Self-excited reverse precessional full rubs, known as dry whip, are more severe than synchronous full annular rubs.

The synchronous rub regime is marked by interaction from the seal, with the latter acting as an additional bearing to yield a coupled rotor/seal system with increased stiffness and natural frequencies. The reduced friction force permits the shaft to slide along the seal's surface without rolling at a particular speed. Figure 6.20 shows a rotor's horizontal (X-direction) and vertical (Y-direction) response during a run up and down sequence with and without rubbing. Differences in the angular acceleration cause the run up and down curves to differ slightly without contacting the seal. The seal has 500 μm clearance, and 3.3 g-cm unbalance is attached to the disk. The resonance range of speeds is wider during the run up than during the run down. In the range of original resonance the response reduces with rub at the seal since it acts as a bearing, even though the shaft is not in continuous contact with the seal. The orbit (shown as insets in the

Fig.) indicates the rotor is bouncing inside the seal, but it maintains the forward precession. As the rotor speed increases or decreases, the operating condition changes from non-contact to contact and then to non-contact again, and is accompanied by jumps in the amplitude. A sudden decrease in the 1× amplitude, accompanied by an onset of other frequency components, is observed in the X-direction when the rotor just contacts the seal during run down to indicate a hammer impact form of external disturbance. A jump-up is observed in the Y-direction during the run down, and is caused by insertion of lubrication oil on the contact surface to avoid severe reverse precession rubbing. Unlike the case in the X-direction where the contact surface is not fully lubricated, the rotor's orbit in the Y-direction is smooth, suggesting the shaft's bounce pattern is considerably influenced by the friction coefficient.

Reverse precession full annular rubs can occur when the ratio of radius to radial clearance ranges between 10 and 40 without an external disturbance.

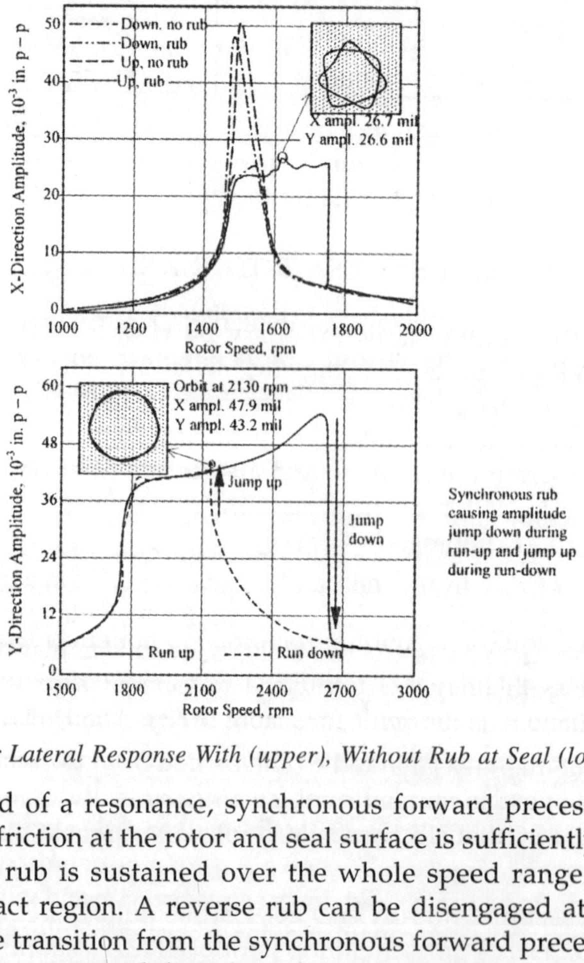

Fig. 6.20. *Rotor Lateral Response With (upper), Without Rub at Seal (lower) (Yu, 2000)*

In the neighborhood of a resonance, synchronous forward precession rub turns into the reverse direction when friction at the rotor and seal surface is sufficiently high. Once generated, the reverse precession rub is sustained over the whole speed range as long as slippage is maintained at the contact region. A reverse rub can be disengaged at a non-resonant speed. Figure 6.21 provides the transition from the synchronous forward precession rub to the reverse precession type without an external disturbing force. As speed approaches the first resonance, the rotor starts bouncing inside the seal as it develops elliptical orbits gradually rotating their major axes due to lateral stiffness anisotropy effects. Intermittent contact at multiple points also appears at this stage. Reverse-dominant precession initiates between 1550 and 1650 rpm, and around 1700 rpm suddenly the full annular rub appears at a frequency of 2300 cpm.

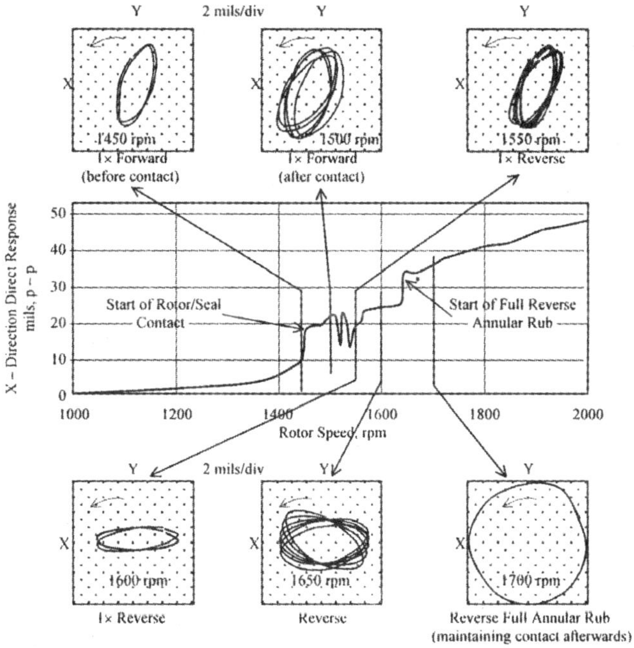

Fig. 6.21. *Transition from Forward to Reverse Precession (Yu, 2000)*

Considerably higher vibration amplitudes occur in the reverse mode at -2300 cpm than in the forward frequency at 2300 cpm, also indicating an almost circular reverse orbit motion. Further increase in shaft speed does not affect the rubbing frequency, though the amplitude increases slightly due to thermal growth of the seal.

The reverse precession full annular rub can also trigger during the run down after the rotor contacts the seal close to the resonance speed, as shown in Fig. 6.22. The rub is maintained until shaft speed is low, and replaces the synchronous 1× motion. When the applied unbalance is reduced to 1.2 g-cm for the same experimental setup the reverse precession does not develop, eventhough the rotor bounces inside the seal at resonance speed. The reduced unbalance also affects the starting point of the rub, as indicated in Fig. 6.23, where seal clearance is 250 μm. The first resonance speed of the original uncoupled rotating system is near 2000 rpm. The reverse rub develops after the resonance speed, and represents a switch from before resonance speed with the higher unbalance, with the same precession frequency of 3120 cpm.

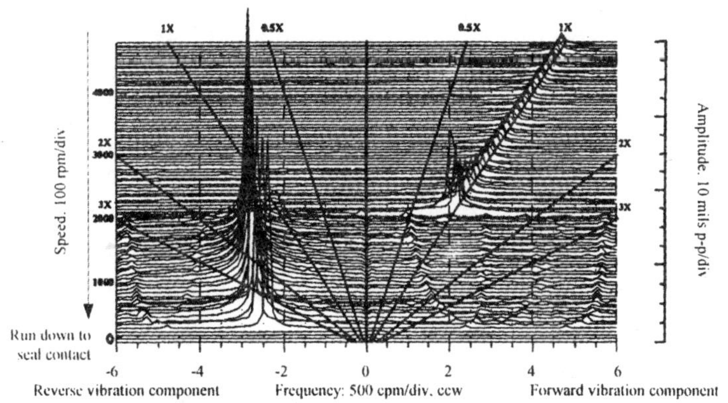

Fig. 6.22. *Reverse Precession Rub Triggered During Run Down (Yu, 2000)*

Although the rotor with higher unbalance initiates contact with the seal at about 1600 rpm, the reverse rub is not engaged quickly until around 1900 rpm, where an increasing contact force due to the synchronous 1× rub motion is able to trigger it.

Considering the effects of shaft speed, once generated, the reverse precession full annular rub does not disengage by reducing the speed until it is very low, when slippage cannot be maintained. The ratio of reverse rub frequency to shaft speed ω/Ω is always less than the ratio of radius to radial clearance in the seal. Assuming the seal vibrates harmonically at the same frequency ω, the relative slip equals $(r\Omega - C_r\omega)$.

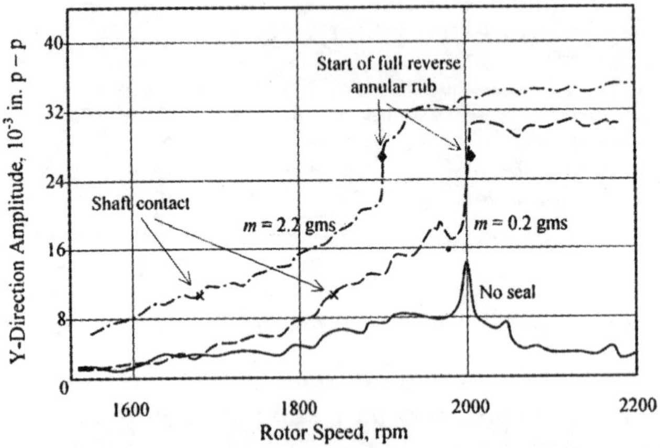

Fig. 6.23. *Effect of Mass Unbalance on Start of Reverse Rub (Yu, 2000)*

The reverse rub disengages when this slippage velocity is near zero, so the friction force, also approaching zero, cannot balance the damping force. The reverse direction rubbing ceases when the slip velocity and the shaft speed are low.

6.10 STABILITY CONSIDERATIONS

Rotating machines experience vibrations arising from a number of causes. Rotor unbalance is just one of them, although a more persistent one. Fluid flow forces, misalignment and maneuver loads in aircraft gas turbine engines are other important sources. There is also another class of vibration that mostly takes the form of flexural motion in the transverse direction at a rotor's natural frequency that is different from the running speed. The vibrations may be sub-synchronous, occurring at a frequency below the shaft speed, or super critical, when the frequency is above the shaft speed. A major feature of these non-synchronous modes of vibration is that the amplitude increases sharply with an increase in speed, and further acceleration may need to be aborted to avoid damage to the machine's components. Thus, such unstable vibrations are damaging and expensive, since they force delays in operation and production. Modification in engineering design may be required to overcome the problem, since they constitute a restraint on the capabilities of the rotating machine.

The mechanisms of instability may be classified by their characteristics, and by their effects on machine operation. Whirling and whipping of the shaft, stick-slip rubs and chatter are some commonly observed phenomena.

When a shaft experiences whirling and whipping, a tangential force normal to the direction of motion is developed. The magnitude of the force is proportional to the deflection, developing at some threshold speed when the force is large enough to offset the stabilizing external damping forces (*ref*: Ehrich, 1999). Figure 6.24 shows the forces developed in a whirling shaft. In polar coordinates the tangential force may be written as

$$2m\omega\,(dr/dt) + c\omega\, r - F_\theta = 0 \tag{6.30}$$

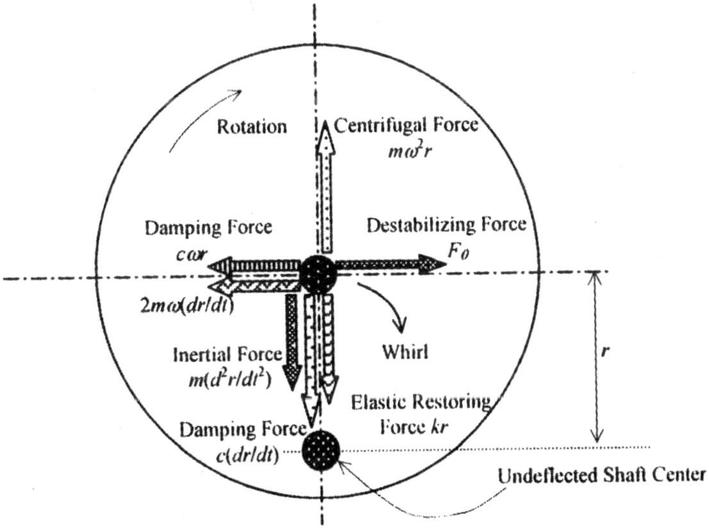

Fig. 6.24. *Forces in a Whirling Shaft (Ehrich, 1992)*

where a constant rate of whirl is assumed. The whirling is based on the premise that some physical feature in the machine creates an exciting force that is normal to the radial deflection of the shaft and opposes the damping force, which tends to diminish the whirling motion. Often this disturbance force is proportional to the radial deflection. As the shaft speed increases to a threshold limit, the increased shaft radial deflection causes the disturbance force to reach a point where it balances the damping force to set up the condition for the onset of instability. Irrespective of the shaft spin speed, the whirling speed at the onset of instability equals the shaft's natural or critical frequency. The direction of the whirl may be in the same direction as the shaft rotation (called the forward whirl), or opposed to it (backward whirl), depending on the direction of the destabilizing force. In the unstable condition the shaft center trajectory can be proved analytically to take the form of an exponential spiral up to a limit. Non-linear characteristics in the machine system will limit the growth of amplitude in the form of additional dissipating energy. Thus, the amplitude of whirl increases sharply with time before stabilizing at a steady state limit, and may be accompanied by damage to the machine's internal components.

Stability problems vary considerably in intensity. An operating unit may be marginally stable and operate for a long time without experiencing excessive vibrations. At many installations a fleet of identical machines may be operating satisfactorily from vibration considerations for many years, with the exception of a single machine. The machines may be operating basically on similar load cycles, handling the same fluid and work identical schedules. Under the circumstances it would be difficult to explain the dynamic experiences of the single machine with the high vibration signature. In such a unit, comparatively small changes often suffice to reduce the size of the problem. Changing the operating temperature of the fluid, going from one type of bearing to another or changing bearing preload may be enough to resolve the problem.

6.11 SELF-EXCITED VIBRATIONS

Self-excited vibrations in a system begin by themselves without any warning, initiating spontaneously. The amplitude increases until some nonlinear effect limits any further increase. The excitation source for these vibrations is a uniform source of power associated with some inherent mechanism within the machine system. The destabilizing force sustaining the motion is derived from and controlled by the motion itself, and the force disappears when the motion stops.

Whirling and whipping instability arises from many sources:

- Internal rotor damping (hysteretic whirl), first reported by Newkirk (1924) and Kimball (1924), can be explained by using the force diagram of Fig. 6.25. Bending in the shaft due to some radial deflection sets up elastic restoring forces in a direction parallel and opposed to the direction of deflection. Internal friction in the shaft material also causes a phase shift in the developed stress, which results in a tangential force component. This sets up the condition for the shaft to whirl. The out-of-phase force from the bending stress increases as the shaft accelerates and reaches a point when the forces from the machine's damping system is exceeded to set up the condition for the onset of instability. Besides friction arising from relative motion between the molecules of the material, loose mating components also are responsible for this phenomenon. Bolted joints, shrink fitted assemblies and rabbets are prime examples. If the bolts are not tightened to the proper torque setting, or adequate interference fit does not exist, relative motion between the components will provide cause for hysteretic whirl in the system. Studies by Gunter (1966), Bolotin (1964) and Vance *et al.*, (1973) have established that hysteretic whirl occurs at shaft speeds above the first critical speed. But once whirl sets in, the induced critical whirl frequency is about half the onset speed.
- Hydrodynamic journal bearings have been a major cause of unstable operation in rotating machines (Newkirk *et al.*, 1925). Referring to Fig. 6.26, a shaft rotating in a journal bearing filled with a lubricant has a nominal radial deflection. Because of the shaft's eccentricity and the average fluid velocity (about half that of the shaft surface), the lubricant develops pressure, which is not symmetric about the line of radial deflection. Also, fluid pressure on the upstream side of the clearance region is higher than on the downstream side. This causes the resultant hydrodynamic force on the shaft to have a tangential component in the direction of rotation of the shaft, tending to whirl the shaft in the same direction. When this force exceeds the stabilizing damping forces, unstable operation will be the consequence. The stiffness and damping characteristics of the fluid film play a major role in determining the development of whirl.

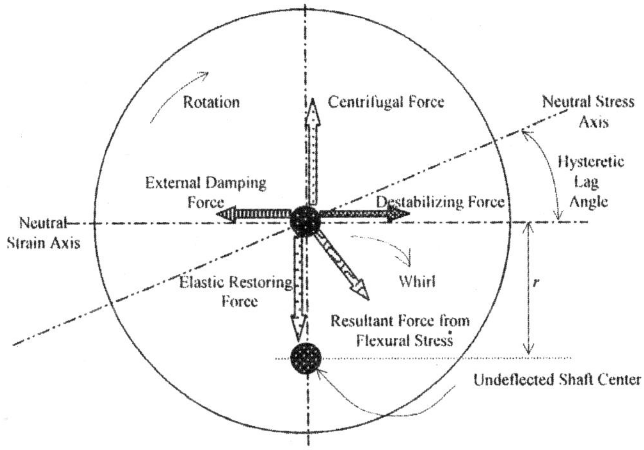

Fig. 6.25. *Internal Rotor Damping Causing Hysteretic Whirl (Ehrich, 1987)*

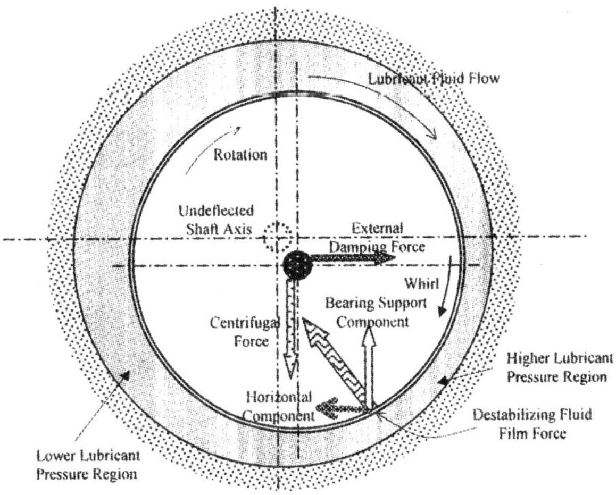

Fig. 6.26. *Bearing Lubricant Film Causing Shaft Whirl (Ehrich, 1992)*

As the speed increases and the eccentricity ratio gradually becomes smaller, the cross coupling stiffness and damping terms become about equal. The speed ratio at which whirl will develop is given by $k_{r\theta}/c\omega$, where $k_{r\theta}$ is the film stiffness in the cross coupled radial and tangential directions. For plain journal bearings this ratio is about one-half. For a flexible rotor, the threshold speed at which the instability occurs is approximated by $\omega_n/(k_{r\theta}/c\omega)$, where ω_n is the rotor critical speed. Thus, a flexible rotor supported in plain journal bearings usually experiences sub-synchronous unstable motion when shaft speed is greater than twice the lowest rotor critical speed.

Steam and gas turbines require close control of radial clearance between the tip of the blades and the casing. The objective behind it is to prevent leakage of steam past the rotating blades without permitting the blades to rub against the case, and thus maximize operating efficiency. In many cases, however, it is difficult to maintain concentricity between the rotor and the stator, with the result that the tip clearance is not uniform along the full circumference. This can happen when the rotor lacks adequate stiffness, or due to larger than normal shaft span between the supporting bearings, causing the rotor to sag downwards. Overhung stages past a bearing support (aircraft jet engines are an example) also tend to exaggerate this situation. From aerodynamic considerations the result of the unequal space around the circumference is to set up a tangential force at the blade tip in the direction of rotation of the rotor. As the clearance distribution becomes more pronounced due to changes in speed or power setting, tangential forces may become large enough to overcome the energy dissipating forces of the damper, with the rotor whirling as a consequence. To overcome the problem, during the design stages bearing and shaft stiffness must be adequately provided for to avoid excessive radial deflection of the shaft. Figure 6.27 illustrates the forces acting on the rotor as a result of uneven tip clearance. Thomas (1958) offered an explanation for the destabilizing mechanism, with the expression for the reaction force $R_{r\theta}$ taking the form:

$$R_{r\theta} = T\beta/(DL) \tag{6.31}$$

where T = torque developed at stage

D = mean blade diameter

L = blade length

β = change in thermodynamic efficiency per unit change in clearance.

Subsequently, Alford (1965) offered similar explanations on stability of compressor rotors. Childs (1986) has done analysis of forces developed on a pump's impeller shroud.

Many examples of unstable rotor vibrations besides those given here can be cited. Aircraft engine propeller whirl has been recognized as a direct consequence of shaft whirl. When the plane of the propeller rotates because of the slope of the shaft's centerline, incremental axial velocities are induced (Taylor *et al.*, 1984; Houbold *et al.*, 1981). At high air speeds destabilizing forces are set up, and can overcome damping forces to cause destructive whirling of the system.

Fluid trapped in a hollow rotor can act as a high-speed centrifuge to set up destabilizing forces. Oil from bearing sumps and steam condensate may inadvertently collect and be trapped in the cavity of the rotor.

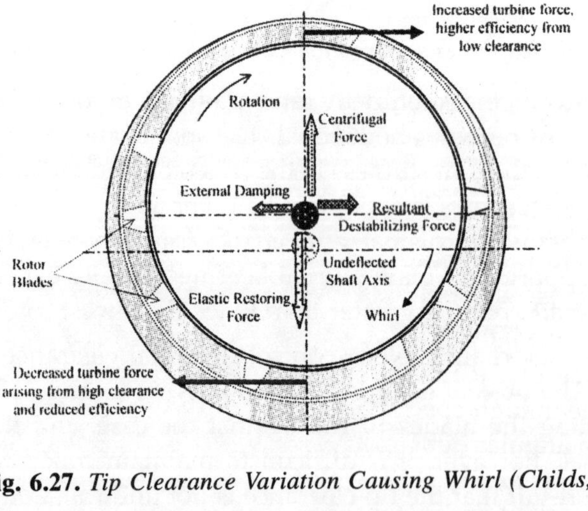

Fig. 6.27. *Tip Clearance Variation Causing Whirl (Childs, 1987)*

When the shaft is operating with some nominal radial deflection, the fluid flies out, but does not remain in one fixed location. The surface of the shaft carries along some fluid with it in the direction of rotation. This advance sets up a force component in the tangential direction; the force is the source of instability because it induces forward whirl. The threshold speed for instability will be above the critical rotor speed.

Rubbing of a shaft against the stator case takes place in an environment where there is no oil or other fluid lubricant, so the rubbing is essentially dry. The dry rub sets up conditions for the shaft to whirl. Dry rub can also take place in a bearing that is starved of the lubricant, or with the loss of clearance in a journal bearing, or due to a loss of radial clearance in a labyrinth seal (see section 12.12), or between the blade tip and casing shroud of a turbo-machine. The mechanics of the rub and the consequent shaft whirl can be understood from Fig. 6.28. Coulomb friction due to rub between the rotating and static parts introduces a tangential force component equal to the product of the radial force and the coefficient of friction. The force induces the shaft into a whirling motion, and the associated centrifugal force produces greater radial deflection, which in turn causes an even greater radial force and whirl producing tangential friction force. Note that the friction force will be in the opposite direction of shaft rotation, thus causing a backward whirl of the shaft. Begg (1983) indicates that the frequency of whirl is generally less than the critical speed. A similar situation is encountered in an axial thrust bearing, where due to the slope of the shaft contact takes place between the rotating and stationary disks at a pivot point. Remedies to alleviate the shaft whirl due to rubbing will range from stiffening of the shaft to reduce radial deflection to improving assembly techniques for maintaining concentricity between rotor and stator.

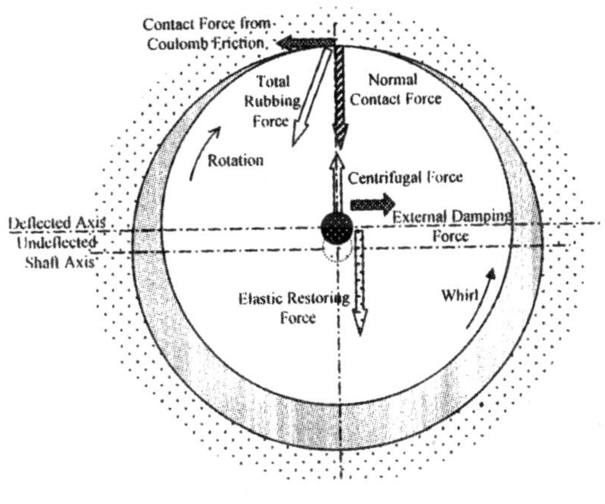

Fig. 6.28. *Dry Friction Rub Causing Shaft Whirl* (ref. *Ehrich, 1987*)

If the rub persists during operation, lubrication of the contact surfaces should alleviate the problem. If lubrication is not possible, another remedy would be to use an abradable lining material on the stator to permit disengagement between the interfering components.

Yet another source of excessive rotor vibration has been suggested by Vance (1988). Transmission of torque in high-speed rotors is accompanied by a rotational twist in the shaft. If that twist results in an angular deflection of the torque vector, then the torque will have a small radial deflection, which is nearly proportional to the radial deflection. The radial torque vector induces bending in the shaft and a deflection that is normal to the radial deflection. When this tangential component is large enough to overcome the energy dissipating forces in the system's damping mechanism, the threshold of instability is crossed and the shaft whirls.

6.12 PERIODIC SYSTEM CHARACTERISTICS

An analytical method to investigate the problem of dynamic instability in rotating machines is based on a system of linear differential equations with periodic coefficients, with the period in time t. A specialized form of this equation, called Mathieu's equation, can be used to represent many different physical problems in rotating machines:

$$\frac{d^2y}{dt^2} + [a - 2q\cos(2t)]y = 0 \qquad (6.32)$$

where a and q are constants. Treatment of the theory behind this equation is given by McLachlan (1947). Stability induced by periodic variation in a rotating machine's system parameters such as stiffness and inertia can be evaluated using this approach. In particular, asymmetric shaft cross-sections and bearing characteristics, pulsating torque and fluctuating axial compressive loads affect the lateral vibration stability of a rotating shaft.

When the stiffness characteristics of a shaft or stator are significantly different in two orthogonal planes, self-excited vibrations are likely to be a problem. An asymmetric shaft rotating at a speed away from whirl speed acts as a periodic change in stiffness in the plane of its radial deflection, causing instability in specific speed ranges determined by the degree of asymmetry. Unstable operation may be expected at one-third and one-half of critical speed, and at critical speed. But with even larger asymmetries in the shaft (or the support) stiffness, generalizations may not be possible.

Eshelman *et al.*, (1980) carried out experiments to determine the effect of cyclically pulsating shaft torque and its effects on rotor stability. A varying torque was superimposed on a constant level torque in a rotating shaft. This caused the natural frequency of the shaft in the transverse direction to shift. The researchers investigated conditions under which the shaft whirled at the first critical speed, varying speed and pulsating torque. Certain combinations of the steady and pulsating torques caused instability in narrow speed bands at discrete fractions of the critical speed.

In a comparable manner, pulsating axial loads are also known to cause lateral instability in a rotating shaft. When the axial load approaches the critical buckling load for the rotating shaft, the natural frequency of vibration in the transverse direction reaches zero. Since the pulsating compressive load causes a loss in lateral stiffness in the shaft, a corresponding reduction in the frequency is the consequence, as noted by McLachlan (1947).

Dry friction was mentioned before, when the shaft rotating at a high velocity encounters instability due to contact between the rotating and static components of the machine. When the shaft is just beginning to turn, breakaway friction produces higher level of friction to cause another group of instability in shafts – chatter, or stick-slip. Periodic motion of the rotor's point of contact is self-excited as a result of negative damping, because motion is taking place in the direction of rotation, and increases the relative contact velocity. Rotor motion counter to contact velocity will reduce relative velocity and increase friction force, thus again reinforcing the periodic motion. In the absence of proper flow of cooling fluid at the cutting bit, dry friction in machine tools exhibit a similar characteristic, and is known as chatter. In both cases the vibration is more torsional in nature than lateral, but vibrations in the transverse direction can also ensue.

6.13 METHODS TO ALLEVIATE VIBRATION INSTABILITY

Rotating machines experience instability due to the presence of a tangential force set up by some physical characteristic related to the machine's design or operation. The tangential force arises due to a deflection of the rotor in the radial direction, thus providing a coupling between the two directions. If adequate steps are taken in the design stage of the machine, the tendency towards unstable operation will be diminished.

Stiffness of the shaft is an important consideration. Destabilizing forces generally become more pronounced when machine operation takes place above the first critical speed, since their sensitivity towards unstable operation increases with the shaft's flexibility. Thus, one possibility is to maintain the operating speed of the machine below the first critical. As the speed ratio ω/ω_n increases, the tendency for the shaft to whirl goes up. Keeping shaft span between the bearings low helps to increase the critical frequency of the rotor, and consequently the operating speed of the machine can be raised without encountering instability problems.

External damping plays a significant role in eliminating unstable operation in a rotating machine. Care must be taken to ensure that damping is external to the rotating system, so the energy dissipating forces in the damper depend on the relative motion between the rotor and the stator. Adequately designed bearing supports play a major role in this respect. In the preliminary design system parameters require adjustment in order to position the natural frequencies of whirl as required. Bearing and support stiffness plays a crucial role, together with support location and mass and stiffness distribution of the rotating components. For this purpose a plot of the natural frequencies of vibration against a particular design parameter provides useful information about the sensitivity of that parameter. Figure 6.29 illustrates a plot of critical speeds as a function of identical support stiffness at both ends. When a number of such plots are generated for variables such as distance between bearings and weight of rotor, a comparison of the design from shaft vibration considerations is possible.

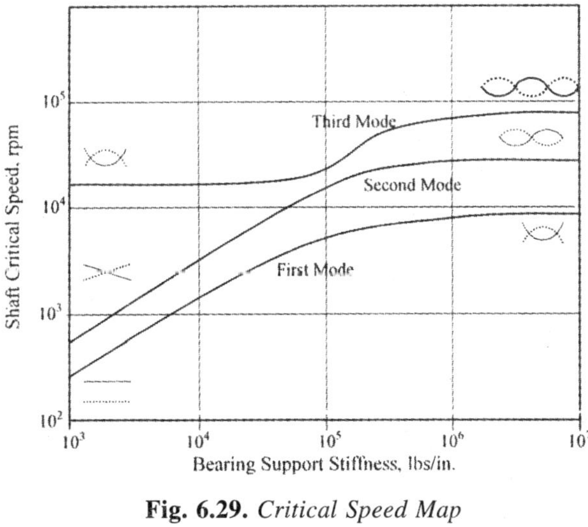

Fig. 6.29. *Critical Speed Map*

The comparison can also be made with test measurements during machine operation to improve the mathematical model with the help of critical speed maps. Use of energy maps for design evaluation has also been suggested (Ehrich, 1987). Potential (or strain) energy and kinetic energy distributions in major sub-assemblies in the machine for a particular vibration mode provide useful information. Since strain energy varies with the square of displacement and kinetic energy varies with the square of velocity, the energy distribution gives important data about the level of dynamic activity in different portions of the machine system. When a system design parameter is changed, its impact on the vibration characteristics can be studied from the energy distribution.

6.14 ROTOR DYNAMIC DESIGN OF HIGH PRESSURE COMPRESSOR

The reliability of centrifugal compressors is by far the most important aspect for many users in offshore petroleum platforms and pipeline pumping industries. With the high power density, aerodynamic forces arising in seals sometimes tend to endanger continuous operation. High sub-synchronous vibrations make it necessary to perform alterations, incurring considerable cost and downtime. The problems indicate lack of adequate knowledge of the forces because computational fluid dynamics models are not able to deal with the complexity of flow through the seals.

A comprehensive experimental investigation for determination of stiffness and damping constants of the labyrinth seals must be a part of the design phase for simulation in the rotor dynamic analysis of the system (Wagner, 1999). Thus, the test rig must be capable of producing and measuring significant gas forces encountered in the compressor. Active magnetic bearings are suitable in this respect because it is possible to set the static eccentricity independently of the rotor speed, and also to superimpose a circular motion of the shaft at adjustable frequency, amplitude and direction of precession.

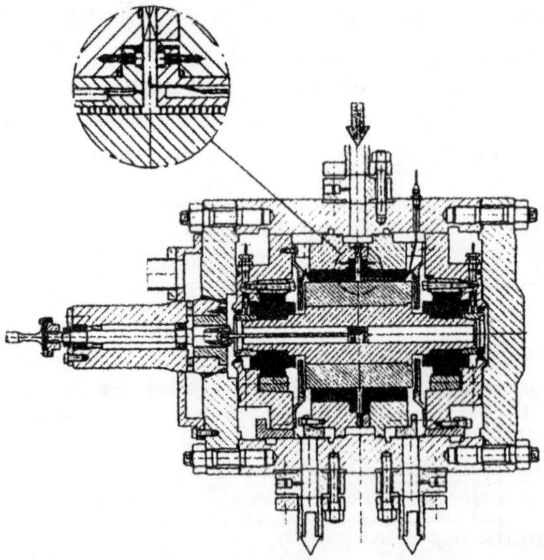

Fig. 6.30. *Test Rig Sectional View (Wagner, 1999)*

At the same time, force can also be measured from the bearings, permitting direct calculation of the transfer function without transducers. The dynamic seal coefficients may then be extracted from the transfer functions. Straight-through, stepped and staggered labyrinth and honeycomb seals of precise geometry as used in the compressor can be used in the test rig shown in Fig. 6.30, which offers the advantage of eliminating extrapolation with reference to geometrical dimensions. Besides casing inlet and outlet pressures, static and total pressures are measured directly in front of the inlet to the labyrinth. The precise swirl at the labyrinth's inlet can then be determined, a primary influencing variable for the objectives of the test.

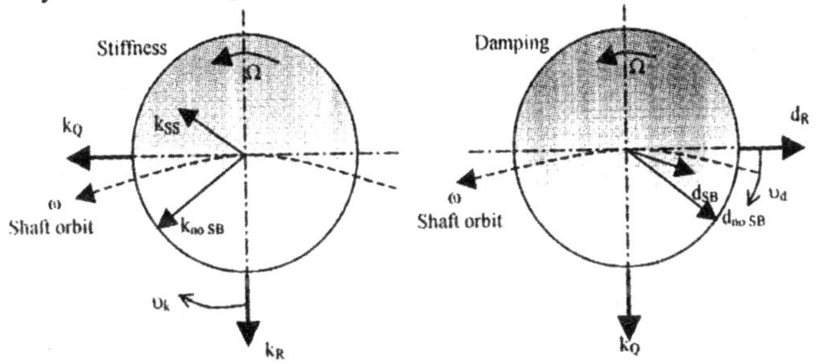

Fig. 6.31. *Stiffness and Damping Vector With and Without Swirl Brake (Wagner, 1999)*

Forces acting in the labyrinth are a combination of linear functions of displacement and velocity, and the stiffness and damping matrices become skew-symmetric only in the special case when eccentricity in the seal is zero. The equations of motion may be separated into real and imaginary components after transformation into the frequency domain, yielding two real equations for determining the four speed dependent labyrinth coefficients to be acquired. The identification is then reduced to a linear regression with speed ω, with the terms relating to stiffness obtained directly at the intersections and damping terms corresponding to the linear gradients. A number of seal types with different diameters, steps, rotating and stationary sealing tips, swirl brakes and abradable coatings on the rotor are examined in the test.

A linear relationship between labyrinth forces and displacement and velocity in the frequency domain is valid for labyrinth seals, but some restrictions are necessary for honeycomb seals. Thus, the labyrinth coefficients may be used as direct and cross-coupled variables for the restoring force $F_R = k_R + \omega\, d_Q$, but contradictions appear for the tangential force $F_Q = k_Q - \omega\, d_R$ in the form of sign changes. To overcome the problem, changing to the polar representation in the form $k = (k^2_R + k^2_Q)^{1/2}$ and $\tan \theta_k = k_Q / k_R$ alleviates the problem. The representation of swirl brakes is shown in Fig. 6.31. Swirl brakes reduce destabilization by reducing total stiffness and damping by turning the vector through a positive or negative angle. As opposed to stunt holes, swirl brakes have the advantage of not requiring additional bearing span and of having their effect precisely quantified during the test.

Figure 6.32 provides a comparison of the dynamic coefficients for the different seal types. The seals are 200 mm in diameter, 80 mm long, have a radial gap of 0.2 mm and operate at inlet density of 200 kg/m³ and pressure ratio of 0.45. Variations from operation at different shaft speeds and swirl ratios are indicated in the diagram. The staggered labyrinth produces the lowest stiffness and damping values. In comparison, forces arising in the honeycomb seal are greater by one order of magnitude. The restoring force in honeycomb seals is several times higher than the tangential force, and owing to the high cross-coupled damping, is heavily dependent on the oscillation frequency. The tangential force has a stabilizing effect in the examined honeycomb seals. The inlet swirl tends to have a larger impact on stiffness than on damping.

In rotor dynamic calculations honeycomb seals cause problems when they are employed at high densities because their dynamic coefficients then reach larger magnitudes than those of journal bearings. In the high-pressure range the lighter loading on the compressor rotor, critical speeds and logarithmic decrements rely more strongly on rotor eccentricity. Thus, it is difficult to get good correlation between calculated and measured results when honeycomb seals are employed.

The test results are applied to a re-injection compressor designed to boost pressure from 400 to 600 bar. The four stage unit has a power requirement of 8.5 MW, and density at discharge is 330 kg/m^3. Sealing is provided by a double acting mechanical contact seal with a closed high-pressure loop. The rotor (dimensions shown in Fig. 6.33) runs at 10,000 rpm and is supported in standardized four segment tilting pad bearings. The dimensions of the rotor are obtained for maximum stiffness from a proven concept, with deflection at mid-span limited to 16 times the bearing amplitude in the first mode. This achieves a high level of logarithmic decrement, and sensitivity to excitation from the labyrinth is lower than in an elastic rotor.

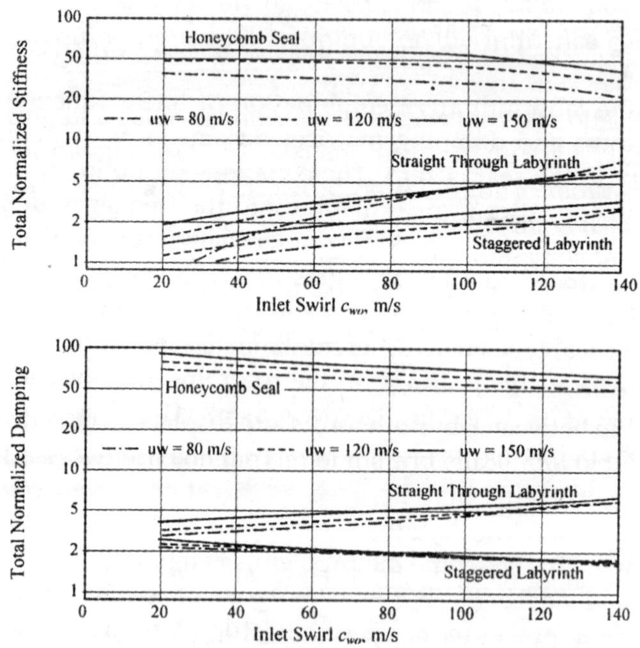

Fig. 6.32. *Seal Dynamic Coefficients (Wagner, 1999)*

A critical speed is in the vicinity of the operating speed, but in most applications operation of the compressor close to resonance is avoided, even though with adequate damping the rotor can operate near resonance without risk. Stability under all operating conditions is an essential requirement, since it may cause a shutdown of the installation besides damaging the machine.

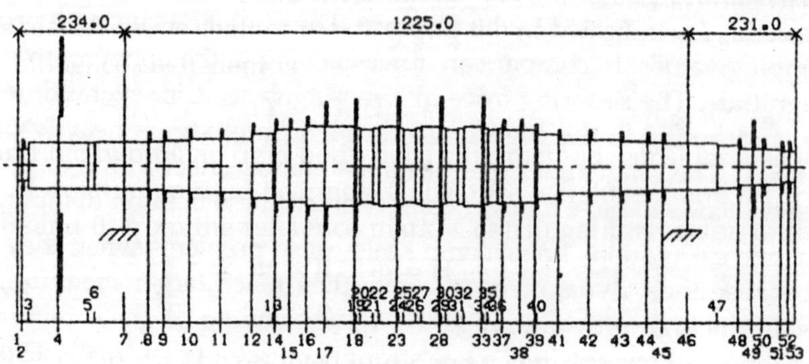

Fig. 6.33. *Rotor Details (Wagner, 1999)*

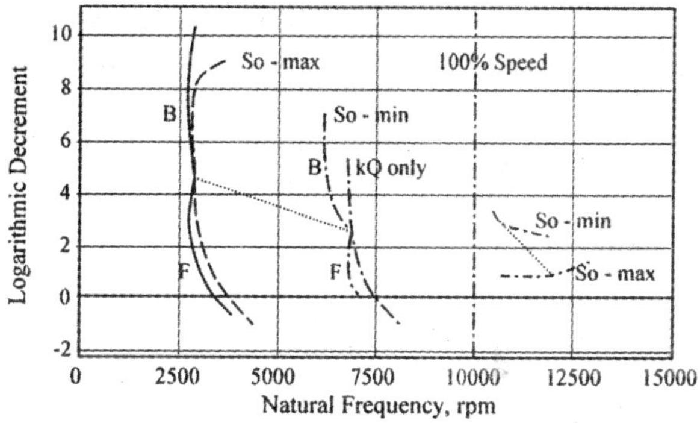

Fig. 6.34. *Effects of Gas Density on Eigen-values (Wagner, 1999)*

Manufacturing tolerances of the journal bearings play a major role for this relatively lightly loaded high-pressure compressor. Permissible variations in the lubricant oil for the bearings must also be taken into account.

Figure 6.34 shows variation of eigen-values in the complex plane when density of the gas is increased from zero to double the design value. Calculations for minimum (0.05) and maximum (0.08) Sommerfeld (So) numbers indicate the impact of bearing tolerances and oil temperature. Lower eigen-values tend to react more sensitively to the Sommerfeld number than upper ones, and the same is true of the effect of gas forces. In the case of the upper eigen-values the gas forces cause the natural frequencies to shift, yet the log decrement sees little reduction. The margin of separation required in API 617 is observed even with double the design density. The curved branches in the Fig. are based on calculations using complete stiffness and damping matrices, and also with only the cross-coupled stiffness k_Q. The complete matrix solutions result in a lower degree of destabilization, and also show the larger influence of gas forces on the natural frequencies.

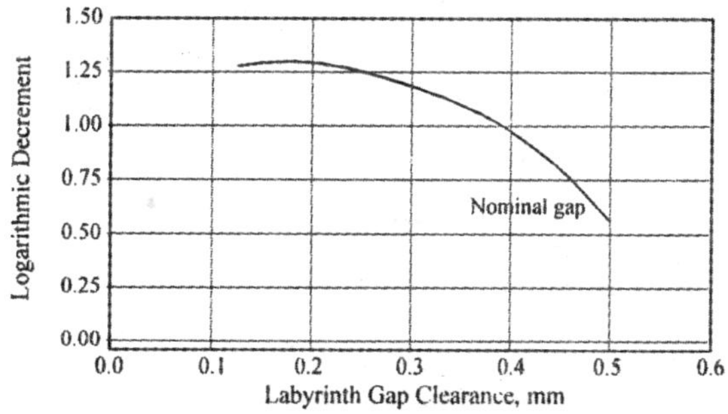

Fig. 6.35. *Effect of Labyrinth Gap Clearance On Rotor Stability (Wagner, 1999)*

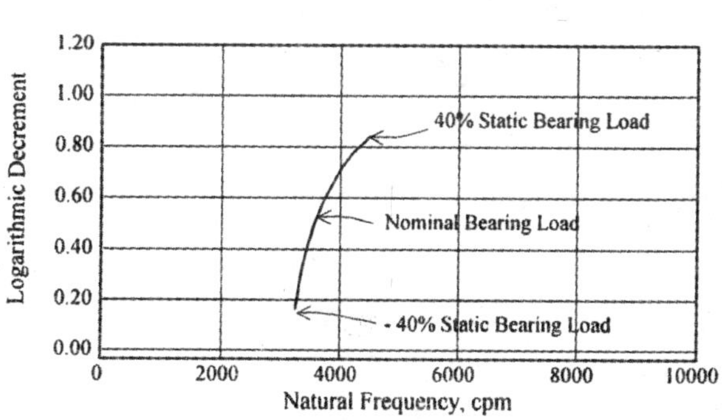

Fig. 6.36. *Effect of Side Loads On Rotor Stability (Wagner, 1999)*

Dynamic coefficients for the seals at the impeller eye and at the balance piston labyrinth are calculated for various clearance widths from available experimental data (*ref*: Zeidan *et al.*, 1993). The importance of examining the influence of clearance widths arises from observations that when the clearance is enlarged through rubbing a compressor exhibits high sub-synchronous vibrations. The theory fails when the labyrinth excitation is calculated for large clearance widths.

The course of the minimum logarithmic decrement due to variations in the labyrinth gap clearance can be observed from Fig. 6.35, using the least favorable combination of bearing geometry and oil temperature. Adequate damping reserve is still present even after the labyrinth clearance is increased to 200 percent of its nominal value.

One other factor to be considered is an asymmetric distribution of pressure development in the volute, causing the static force in the bearing to change. Figure 6.36 illustrates the influence of the static gas forces on the eigen-values on the complex plane, using minimum damping. Here, too, the assumption of most unfavorable bearing geometry and oil temperature and of damaged labyrinth is used. Adequate damping reserve is still noticeable.

Triple dry gas seals are not favorable from rotor dynamic considerations, mostly because the relatively large bearing span results in insufficient stability. One solution is to install swirl brakes on the shroud and the balance piston labyrinths. The swirl brakes have the effect of reducing the cross-coupled stiffness for the prescribed operating conditions by 70 percent. The spring and damping coefficients of the labyrinths with and without the swirl brakes are obtained from test measurements. When a hydrocarbon mixture is envisaged for the compressing unit, the reliability of the stability calculations is of significance, and can be obtained from a back-to-back experimental verification using labyrinth seals manufactured and installed in the compressor with and without the swirl brakes.

Verification of the test and analytical results is obtained from measurements on an operating unit, with its design based on the lessons learned. With the swirl brakes installed, no noteworthy sub-synchronous vibrations were observed at any of the operating points. The logarithmic decrement is determined to be 0.3. For the second test run labyrinth seals without the swirl brakes are used. Instability is encountered at 100 percent speed at discharge pressures between 240 and 320 bar, based on supply oil temperature. The frequency of vibration is around 6,000 cpm, also dependent on oil temperature. Based on the investigation, the necessary stability limit of 10 μm peak to peak sub-synchronous vibration limit is achieved by using a three-pronged approach: (*a*) increase machine speed at constant discharge pressure and oil temperature, (*b*) increase discharge pressure at constant speed and oil temperature, (*c*) increase oil temperature at constant speed and discharge pressure. The tests are also useful in ascertaining that the frequency of the sub-synchronous vibrations increases after the onset of instability. This effect is explained by the rub at the balance piston to effectively reduce the bearing span and raise the frequency of vibration as a direct consequence.

6.15 EFFECT OF SQUEEZE FILM DAMPER ON ROTOR RESPONSE AND STABILITY

The squeeze-film damper has the ability to attenuate vibration and stabilize a rotating assembly system by providing multi-directional damping. Broadly speaking, there are two different configurations of the squeeze-film damper. One is the arrangement where a centralizing retainer spring in the form of a squirrel cage is connected to the outer race of its rolling-element bearing, and acts as a parallel element within the squeeze-film annulus. The second is a structure without a re-tainer spring, where, for a horizontal machine say, the outer race of the rolling-element bearing is assumed to remain at the bottom of the damper clearance circle until the unbalance forces become large enough to create lift.

One of the advantages of the configuration with a retainer spring is the ability to tune the critical speeds of the system away from the operating range to ensure smooth running. However, when different unbalances, static eccentricity ratios and supply pressures are applied, the system resonance will often migrate from their expected positions. In particular, if the damping is excessive, which can sometimes occur with increase in the oil supply pressure or in the static damper eccentricity ratio, resonant frequencies of the assembly may approach the corresponding frequencies for the model without a retainer spring. This is because the damper locks out the retainer spring, thus rendering it inoperative.

In an experimental investigation aimed at a direct comparison of the vibration performance of two configurations of a flexible rotor-bearing assembly incorporating a squeeze film damper, the subject is examined from the points of view of forced vibration and stability control (Chu *et al.*, 1996). The schematic diagram of the test rig is shown in Fig. 6.37. It consists of a flexible stepped shaft supported in two ball bearings, one of which is selfaligning; the other is flexibly mounted on four supporting bars, constituting a centralizing spring. A squeeze film damper annulus is placed around the outer race of the latter ball bearing to attenuate rotor vibration. Oil is pumped via three holes and a central groove into the annular clearance space of the damper. The shaft carries an overhung disk held by a taper-lock bush. The drive shaft is connected to the main shaft by a flexible rubber coupling to isolate the test shaft from any vibration influence from the motor.

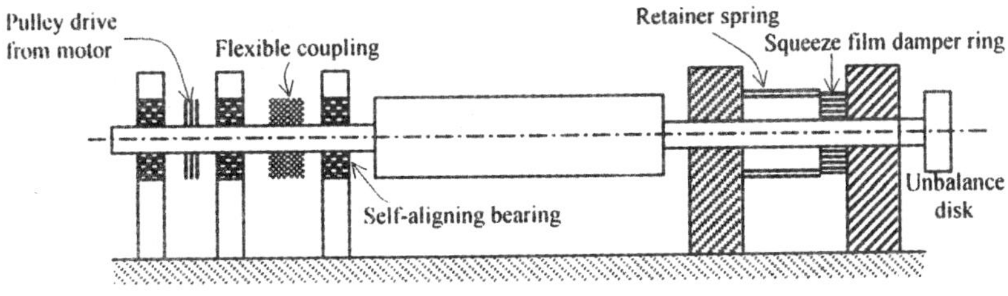

Fig. 6.37. *Schematic Diagram of Experimental Rig (Chu, 1996)*

The squeeze film damper has a two-land geometry and is unsealed. Journal radius is 50 mm, radial clearance is 0.1 mm, land length is 10 mm and oil viscosity is 0.00375 N-s/m^2. The rig is first impulse tested to obtain its undamped natural frequencies. For the configuration with a retainer spring, in order to eliminate the influence of oil damping, the damper ring is temporarily detached from the rig. The test result at the damper ball bearing gives a first undamped natural frequency of 14 Hz and a second of 41 Hz. The retainer spring is then made inoperative by shimming the damper annulus within the damper ring. The corresponding impulse test result indicates the first two undamped natural frequencies of 31.2 Hz and 88.0 Hz.

The shaft and the overhung disk are first balanced before running, as also the two together. One of several unbalance masses can be screwed into the overhung disk. In order to observe the effects of different levels of unbalance, this is chosen such that the non-dimensional eccentricity u of the mass unbalance Q_c $(= u/c)$ at the position of the overhung disk has values of 1.12, 0.649 and 0.148.

Radial and transverse squeeze film forces (P_r, P_θ) obtained by integrating the dynamic pressure distributions within the annulus are shown in Fig. 6.38, taking into consideration film rupture and the attitude angle α of the shaft. The squeeze film is rather difficult to specify, particularly with respect to the circumferential extent around the rolling element bearing. This is largely due to cavitation in the oil film which arises when negative pressures are produced

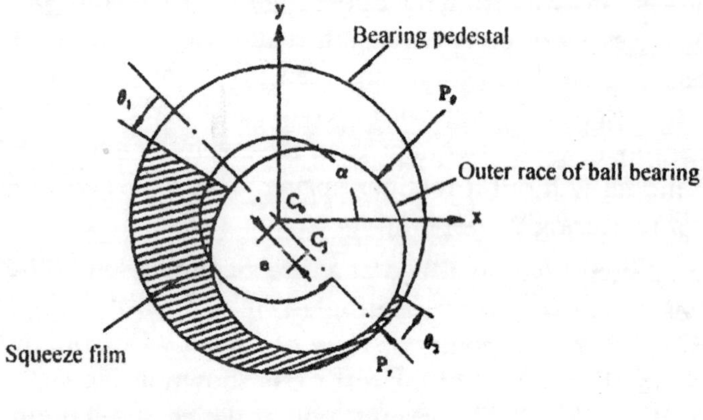

Fig. 6.38. *Details of Squeeze Film Damper (Chu, 1996)*

between rapidly separating surfaces during vibration and which cannot be compensated by sufficient supply pressure. To achieve an understanding of the way in which this oil-film extent governs the vibration performance, consider the two classical extremes – the half (π) and the full (2π) film (*ref*: Humes *et al.*, 1978). The forces (P_r, P_θ) for the short film π are:

$$P_r = \frac{\eta R L^3 \pi}{c^3} \frac{1+2\varepsilon^2}{(1-\varepsilon^2)^{5/2}} e' \tag{6.33}$$

$$P_\theta = \frac{\eta R L^3 \pi}{c^3} \frac{e\alpha'}{(1-\varepsilon^2)^{3/2}} \tag{6.34}$$

and for the full film by:

$$P_r = \frac{\eta R L^3 \Omega}{c^2} [\alpha' \varepsilon g_1 + \varepsilon' g_2] \tag{6.35}$$

$$P_\theta = \frac{\eta R L^3 \Omega}{c^2} [\alpha' \varepsilon g_3 + \varepsilon' g_1] \tag{6.36}$$

where:

$$g_1 = -2\varepsilon \cos^3 \theta_1 \, (1 - \varepsilon^2 \cos^2 \theta_1)^{-2}$$

$$g_2 = \varepsilon \sin \theta_1 \, [3 + (2 - 5\varepsilon^2) \cos^2 \theta_1] \, (1 - \varepsilon^2)^{-2} \times (1 - \varepsilon^2 \cos^2 \theta_1)^{-2} + \psi \, (1 + 2\varepsilon^2) \, (1 - \varepsilon^2)^{-5/2}$$

$$g_3 = \varepsilon \sin \theta_1 \, (1 - 2 \cos^2 \theta_1 + \varepsilon^2 \cos^2 \theta_1] \, (1 - \varepsilon^2)^{-1} \times (1 - \varepsilon^2 \cos^2 \theta_1)^{-2} + \psi \, (1 - \varepsilon^2)^{-3/2}$$

$$\tan \theta_1 = -\varepsilon'/\varepsilon\alpha'$$
$$\psi = \pi/2 + \tan^{-1}[\varepsilon \sin \theta_1 (1 - \varepsilon^2)^{-1/2}]$$
ε = eccentricity ratio = e/c = eccentricity/clearance in damper
Ω = angular velocity of rotor
η = oil viscosity
θ = circumferential coordinate and $' = d/dt$

For circular orbits concentric about the center of the damper and the given damper values, the 2π film has an effective linear damping coefficient from the two lands, and is $c_0 = 2\pi\eta R(L/c)^3 = 1.178 \times 10^3$ N-s/m. For the π film the value is $c_0 = \pi\eta R(L/c)^3 = 0.589 \times 10^3$ N-s/m. The linear values relate only to small vibrations, with $\varepsilon \to 0$.

For a static eccentricity ratio of zero a higher supply pressure produces higher system damping, and such damping suppresses vibration. This is almost certainly due to maintaining the clearance full of oil. For concentric vibration ($\varepsilon_0 = 0$), as the supply pressure is increased from 0 Pa to 34.5 kPa, there is a shift in resonance position from 40.0 rev/s to around 34.0 rev/s. It is expected that if the supply pressure is further increased, the resonance will finally reach a value of 31.8 rev/s, the value without retainer spring. For eccentric vibration with a static damper eccentricity ratio of unity, increasing the supply pressure does not lead to change in the position of the resonance, remaining at around 31.8 rev/s.

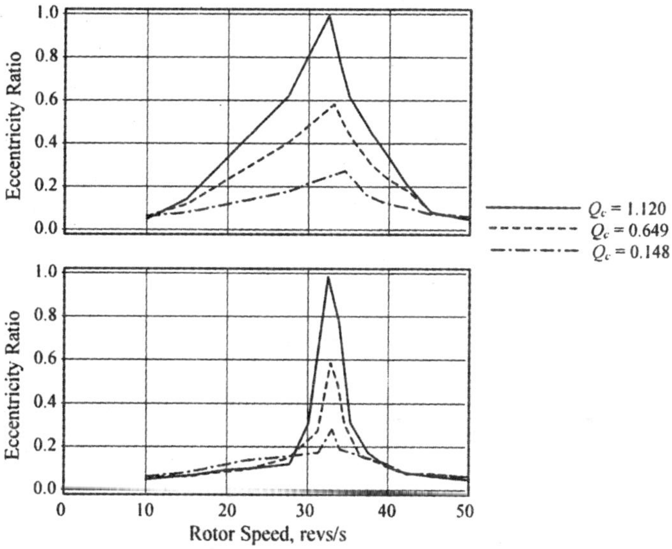

Fig. 6.39. *Vibration Response With* (upper) *and Without* (lower) *Retainer Spring (Chu, 1996)*

The reason for this is that the system damping is always high and is caused by the high static eccentricity ratio. As supply pressure and hence damping is increased, it can only reduce the peak value of the vibration and make no further shift in the resonant frequency.

For a high-speed rotating assembly, it is important to ensure that the shaft is well balanced and runs in a quiet regime. Unfortunately, there will always be some residual unbalance, which, if large enough, will cause severe vibration. This can sometimes lead to fracture of the shaft and damage to the foundation. Also, even relatively low amounts of unbalance may change the position of the resonance for the configuration with retainer spring and make it unpredictable at the design stage. Figure 6.39 shows three response curves for concentric vibration at the squeeze film damper location for the configuration with and without the retainer spring. For low unbalance, say $Q_c = 0.148$, as shown by the lowest curve, the position of the resonance is at about 34 rev/s. If the unbalance is increased to $Q_c = 0.649$, a resonance is observed at around 33 rev/s and if unbalance is $Q_c = 1.12$, the position is at 32 rev/s. At the overhung disk location the shaft vibration with no supply pressure is nearly half that of the response when the supply pressure is 34.5 kPa.

For a given operating condition, the system will be stable if all the eigen-values λ that enable the eigen equation $D(\lambda) = 0$ to be satisfied are such that their real parts are negative. Even one root with a positive real part is sufficient to indicate unstable conditions. If the relation between the real part of the eigen -value s and the rotating speed is obtained, operating stability of the system can be assessed. A numerical search for all the roots of the eigen equation can be very time-consuming, and tends to lead to a loss of physical insight. For the eigen-equation $D(\lambda) = 0$, the Leonhard locus (Naslin, 1965) retains a feel for the underlying principles. Consider the complex λ plane and the complex D plane. The function $D(\lambda)$ maps each point in the λ plane to a corresponding point in the D plane. The roots of the eigen equation can be regarded as those points in the λ plane that map to the origin of the D plane. A point traveling along the imaginary axis in the λ plane from $\omega = 0$ to $\omega = \infty$ will map to a curved locus in the D plane, and is known as the Leonhard locus. For a stable system all the roots will lie to the left-hand side of the locus in the λ plane. This implies that the Leonhard locus will be such that the origin is always to its left-hand side. Thus, for stability, the Leonhard locus must encircle the origin, in an anti-clockwise direction, as ω increases from zero. At the threshold of stability the Leonhard locus will pass exactly through the origin of the D plane. As an example of this theory, consider the instability problem for the chosen rotor dynamic system. The combined rotor-damper-seal system is assumed to be supported by a linear damper with damping coefficient $c_o = 0.589 \times 10^3$ N-s/m say, to simulate the linear π model for the squeeze film damper. Figure 6.40 shows the two Leonhard loci for this case. It is seen that as the frequency ω increases the curve for $\Omega = 65.4$ rev/s encircles the origin in an anticlockwise direction and the curve for $\Omega = 65.5$ rev/s does not. Therefore, 65.5 rev/s can be reasonably regarded as the onset speed for instability of the system.

As an alternative, the dynamic stiffness method may also be used to analyze the stability of the rotor dynamic system, incorporating, say, the linear π model of the squeeze film damper. Consider a composite system consisting of two subsystems A and B, where in this case, the subsystem A is the rotating assembly consisting of rotor and linear damper and the subsystem B is the seal.

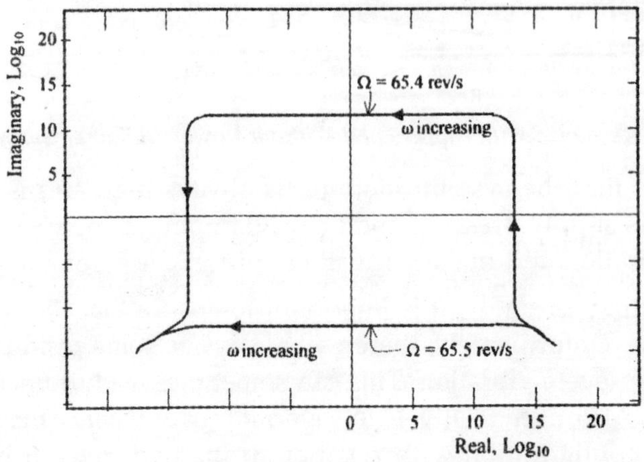

Fig. 6.40. *Leonhard Loci for Linear Analysis with $c_0 = 0.589 \times 10^3$ N-s/m (Chu, 1996)*

Subsystem A is isotropic. Apply forces of amplitudes F_x and F_y at the overhung disk. Let the amplitudes of the response be X and Y, also at the disk. Then the dynamic stiffness Z_x and Z_y for the shaft at the disk may be defined as follows. Since the shaft is circular and isotropic and the damper is also iso-tropic, these two dynamic stiffness coefficients will be complex and equal. They can be written as $Z_x = F_x/X$ and $Z_y = F_y/Y$. Subsystem B (the seal) has terms that include direct and cross-coupled forcedeflection relationships, and some of the terms are functions of the rotating speed Ω. Now connect the two subsystems together at the overhung disk mass. At the borderline between stability and instability the system will vibrate at a natural frequency and no force will be required. This condition can be expressed to determine an expression for the borderline relation between stability and instability. The procedure calls for selecting trial values of rotor speed Ω, determining the real and complex portions of the two stiffness expressions and finding a value for the frequency ω such that the stiffness components come close to zero. A graphical approach provides the value of the instability onset frequency ω and the corresponding speed Ω at the intersection of the two curves. An added advantage of this procedure is that one subsystem, say the seal, can be maintained unaltered and the other, the rotator with the damper, revised as required to provide the requisite instability onset speed.

6.16 FULL LOAD TESTING FOR ROTOR DYNAMICS IMPROVEMENT

API 617 provides guidelines for rotor dynamics characteristics of compressors during shop tests. However, the requirements deal mainly with synchronous unbalanced response of the rotor, and do not relate to sub-synchronous vibrations induced by the flow of gas, or in other words, stability. In multi-stage centrifugal compressors when the gas crosses through the channel, distortions take place in the velocity distribution along the impeller, diffuser, channel return, inlet guide vanes and volute. This induces a complex pattern of dynamic forces on the rotor, resulting in unstable operation of the machine.

The aerodynamic forces may be split into two categories. The first type is forced excitation generated directly by the flow itself at low flow rates, and manifests in the form of a rotating stall. Prediction criteria for this event are not accurate when the flow coefficient is low in high-pressure centrifugal compressors. The second type of force arises from interaction between the fluid and the structure, and develops in regions where the fluid is confined. The displacement of the rotor creates local disturbances in the pressure and velocity, and appears as a special reaction. The interaction is of considerable significance when the compressor has several impellers, and hence a number of inter-stage and piston labyrinths, where the gas is confined.

The vibration problems are illustrated in the full load testing of two natural gas centrifugal compressors of the type shown in Fig. 6.41 (*ref:* Gelin *et al.*, 1996). One unit is used for re-injection of natural gas driven by a variable speed electric motor, and the other is dedicated to a gas lift application powered by a fixed speed motor. Both machines are fitted with tilting pad bearings and dry gas seals. The units are split into two sections for intermediate cooling. The rotor for the re-injection application has higher stiffness since it handles gas at a greater pressure, and has a part of the gas path machined in the shaft at the impeller eye locations. The initial design called for vaneless diffusers.

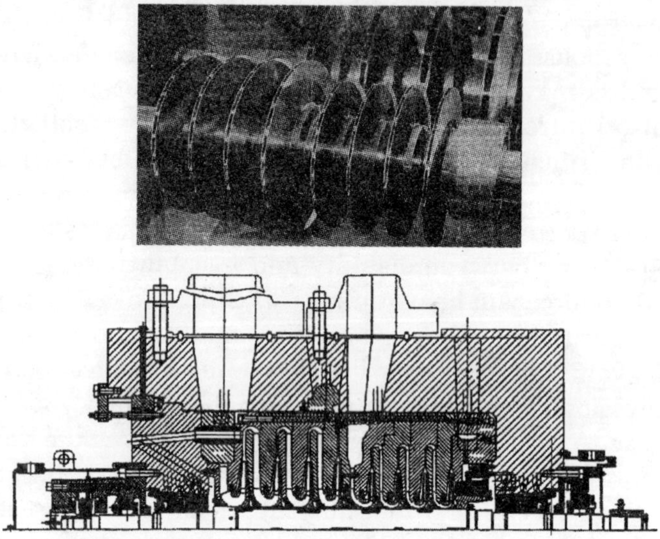

Fig. 6.41. *Compressor Rotor (upper) and Cross Section (lower) (Gelin, 1996)*

Low-pressure performance tests were carried out according to ASME PTC 10 Class III requirements, and the units exhibited good vibratory characteristics. Sub-synchronous vibrations appear at 7 percent speed, accompanied by unacceptable levels at the first natural frequency of the rotor, in the full load test of the re-injection unit. The rotating stall problem is overcome by using vaned diffusers in the second section and for the last impeller of the first section. The first natural frequency peak was diminished by installing a honeycomb seal in place of a toothed labyrinth on the balance piston, and by modifying the tilting pad bearings to provide increased stiffness.

The gas lift compressor exhibited high synchronous vibrations at the first bending mode while operating at 13,800 rpm, 17 bar suction pressure, 125 bar discharge pressure and 100 percent flow in each section inlet. On dismantling the machine the thrust bearing pads, thrust collar and the dry gas seal were found to sustain extensive damage. Contact was noted on the impeller shrouds, but the radial bearings and the internal labyrinths were intact. The instability in the radial direction led to a major incident in the axial direction, with notable radial damage that was inexplicable. Investigating the rotor dynamics and thermal and aerodynamic excitation of the flow in the dry seals revealed two different and uncorrelated events: instability arising from the first bending mode under the prescribed operating conditions and a thermal problem with the gas seals. The problem was resolved by stiffening the shaft in a similar manner as on the other unit, installing a honeycomb seal on the balance piston, adjusting the pads on the radial bearings and redesigning the inlet and discharge volutes of the second section. Subsequent results form testing at full load were satisfactory, and permitted installation of the unit at the site.

Diagnosis of synchronous vibrations is obtained from a rotor dynamic analysis using the elastic beam representation of the shaft, each node having two radial displacement and two rotational degrees of freedom. Impellers, balance piston and gas seals are simulated as pure mass and inertia without stiffness. Radial tilting pad bearings are modeled by complete stiffness and damping matrices that include cross-coupling terms arising from the pad inertia. Fluid structure interaction due to the impellers and internal seals are simulated like the bearings, by stiffness and damping matrices of the form:

$$[K_f] = \begin{bmatrix} K & k \\ -k & K \end{bmatrix} \quad \text{and} \quad [C_f] = \begin{bmatrix} C & c \\ -c & C \end{bmatrix} \tag{6.37}$$

Influence from the interaction between the fluid and the structure can be obtained from the difference $(k - C\omega)$ because it represents the destabilizing cross-coupled force intensity. Dynamic characteristics of the radial tilting pad bearings come from the solution of the Reynolds equation with introduction of a perturbation (Frene *et al.*, 1990). Thermal equilibrium on each pad is verified to estimate the actual oil viscosity and clearance and preload in the bearing. The numerical resolution uses an iterative procedure based on a finite difference scheme. Natural frequencies and the associated damping in the form of logarithmic decrement are then computed by a parametric analysis from the rotor model at nominal speed. Figure 6.42 gives the corresponding chart for the first lateral mode in the forward direction for the re-injection compressor, where the diametral clearance is expressed as a ratio of bearing diameter and preload expressed conventionally.

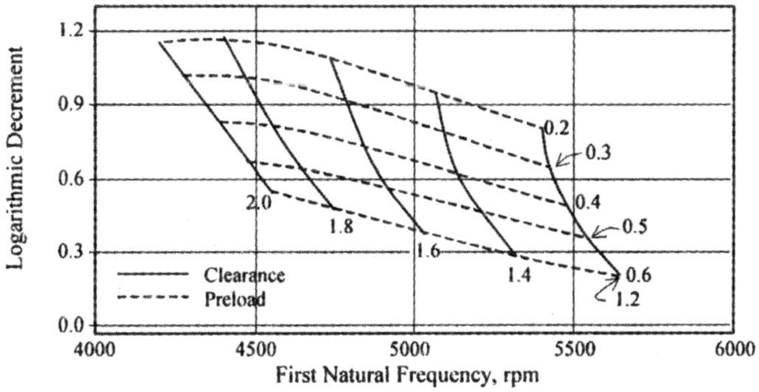

Fig. 6.42. *Bearing Optimization Chart for Re-injection Compressor (Gelin, 1996)*

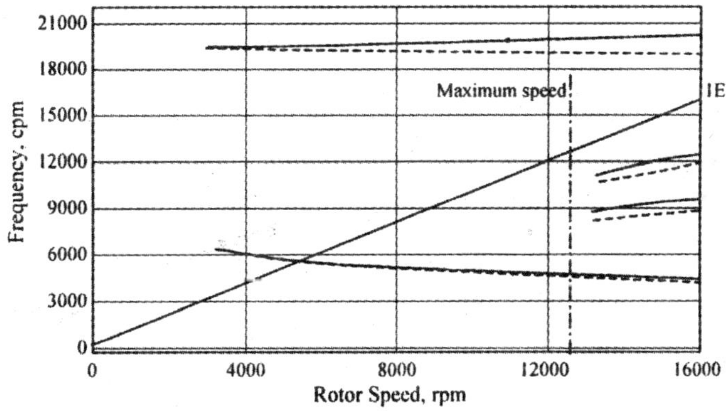

Fig. 6.43. *Campbell Diagram for Re-injection Compressor (Gelin, 1996)*

The diagram permits selection of the optimum bearing geometry to obtain a high natural frequency and good damping. Thus, paying special attention to clearance and preload definition achieves noticeable changes in the rotor dynamic characteristics. The Campbell diagram for the re-injection compressor is shown in Fig. 6.43.

Even after optimization in the bearings both compressors exhibited sub-synchronous vibrations at the first natural frequency during full load tests. Attention was turned to the cross-coupled forces from the internal seals where the gas tends to confine. The seals on the impeller eyes are of teeth-on-stator labyrinth type, with only a shunt hole on the balance piston and a separation wall between the last impeller and the balance piston. With the higher flow rate the shunt hole became ineffective since pressure in the discharge volute was lower than the discharge pressure from the last impeller, and this caused the gas to enter the toothed labyrinth directly from this impeller's discharge. The undesirable condition was rectified by improving the feed into the shunt hole, installing a separation wall between the last impeller and the balance piston and using a honeycomb seal to improve the overall global damping.

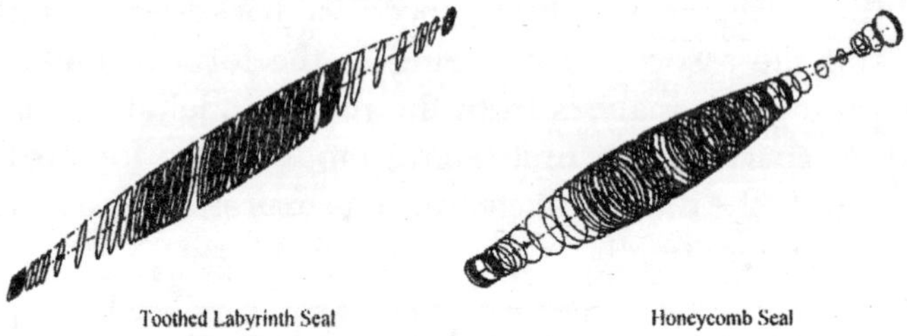

Fig. 6.44. *Differences in First Mode due to Seal Type (Gelin, 1996)*

The honeycomb seal conferred two interesting characteristics of increased natural frequencies. The first natural frequency increases mainly due to the mode becoming less sensitive to fluid-structure interaction. This increases the direct stiffness of the honeycomb seal. At the same time, a significant shift in the logarithmic decrement occurs due to the higher damping of the honeycomb seal. The corresponding mode shapes are shown in Fig. 6.44. The difference in shape is attributed to the high direct stiffness of the honeycomb seal relative to the toothed seal.

6.17 LM 2500 ENGINE VIBRATION CHARACTERISTICS

Among the many deleterious effects of vibrations, acceleration of wear in gas turbine engines is probably the most objectionable since it directly affects the durability characteristics of the components and the overall system. Critical speeds and associated mode shapes of the complete structure of a gas turbine engine must be understood to deal with vibration related problems. A multi-level, multi-rotor mathematical model that includes all rotating and stationary parts, including the support system, is developed to calculate the effects of rotating unbalance. Analytical predictions of natural frequencies of vibration and structural response must be correlated with actual engine vibration measurements.

Engines derived from aircraft power plant technology tend to have structurally flexible casing and support members because of thinner wall sections. Thus, many of the natural frequencies are associated with considerable dynamic activity in the stationary sections of the engine, driven by imbalance in the rotors and misalignment in the bearings. In the case of derivative engines used to power heavy military vehicles (see section 4.5), external excitation also plays a major role. For instance, when an army tank traverses a ploughed field at a speed of 20 miles an hour, the vehicle and the propulsion system are subjected to a spectrum of acceleration levels spread over a considerable frequency band. Vibration modes that are mostly associated with shaft bending are substantially affected by the casing and support system's geometric characteristics. Understanding the analytical predictions for mode shapes provides useful insight in the matter of locating vibration measurement sensors and to determine vibration severity during engine testing.

To illustrate the process of analyzing vibration characteristics of an engine system, General Electric Company's LM2500 marine gas turbine (shown in Fig. 6.45) will be reviewed (Thompson *et al.*, 1988). The US Navy has accumulated extensive fleet experience with this power propulsion engine for ships. Two gas generator rotor and two power turbine rotor critical speeds exist in the turbine's operating range. One mode, identified with the compressor rotor and observed between 8,100 and 8,500 rpm, exhibits significant dynamic activity at the compressor rear frame, a standard vibration monitoring location. Another mode attributed to the high-pressure turbine displays high activity in the 5,000 and 7,000 rpm range, and also is predominantly observable at the compressor rear frame. In the low pressure rotating system one of the natural frequencies occurs between 1,500 and 1,700 rpm and the other between 3,200 and 3,500 rpm, and both modes are driven by the low-pressure turbine. Interestingly, both modes are also seen primarily by the compressor rear frame accelerometer.

To understand the wear accelerating effects of vibrations in General Electric Company's LM2500 engine, the US Navy's engineers developed an analytical model for calculating the rotor dynamics and structural response (Thompson *et al.*, 1988). The mathematical procedure is a multi-level extension of the classical Myklestad-Prohl method (Mykelstad, 1944, Prohl, 1945), with the high and low pressure rotating systems and the engine's casing and support structure included in the model. Previous efforts in evaluating the vibratory characteristics of the marine engine during the design phase were restricted to determining the critical speeds. The dynamic model is now adjusted to reflect the actual behavior of the engine during operation. The gas turbine engine, shown in the upper half of Fig. 6.45, provides the main propulsion power for the Navy's frigates, destroyers and cruisers, and is considered a reliable workhorse of the fleet. But during certain conditions the engine exhibits vibrations that tend to accelerate wear in the components.

A number of parts cannot be modeled accurately with constant cross-section beam elements, so finite element techniques are necessitated to define deflection and rotation coefficients due to the action of shearing force and overturning bending moment in the beam. The development of expressions for the coefficients then permits their introduction into a transfer matrix form. The matrix multiplication process than proceeds normally across the hybrid elements. Transitional elements occur at joints between different levels of the structure and at points where an overhanging structural element is attached to a support point. Some examples of the latter are cantilevered segments on compressor and turbine rotors. Compatibility equations are then written for these elements to define the transfer of shear, bending moment, slope and deflection across the element or from one structural level to another.

Polar moment of inertia of the components tends to cause varying effects due to the difference in speed of the multiple rotors. This aspect is compensated by specifying the reference rotor to operate at synchronous speed, and by defining the other rotor speeds as a function of the reference rotor. In this respect the non-rotating components are assumed to be operating at the same speed as the reference rotor, so the results may be interpreted to say that the stationary parts execute dynamic motion that is synchronous with the reference rotor. The stiffness of flexible couplings, bearings and squeeze film damper bearing are defined in a similar manner. Offline calculations are required to develop the functional relationship between rotor speed and bearing load for the squeeze film damper bearing.

Model definition starts with sectioning the full structure into a series of connected elements, referred to as a 'Line'. Starting with the gas generator rotor, the lower half of Fig. 6.45 shows it as 'Line A'. Interconnections representing bearings and their supports, struts and mounts are defined by force and moment springs to structurally couple the line members. The springs may even be provided with viscous and damping properties for a forced response analysis. Boundary conditions for the rotating and static components with a free end take the form of zero moment and shear force at the end. Upon completion of the matrix multiplication process a square matrix is obtained, the determinant of which is zero for the natural frequencies.

The mathematical model is first used in the calculation of critical speeds and their associated mode shapes in the turbine's operating speed range. The high and low pressure rotating systems

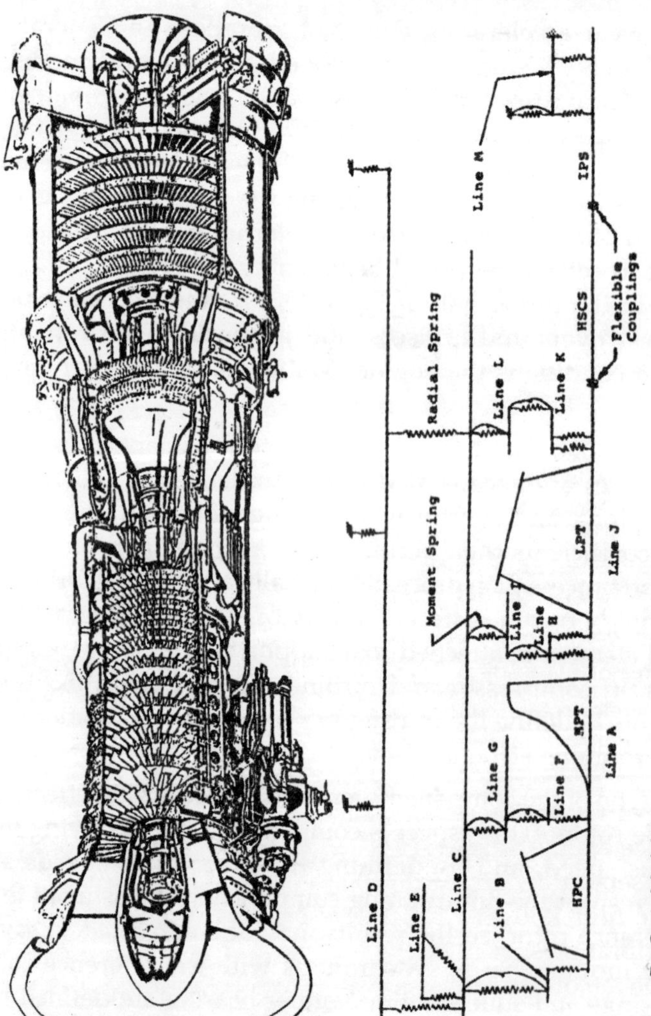

A Gas Generator D Outer Module G Combustor Inner Liner J Power Turbine/HSCS/Pinion M Reduction Gear Case
B A – Sump E Accessory Gearbox H C – Sump K D – Sump
C Outer Casing F B – Sump I Turbine Mid Frame L Turbine Rear Frame

Fig. 6.45. *Switch Position and Direction*

Table 6.3. ***LM2500 Gas Turbine Engine Critical Speeds***

Mode #	*Reference Rotor System*	*Speed RPM*	*Mount Stiffness Direction*	*Description of Associated Mode Shape*
1	PT	331	Vertical	System rocking Mode
2	PT	346	Vertical	System rocking Mode
3	PT	835	Vertical/Horizontal	1st stage HPT nozzle
4	PT	1090	Horizontal	Accessory gearbox
5	PT	1400	Vertical	Accessory gearbox
6	PT	1632	Horizontal	Gas turbine bending/ Accessory gearbox
7	PT	1916	Vertical	Gas turbine bending/ Accessory gearbox
8	PT	2565	Horizontal	PT rotor
9	PT	3520	Vertical/Horizontal	PT rotor/Case
10	GG	4400	Vertical/Horizontal	PT rotor/Diffuser
11	GG	5400/9000	Vertical/Horizontal	HPT rotor/Case
12	GG	5227	Vertical	PT rotor/CRF
13	GG	5615	Horizontal	Drive shaft/CRF
14	GG	5997	Vertical	PT rotor/CRF
15	GG	6226	Horizontal	Drive shaft/CRF
16	GG	6558	Horizontal	PT case/GG rotor/CRF
17	GG	7000	Vertical	PT case/GG rotor/CRF
18	GG	7500	Vertical	PT case/GG rotor/CRF
19	GG	8400/8600	Vertical/Horizontal	CRF/Compressor rotor
20	GG	9500/9750	Vertical/Horizontal	CRF/Compressor rotor

Notes:
1. Modes 11, 19 and 20 are observed critical speed, cannot be readily identified in analysis, response remains high in given speed range.
2. Modes 3, 9, 10, 11, 19 and 20 are not affected by mount stiffness.
3. See text for discussion on mount stiffness direction.

Legend: CRF – Compressor rear frame
GG – Gas generator rotor
PT – Power turbine rotor

operate in their own speed range, hence two executions are required for each rotor considered as the reference. And since engine mount system stiffness properties are determined for motion in the horizontal plane separately from motion in the vertical plane, a total of four sets of results are obtained. The four known critical speeds were predicted, and their mode shapes also correlated with the vibration survey measurements. The mode shapes provide added insight for further refinement of major structural members participating actively in the modes. The mode shapes also confirm the most sensitive locations on the outer engine case to measure the engine's response. Mode shapes for three of the critical speeds are shown in Fig. 6.46.

A number of the calculated modes are complementary pairs in the horizontal and vertical planes, and can be readily identified. But many other modes do not have a resembling complementary mode in the other direction, mostly because the effective mount stiffness values are substantially different, and this causes the pair's natural frequencies and mode shapes also to differ.

An investigative analysis of each of the critical speeds and associated mode shapes has been carried out (table 6.3). Many of these critical speeds turn out to be duplicates, representing a complementary pair of modes. The engine's internal components are practically symmetric about the centerline, but the mount system is asymmetric. The engine is supported by a system of links and brackets from yokes mounted to the base plate, with discrete attachment points between the engine and the yoke. The upshot of this mounting arrangement is that the stiffness property in the horizontal and vertical directions assumes differing values. In other words, a given horizontal force applied on the engine produces a horizontal deflection, but it does not match the vertical deflection if an identical force were applied in the vertical direction. But since most of the activity for a mode takes place in a selected portion of the engine, the analytical model calculates two separate modes of vibration, one of them corresponding to the vertically oriented mount stiffness and the other to the horizontal direction stiffness. To some extent this aspect of the mounting system of the engine is also responsible for the spread in the engine speed observed for any particular mode of vibration during engine operation. Note also that if a vibration mode does not show participation of the engine mount the variance in the horizontal and vertical critical speeds will diminish.

For analyzing dynamic characteristics of the mount system careful attention must be paid to the pair of horizontal and vertical vibration modes. Dynamic loads and stresses in the components will vary for the modes. Finite element evaluation is then required to establish and evaluate the suitability of the component's design traits.

Stiffness, or spring, values used in the analytical model needs some elaboration. For a component such as the compressor rear frame a three dimensional finite element model is a prerequisite. Boundary conditions reflecting the physical realities of the component as it is assembled and operates in the engine system must be applied in the model. A static force is then applied, and deflections calculated at appropriate locations of the model for deriving the stiffness coefficients. Derivation of mass and inertia values, on the other hand, may not be as complex. As a first check the component may be pan weighed. Location of the component's center of gravity and mass inertia properties about the major axes is generally provided by most structural mechanics computer programs as a matter of routine.

In the calculation sequence for system response, damping properties must be described. Aside from damping associated with the lubricant film in the bearings and material damping because of motion between the molecules in the metal, frictional damping in the structure plays a major role in the determination of response. Thus, if a bolted flange joint is not adequately tight as called for in drawing specifications, slippage between the mating parts will be the result. Damping values are generally obtained from bench test on components, and from observations during engine test.

Table 6.3 gives a listing of critical speeds for the LM2500 engine. The speed parameter applies to the high or low-pressure rotor to which the associated mode is ascribed. The ratio of the power turbine speed to the gas generator speed introduces important and substantial inertial effects, which is maintained at a constant level in the calculation sequence. However, this ratio assumes varying values in the full operating speed range, hence the power turbine speed can change both the frequency and the mode shape of the gas generator's critical speeds.

A series of vibration modes involving components in the back end of the engine, such as the turbine rear frame, power turbine case, diffuser and coupling shaft, may hold significance. Higher than usual wear patterns have been detected in these components, possibly because of the large dynamic movement. To investigate the situation, a sudden increase in activity was uncovered at the predicted modes. A detailed analysis of the vibration signature of the accelerometers mounted on the turbine rear frame and the power turbine case confirmed the mathematical predictions. Primary exciting force behind this mode was determined to arise from unbalance in the gas generator rotor. Additional trim balancing of this rotor was successful in reducing the response in the back end of the engine.

Evaluation of the previously unknown modes indicated many of them display considerable participation of components in the back end of the engine. The turbine rear frame, power turbine case, diffuser and the high speed coupling shaft play a major role in two of the modes shown in Fig. 6.46. They provide some reasons for the wear pattern observed in the Navy's operating LM2500 engines, and are corroborated by a sudden increase in response at the turbine rear frame at speeds predicted in the analysis. For further confirmation, detailed analysis of vibration data from an engine experiencing a severe case of the wear out modes gave the right correlation with the mathematical predictions. The primary excitation behind this mode comes from unbalance in the gas generator rotor. In-place trim balancing of the high pressure compressor was successful in lowering the response at the compressor rear frame from 6.0×10^{-3} in. double amplitude at 8250 rpm and 4.5×10^{-3} in. double amplitude at the turbine rear frame between 7000 and 7500 rpm down to 2.7×10^{-3} in. at the compressor frame and 2.2×10^{-3} in. at the turbine frame.

The multi-level dynamic model also offers the potential for evaluating deflections in the engine casing and internal structural components and relative motion between the elements, referred to as a 'Line'. Starting with the gas generator rotor, the lower half of Fig. 6.45 shows it as 'Line A'. Interconnections representing bearings and their supports, struts and mounts are defined by force and moment springs to structurally couple the line members. The springs may even be provided with viscous and damping properties for a forced response analysis. Boundary conditions for the rotating and static components with a free end to take the form of zero moment and shear force at the end. Upon completion of the matrix multiplication process a square matrix is obtained, the determinant of which is zero for the natural frequencies.

The mathematical model is first used in the calculation of critical speeds and their rotor and the casing. Several engines removed for high vibrations were noted to have broken combustor retaining pins. High dynamic activity of this component in most of the high-speed modes was able to explain the cause of the failure in the retaining pins.

Four of the modes in the power turbine's low speed range between 1000 and 2000 rpm are of special interest, where the accessory gearbox is the most active part on the engine. Two of the modes see the gearbox moving in phase with the engine casing, while the other two modes have the gearbox motion out of phase with the case.

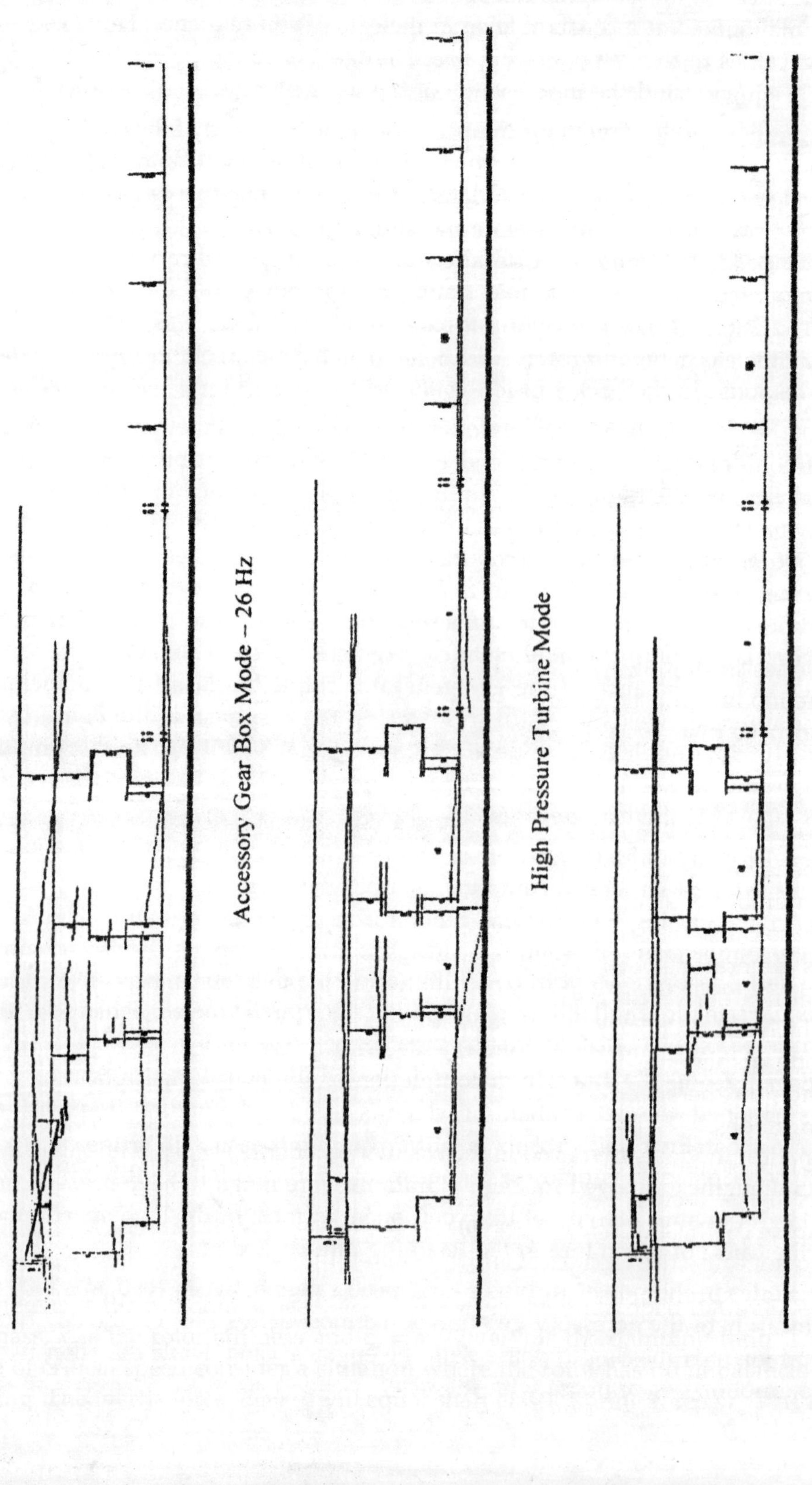

Fig. 6.46. *Mode Shapes for Known Critical Speeds of LM2500 Engine (Thomson, 1988)*

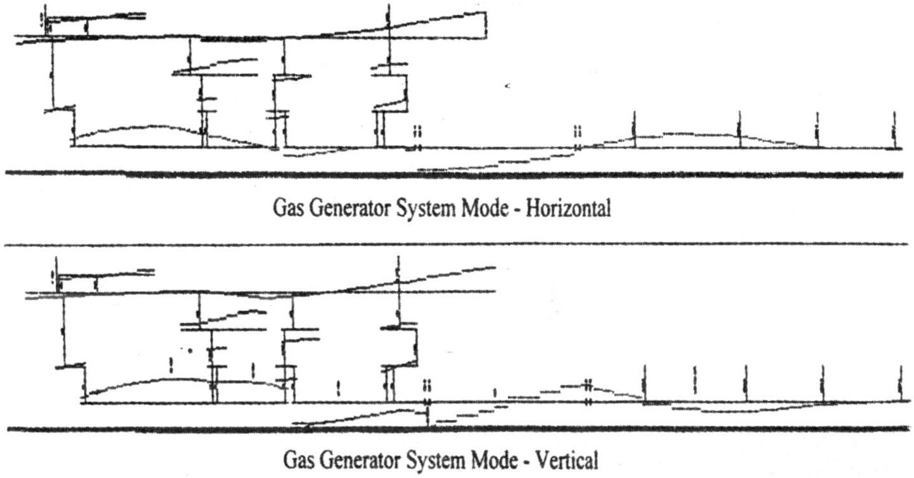

Fig. 6.47. *Newly Identified Mode Shapes of LM2500 Engine (Thomson, 1988)*

The analytical results confirm and add clarity to observations on the fleet engines: the mount links of the gearbox wear, become loose and permit the assembly to move relative to the engine. Since the gearbox supports the main fuel pump and its control, the lubricating oil pump, the starter, the air/oil separator and the transfer gearbox, redesign of the mounting system is under study.

Two other modes in the low speed range involve translation and rocking motion of the engine's enclosure module, mostly due to the stiffness values of the mount linkages, but engine operation does not take place at the speeds.

Vibration surveys on a number of the engines indicated that response of the high-speed modes in the gas generator running range was to some extent dependent on the power turbine speed, and occurred during pier side operation when the power turbine is unloaded. To understand this behavior, note that the two rotating systems in the engine are not mechanically connected, but a number of structural members in the bearing housings, support system and casing are present to cause the two rotating systems to interact. The ratio of the power turbine speed to the gas generator speed introduces considerable inertial effects. A given power turbine speed defines the stiffness of the flexible diaphragms in the high speed coupling shaft, through which engine power is transmitted to the reduction gears. Thus, the frequency at which the mode displaying high response at the turbine rear frame due to compressor unbalance depends on the speed at which the power turbine runs. Figure 6.47 shows gas generator system modes.

The dynamic representation of the engine shown in Fig. 6.45 includes four input pinion shaft bearings for the reduction gear. The bearings are of the plain journal type, and their stiffness and damping coefficients rely on shaft speed and imposed load. The flexible diaphragms in the coupling do not cause the pinion shaft to affect the high-speed modes, but the effect from the bearings on the low speed modes can be pronounced. The coupling shaft tends to isolate the power turbine rotor from excessive loads and motion arising in the drive train. But if a condition arises where the pinion shaft bearings run with little load, their reduced effective stiffness and damping could cause damaging vibrations in the coupling.

6.18 PREVENTING SUB-SYNCHRONOUS VIBRATIONS IN LARGE STEAM TURBINES

Low frequency vibrations in turbine rotors are an important issue in developing higher capacity and advanced steam condition units. The non-synchronous vibrations set in as the load is increased, and are caused by steam whirl, flow disturbance in the control stage or rubbing between the rotor and the casing. Whirling of the shaft from the steam is a self-excited vibration of the first vibration mode at high loads, and is similar to, but not as severe as, oil whip. In addition, steam whirl can occur even when tilting pad bearings are used to control the oil whip. Exciting forces generated by the blades and seals have been investigated but cannot be accurately estimated during the design stage (McClosky *et al.*, 1992). On the other hand, vibrations caused by flow disturbance in the control stage and by rubbing are forced vibrations that result in exciting the first rotor mode (Wu *et al.*, 1993).

To solve the problem an analytical and vibration testing of the turbine's rotor under actual operating conditions has been carried out by Mitsubishi Heavy Industries of Japan (Kanki *et al.*, 1997).

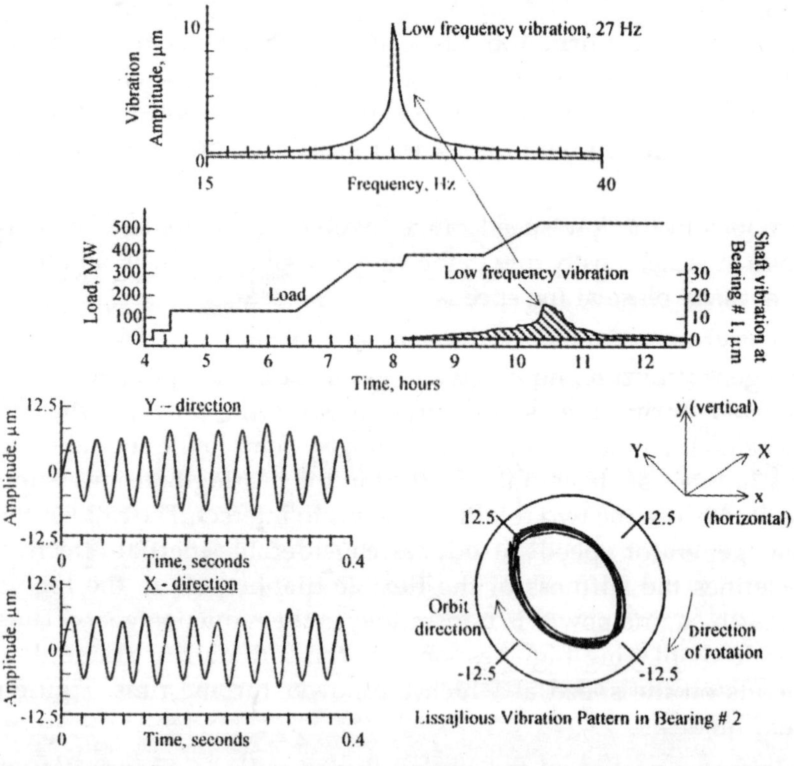

Fig. 6.48. *Shaft Low Frequency Vibrations* (upper) *and Trajectory* (lower) *(Kanki, 1997)*

The results indicate that the low frequency vibrations are due to a reduction in the available damping at high loads, and the falloff in the damping is caused by the presence of unstable forces in the control stage. As a counter-measure, a squeeze film damper has been designed and tested to verify its damping properties. The damper mechanism is then installed on the turbine unit and the test sequence performed at actual operating conditions. The field tests confirmed the increased damping in the system introduced from the squeeze film damper, and the low frequency vibrations have been eliminated.

The rotor in question comes from a 450 MW high-pressure steam turbine operating at 3000 rpm. Eight governing valves connected to the inlet adjust the flow of steam. The unit experienced a strong low frequency (27 Hz) vibration. Figure 6.48 provides the shaft vibration signature in bearing # 1 and low frequency trajectory in bearing # 2 at full power output during a typical run. The direction of the whirl is forward. The occurrence and threshold of the low frequency vibration change if the opening sequence of the eight valves is altered. The damping ratio of the first mode is measured using an inertia type of exciter placed on the bearing through a jig, and the natural frequency and damping ratio are obtained by measuring the frequency response. The stability of the rotating system is evaluated by using a swept frequency sine wave to excite the bearing pedestal, then measuring the response through an appointed natural frequency. The procedure is repeated for a number of speed and load points. A gradual reduction in the damping ratio of the rotor is noted with the increase in turbine output up to about 300 MW, when the damping coefficient suddenly decreases at points where the load exceeds 300 MW.

The test thus clarifies that the low frequency vibration is self-excited by the steam whirl when the damping is low. The destabilizing force is generally considered to arise from stage excitation, also known as the Alford force, and from the labyrinth seals. These forces may be represented by eight stiffness and damping coefficients. To determine the effects of these forces on the stability of the system, the dynamic coefficients are incorporated into the complete rotor model in the same manner as bearing coefficients. The analytical model for the full high and low-pressure turbine and generator rotors is shown in Fig. 6.49. The vibration analysis uses the transfer matrix method. The objective of this stability analysis is to determine the quantitative effects of partial admission of the control stage and alignment between the control stage and the nozzle.

Improvement in the damping ratio for the first rotor mode is achieved by replacing the # 2 bearing with a squeeze film damper bearing, shown on the left in Fig. 6.50. The inner ring of the bearing is fitted to the outer ring by 36 pins. The design procedure for the squeeze film damper aims at placing its stiffness and damping characteristics in between those for the bearing pedestal and the tilting pad bearing. The spring constant for the pins and the film's damping coefficient are selected by taking into consideration the need for maximum damping ratio and to avoid contact between the inner and outer rings under the combined action of the rotor weight and partial admission at the control stage. Setting the bearing clearance to 0.25 mm provides the optimum damping coefficient, even when the static displacement (calculated from the three-dimensional finite element model shown on the right in Fig. 6.50) produced in the housing is taken into account and shaft eccentricity becomes zero.

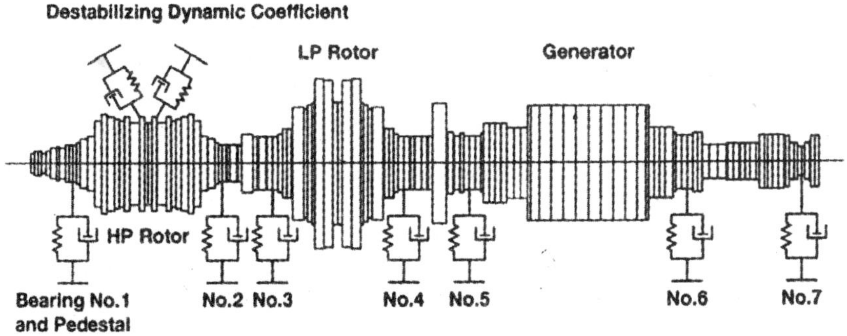

Fig. 6.49. *Rotor Dynamics Model with Destabilizing Dynamic Coefficients (Kanki, 199?)*

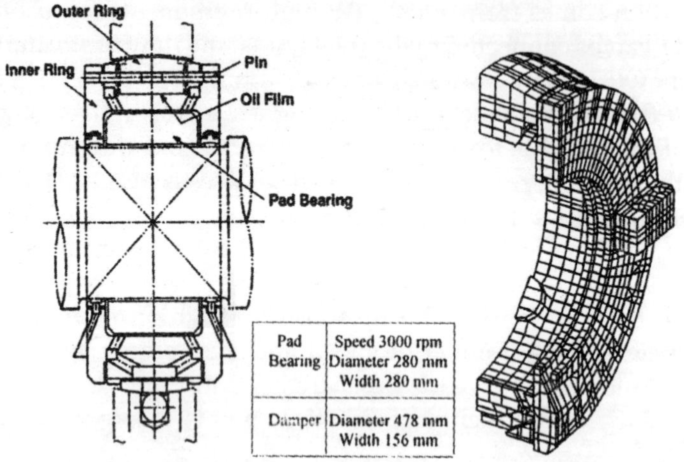

Fig. 6.50. *Squeeze Film Damper Bearing: Design* (left), *Finite Element Model* (right) *(Kanki, 1997)*

Verification of the damping effect developed in the squeeze film damper bearing is obtained from shop tests. The required increment in the damping coefficient can be seen in Fig. 6.51 (*left*). The high-pressure rotor is then installed at the site and the vibrations measured at different operating conditions. The first mode's damping ratio is also measured. Figure 6.51 (*right*) shows the spectrum of shaft vibration at rated load.

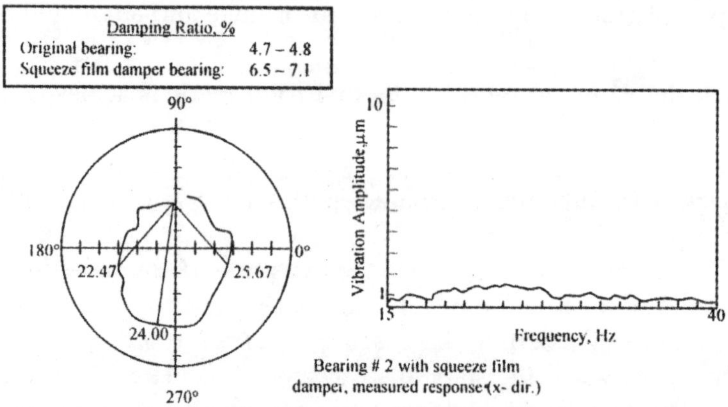

Fig. 6.51. *Measured Damping Ratio* (left), *Shaft Vibrations in Bearing* (right) *(Kanki, 1997)*

The results provide evidence that the low frequency vibration is fully eliminated under all conditions as a consequence of the increased damping by using the squeeze film damper bearing. The turbine has been satisfactorily operating free of the low frequency vibrations for two years after the installation of the squeeze film damper.

6.19 EXAMPLE PROBLEMS

Example Problem # 6.1: A 24.0 in. diameter pulley *B* weighs 145 lbs and receives 33.5 HP at 325 rpm from a shaft located at an angle of 45° below the pulley (Fig. 6.52). A 19.25 in. pitch diameter gear *C* weighing 57.5 lbs. delivers 35 percent of the power horizontally at the top to gear *G*, and a 12.375 in. pitch diameter gear *E* weighing 28 lbs. delivers the remaining power to another gear *H* downward to the left at an angle of 30° below the horizontal. Both gear sets have 20° pressure angle. The system is required to have corrosion protection for long-term effects. The shaft is supported in bearings at *A* and *D*. Determine the torque transmitted by the gears and the pulley, and obtain the bending moment distribution in the shaft.

***Solution*:** The torque acting on the shaft between *B* and *C* is:

$$T_B = 63025(\text{hp})/\text{rpm} = 63025 \times 33.5/325 = 6496 \text{ in-lbs.}$$

$$T_c = 63025 \times .35 \times 33.5/325 = 2274 \text{ in-lbs.}$$

The remaining power is delivered by gear *E*. Hence torque between *C* and *E*:

$$T_E = 63025 \times .65 \times 33.5/325 = 4223 \text{ in-lbs.}$$

The torque on the pulley is: $(F_1 - F_2) \times D_B/2 = 6496$ in-lbs.

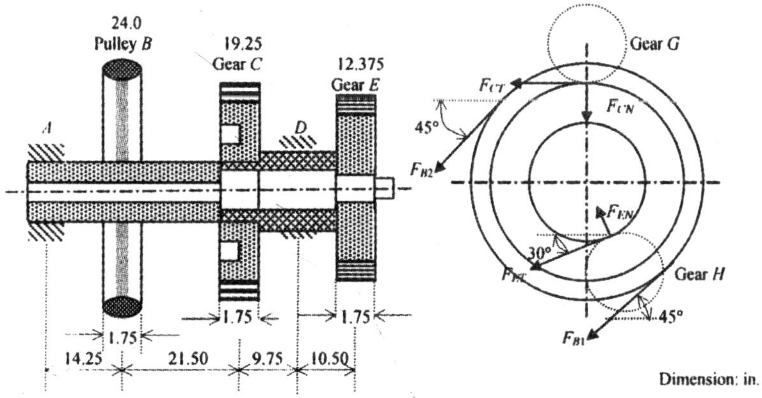

Fig. 6.52.

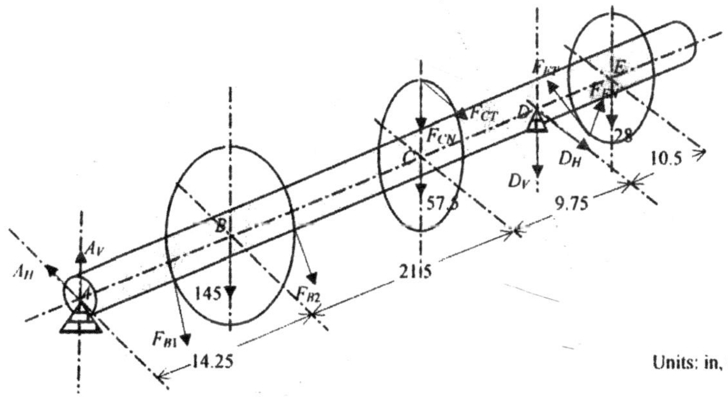

Fig. 6.53.

Hence, $F_1 - F_2 = 6496/12 = 541.4$ lbs.

Assume belt tension ratio:

$F_1/F_2 = 3$ for a flat belt, with adequate initial tension.

Then: $F_1 = 812.1$ lbs., and $F_2 = 270.7$ lbs.

Total bending force at pulley B is:

$F_B = F_1 + F_2 = 1082.7$ lbs. Reactive tangential forces on gears C and E are:

$$F_{CH} = F_{CT} = T_C / r_C = 2274/9.625 = 236.2 \text{ lbs}$$
$$F_{ET} = T_E / r_E = 4222/6.1875 = 682.5 \text{ lbs}$$

The normal force tending to separate contact at the gear teeth in gears C and G and mating gears E and H are:

$$F_{CV1} = F_{CN} = F_{CT} \times \text{Tan}(20°) = 236.3 \times 0.364 = 86.0 \text{ lbs}$$
$$F_{EN} = F_{ET} \times \text{Tan}(20°) = 682.3 \times 0.364 = 248.4 \text{ lbs.}$$

Referring to Fig. 6.53, sum the forces in the horizontal and vertical planes:

$$A_H - F_B\text{Sin}(45°) - F_{CT} - D_H + F_{ET}\text{Cos}(30°) - F_{EN}\text{Cos}(60°) = 0$$
$$A_V - F_B\text{Cos}(45°) - 145 - F_{CN} - 57.5 - D_V + F_{EN}\text{Cos}(30°) + F_{ET}\text{Cos}(60°) - 28 = 0$$

Bearing reactions are obtained from static equilibrium conditions (sum of forces and moments is zero) in each plane. Bending moment and shear force diagrams can now be prepared for the horizontal and vertical planes, shown in Fig. 6.54.

Maximum bending moment occurs at pulley B. Total bending moment at B is:

$$M_B = [(9748.4)^2 + (11086.5)^2]^{1/2} = 14764.6 \text{ in-lbs}$$

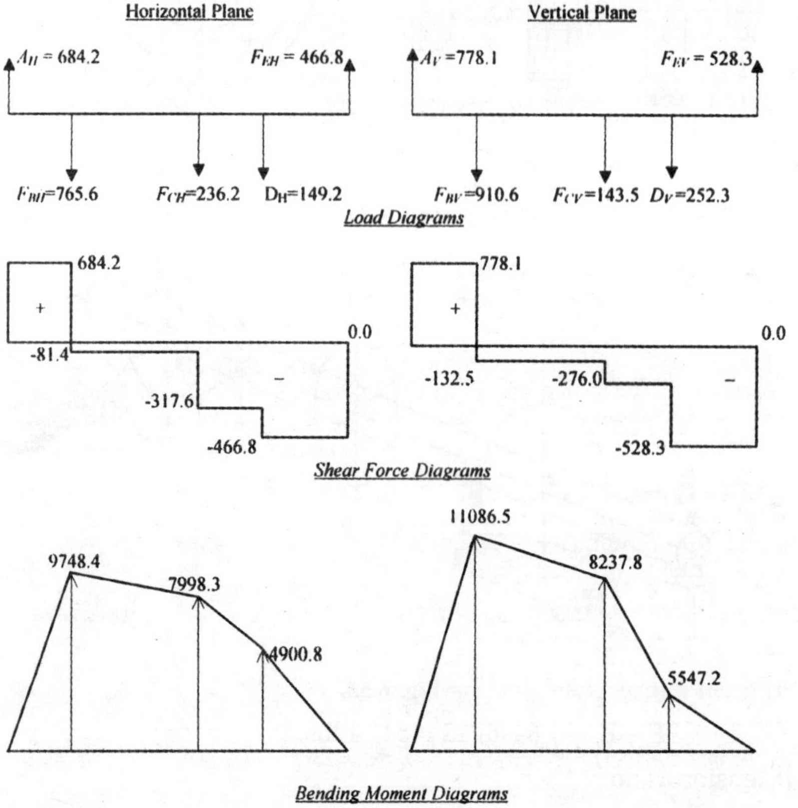

Fig. 6.54.

Note that at pulley B the shaft also transmits a torque load of 6496 in-lbs.

Example Problem # 6.2: In problem # 6.1 assume the shaft is to be designed for maximum loads. The shaft is to be machined from C1040 steel. Determine the shaft diameters. Assume the shaft is hollow, with $D_i/D_o = 0.5$.

Solution: For a hollow shaft the normal, shear and combined stresses are:

$$\sigma_x = \frac{32M}{\pi D_o^3[1-(D_i/D_o)^4]} \tag{6.38}$$

$$\tau = \frac{16T}{\pi D_o^3[1-(D_i/D_o)^4]} \tag{6.39}$$

$$\tau_{max} = \frac{0.5S_{yp}}{N} = \frac{16\sqrt{M^2+T^2}}{\pi D_o^3[1-(D_i/D_o)^4]} \tag{6.40}$$

where N is a safety factor.

Normalized C1040 steel has the following material properties:

$$S_u = 85{,}000 \text{ lbs/in}^2$$

$$S_{yp} = 50{,}000 \text{ lbs/in}^2$$

Endurance limit for C1040 may be assumed to be: $S_n{}' = 0.5S_u = 42{,}500$ lbs/in^2. If the shaft has profiled keyways at B and C, K_f (bending) = 1.6 and K_{fs} (torsion) = 1.3. For the keyway at E assume K_f (bending) = 1.3 and K_{fs} (torsion) = 1.3. For operation in a non-corrosive environment and a survival rate of 95 percent, the corrosion factor will be assumed at $C_R = 0.869$, and for a machined surface $C_F = 0.77$. Also, since the shaft will be larger than ½ in. diameter and smaller than 2 in. diameter, assume $C_S = 0.85$. The appropriate endurance limit is then obtained from:

$$S_e = (1/K_f)C_RC_SC_FS_n{}' = (1/1.6) \times 0.869 \times 0.85 \times 0.77 \times 42{,}500$$

Or,

$$S_e = 15{,}108 \text{ lbs/in}^2$$

The factor of safety N is assumed to be 3.75. Since the load is to be gradually applied, $K_{sb} = K_{st} = 1.0$. Stress concentration at the pulley is due to the profiled keyway and light press fit will be $K_t = 1.6$ for shafts subjected to fluctuating bending loads, and $K_f = 2.5$ for fretting corrosion. Thus, for the keyway and press fit combination,

$$K_fK_tM_B = 2.5 \times 1.6 \times 14764.6 = 59058 \text{ in-lbs.}$$

$$\tau_{max} = (.5 \times 15108)/3.75 = 2014.4$$

$$= [(16/\pi D_o^3) \times (M^2 + T^2)^{1/2}]/[1 - (D_i/D_o)^4]$$

Substituting the values for moment and torque at location B, then $D_o^3 = 160.2$, outer diameter $D_o = 5.432$ in. Select $D_o = 5.50$ in. and inner diameter $D_i = 2.75$ in. Shaft segment between A and B experiences $M_{AB} = [(9748.4)^2 + (11086.5)^2]^{1/2} = 14764.6$ in-lbs and $(M^2 + T^2)^{1/2} = 16131$ in-lbs. using $K_f = 1.6$ for corrosion. Then $D_o^3 = 69.60$, outer diameter $D_o = 4.113$ in. Select $D_o = 4.125$ in. and inner diameter $D_i = 2.0625$ in.

Example Problem # 6.3: Shaft deflections must be calculated in order to establish minimum permissible clearance between pulleys, gears and assembly enclosure. Explain the numerical integration method. How are shaft deflections related to the fundamental natural frequency of vibration?

Solution: Many methods are available for determining the deflection along the length of a shaft subject to a set of operating loads. The numerical integration approach is based on the following differential equations:

Load equation: $$\frac{d^4y}{dx^4} = \frac{w}{EI} \qquad (6.41)$$

Shear force: $$\frac{d^3y}{dx^3} = \frac{V}{EI} \qquad (6.42)$$

Bending moment: $$\frac{d^2y}{dx^2} = \frac{d\theta}{dx} = \frac{M}{EI} \qquad (6.43)$$

Slope: $$\frac{dy}{dx} = \theta \qquad (6.44)$$

By considering finite distances along the shaft length (in the x direction), the above equations can be integrated through numerical summation. As an example, $y = \Sigma\theta\Delta x$. The deflections are obtained by using the following step-by-step procedure, using a tabular format:

- Place a station number at each shear force location in the shaft, as also at each change in its cross-section. When long sections of the shaft with constant section are encountered, place stations at intermediate points. (1st column).
- Enter station numbers on alternate lines of the table, and force on the same line as the station number. Upward forces are positive, downward forces are negative. (2nd column).
- Calculate vertical shear at each station by summing forces in column # 2, and enter shear values in 3rd column.
- Enter the axial distance of the station from the previous station on the same line as station number. (4th column).
- Calculate bending moment at each station by multiplying shear force with axial distance. The product when added to the previous entry in 5th column gives bending moment at that station.
- In 6th column, on a line between stations enter area moment of inertia I at the section between the two stations.
- In 7th column enter product EI on the same line as I (E is material Young's modulus).
- In 8th column enter the value of (M/EI). The values are entered both at the station location as well as between stations.
- In 9th column enter the average value of M/EI on the line in between stations.
- Calculate slope at each station by entering running total of the product of 9th column and shaft segment length (4th column). Enter in 10th column on same line as station numbers.
- In 11th column enter arithmetic average of slope from 10th column, in between stations.
- Calculate deflection increment by multiplying 11th column with 4th column, and enter on the next lower line from average slope value. Enter in 12th column on same line as average slope.
- Total all deflection increments between bearing reactions, reverse the sign and divide by distance between the bearings. This gives the integration constant. Multiply this

constant by shaft spans of 4th column and enter in 13th column on same line as deflection increments.

- Add deflection increments of 12th column and the integration constant in 13th column. Enter the sum as part of a running total in 14th column on the same line as station number.

In this sequence of calculation the bearings are assumed to be rigid, hence the deflection at the bearing supports is zero. When shaft loads are separated into horizontal and vertical components, total shaft deflection is obtained by calculating the square root of the sum of the squares of the horizontal and vertical deflections at each station.

Practical difficulties in the manufacturing and assembly of a built-up rotor do not permit the center of mass of a symmetric rotating system to coincide with the actual center of rotation. Additionally, static deflections due to weight of the components and constant lateral loads cause a further shift of the mass center from the axis of rotation. As the rotating shaft's speed increases, the kinetic energy of the attached masses also increases. When the kinetic energy reaches the potential energy of the deflected shaft, the vibrations in the shaft will be at a peak. The rotor speed at which this disturbance occurs is called the fundamental frequency of vibration, or critical speed, of the shaft. Equating the kinetic energy of the rotating masses to the potential energy of the deflected shaft results in an equation that will define the critical speed of the shaft. This equation is called Rayleigh's equation. Thus the lowest, or fundamental, frequency of vibration for a shaft simply supported at two points is given by Rayleigh's method:

$$f_n = 187.7\left[\frac{W_1y_1 + W_2y_2 + W_3y_3 + ... + W_ny_n}{W_1y_1^2 + W_2y_2^2 + W_3y_3^2 + ...W_ny_n^2}\right]^{1/2} \tag{6.45}$$

where W_n is the weight of the rotating mass in pounds and y_n is the static shaft deflection due to W_n. The constant factor = $(60/2\pi) \times \sqrt{386} = 187.7$ and frequency is in cycles/min.

Example Problem # 6.4: Shafts are used in many different ways in all types of mechanical equipment as power shafts, line shafts and camshafts, among others (Fig. 6.55). An axle is a stationary member used to support rotating wheels, idler gears and pulleys. A stub, or head, shaft is generally integral with an engine, motor or other prime mover, and has a projection to facilitate connection with other shafts. A jackshaft, or counter shaft, connects a prime mover with a line shaft or a machine. Flexible shafts may be used to transmit motion between two components where the rotating axes are at angle with each other.

Typically a shaft is subjected to fluctuating loads, so considerations for stress concentration and endurance strength play a major role in its design. Of equal importance, and sometimes more so, is the consideration of shaft rigidity, or stiffness. Excessive lateral deflections can cause premature wear and failure of the bearings, while also reducing the critical speed, which can cause unacceptable shaft vibrations when operating at or near this speed.

Generally, shafting of 3 to 3½ in. diameter is made of cold drawn carbon steel round bars. If toughness, shock resistance and strength are of prime concern, alloy steel bars (for example, 1347, 3140, 4150, 4340, 8650) may be more suitable. Heat treatment may also be specified to produce desirable properties. Surface wear resistance requires a carburized grade steel (1020, 1117, 4820, 8620, among others).

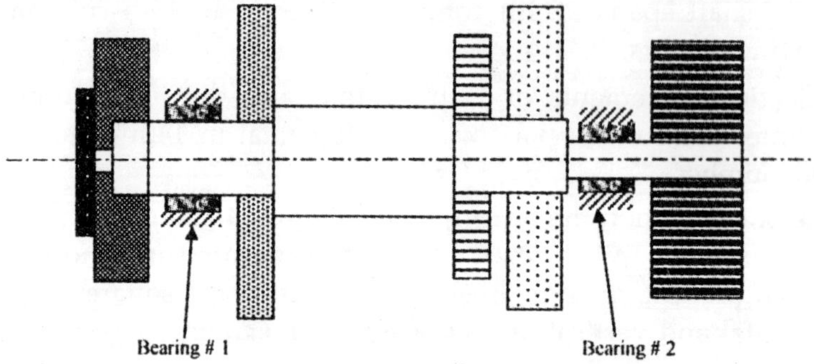

Fig. 6.55. *Rotor-bearing System Design Example*

Cost control can be obtained by using a plain carbon steel, if possible. Nitriding, cyaniding, flame hardening and other case hardening methods provide greater surface wear resistance.

Shafts larger than 3½ in. diameter call for machining from hot rolled carbon steel. Adequate stock must be removed in the machining process to eliminate decarburized scale caused by the hot rolling process. Larger diameter shafts are often times forged and machined to the required size from .45 percent carbon steel.

Shaft diameter of a rotor system can be selected once the distribution of bending moments and torque along its axial length is available. The diameter of each section of the shaft will depend on stresses due to the combination of bending moment and torsion. The location of maximum stress in an application is sometimes not difficult to tell. In general, however, a systematic approach calls for drawing shear and bending moment diagrams to locate the points of maximum moments.

Example Problem # 6.5: Table 6.4 provides the geometric definition of a rotor and bearing system. A number of factors are to be taken into account in the computation of flexural (or transverse) natural frequencies and their corresponding modes of vibration of a rotating system. First, the deformation of the rotor is due to a combination of bending and shearing, which is inherently more complex than simple twisting or stretching. Bending is generally not confined to a single plane, so two orthogonal planes (for example, xz and yz) must be considered. Within each plane it is necessary to characterize the motion of each point i by two independent coordinates: a transverse displacement and an angular rotation. Finally, when the rotor is spinning, the motions in two orthogonal planes are coupled by gyroscopic and other asymmetric effects. The location and nature of the bearings play a major role in flexural vibrations, and requires the development of more detailed models of bearings and their support structures.

It will be assumed that all deformation takes place in the massless connecting shafts. The axially symmetric rigid disks have mass and axial and transverse moments of inertia, but are of negligible axial length. Each disk has four degrees of freedom; the center may displace transversely through small displacements, and the disk may rotate through small angles about the two lateral axes. Stiffening effects due to out-of-plane gyroscopic moments are ignored.

Assume the shaft spin speed is 3600 rpm. Station 0 is a dummy station, located at the left hand end of shaft.

Table 6.4. ***Rotor-bearing System Design Data***

Stn. #	Length in.	Diameter		Shape Factor κ	Material Properties			Bearing Stiffness		Added Weight	
		Outer in.	Inner in.		E (10^6 lbs/in^2)	G (10^6 lbs/in^2)	ρ lbs/in^3	K_{xx} (10^6 lbs/in.)	K_{yy} (10^6 lbs/in.)	Weight lbs	WR^2 lb-in^2
0	0	0	0	1	30	12	0.28	–	–	–	–
1	8.06	8.93	7.13	1	27.25	11.09	0.28	1.38	1.45	–	–
2	3.77	9.76	7.85	1.25	27.13	10.88	0.28	–	–	43.2	11057
3	12.44	11.12	9.38	1.38	26.89	10.83	0.28	–	–	111.7	45092
4	9.48	12.29	10.54	1.13	26.75	10.77	0.28	–	–	–	–
5	11.33	14.13	12.51	1.13	26.5	10.27	0.28	–	–	124.8	109000
6	9.48	12.89	10.91	1.13	25.5	9.85	0.28	–	–	–	–
7	12.44	11.74	9.9	1.13	26.7	10.32	0.28	–	–	179.3	187050
8	10.38	10.72	9.15	1	27.25	11.09	0.28	–	–	38.7	5852
9	8.87	9.13	7.49	1	27.5	11.13	0.28	1.43	1.35	–	–
10	9.24	8.53	6.56	1	27.5	11.13	0.28	–	–	84.2	23400
11	0	5	0.00	1	30	12	0.28	–	–	–	–

Table 6.5. *Geometric Properties of Rotor*

Stn. #	Cross sectional		Total mass at station lbs.	Total WR^2	
	Area, A in.	Inertia, I_{diam} In		Shaft axis lbs-in²	Diameter lbs-in²
0	0	0	25.803	421.2	246
1	22.704	186.298	39.826	696.2	387
2	26.417	259.018	106.352	12631.7	6478
3	28.015	370.568	202.749	47765.3	24100
4	31.379	514.09	95.055	3783.7	2110
5	33.895	754.502	228.38	113107.7	56800
6	37.011	659.677	104.286	3379.3	1960
7	31.273	460.955	269.983	189556.6	95000
8	24.501	304.184	101.319	7208.7	3730
9	21.407	186.589	57.179	906.8	551
10	23.348	168.971	114.602	23840.1	11974
11	19.635	30.68	0	0	0

Table 6.6. *Shaft Element Field Matrix Parameters*

Station #	L	L/(EI)	$L^2/(2EI)$	$L^3/(6EI)$	ε
0	0.00	0.00	0.00	0.00	0.00
1	8.06	1.60×10^{-9}	6.43×10^{-9}	1.73×10^{-8}	2.58×10^{-7}
2	3.77	5.36×10^{-10}	1.01×10^{-9}	1.27×10^{-9}	3.95×10^{-8}
3	12.44	1.25×10^{-9}	7.76×10^{-9}	3.22×10^{-8}	3.71×10^{-7}
4	9.48	6.89×10^{-10}	3.26×10^{-9}	1.03×10^{-8}	2.36×10^{-7}
5	11.33	5.67×10^{-10}	3.21×10^{-9}	1.21×10^{-8}	3.28×10^{-7}
6	9.48	5.63×10^{-10}	2.67×10^{-9}	8.43×10^{-9}	2.19×10^{-7}
7	12.44	1.01×10^{-9}	6.28×10^{-9}	2.61×10^{-8}	4.26×10^{-7}
8	10.38	1.25×10^{-9}	6.49×10^{-9}	2.25×10^{-8}	3.96×10^{-7}
9	8.87	1.73×10^{-9}	7.67×10^{-9}	2.27×10^{-8}	3.30×10^{-7}
10	9.24	1.99×10^{-9}	9.18×10^{-9}	2.82×10^{-8}	3.28×10^{-7}
11	0.00	0.00	0.00	0.00	0.00

External masses in the form of disk, coupling or gear may be located here, as also a bearing. Shaft data – length, diameters, material properties – at station 1 correspond to shaft geometric properties between stations 0 and 1, other station data following the same format. Shear shape factor is used for shear compliance, as defined in Timoshenko's beam theory (1974). The x and y directions are located in a plane normal to shaft axis, and are mutually perpendicular to each other. Ignoring material and bearing damping, determine the first critical frequency and mode shape.

***Solution*:** The dimensional data provided at the shaft stations must first be converted into geometric properties such as cross-section area, area and mass moments of inertia. This data is shown in table 6.5. Equations (6.19) and (6.20) require parameters L, $L/(EI)$, $L^2/(2EI)$, $L^3/(6EI)$ and factor $\varepsilon = L^2/(\kappa GA)$ for all elements, from which $[F_i]$, $[S]$, $[T]$ and $[W]$ matrices are formed. The required elements are shown in table 6.6. The assembled field matrix $[F_1]$ for the first shaft segment is provided below:

$$[F_1] = \begin{bmatrix} 1 & 0 & 0 & 8.057 & 6.40\times10^{-9} & 0 & 0 & 3.82\times10^{-8} \\ 0 & 1 & -8.057 & 0 & 0 & -6.4\times10^{-9} & 3.82\times10^{-8} & 0 \\ 0 & 0 & 1 & 0 & 0 & 1.6\times10^{-9} & 6\times10^{-9} & 0 \\ 0 & 0 & 0 & 1 & 1.6\times10^{-9} & 0 & 0 & -6\times10^{-9} \\ 0 & 0 & 0 & 0 & 1 & 0 & 0 & -8.057 \\ 0 & 0 & 0 & 0 & 0 & 1 & 8.057 & 0 \\ 0 & 0 & 0 & 0 & 0 & 0 & 1 & 0 \\ 0 & 0 & 0 & 0 & 0 & 0 & 0 & 1 \end{bmatrix}$$

The procedure for assembling field matrices for the other shaft segments is identical, and the remaining matrices for the shaft segments of the rotor are not listed here.

Similarly, the point matrix $[P_i]$ defined in equations (6.14) and (6.15) can be assembled from its elements. Material and bearing damping coefficients are disregarded, as also out of plane gyroscopic effects. Since damping and gyroscopic effects constitute the imaginary elements, the point matrix will only have real terms. The point matrix for the first station on the shaft is assembled from the $[B_i]$, $[I_i]$ and $[O_i]$ sub-matrices, where $[I_i]$ is the identity matrix and $[O_i]$ is the null matrix. The values shown are for the trial frequency of 5495.5 rpm, which is determined to be the first natural frequency of the rotor. The assembled point matrix is shown below:

$$[P1] = \begin{bmatrix} 1 & 0 & 0 & 0 & 0 & 0 & 0 & 0 \\ 0 & 1 & 0 & 0 & 0 & 0 & 0 & 0 \\ 0 & 0 & 1 & 0 & 0 & 0 & 0 & 0 \\ 0 & 0 & 0 & 1 & 0 & 0 & 0 & 0 \\ 0 & 0 & -5249 & -331827 & 1 & 0 & 0 & 0 \\ 0 & 0 & -331827 & 5249 & 0 & 1 & 0 & 0 \\ 0 & 1.42\times10^{-6} & 0 & 0 & 0 & 0 & 1 & 0 \\ 1.34\times10^{6} & 0 & 0 & 0 & 0 & 0 & 0 & 1 \end{bmatrix}$$

The system transfer matrix $[M]$ as defined in equation (6.22) is obtained with the help of the point and field transfer matrices, which sequentially transfer the rotor's state from the left end at station 1 to the right end at station n. Sequential multiplication of the state and field transfer matrices gives the following matrix $[M]$:

1	0	0	0	0	0	0	0
0	1	0	0	0	0	0	0
0	0	1	0	0	0	0	0
0	0	0	1	0	0	0	0
0.0014	0	-179750	-10262969	1	0	0	0
0	-0.0014	-10262969	179750	0	1	0	0
0	-98226	0	0	0	0	1	0
-98226	0	0	0	0	0	0	1

Boundary conditions at the left and right ends of the shaft are to be imposed next. The left end of the rotor is supported in a bearing, while the right end is free. For the pinned end the required conditions are $u = 0$ and $M_y = 0$, and at the free end the conditions are $V_x = 0$ and $M_y = 0$. Similar conditions are set up for the vertical plane. Applying the boundary conditions as described in equations (6.23) or (6.24), a non-trivial solution for $\{z^L\}$ exists only if the coefficient determinant vanishes. For the trial frequency of 5495.5 rpm, the coefficient determinant is calculated to have the minimum value of 1.266×10^{-24}. The corresponding relative deflections in the horizontal and vertical planes are shown in table 6.7. Since the bearing support stiffness K_{yy} at both ends of the shaft is substantially less than K_{xx}, the dynamic motion in the yz plane is also higher than in the xz plane, thus identifying this mode with the y-direction. The relative deflections in the y-direction are plotted in Fig. 6.56. Note that there is little flexural motion in the shaft since most of the activity is in the bearings, thus identifying this mode as a rigid body (or, rotor bounce) mode of vibration. Also, activity in the left end bearing is considerably more than in the right end bearing located 9.2 in. from the shaft's free end.

Table 6.7. *Relative Shaft Deflections*

Station #	*Deflection u, xz plane*	*Deflection v, yz plane*
0	27	-42.7
1	27.2	-42.7
2	24.7	-32.6
3	26.3	-32.5
4	36.6	-38.7
5	41.1	-41.0
6	44.1	-42.1
7	42.5	-39.3
8	35.4	-31.8
9	23.0	-20.1
10	10.1	-8.2
11	1.0	1.0

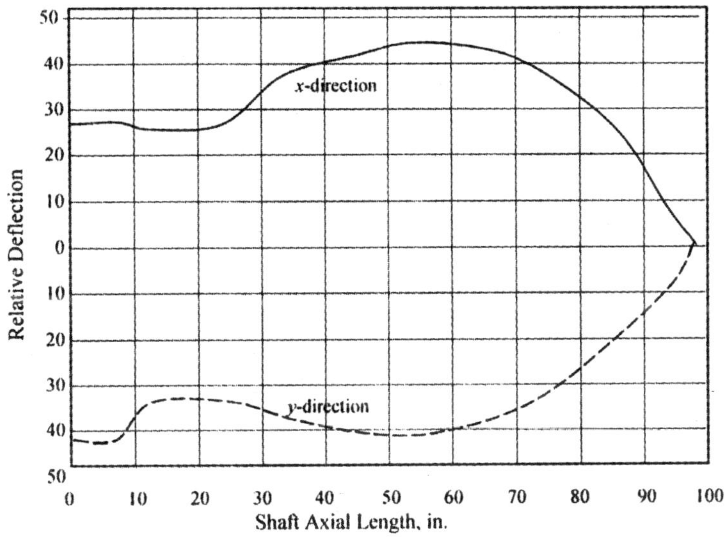

Fig. 6.56. *Shaft Mode Shape at 5495.5 rpm Frequency*

REFERENCES & BIBLIOGRAPHY

Alford, J., "Protecting Turbo-machinery from Self-excited Rotor Whirl", ASME Journal of Engineering for Power, 87(4):333-334, New York, 1965.

American Petroleum Institute, "Centrifugal Compressors for Petroleum, Chemical and Gas Service Industries", API Standard 617, 6th edition, 1995.

American Society of Mechanical Engineers, "Compressors and Exhausters", Power Test Code # 10, 1965 (reaffirmed 1979), ANSI PTC # 10, 1979.

Balda, M., "Dynamic Properties of Turbo-set Rotors", Symposium on Dynamics of Rotors, IUTAM, Lyngby, Denmark, 1975.

Barrett, L.E., Allaire, P.E., Gunter, E.J., "Proceedings of the Conference on the Stability and Dynamic Response of Rotors with Squeeze Film Bearings", University of Virginia, Charlottesville, May 1979.

Barrett, L.E., Allaire, P.E., Gunter, E.J., "Optimum Bearing and Support Damping for Unbalance and Response and Stability of Rotating Machinery", Journal of Engineering for Power, 100(1):89-94, 1978.

Begg, I.C., "Friction Induced Rotor Whirl–A Study in Stability", ASME, paper # 83-DET-10, New York, 1983.

Bentley, D.E., "The Re-excitation of Balance Resonance Regions by Internal Friction", 82-PET-49, ASME, New York, September, 1982.

Bolotin, V.V., "Non-conservative Problems of the Theory of Elastic Stability", Pergamon Press, New York, 1964.

Booker, J.F., Huebner, K.N., "Application of Finite Element Methods to Lubrication, An Engineering Approach", Journal of Lubrication Technology, 94(4):313-323, October, 1972.

Booser, E.R., "Handbook of Lubrication", Vol. 2, CRC Press, Boca Raton, FL. 1973.

Brown, R.D., Hart, J.A., "A Novel Form of Damper for Turbo-machinery", CP2443, NASA, Washington, DC, June, 1986.

Cataford, G.F., Lancee, R.P., "Oil Free Compression on a Natural Gas Pipeline", 86-GT-293, ASME, New York, June, 1986.

Childs, D.W., "Force and Moment Dynamic Coefficients for Pump Impeller Shroud Surfaces", Proceedings of 4th Workshop on Rotor Dynamic Instability Problems in High Performance Turbo-machinery, NASA, CP, Texas A & M University, College Station, TX, June, 1986.

Childs, D.W., "Identification and Avoidance of Instabilities in High Performance Turbo-machinery", Von Karman Institute for Fluid Dynamics, lecture, Rhode-Saint Genese, Belgium, January, 1987.

Choi, Y.S., Noah, S. T., "Non-linear Steady State Response of a Rotor Support System", Journal of Vibrations, Acoustics, Stress and Reliability in Design, 169(3):255-261, July, 1987.

Choy, K.C., "Dynamic Analysis of Flexible Rotor Bearing Systems Using A Modal Approach", Ph.D. Thesis, University of Virginia, 1978.

Chu, F., Holmes, R., "The Effect of Squeeze Film Damper Parameters on the Unbalance Response and Stability of a Flexible Rotor", ASME paper # 96-GT-377, New York, 1996.

Crede, C.E., Ruzicka, J.E., "Theory of Vibration Isolation", Shock & Vibration Handbook, 3rd edition., Mc-Graw Hill, New York, 1988.

Den Hartog, J.P., "Mechanical Vibrations", Dover Publications, New York, 1985.

Dennis, A.J., Erikson, R.H., Seitelman, L.H., "Transient Response Analysis of Damped Rotor Systems by the Normal Modes Method", ASME Paper # 75-GT-58, New York, 1975.

Deutschman, A.D., Michels, W.J., Wilson, C.E. "Machine Design–Theory and Practice", MacMillan Publishing Co., New York, 1985.

Dimarogonas, A.D., "Newkirk Effect: Thermally Induced Dynamic Instability of High Speed Rotors", ASME paper # 83-GT-26, New York, April, 1983.

Ehrich, F.F., O'Connor, J.J., "Stator Whirl with Rotors in Bearing Clearance", Journal of Engineering for Industry, 89(3):381 – 390, August, 1967.

Ehrich, F.F., "The Dynamic Stability of Rotor/Stator Radial Rubs in Rotating Machinery", Journal of Engineering Ind., Pages 1025–1028, 1969.

Ehrich, F.F., "Self-excited Vibration", Shock & Vibration Handbook, 3rd edition, Mc-Graw Hill, section 5, 1987.

Ehrich, F.F., "The Role of Bearing Support Stiffness Anisotropy in Suppression of Rotor Dynamic Instability", DE-Vol. 18-1, ASME, New York, 1989.

Ehrich, F.F., "Handbook of Rotor Dynamics", Krieger Publishing Co., Malabar, FL, 1999.

Eshelman, R.L., Eubanks, R.A., "Effects of Axial Torque on Rotor Response: An Experimental Investigation", ASME paper # 80-WA/DE-14, New York, 1980.

Fowlie, D.W., Miles, D.D., "Vibration Problems with High Pressure Compressors", ASME paper # 85-Pet-28, New York, 1985.

Frene, J., Nicolas, D., Degueurce, B., Berthe, D., Gaudet, M., "Lubrification Hydrodynamique", Eyrolles Edition, Lecreusot, France, 1990.

Gelin, A., Pugnet, J.M., Bolusset, D., Friez, P., "Experience in Full Load Testing of Natural Gas Centrifugal Compressors for Rotor Dynamics Improvement", ASME paper # 96-GT-378, New York, 1996.

Gunter, E.J., "Dynamic Stability of Rotor Bearing Systems", SP-113, NASA, Chapter 4, Washington, D.C., 1966.

Gunter, E.J., Barrett, L.E., Allaire, P.E., "Design of Non-linear Squeeze Film Dampers for Aircraft Engines", ASME Journal of Lubrication Technology, 99(1):57-64, January, 1977.

Hagg, A.C., Sankey, G.O., "Some Dynamic Properties of Oil Film Journal Bearings with Reference to Unbalance Vibration of Rotors", Journal of Applied Mechanics, 23(2):302-306, June, 1986.

Hildebrand, F.B., "Advanced Calculus for Applications", Prentice-Hall, New Jersey, 1986.

Houbold, J.C., Reed, W.H., "Propeller Nacelle Whirl Flutter", Institute of the Aerospace Sciences, 61 – 34, New York, January, 1981.

Humes, B., Holmes, R., "The Role of Sub-atmospheric Film Pressure on the Vibration Performance of Squeeze Film Bearings", Journal of Mechanical Engineering Science, Volume 20, No. 5., Pages 283 – 289, 1978.

ISO 1940/1, "Mechanical Vibrations-Balance Quality Requirements of Rigid Rotors-Part I, Determination of Permissible Residual Unbalance", International Organization for Standardization, Geneva, Switzerland, 1986.

Kanki, H., Kaneko, Y., Kurosawa, M., Yamamoto, T., "Prevention of Low Frequency Vibration of High-capacity Steam Turbine Units by Squeeze Film Damper", ASME paper # 97-GT-11, New York, 1997.

Kirk, R.G., "Labyrinth Seal Analysis for Centrifugal Compressor Design – Theory and Practice", Proceedings, International Federation for the Theory of Machines and Mechanisms 1986 International Conference on Rotor Dynamics, Pages 589-596, Japan Society of Mechanical Engineers, Tokyo, September, 1986.

Kirk, R.G., Mondy, R.E., Murphy, R.C., "Theory and Guidelines to Proper Coupling Design for Rotor Dynamics Consideration", ASME paper # 83-DET-93, New York, 1983.

Kimball, A.L., "Internal Friction Theory of Shaft Whirling", General Electric Review, 27(4): 244, April, 1924.

Kramer, E., "Instabilities of Rotating Shafts", Proceedings of the Conference on Vibration of Rotating Systems, Institution of Mechanical Engineers, London, Pages 230-246, 1982.

Klosterman, A.L., "On the Experimental Determination and Use of Modal Representations of Dynamic Characteristics", Ph.D. Thesis, University of Cincinnati, 1971.

Lawen, J.L., Flowers, G.T., "Interaction Dynamics Between a Flexible Rotor and an Auxiliary Clearance Bearing", ASME Journal of Vibration & Acoustics, Vol. 121, Pages 183–189, New York, 1999.

Li, D.F., Gunter, E.J., "Component Mode Synthesis of Large Rotor Systems", ASME paper # 81-GT-147, New York, 1981.

Li, D.F., Gunter, E.J., Barrett, L.E., Allaire, P.E., "The Dynamic Analysis of Multi-level Flexible Rotor Systems Using Transfer Matrices and Component Mode Synthesis", Final Report UVA/628145/MAE81/102, University of Virginia, Charlottesville, 1991.

Linn F.C., Prohl, M.A., "The Effect of Flexibility of Support upon the Critical Speeds of High Speed Rotors", Transactions, Society of Naval Architects and Marine Engineers, 59:536-553, 1951.

Loomis, A.L., "Compressed Air & Gas Data", Ingersoll Rand Co., Woodcliff Lake, NJ, 1989.

Lund, J.W., "Spring and Damping Coefficients for the Tilting Pad Journal Bearing", Transactions American Society of Lubrication Engineers, 7(4):342-352, 1964.

Lund, J.W., Thomsen, K.K., "A Calculation Method and Data for the Dynamic Coefficients of Oil Lubricated Journal Bearings", Topics in Fluid Film Bearing and Rotor Bearing System Design and Optimization, ASME, New York, Pages 1–28, 1978.

Mabie, H.H., Reinholtz, C.F., "Mechanisms and Dynamics of Machinery", John Wiley & Sons, New York, 1987.

Marmol, R.A., Vance, J.M., "Squeeze Film Damper Characteristics for Gas Turbine Engines", ASME Journal of Mechanical Design, 100(1):139-146, January, 1982.

Maslen, E.H., Sortore, C.K., Vazquez, J.A., Knopse, C.R., "Synchronous Response Estimation in Rotating Machinery", ASME paper # 00-GT-397, New York, 2000.

McLachlan, N.W., "Theory and Applications of Mathieu Functions", Oxford University Press, New York, 1947.

McCloskey, T.H., Adams, M.L., "Troubleshooting Power Plant Rotating Machinery Vibration Problems Using Computational Techniques: Case Histories", Institute of Mechanical Engineers, # C432/137, Pages 233 –244, London, 1992.

Miyachi, T., "A Parametric Study of the Unbalance Vibration of a Rotor Supported in a Light Flexible Casing", Gas Turbine Society of Japan paper # 83-TOKYO-IGTC-125, Tokyo, 1983.

Muller, P.C., Bajkowski, J., Soffker, D., "Chaotic Motions and Fault Detection in a Cracked Rotor", Non-linear Dynamics, Volume 5, Pages 223–254, 1994.

Muszynska A., "Partial Lateral Rotor to Stator Rubs", paper # C281/284, Institution of Mechanical Engineers, London, 1984.

Muszynska A., "Whirl and Whip–Rotor/Bearing Stability Problems", Journal of Sound & Vibration, 110(3):443-462, 1986.

Myklestad, N.O., "A New Method of Calculating Natural Modes of Uncoupled Bending Vibration of Airplane Wings and Other Types of Beams", Journal of Aeronautical Sciences, Vol. 12, Pages 153 –162, 1944.

Myklestad, N.O., "Fundamentals of Vibration Analysis", Mc-Graw Hill, New York, 1956.

Naslin, P., "The Dynamics of Linear and Non-linear Systems", Blackie, 1965.

Nayfeh, A.H., Mook, D.T., "Non-linear Oscillations", John Wiley & Sons, New York, 1979.

Newkirk, B.L., "Shaft Whipping", General Electric Review, 27: 169 – 178, 1924.

Newkirk, B.L., Taylor, H.D., "Shaft Whipping due to Oil Action in Journal Bearing", General Electric Review, 28(8): 559–568, 1925.

Orcutt, F.K., "The Steady State and Dynamic Characteristics of the Tilting Pad Journal Bearing in Laminar and Turbulent Flow Regimes", Journal of Lubrication Technology, 89(3):392 – 404, 1967.

Pincus, O., Sternlicht, B., "Theory of Hydrodynamic Lubrication", Mc-Graw Hill, New York, 1984.

Prohl, M.A., "A General Method for Calculating the Critical Speeds of Flexible Rotors", Journal of Applied Mechanics, Pages 142–147, 1945.

Rangwala, A.S., "Turbo-machinery Dynamics: Design and Operation", McGraw-Hill, New York, 2005.

Roark, R.J., Young, W.C., "Formulas for Stress and Strain", 5th edition, Mc-Graw Hill, New York, 1989.

San Andres, L.A., Vance, J.M., "Effect of Fluid Inertia on Finite Length Squeeze Film Dampers", 86-TC-4D-1, American Society of Lubrication Engineers, Park Ridge, IL, October, 1986.

Sawyer, T., "Gas Turbines – Volumes I, II and III", International Gas Turbine Institute Atlanta, ASME, 1984.

Schollhorn, V.K., Ebi, G., Steigleder, K., "Frettinganrisse in Einen 936 MW Generator", VGB Kraftwerkstechnik, Volume 73, No. 4, Pages 340–344, 1993.

Sortore, C.K., "Observer Based Critical Response Estimation in Rotating Machinery", Ph.D. Dissertation, University of Virginia, Charlottsville, VA, 1999.

Taylor, C.F., "The Internal Combustion Engine in Theory and Practice", Volume 2, The M. I. T. Press, Cambridge, MA., 1997.

Taylor, E.S., Browne, K.A., "Vibration Isolation of Aircraft Power Plants", Journal of the Aeronautical Sciences, 6(2): 43–49, 1984.

Thomas, J.J., "Instabile Eigenschwingungen von Turbinenlaufern, Angefacht durch die Spaltstromungen Stopfbuschen un Beschaufelungen", AEG-Sonderdruck, 1958.

Thompson, B.D., Badgley, R.H., "Application of an Advanced Hybrid Rotor Dynamics Model to the Complete Structure of a Marine Gas Turbine Engine", ASME paper # 88-GT-123, New York, 1988.

Timoshenko, S., Young, D.H. and Weaver, W., "Vibration Problems in Engineering", John Wiley & Sons, Inc., New York, 1974.

Vance, J.M., Lee J., "Stability of High Speed Rotors with Internal Friction", ASME paper # 73-DET-127 , New York, 1973.

Vance, J.M., Laudadido, F.J., "Experimental Measurement of Alford's Force in Axial Flow Turbomachinery", 84-GT-140, ASME, New York, June, 1984.

Vance, J.M., "Torque Whirl – A Theory to Explain Non-synchronous Whirling Failures of Rotors with High Torque Load", Journal of Engineering for Power, 100(2): 235-240, April, 1988.

Vance, J.M., "Rotor Dynamics of Turbo-machinery", John Wiley & Sons, New York, 1988*a*.

Wachel, J.C., Tison, J.D., "Vibrations in Reciprocating Machinery and Piping Systems", Proceedings of the 23rd Turbo-machinery and Piping Systems, Texas A & M University, College Station, TX 1994.

Wachel, W.D., "Non-synchronous Instability of Centrifugal Compressors", ASME paper # 82-Pet-22, New York, 1982.

Wagner, N.G., "Reliable Rotor Dynamic Design of High Pressure Compressors Based on Test Rig Data", ASME paper # 99-GT-150, New York, 1999.

Wang, Z., Lund, J.W., "Calculations of Long Rotors with Many Bearings on a Flexible Foundation", C291/84, Proceedings of 3rd International Conference on Vibration in Rotating Machinery, Institute of Mechanical Engineers Conference Publication, York, England, 1984.

Wu, F., Flowers, G.T., "A Study of the Influence of Rubbing on the Dynamics of a Flexible Disk Rotor System", ASME paper # 93-GT-35, New York, 1993.

Yu, J.J., Goldman, P., Bently, D.E., Muzynska, A., "Rotor/Seal Experimental and Analytical Study on Full Annular Rub", ASME paper # 00-GT-389, New York, 2000.

Zeidan, F.Y., Perez, R.X., Stephenson, E.M., "The Use of Honeycomb Seals in Stabilizing Two Centrifugal Compressors", Proceedings, 22nd Turbo-machinery Symposium, Texas A & M University, College Station, Texas, 1993.

Zhou Ren-Mu, "Instability of Multi-stage Compressor K 1501", CP2443, NASA, Washington, DC, June, 1986.

Part III
Engine Operational Characteristics

CHAPTER 7

Experimental and Test Measurement

7.1 INTRODUCTION

Vibration measurement in turbo-machines is a prerequisite for fault diagnosis, performance verification and condition monitoring. Advances in sensors, signal-processing procedures and data-storage devices have simplified evaluation of dynamics of motion of the rotating shaft. Improvements in electronics and application of microprocessors to vibration analysis have further enhanced capabilities of engineers to obtain vibration free operation of machines. The Fast Fourier Transform (FFT) is the foundation of most vibration data processing, and diagnostic and modal analysis techniques have taken advantage of FFT capabilities for widespread use in data reduction of vibration signature of rotating machines.

Higher and lower harmonics of the rotor speed, where characteristic frequencies are an integral or fractional multiple of the machine's operating speed, are typical of all forms of turbine and reciprocating engines, compressors and pumps. For instance, 'ghost' components are sometimes observed in vibration spectra obtained from measurements in machines coupled to a gearbox, where vibration frequencies occur in multiples of the number of teeth on a gear. The presence of a number of blades, support struts and obstructions in the flow path of a turbine or compressor has similar effects on the vibration signature of a machine component at resonance. Multiple cylinder engines experience firing of the air-fuel mixture at various crank angle locations in an engine, causing vibration peaks at several frequencies in the crankshaft.

A major goal in the development process of turbine blades is to prevent their failure from high cycle fatigue. Maximum dynamic stresses in the blade are obtained from a correlation between the numerical model's free vibration results and dynamic strain measurements. These vibratory stresses are transferred from the gauges either by slip rings or telemetry during engine tests with the blade prototype. Vibration of the blade can be characterized with an adequate number of gauges placed on the airfoil contour close to the peak stress location for the mode of interest. Measuring vibratory stresses in a turbine or compressor blade is a substantially harder task than it is for most other structures. The difficulties encountered stem from the harsh operating conditions such as rotation, fluid flow and elevated temperature. Structural peculiarities in the form of sharply peaked modal strain and mistuning effects go on to compound the task.

A variety of devices exist to measure the dynamics of the blade. Strain gauges, accelerometers, proximity probes, laser vibrometer and LVDT's are available, but because of the need to not perturb the aerodynamics around the blades and the high centripetal loading, the choice is limited to strain gauges, and more recently, light probe systems. Further, the rotating system necessitates the use of a slip ring to transmit the strain gauge signals to the non-rotating environment. It thus imposes the restriction of the devices to many fewer than the product of the number of modes of interest and the number of blades on the disk. In principle, the rotational symmetry of the bladed disk should ease the instrumentation process. Indeed, since the modes of vibration of the system consist of harmonic waves of constant amplitude traveling forward and backward around the disk, the resonant response of two different blades should have the same amplitude but be out of phase with respect to each other. For a perfectly tuned disk, then, it would be sufficient to instrument a single blade. Unfortunately, manufacturing variability and in-service wear generally create a small level of blade-to-blade variation in their structural and material properties. Referred to as mistuning, this aspect introduces dramatic differences in the amplitudes of the resonant response of different blades. Accordingly, it would be desirable to instrument a series of blades in order to capture the one yielding the largest resonant response. But the limitations of the slip ring preclude this possibility.

Infrared temperature measurement techniques are used when contact with a hot target is not possible. In a moving body the target cannot be reached by traditional instruments, where the surface is at such a high temperature that damage to the contacting sensor may occur, or when the measuring instrument modifies the target temperature. A radiometric thermometer in gas turbine monitoring offers the advantage over the contacting sensor of high sensitivity, accuracy and working frequency, while being a non-intrusive scanning system. But there are disadvantages of a more complex operation and management of the system, the temperature range is limited and the equipment is expensive when compared with a traditional system.

Errors in metal temperature measurements of the components arise mostly from the hostile environment in which the instrument must operate. Contamination of the signal comes from emitted and reflected radiation coming from the hot gases of combustion, interference due to transient events and dumping of the signal due to inclination of the axes between the target surface and thermometer axis. The sight of the flame coming from the combustion chamber and carbon particles in exhaust gases can pose considerable errors in radiometric scanning for temperature measurements of the combustion liner. The presence of steam, CO_2, CO, SO_2, unburned particles and dust cause a considerable number of emission, reflection and scattering phenomena in the first stages of a turbine.

Torsional vibration velocity of a rotating shaft can be measured using a toothed wheel and a fixed sensor. The signal generated by the teeth of the wheel passing the fixed sensor has a frequency that equals the number of teeth multiplied by shaft speed. If the shaft is experiencing torsional vibrations, the carrier frequency will exhibit frequency modulation, or a change in frequency, since the time required for each tooth to pass the fixed pickup varies. Other methods for indirect measurement of torsional vibrations use strain gauges to indicate instantaneous torque. The strain gauges are usually mounted at an angle of 45° to eliminate influence of shaft bending on torque measurements. Such techniques, however, are not common diagnostic tools, since strain gauge signals are difficult to take from a rotating shaft. Measurement of torsional vibrations is, at best, a tricky proposition, because the oscillatory motion is superimposed on the steady rotational motion of the shaft. In comparison, instruments that convert shock or vibratory motion into an optical, mechanical or, most commonly, an electrical signal that is proportional to the experienced motion can more readily measure lateral vibrations of a rotating shaft.

Analysis of lubricating oil for predictive maintenance of steam and gas turbines is a popular practice for monitoring machine condition. Checking the debris from oil filters and magnetic plugs aids in detecting metallic wear in bearings, seals and at locations where rub can occur. The system is effective in the sense that tests may be performed to determine if a system is nearing or has reached a failure mode. Further damage can then be contained or avoided through immediate shut down or corrective action. Spectroscopy is widely applied to provide a quantitative and multi-elemental analysis of wear debris in the oil. Wear metals such as iron, aluminum, chromium, copper, tin, lead, silver, titanium and nickel are detectable. Concentration trends are established through routine monitoring to indicate if a continuing wear-related condition exists, the rate of wear, and as a consequence, the immediacy of the wear problem. If the component experiencing the problem is not extensively used in other areas of the machine, it may also be possible to tell the exact source of the problem from the chemical analysis of the debris.

7.2 VIBRATION SENSORS

Vibration measurement instruments fall into three broad categories: transducers to measure the vibration set up in the machine, electronic devices to process the signal and analyzers to calculate spectral blocks of data and display the results. Within each category, associated components are needed to condition, process, record and store the signal.

Piezoelectric transducers, accelerometers and gauges are commonly used to measure vibration signals. Generally exhibiting high-elastic stiffness, a piezoelectric material electrically responds to force, and produces an electric charge proportional to the force applied to it in its linear elastic range. Motion transducers (or pickups or sensors) come in two basic types. A fixed reference type has one terminal attached to a fixed point, the other to the point whose motion is to be measured. A mass-spring (or seismic) transducer has a mass and spring system attached to a base, and the base is mounted on the vibrating structure; dynamic movement is inferred from motion of the mass relative to the base.

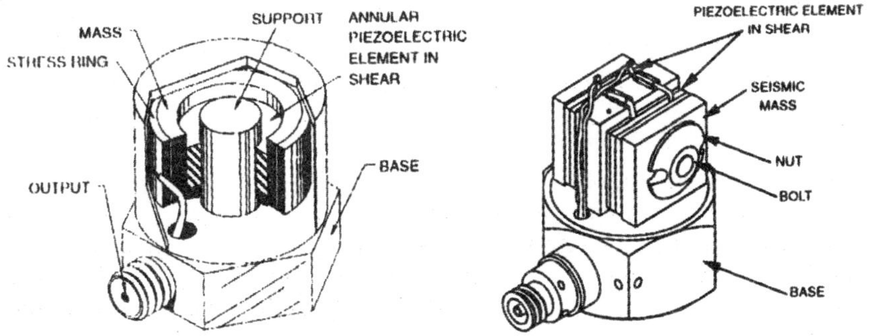

Fig. 7.1. *Seismic Piezoelectric Accelerometers* (Courtesy: *Bruel & Kjaer, Endevco*)

A piezoelectric accelerometer (fig. 7.1) is a linear seismic transducer that produces an electric charge proportional to the applied acceleration through the base. The piezoelectric crystal producing the charge acts as the spring. In a properly designed accelerometer, the mass and the frame have high stiffness coefficient, the crystal has relatively little mass, and viscous damping exists only between the frame and the mass. For a given application practical considerations may not fulfill all of these assumptions. At higher frequencies of excitation, for example, inertial forces may deform the housing and change the performance of the seismic system by distorting the indicated frequency.

A force gauge is a mechanical impedance measurement device to relate the force applied to a structure of an operating machine. Force gauges use piezoelectric transducing elements to generate an output charge or voltage proportional to the applied force. Amplitude response, signal conditioning requirement and environmental effects associated with force gauges are similar to those related to piezoelectric accelerometers. Force gauges for use with a vibration excitation are attached at one end to the structure and the other end to a vibration exciter.

For shock excitation force gauges are usually built into the head of a hammer. Interchangeable faces of various materials to control the generated waveform of the shock pulse may be used on the head of the hammer. A hard material produces short duration, high amplitude shocks with fast rise and fall times. Shocks of short duration have a broad frequency spectrum. A softer material, on the other hand, causes longer, lower amplitude shocks with a narrow frequency spectrum. Impedance heads combine the functions of a force gauge and an accelerometer in a single instrument. They are convenient for measuring driving point impedance.

The laser Doppler vibrometer uses the Doppler shift of a laser light which is scattered from a vibrating test object, and produces a real time analog signal output proportional to the instantaneous velocity. Benefits of the system arise from lack of mounting and mass loading effects, and are not affected by environmental factors. Measurement may be done from a remote standoff distance.

The electro-optical measurement system tracks displacement of a light-dark target. The target provides discontinuity in the intensity of reflected light from an object, and the lens of a photo processing equipment forms the image. Electrons are emitted in proportion to the intensity of the light, causing an electronic image to be generated in real time.

Accelerometer preamplifiers are needed to convert a weak signal at high impedance from a typical piezoelectric transducer into a voltage signal at low impedance. Voltage signal is less prone to the influence of electromagnetic noise and other effects. Pickup of noise in a high impedance circuit increases with cable length, so it is advantageous to mount the preamplifier close to the transducer. A line drive amplifier, with its miniaturized thick film circuits, can be attached or even included in the transducer. The output cable from a preamplifier is less subject to noise problems. Line drive preamplifiers have restrictions on dynamic and frequency range compared with a general-purpose charge preamplifier.

Signal conditioning is required to limit frequency band of the signal, to integrate it (to velocity or displacement) and to adjust its gain. High and low pass filters are required to remove extraneous low and high frequency signals and to restrict measurement to the frequency range of interest. Extraneous low frequency signals arise from thermal transient effects, strain in accelerometer base and vibrations transmitted through the foundation from other sources.

At the high end of the frequency spectrum resonance arising from the accelerometer's own geometry must be filtered. The signal in the input amplifier is not affected by this high and low pass filtering.

Many criteria for evaluating operating machinery vibrations are based on velocity, and it is often preferable to express it in terms of displacement. It may also be desirable to determine the total motion of a rotating shaft by adding displacement of the bearing housing, measured with an accelerometer, to the relative displacement of the shaft in its bearing, measured with proximity probes. Acceleration signals can be electronically integrated to yield velocity and displacement; an accelerometer combined with an integrator can produce a velocity signal that is valid over a frequency range of three decades (1000:1).

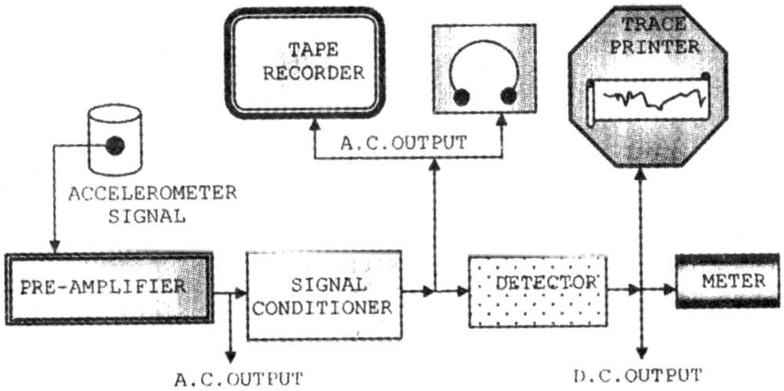

Fig. 7.2. *Vibration Measurement System*

Detectors are used to extract parameters that characterize a signal, typical ones being arithmetic average, mean square and root mean square values. A full wave rectifier obtains the signal magnitude, and with a smoothing circuit, produces the arithmetic average. Mean square and root mean square values are independent of phase relationships, and are preferred for describing stationary signals. Mean square values are directly additive when two signals of different frequency bands are added together.

Vibration meters receive signal from a transducer and process it to display relevant vibration parameters. For measurement on rotating machines a frequency range of 10 Hz to 10 kHz is desirable. The lower frequency limit includes any sub-harmonic components due to factors such as oil whip in plain bearings. The upper limit includes higher order frequencies in multi-cylinder internal combustion engines, tooth meshing frequencies and their harmonics in gears and frequencies associated with faults in rolling element bearings.

Evaluation of vibration signatures can be done at a later date if the signal is stored on instrumentation tape recorders using direct, frequency modulation or digital techniques. Ability to record low frequency signals is the prime difference between the methods. Direct recordings cannot record down to a direct current signal, 25 Hz being the lower limit. In contrast, frequency modulation method can go down to a direct current signal, which represents a constant deviation of the carrier frequency.

Figure 7.2 provides a block diagram of a typical vibration measuring system consisting of a preamplifier, signal conditioner, detector and indicating meter. Vibration meters often combine all of these elements in a single unit.

7.3 MEASUREMENT TECHNIQUES

Careful pretest planning can save a lot of time and aggravation in the measurement of vibrations, and ensures that the most useful data is obtained. Starting with a clear definition of the test's objectives, factors that need important consideration must be identified. For example, estimates of the characteristics of the motion to be measured must be made, including the range of frequency, amplitude, dynamic motion, duration and principal direction of activity. This information provides a basis in the selection of the measurement equipment. Previous experience on a similar type of equipment may provide a guide in the estimation of signal characteristics. Preliminary measurements may also be helpful. Environmental conditions, locations where measurements are to be made, system components and method of data transmission are other factors to be taken into account.

Selection of the parameter to be measured (displacement, velocity, acceleration, strain) may be predetermined by specifications or standards. At high frequencies, acceleration measurements provide the highest signal output. A transducer of small size and mass suffices to measure acceleration, which is a consideration for smaller components. The results are useful in load and stress analysis. Velocity measurement is widely used on rotating machinery because the velocity spectrum is generally more uniform in the intermediate frequency range. If vibration readings are to be correlated with acoustic measurements, sound pressure level is proportional to the velocity of the vibrating surface. Displacement measurements are particularly important if the vibrating part must not come into contact with other components in the machine, and for relative motion between the rotor and the stator. The magnitude of the displacement provides an indication of the stress experienced by the component. Also, at low frequencies acceleration and velocity parameters may be too small to be useful.

A number of factors need to be considered in the selection of transducers, including characteristics of the transducer, motion type and environmental issues. Conflicting situations often arise. As an example, sensitivity requirement would suggest that a transducer of larger size must be selected, but mass considerations would call for a small size in order to reduce inertia loading on the component. In special cases even the smallest transducer may result in an unacceptable load. Then a device that does not come into direct contact with the test surface may be the right choice.

A transducer may be mounted by screw, cement type of epoxy or attached with a bracket to the test surface. The method of mounting affects the resonance frequency, and hence, the useful range of frequency of the transducer. Also, ability to withstand high levels of vibration, operating temperature, stability and repeatability of the method of mounting should be reviewed. A stud-mounted transducer offers the highest resonance frequency of any mounting methods, and accepts high vibration levels without loosening from the test surface. When stud mounting is not possible, a transducer may be bonded to a test surface by a thin layer of cement (epoxy or dental type). The method provides good frequency response when the accelerometer is directly attached to the test surface. Magnetic mounting is useful in measuring low acceleration levels. A hand held transducer provides the poorest performance, but it comes in useful in making a rapid survey of a test surface by changing measurement location. A probe with one end threaded into a transducer is pressed against the test surface. Measurement accuracy is not available with the procedure, and the frequency range is restricted between 20 and 1,000 Hz.

If a transducer is to be used under unusual temperature conditions, calibration must be performed in the temperature range in which it will operate. One method calls for a direct comparison of the generated output with the output from another standard transducer when both transducers are attached to a vibration exciter and subjected to the same excitation. The free-fall calibration method calls for the transducer to fall freely for a short period under the influence of gravity. The peak signal is measured for a value of 1 g. Calibration of a complete vibration measurement system may be done by a field calibrator. This is a portable device containing a small shaker. The accelerometer can be mounted on the shaker and subjected to a known displacement, velocity or acceleration at a fixed frequency. The calibrator is provided with an internal oscilloscope and a stable built-in accelerometer in a feedback loop that controls the electro-dynamic exciter.

Data transmission between the transducer and the electronic instrumentation is generally through a cable, but the aerospace industry relies on telemetry. The choice of a suitable cable depends on the application, amplifier type and environmental conditions. Cable jackets may be made of silicon rubber, poly-vinyl-chloride or Teflon, depending on the temperature range to which it is subjected. The cable must be tied down near the accelerometer and at regular intervals to avoid cable whip and induced cable noise. Light and flexible cables prevent loading the transducer and the structure under test. Cables should be generally treated with the same care as transducers in vibration measurement tests.

7.4 SIGNAL PROCESSING

FFT analysis, digital filtering and optimization are the methods used for processing of the signal. Commercial software programs employing these digital techniques are available and widely used. Analog-to-digital (A/D) converters serve to convert a continuous signal into a sequence of digital numbers to represent the instantaneous value of the signal at specified time increments. It may be possible to regain the original analog signal by the reverse process using a digital-to-analog (D/A) converter. Time increments are normally uniform, representing a constant sampling frequency. Time intervals may also be selected on other bases, for instance, uniform increments in shaft speed for order tracking.

A/D conversion is done through the sample and hold procedure. The instantaneous value of the analog signal is sampled at the instant of each pulse of the sampling clock, and holds that analog voltage constant until the A/D conversion process is complete. The process is then repeated. For multiple channel conversion it is common to use a single A/D converter multiplexing between channels, with the sample and hold circuits for all channels synchronized. The output of the A/D converter is a binary integer number with 2^N possible values, where N is the number of bits. The actual dynamic range of the measurement may be limited by factors other than the least significant bit, such as noise level and linearity in the analog parts of the system.

Discrete sampling in the time domain corresponds in the frequency domain to a periodic repetition of the spectrum, with a periodic spacing equal to the sampling frequency. If it is desired to obtain the correct frequency spectra, it is necessary to ensure the analog signal does not contain frequency components greater than half the sampling frequency, and this is achieved by using appropriate anti-aliasing filters with steep characteristics. Their application makes it possible to use up to 80 percent of the theoretically available spectrum, but it may be accompanied by some phase distortion in the vicinity of the cut-off frequency.

An ideal band-pass filter transmits that part of the input signal within its band pass, and completely attenuates all other frequency components. An analysis may be performed over a frequency range by using a single filter with a tunable single filter swept over the entire frequency range. For general vibration analysis, it is common to use tunable filters, with the center frequency tuned manually, or synchronized with the X position of the pen on a graphic recorder. Alternatively, the center frequency may be synchronized with an external signal, for instance, by a trigger pulse once per revolution of a rotating shaft. The filter then becomes a tracking filter, which may be used to identify the component corresponding to a designated harmonic, or a multiple of the synchronizing signal. An adapter to multiply or divide the frequency is needed to generate a tuning frequency from the original signal. Digital filter analyzers have superseded banks of filters, each with their center frequency. A digital filter can process a continuous digitized signal, and provide another modified signal as an output. The relationship between the output and input signals is expressed as a difference equation, with properties similar to those of a differential equation used in describing an analog filter. Figure 7.3 illustrates the difference between a 3 dB bandwidth and the effective noise bandwidth, the first being most relevant when separating discrete frequencies, and the second being used for random signals.

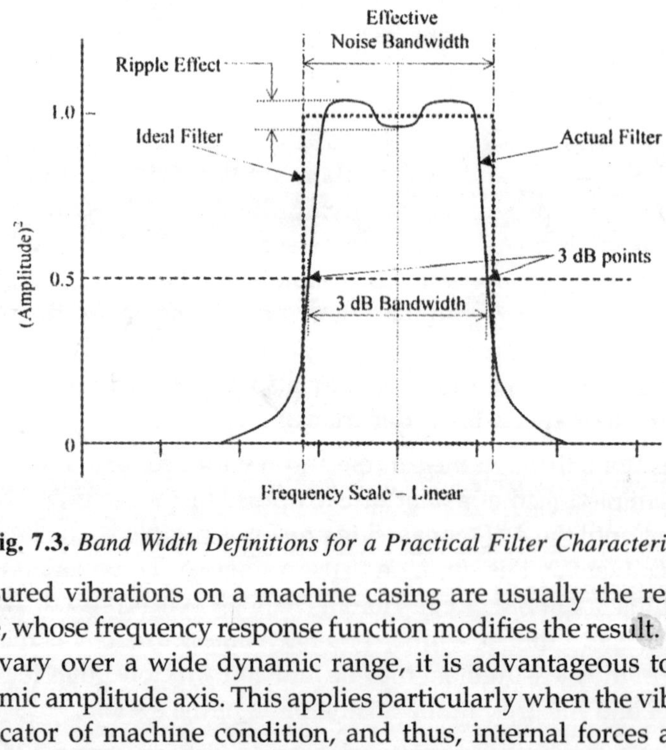

Fig. 7.3. *Band Width Definitions for a Practical Filter Characteristic*

Externally measured vibrations on a machine casing are usually the result of internal forces acting on a structure, whose frequency response function modifies the result. Because the structural response functions vary over a wide dynamic range, it is advantageous to depict the vibration spectra on a logarithmic amplitude axis. This applies particularly when the vibration measurements are used as an indicator of machine condition, and thus, internal forces and stresses. Another aspect to be considered is the dynamic range. The signal from an accelerometer, including the preamplifier, can have a valid dynamic range of 120 dB. A wide range of this magnitude can only be accommodated on a logarithmic amplitude axis. An exception where a linear amplitude scale is preferable to a logarithmic scale is for representation of relative displacement signals using proximity probes. This parameter often needs comparison with rotor dynamic and bearing calculations, and the dynamic range for relative shaft vibrations does not need depiction on a logarithmic scale.

Filter analyzers can cause significant delays to influence the speed of analysis. The filter and the detector introduce delay on the same order as the response and averaging times. In a swept frequency analysis, the delays mean that the recorded spectrum departs from the true spectrum, as shown in fig. 7.4. In the selection of paper speed with an X-Y recorder errors in the level of peaks and valleys and the frequency offset due to the delay must be limited.

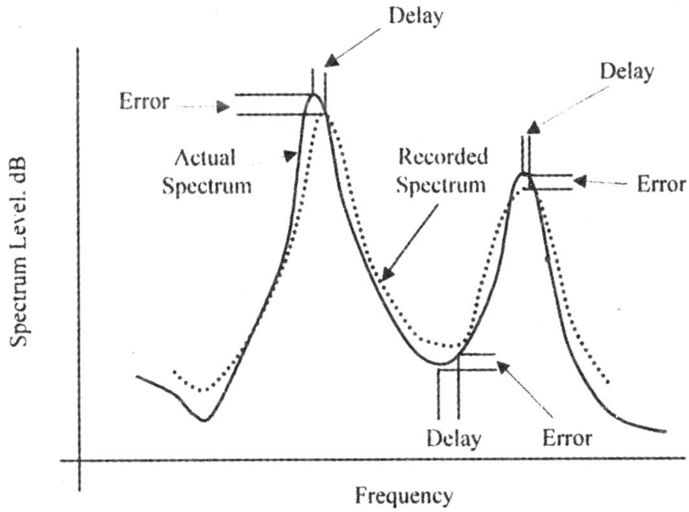

Fig. 7.4. *Delay and Error between Actual and Recorded Spectra in Swept Frequency Analysis*

The FFT algorithm is an efficient way to calculate the Discrete Fourier Transform (DFT), which is a finite approximation of the Fourier integral transform. For real valued time signals, the FFT algorithm makes use of the following form of the equation:

$$S_x(m) = \Delta t \sum_{n=0}^{N-1} x(n\Delta t) \exp(-j2\pi m \Delta f n \Delta t) \tag{7.1}$$

$$x(n) = \Delta f \sum_{m=0}^{N-1} S_x(m\Delta f) \exp(j2\pi m \Delta f n \Delta t) \tag{7.2}$$

The equations shown above give spectrum values $S_x(m)$ at N discrete frequencies $m\Delta f$, and give the time series $x(n)$ at N discrete time points $n\Delta t$. Fourier transforms are integrals of continuous functions, while DFT equations are finite sums, but otherwise have similar properties. Figure 7.5 graphically illustrates the difference between the Fourier integral and the corresponding DFT. The DFT concept is used to generate from a physical signal a unique set of sine and cosine terms at appropriate frequencies. The set is unique because of the orthogonal nature at different frequencies of the sine functions, of the cosine functions and of the sine functions compared to cosine functions.

A typical non-stationary signal resulting from measurements made during a machine run-up or coast-down can be analyzed by dividing it into a series of short quasi-stationary time periods (often overlapping), with the speed roughly constant in each step. The signal may be analyzed with a tracking filter tuned to a specific harmonic of the operating speed, and the results recorded as a function of speed. If a phase meter is included between the filtered and the tracking signals, phase and amplitude may be recorded against speed to provide a Bode plot.

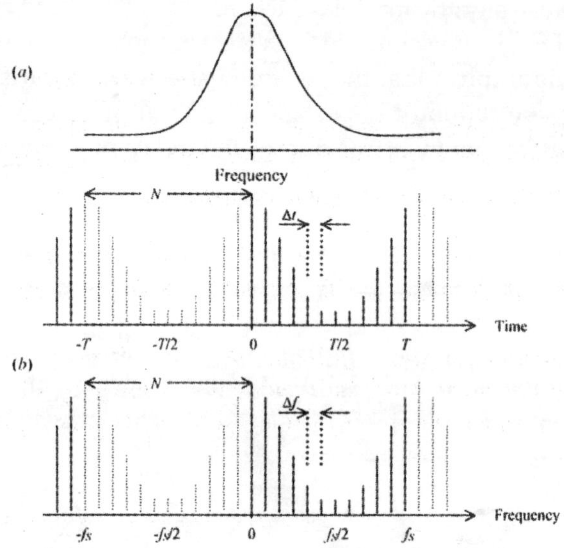

Fig. 7.5. *Comparison of (*a*) Fourier and (*b*) Discrete Fourier Transforms*

Using an FFT spectrum analyzer, the behavior of several harmonics may be studied simultaneously. A scan analysis on an FFT analyzer with a long memory permits this operation; a short Hanning window is scanned through the record to view successive instantaneous spectra on the display screen. Varying the step length may change scan speed. An effective way of representing a scan analysis is by a 'water fall' or 'cascade' plot, as shown in fig. 7.6. The third dimension of the plot may be time or speed; for a simple scan analysis it is usually time. If the spectra are arranged at equal intervals of shaft speed, harmonically related components (their base following radial lines) can be readily separated from constant frequency components (with base line parallel to speed axis). Note that the cascade plot has characteristics of a Campbell diagram. During a slow startup or coast down of a rotating machine it may be possible to obtain vibration spectra at constant shaft speed. Each spectrum is then a windowed section of a continuously varying signal with a small speed change within the window length. As a consequence, the peak corresponding to each harmonic is not always localized in one line; in particular, the higher harmonics are likely to be spread over progressively more lines.

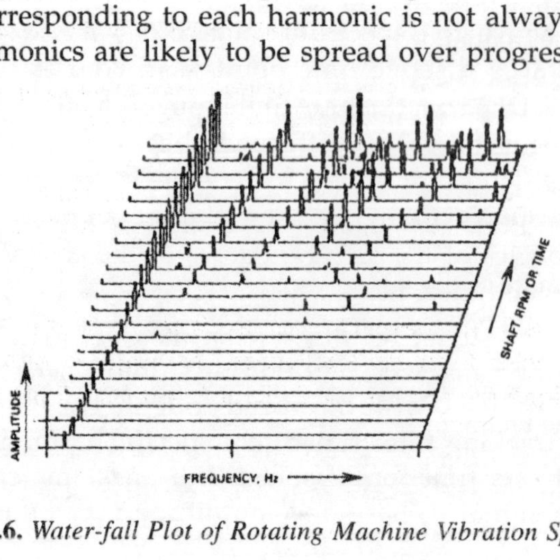

Fig. 7.6. *Water-fall Plot of Rotating Machine Vibration Spectra*

7.5 MODAL ANALYSIS

Natural frequencies, damping factors, modal vectors and modal scaling of a linear time invariant system can be determined by the process of modal analysis. Modal parameters can also be determined by finite element analysis. Experimental methods provide verification and correction of the calculated results. If a finite element model does not exist, experimental results serve as the model for future evaluation for structural modifications.

The success of an experimental modal analysis process depends on having specific goals for the test. Every phase of the process is affected by the established goals. Modal analysis theory has not changed for a long time; however, its application to experimental measurements has changed substantially. Certain aspects of transform relationships have taken on renewed importance, since digital forms of integral transforms are in constant use. The theory uses a more thorough understanding of how the structural parameters of mass, damping and stiffness relate to the impulse response function (time domain), frequency response function (Fourier or frequency domain) and transfer function (Laplace domain). The primary concern in the acquisition of data for the formulation of a modal model relates to digital signal processing. Conversion of analog signals into a corresponding sequence of digital values must accurately describe the time varying characteristics of the inputs to, and responses from, a system. Digital signal processing helps to condense the amount of measured data, estimate the modal parameters and reduce the influence of noise in the measurement procedure.

The frequency-response function is the prime measurement to be made in an experimental modal analysis. A measurement model is needed to permit estimation of the frequency response function from measured input and output data in the presence of noise (representing errors). The Fourier transform is used for this computation, and is performed digitally using a fast Fourier transform algorithm. Several factors contribute to the quality of actual measured frequency response function estimates. Errors arise from overloading the input, extraneous signal input and strong electric or magnetic fields nearby. Non-linear effects shift energy from one frequency to many new frequencies in a manner that may be difficult to recognize. One way to reduce this effect is by choosing a randomly different input signal for each of the contributing averages.

Signal averaging is widely used in structural signature analysis where the event is a revolution of a rotating shaft. Since the frequency response function is expressed in terms of system properties of mass, damping and stiffness, the functions may also be considered to be constant. Hence, when formulating the frequency response function by using appropriate algorithms, the estimate is intrinsically unique, as long as the system is linear and noise is eliminated. The estimate will be valid whether the input is stationary, non-stationary or deterministic. The uniqueness aspect of the frequency response function permits greater flexibility in the testing procedure. Each function is derived as a result of a separate test, or as the result of different portions of the same continuous test. In either case the estimate of the frequency response function is the same, as long as the time-history data for the auto and cross-power spectra are acquired simultaneously. Since the Fourier transform is a linear function, signal averaging may be of synchronous, asynchronous or cyclic type. Input excitation to the system may be of the random signal or deterministic type. The choice of input depends upon the characteristics of the system and of the parameter estimation, as also the expected utilization of the data. For a linear system all input forms should give the same expected value. Many signals are appropriate for use in modal analysis. Figure 7.7 shows a typical test configuration when an impact form of excitation is used on a component.

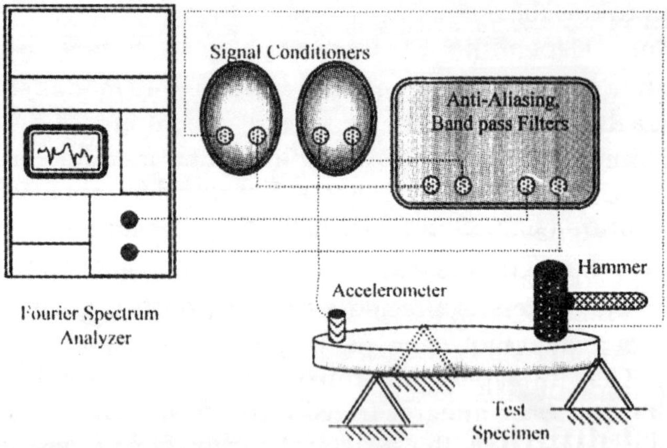

Fig. 7.7. *Test Configuration for Impact Excitation*

Modal parameter estimation, or identification, uses numerical techniques to separate the contributions of individual modes of vibration in measurements such as frequency response functions. The concept calls for estimation of individual single degree of freedom contributions to the multiple degree of freedom measurement. Modal participation vectors are a result of the estimation algorithms, and relate how well each modal vector is excited from each of the reference locations in the measured data. The process is done in multiple stages. The common characteristics of different modal parameter estimation algorithms can be more readily identified by using a matrix polynomial model rather than a physically based mathematical one. Using a polynomial form for the frequency response function, a linear equation with unknown coefficients can be developed which is valid at each measured frequency. In parallel, a time domain representation is also formulated. Most of the modal parameter estimation starts from a general matrix polynomial formulation that is justifiable based upon the underlying matrix differential equation. The process yields essentially the same characteristic matrix polynomial equation for both time and frequency domain data. The roots, or modal frequencies, of the matrix characteristic equation in the time domain are in the z domain and must be converted to the frequency domain, while the roots in the frequency domain are already in the desired domain. The modal frequencies can be found using a number of numerical techniques. Using a companion matrix approach the roots of the matrix characteristic equation can be found as eigen-values. Eigen-vectors that are found from the solution utilizing the companion matrix might be useful for modal parameters, unless scalar coefficients are used. Modal participation vectors from the eigen-vectors of the companion matrix approach are generally found in a second-stage solution of the process. For nearly all modal identification problems, a large amount of redundancy exists, because of which the data space is filtered, or sieved, in order to obtain a reasonable solution in the minimum amount of time.

7.6 ANALYZING TRANSIENT MACHINERY VIBRATIONS

Fast Fourier Transform (FFT) analyzers are conventionally intended for use with harmonic signals that are stationary in time, so the signals are in the center of the FFT bin. But extracting useful information from transient vibration data containing noise poses some problems. An understanding of the FFT process and how it may be used to provide good approximations of transient data can provide information about problems as varied as detection of bad bearings, oil whip and steam whirl.

Accelerometer data in the form of a 400-line spectrum and 8 averages is a reasonable first choice for new machines. Depending on the character of the data, the number of lines and averages may be modified to produce a spectrum where the noise level will impact the discrete frequency signal by an acceptable amount. The random noise level in the same bin as a discrete signal may add or subtract from the amplitude of the signal in the real and imaginary spectra. The number of averages smoothens the noise level to its average value, but averages the magnitude of the discrete signal to a value nearer its actual value.

Figure 7.8 provides a comparison of two-time histories and their spectra, where the peak amplitude of both traces is about 200 mV (Frarey, 2002). However, the upper spectrum associated with the upper time trace only shows a peak amplitude of 180 mV, and the lower spectrum shows the correct amplitude of 200mV.

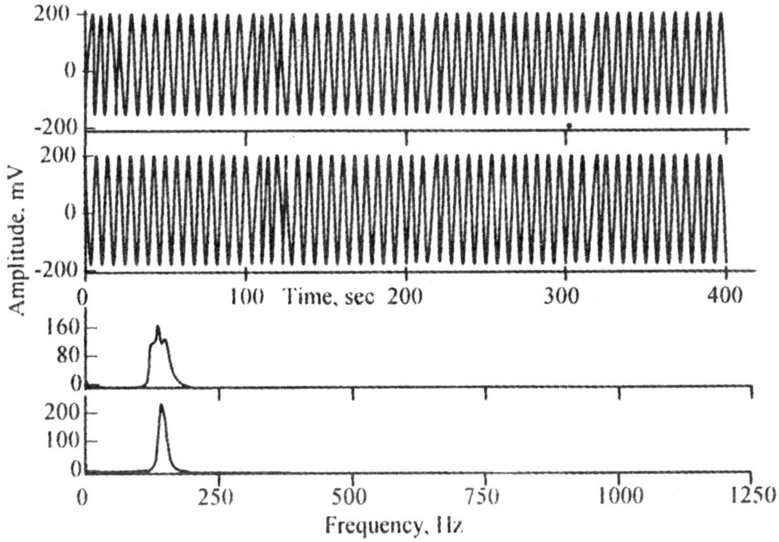

Fig. 7.8. *Two Traces of Equal Amplitude and Their Spectra (Frarey, 2002)*

Another difference between the two spectra is the width of the peak. Close inspection of the upper trace will show that the time per cycle is varying, implying that the signal is frequency modulated and thus moving through several bins during the data acquisition period. In a real case this can happen when the machine speed varies during the data acquisition time, or in the case of instability such as oil whip or steam whirl, the frequency of the signal has varied in that time. The true amplitude of a fat peak can be partially corrected by taking the square root of the sum of the squares of the data in each bin. Complete correction is not obtained since the signal resides for some time of the bin, introducing a picket fence form of error in the reconstruction of the signal. A solution for this would be to reduce the resolution of the spectrum by either selecting a higher maximum function value or by reducing the number of lines of resolution.

The peak hold option of an FFT analyzer stores the maximum amplitude achieved in a bin during the input time. It is touted as a way to get the amplitude portion of the Bode plot during coast-downs or startups. But using the FFT analyzer in this mode violates the rule that the data must be stationary. It is still possible to get a good amplitude versus speed plot if the analyzer is set up to give it a chance to gather accurate data. This can be done by capturing the complete coast-down in the input buffer and then deciding how to set up the analyzer to handle the data.

The number of lines of resolution in the spectrum may be set between 200 and 1600, but the highest resolution may not be the right choice to exactly determine the critical speed. High resolution requires a lot of data to fill the inner buffer, which takes time. The time to fill the buffer must be minimized. Thus, if it takes 8 seconds to fill a 1600 line spectrum, a 200 line spectrum would require only 1 second. But during that same time both the amplitude and frequency can change dramatically. The peak hold value saved in the spectrum is low because the data resides in the bandwidth of the bin for a very short time. The peak-to-peak amplitude of the data in the input buffer also changed during the time to fill the buffer. For a slow rate of change in rotor speed, this can result in a significant difference in the maximum recorded amplitude. The 200-line spectrum will be closer to the actual peak-to-peak value at the critical speed than the one obtained from a 1600-line spectrum. The actual rpm may then be determined by measuring the period of the data at the critical speed.

Detecting a defective rolling element bearing is based on the principle of calculating frequencies dependent on the location of the defect, the geometry of the bearing and its speed. The vibration spectrum may then be examined for a signal at these frequencies, and if the signal exists, the bearing is judged defective. The problem with this approach is that in a number of situations a signal is not noted even when the bearing is defective, or worse still, the bearing soon fails. The background noise in the spectrum may be very low, or the defect in the bearing not large enough, to cause this situation. The reason for the poor success rate of this method lies in the nature of the generated signal. Figure 7.9 shows the spectra generated for several types of signals. The spectrum in the top pair is generated by a sine wave. The next one is for a square wave, where the energy is shared by all the odd harmonics of the repetition rate of the square wave. Energy from the signal is shared by the odd harmonics of the repeating rate of the square wave. The third pair shows the spectrum for a much shorter duration pulse, where the fundamental is still the largest spectrum in the signal and more energy is shared with the higher harmonics. The fourth pair of signals has a very short impact in the time trace. There are many more harmonics of the fundamental mode of low amplitude, compared with the time trace amplitude, and is a typical spectrum generated by a hammer in modal analysis testing.

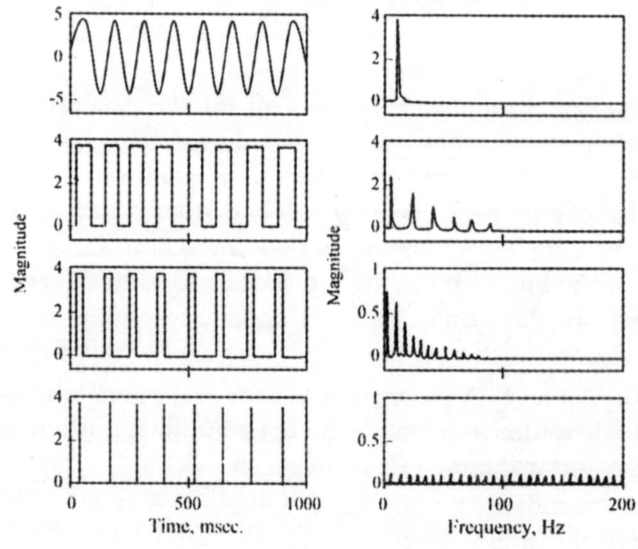

Fig. 7.9. *Spectrum Energy Distribution as Function of Pulse Shape (Frarey, 2002)*

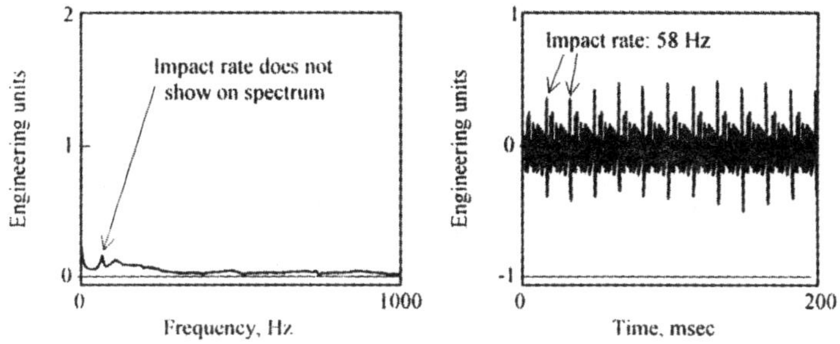

Fig. 7.10. *Time Trace and Resulting Spectrum of Defective Bearing (Frarey, 2002)*

If the shape of a defective bearing's time trace is sinusoidal, the defect signal would be readily identifiable. But the impact from a ball impacting a pit in the raceway produces a signal akin to a hammer impact. Figure 7.10 shows the spectrum and the time trace for a defective bearing. The time trace contains a sharp impact, meaning that the energy is spread among many harmonics, as the spectrum indicates. But the fundamental defect frequency is barely detectable, and if a little background machinery noise is added it would be obscured. A bearing defect pulse can excite a structural response at high frequencies, where small motions can generate high acceleration levels.

The detection can be facilitated by enlarging the impact impulse and by removing the ring down frequency with a software envelope detector using the Hilbert transform technique. The transform shifts the phase of every signal in the input by 90°, creating a situation analogous to real and imaginary signals. The amplitude of the signal is equal to the square root of the sum of the squares at each sample of the data. Figure 7.11 shows the same sequence of signals.

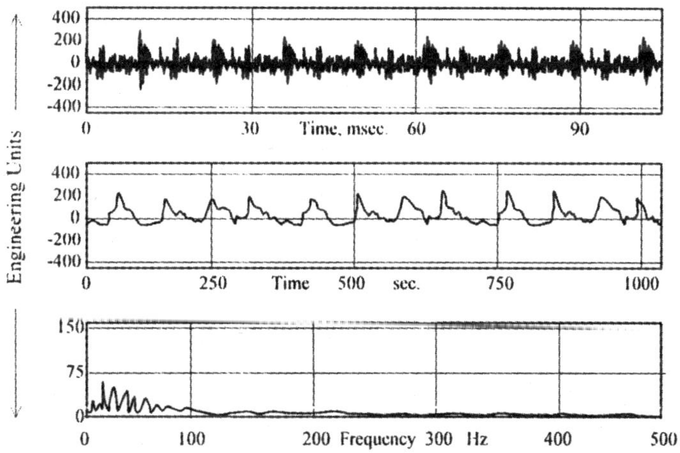

Fig. 7.11. *Defective Bearing Signal Analysis Using Hilbert Transform (Frarey, 2002)*

The top trace shows the raw data from a defective bearing. The second trace is the envelope of the defect signal, and the bottom shows the spectrum of the envelope. To examine the spectrum of the defect signal, a zoom Hilbert transform could be performed. This has the effect of moving the defect signal to the right in the spectrum, but it destroys the typical defect type of signal shown in the top two traces.

7.7 POSITIONING STRAIN GAUGES ON BLADES

Strain gauges on a rotating turbo-machine airfoil are used to obtain vibration measurements to prevent failure from fatigue. Maximum dynamic blade strains are transferred from the gauges either by slip rings or by telemetry during engine tests. Vibrations of the blade are characterized with an adequate number of strain gauges placed on the airfoil's contour, with their location determined by numerical mode shapes (Kielb *et al.*, 2001) or from experimental stress analysis (Purcell *et al.*, 1996). The strain gauge is instrumented close to the peak stress location for the mode of interest. But a qualitative technique may lead to inaccuracy because of ambiguous mode identification occurring frequently for shrouded disk and blades assembled on a flexible rotor due to high engine orders. Sensemeier *et al.*, (1998) provide a systematic investigation for application of strain gauges in a blade, and Griffin (1992) extends the procedure for a complete bladed disk using a model of mistuned masses and springs.

With a small number of strain gauges it is not possible to obtain an accurate perspective of the strain fields from measurements alone. The issue is resolved by relying on finite element analysis to obtain the overall shape of the strain field and use the measurements as a scale factor for the computational results. The peaks of the strain distribution are often very sharp, and a slight error in positioning of the gauge, either in location or direction, may result in a measured strain much less than a maximum value.

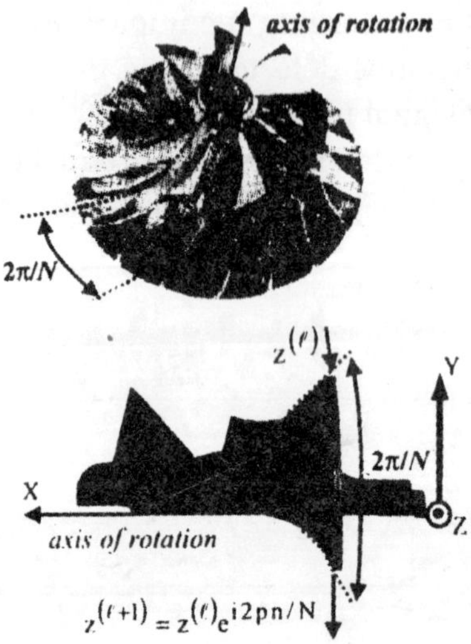

Fig. 7.12. *Periodic Finite Element Model of Impeller (Szwedowicz, 2002)*

The location of the gauges thus appears as a compromise between high-strain values and low sensitivity to mis-positioning. Test measurement efforts may be minimized by optimizing the placement of gauges using four criteria: mode-gauge orthogonality, strain-gauge sensitivity, low strain gradient and distance principle among the gauges (*ref.* Szwedowicz *et al.*, 2002). An objective function is introduced to fulfill these criteria, which is then optimized through an algorithm. The procedure is illustrated for a radial compressor disk, with its dynamic behavior obtained from the finite element computation of the cyclic sector of the impeller containing $N = 9$ sectors (fig. 7.12). The eigen frequencies and mode shapes are computed using the general purpose finite element computer program. The blade's geometry is simulated using a parabolic mesh, and the governing equations are in a piece-wise approximation method. Results from the evaluation provide displacements and strains at the nodes, expressed in the spatial Cartesian system. In the absence of an external force on the structure, the relative dynamic stress is absent in the normal direction to the local airfoil contour around the node, assuming element size approaches small dimensions.

In a cyclically symmetric structure the mode shapes are orthogonal relative to each other. Thus, if vibrations measurements on the rotating impeller are carried out using a certain number of strain gauges, reduced vibration vectors of the assembly are then obtained for all measured mode shapes. The measured mode shapes can be identified for the applied configuration of strain gauges. In the best case the gauges are applied to the peak stress location for a given mode. Because of the many different nodal diameters the disk mode number of interest is mostly higher than the number of applied gauges. On the other hand, dynamic peak stresses can be found in regions of high temperature or on disk parts with non-regular contour geometry that are not accessible for gauge instrumentation. Hence, the strain gauges have to be applied in the airfoil region of dynamic strains where magnitudes are high enough in relation to the strain peak of the analyzed mode shape. This relation may be expressed as gauge sensitivity.

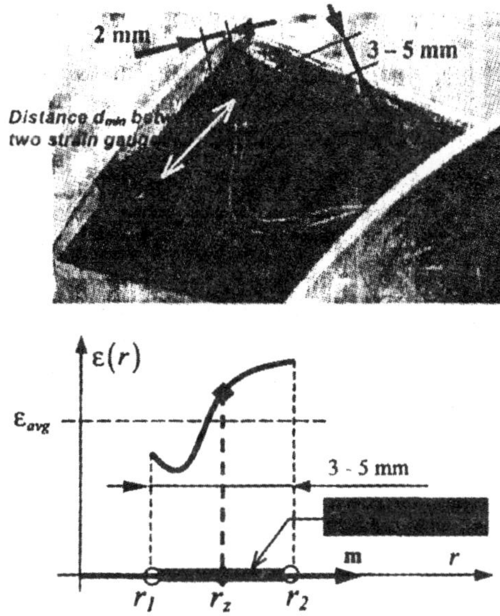

Fig. 7.13. *Instrumented Impeller and Variation of Strain Along Gauge (Szwedowicz, 2002)*

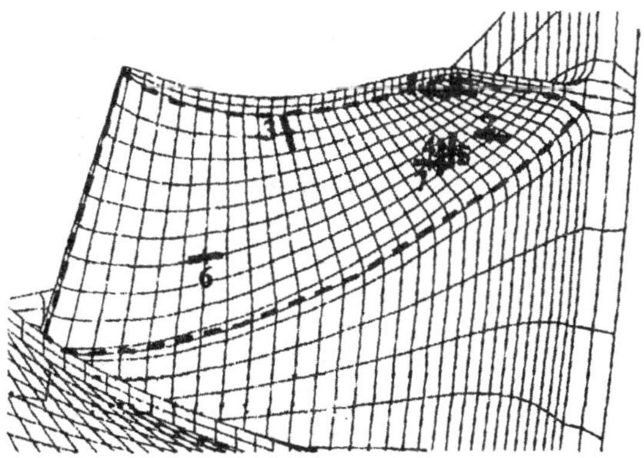

Fig. 7.14. *Optimal Strain Gauge Placement on Impeller Blade (Szwedowicz, 2002)*

The length of the strain gauges is generally between 3 and 5 mm, measuring average strain along the instrumented direction between two points. To avoid unexpected measurement errors, sharp strain gradients should not occur within the region of the application of the gauge. The vector length between the corner and mid-side nodes are calculated from their coordinates. If the relative relations of the dynamic strains among the nodes in the time domain are assumed to be constant, the gradient of the strain distribution around the node can be evaluated. As a check, for each mode and contour node the relative strain gradient may be calculated. Figure 7.13 shows the example of the impeller instrumented with gauges and variation of the strain along the gauge.

A minimal distance, based on the specific gauge instrumentation procedure, must be maintained among adjacent strain gauges. But the number of blades N in the assembly is usually larger than the number of available gauges, so this aspect may not be relevant, although the effects of variation between blade geometries also needs consideration. Optimal positions of the gauges may then be searched based on the mentioned criteria. To demonstrate the approach, 8 strain gauges are shown in fig. 7.14 on the impeller blade's pressure side for evaluating nodal diameters from 0 to 4 and all vibration modes up to 6. The finite element model's 636 mid-side nodes and 224 corner nodes are used in the numerical analysis.

In the absence of gauge failure their locations are dictated by the modal strain distribution and by the measurement noise arising from intrinsic gauge error, quantification error, electrical interference and potential mis-positioning. While the noise and mis-positioning do not affect the measurements to the same extent, they have the common effect of producing variability in the obtained results. The variability may be quantified through the signal to noise ratio of the mean value of the strain and the standard deviation of the contributions of noise and mis-positioning. The application of robust positioning of strain gauges is demonstrated for a fan blade from a Honeywell engine (*ref*: Mignolet *et al.*, 2002).

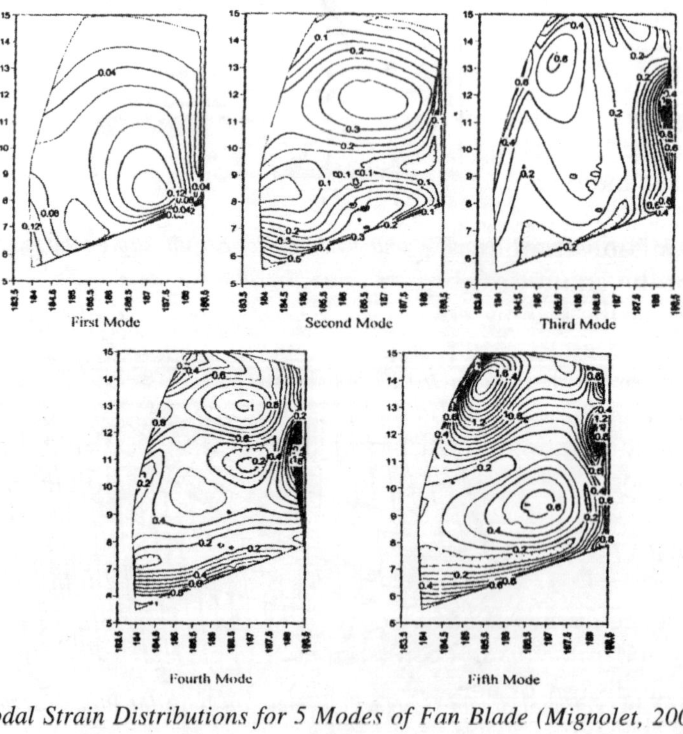

Fig. 7.15. *Modal Strain Distributions for 5 Modes of Fan Blade (Mignolet, 2002)*

Figure 7.15 shows the first five modes on the pressure side of the blade, calculated from an ANSYS finite element model consisting of 4380 8-noded brick elements and 7650 nodes. A discrete set of gauge locations and orientations was formed by 4624 nodes on the pressure and suction sides for 36 different blades set 5° apart, giving a total of 166,464 potential locations and angles for the strain gauges. The measurement noise is assumed to have a standard deviation of 1 percent of the peak strain value for each mode. For mis-positioning the deviation is varied from 0.01 in. to 0.12 in. For modes 1 and 5 the peak signal to noise ratio decreases slowly as a function of mis-positioning level.

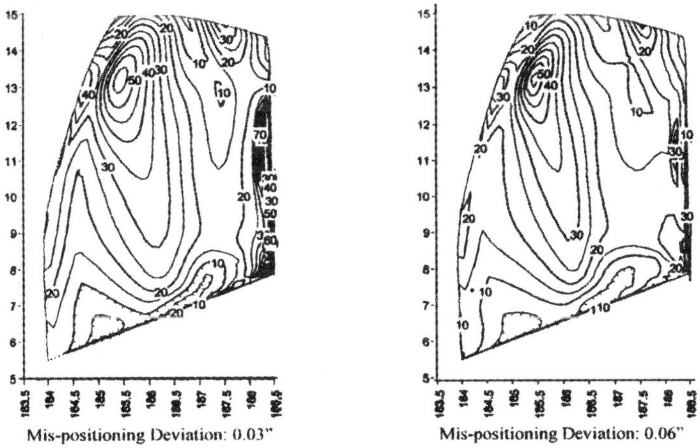

Fig. 7.16. *Distribution of Signal to Noise Ratio for 3rd Mode (Mignolet, 2002)*

On the other hand, modes 2 and 3 experience an initially rapid decrease of the ratio, but the trend suddenly changes and an almost constant value is noted. The trend can be understood from the distribution of the ratio in figs. 7.16 for the pressure side of the blade for the third mode, using standard deviation in location mis-positioning of 0.03 in. and 0.06 in. The figs. are obtained by keeping for each location on the blade the largest value of the signal to noise ratio as the angle is varied through its 36 different values. Two dominant peaks are noted, one close to the trailing edge and one toward the leading edge. For the smaller mis-positioning level the sharp trailing edge peak yields the largest value of the ratio, but the situation is reversed with the increase in mis-positioning. The different rate of decay with mis-positioning can be understood from the strain distribution plot of the third mode in fig. 7.15, where the strain builds up rapidly at the trailing edge. But the leading edge is quite blunt, so mis-positioning only produces a mild effect. A similar situation occurs for the first two modes. Indeed, the strain distribution for mode 1 in fig. 7.15 exhibits two almost equal peaks on the pressure side, with the larger one at the root of the blade on the leading edge. The results thus support the choice of signal to noise ratio as the appropriate metric for the selection process of strain gauge location and direction.

7.8 BLADE VIBRATION MEASUREMENT

A major challenge during the design process of a modern low aspect ratio high-speed axial compressor is to find rotor blade geometries that meet both the aerodynamic and mechanical requirements. The design of aero engine rotor blades is an iterative process between the aerodynamic and structural considerations. The standard process is an iterative exchange of more or less intuitive variations of the blade geometries based on past experience. After designing the blades, the hardware has to prove the results of the analytical process under real engine conditions. Compressor rotor blade vibration measurements are crucial to the proper assessment of critical operating conditions and the prediction of blade life. At MTU these measurements are performed using strain gauges or the frequency modulated grid system (FM grid). Both methods are very costly and time consuming. Hence, MTU initiated work on an optical blade vibration measurement (OBM) system to provide an advanced alternative measurement solution to the two standard methods (Frischbier *et al.*, 1996). This system now can be used in any compressor without touching the blades by only mounting optical probes in the compressor casing. As a result, blade resonance frequencies and amplitudes of any blade can be measured precisely.

In the initial aerodynamic design, the analysis of the blade frequencies revealed the frequency of the second bending mode with coupled torsion (2FIT) at zero rpm was just 2 percent below the frequency of the first torsional mode with coupled bending (1T2F). Due to the higher frequency increase of the second bending mode with rotational speed (stiffening from centrifugal loading) a mode shape transfer (coupling) between both modes was analyzed in the range from 60 to 80 percent nominal speed. Mode shape exchanges of this kind, most of the time are aeroelastically unstable and must be avoided.

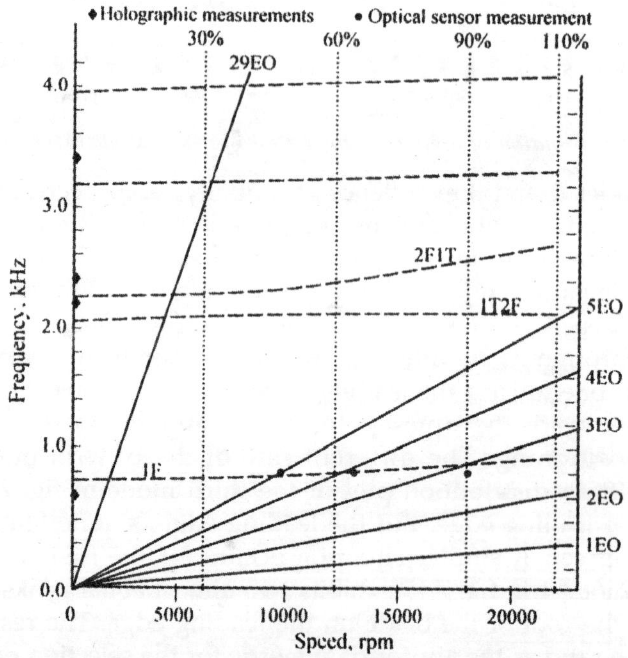

Fig. 7.17. *Campbell Diagram with Analytical and Experimental Results (Frischbier, 1996)*

The geometry of the blades therefore had to be changed in such a way that in the new configuration the frequency of the second bending mode would be distinctively higher in the whole operation range than the frequency of the first torsional mode. The task for the redesign process therefore was to maximize the frequency of the second bending mode (2FIT) with additional constraints on the upper boundaries of the first bending (1F) and the first torsional mode (1T2F). This was accomplished by optimization of the blade profile thickness, with the elements split into groups based on their physical location. For each group the element with the maximum thickness is the variable, and within each group the thickness of all elements are changed with the defined reference element. The thickness of the reference elements is the design relative value. With this procedure the characteristics of the aerodynamic profiles were preserved for all possible variations of the design variables. Additionally, it had to be assured that for all proposals of the optimization code the blade thickness must decrease from the blade root to the tip with due respect to the static stresses at the root. Maximization of the 2nd bending mode called for constraints in the form of upper bounds for the first bending and torsion modes and upper and lower bounds for the thickness of the elements.

The optimization process is an iterative repetition of two program modules. Starting with the current geometry a finite element analysis (normal mode analysis) gives the values of the objective function and of the characteristic constraints and their derivatives (first order). With these linearized sensitivities the approximated optimization problem is solved in one module and a proposal for the new geometry delivered. Based on this new geometry another modal analysis is performed giving the new real frequency values and their new sensitivities. These sensitivities again are the basis for a new optimization run. The procedure is repeated until the proposals of the optimization module converge to a final solution. Figure 7.17 shows the solution in the form of a Campbell diagram for the final milled blisk, together with measured frequencies at zero rpm with a laser holography system.

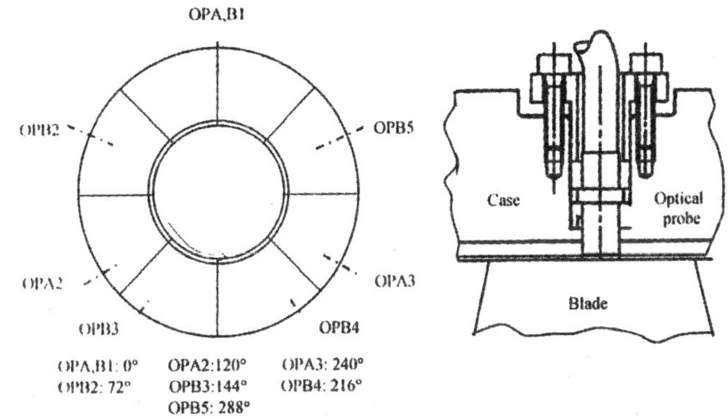

Fig. 7.18. *Optical Probe Setup on Compressor Casing (Frischbier, 1996)*

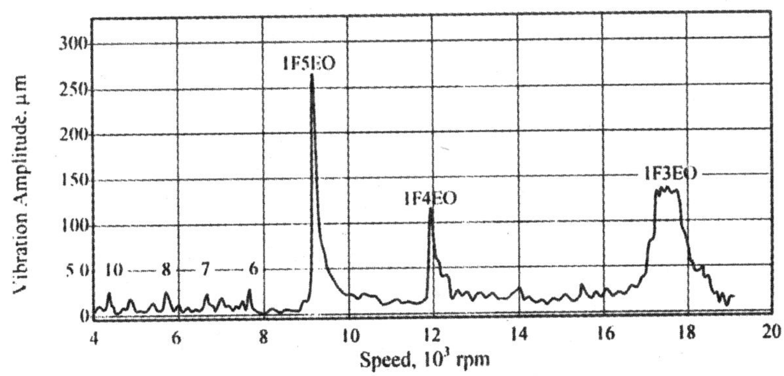

Fig. 7.19. *Measured Vibration Amplitude/Speed Profile (Frischbier, 1996)*

The basic principal of optical blade vibration measurement is known for many years. Optical trigger probes are installed in the compressor casing above the rotor, and the blade transit time between the probes are measured. In the absence of blade vibration the transit times are a function of rotor speed, rotor radius and circumferential probe position. In the presence of disturbing vibrations, transit times deviate from those in the undisturbed condition, the blades passing the probes earlier or later than normal, depending on their momentary deflection. Analysis of these transit time deviations will then identify the amplitudes and frequencies of the blade vibrations encountered. Thus, the blades remain free from instrumentation, the transmission medium between the rotor and the stator is eliminated and the measurements are concurrent on all blades. A total of seven flanges to accommodate the optical probes are attached on the compressor casing above the rotor at the same axial position.

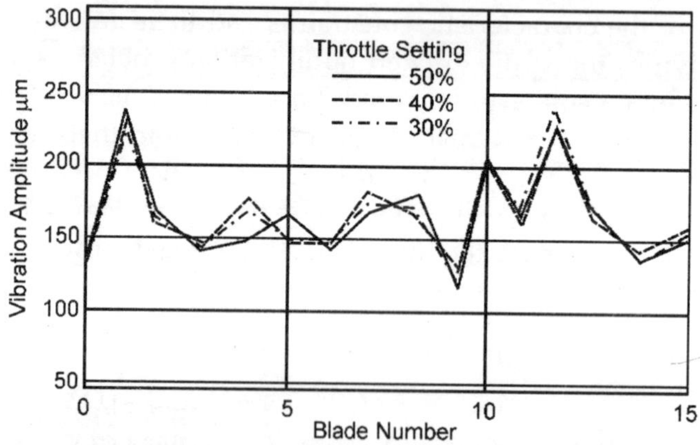

Fig. 7.20. *Individual Blade Amplitudes at 9,186 rpm Resonance (Frischbier, 1996)*

Three of the probes are spaced 120° apart circumferentially, and five 72° apart (fig. 7.18). To reduce the complexity of the setup, a one pulse per revolution signal is omitted, so that the test data could be keyed to relative blade numbers. Initial measurements are taken using three equally spaced probes during slow accelerations and decelerations between 2,000 rpm and 15,000 rpm, and then using five equally-spaced probes between 10,000 rpm and 20,000 rpm. Accelerations and decelerations last about three minutes.

Vibrations of appreciable amplitudes are observed in the first bending mode only. Plotted in fig. 7.19 is the typical resonant vibration amplitude versus speed profile at the 9186 rpm resonance. Amplitude scatter about the mean value is 35%. Differences in the amplitude levels are caused by aerodynamic coupling effects of the mistuned blades. While the rig is passing the resonance band of the 1F bending modes during acceleration or deceleration, at each discrete step in the speed band different sets of combinations of blades are vibrating in resonance. Due to these varying resonating blade packages and due to the corresponding different coupled unsteady air loads also the individual maximum resonance amplitudes of the blades display variations. Vibration amplitudes widen slightly when the throttle is closed. All observed amplitudes are under the theoretical endurance limit of 4.7 μm at 9186 rpm. Individual blade amplitudes across the rotor at the 9186 rpm speed are shown in fig. 7.20.

7.9 HIGH TEMPERATURE PROXIMITY MEASUREMENT

Deflections in the disk and shaft are measured as part of the design development, monitoring and commissioning of all classes of turbo-machinery in a high temperature environment using a number of sensor types. The most common form is the eddy current sensor, but it can operate only up to the curie point of rare earth magnets, about 200°C, and calibrations are often affected by target material property variations. Improvements to the basic concept fail to address the issue of limited temperature capacity.

Many other operating principles are available for the purpose, as shown in the table 7.1. The use of capacitance sensors permits measurement of the gap between the engine casing and rotor blades. The small size and low cost make this sensor a logical selection to apply to high temperature proximity measurement, since it is not susceptible to errors induced by changes in temperature (Muller *et al.*, 1997, Sheard *et al.*, 1997).

Proximity measurement systems consist of three main components. The first is a sensor of the type shown in fig. 7.21. The second is an engine-mounted oscillator, amplifier and filter,

Table 7.1. *Comparison of Clearance Gap Sensors*

Sensor	*Resolution*	*Temperature Limit*	*Target Geometry Dependent*	*Relative Size*	*Cost*
Electro-mechanical	0.05 mm	1200°C	No	Large	High
Optical	0.05 mm	1550°C	No	Large	Very high
Capacitive	5% of range	1100°C	Yes	Very small	Low
Eddy current	5% of range	650°C	Yes	Small	Low
Microwave	0.10 mm	1200°C	No	Medium	Medium
Fluidic	0.25 mm	1400°C	Yes	Small	Low

Additional factors: Optical and capacitive — most mature; Capacitive — least complex; Capacitive and Microwave — most durable

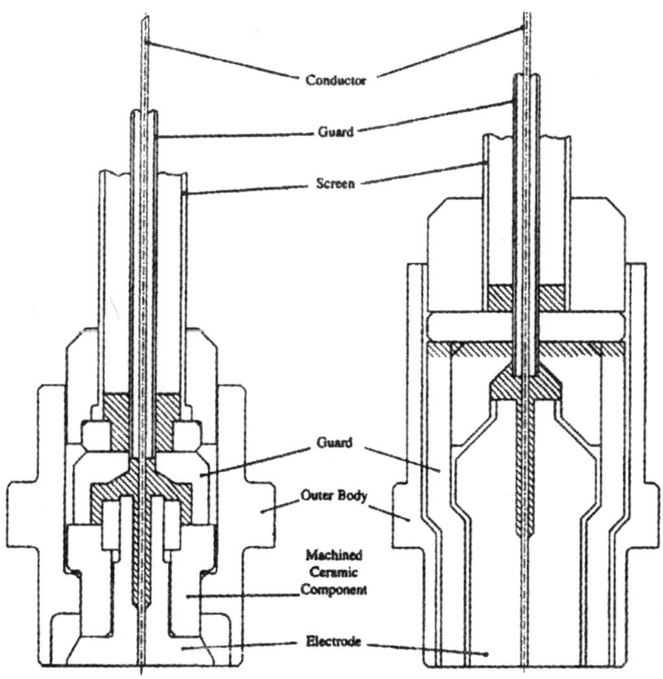

Fig. 7.21. *Capacitive Sensors: Tri-axial* (left); *New Design* (right) *(Sheard, 1997)*

and the third is a rack-mounted receiver linked to the amplifier by coaxial cables. In a tri-axial sensor the center wire is connected to the center tip. A guard surrounds the center wire and the tip, and an outer screen surrounds the guard. The capacitance measured between the sensor's tip and the target is isolated from the surroundings by a high-temperature resistant sheathing, and then encapsulating the cable and the sensor in a third layer. The construction provides isolation of the center wire, and when the temperature falls below 200°C the cable between the sensor and the oscillator takes the characteristics of a low-noise PTFE-coated coaxial cable.

The amplifier drives the guard at 16 kHz and constant amplitude, and is insensitive to low speed thermal changes in cable capacitance to maintain the amplitude level. It induces an identical sinusoidal oscillation in the center wire that is immune from changes in cable capacitance. As soon as a target comes into the range of the sensor's tip, the additional capacitive couple causes a change in the induced voltage on the center wire, manifesting itself at a reduced voltage relative to that on the guard. The amplifier works at high gain/bandwidth product differential, with both the reference guard voltage and the induced center wire voltage fed into it. The output from the amplifier is a 16kHz sine wave with an amplitude proportional to the difference between the reference and induced inputs, and is fed through a 50 ohm drive stage for connection to a distant receiver module. The sine wave is synchronously rectified to eliminate quadrature errors, then passed through a 3-pole low pass filter with a 3-db point at 5 kHz. The DC voltage output is scaled to give a 0 to 10 volt output, proportional to the capacitance between the sensor tip and the target over the working range.

The development of the sensor should aim to be representative of conditions found in current generation turbines. The front face of the sensor is exposed to peak operating temperature of 1100°C, has a body temperature of 800°C, cable temperature of 600° C and the flexible joint temperature of 200°C. A traditional tri-axial sensor is assembled from metallic and ceramic components. A fully guarded sensor has less than 0.1 pF stray capacitance because of air linkage from the electrode to the earth across the guard ring, and may be used towards a solid target or over unshrouded blades using the appropriate circuitry. A newer concept eliminates all ceramic components, maintaining electrical isolation by coating the electrode and the guard with alumina. Elimination of machined ceramic components enables extension of the guard to the tip of the sensor to enhance its suitability for use with proximity and tip clearance electronics. The electrode, guard and housing are conically shaped for easy fitting, and application of a pre-load during assembly assures positive location and durability during engine test. The guard ring is dimensioned to obtain axial coincidence and reduced stray capacitive linkage to the sensor body, thereby maximizing the operating range for a given sensor diameter.

Stray capacitance tends to reduce linearity, and hence operating range, of the sensor. From a comparative study between the diameters of the guard and the electrode, a ratio of 1.4 between the two is found to maximize the range of the sensor. Performance of the modified design sensor is obtained from tests. The critical parameter at elevated temperature is isolation resistance between the components, hence the sensor is tested in an oven. Measured isolation compares favorably with theoretical resistance for a pure alumina coating. With an isolation of 10 k-ohms in the sensor the amplifier performance is unaffected, but the system does not function for 1 k-ohm isolation. At 1100°C the parts maintain over 10 k-ohms isolation, making them suitable for the application. The attached cables exhibit some thermo-electric effects starting at 700°C. Dissimilar metals used in the cable generate a voltage measuring up to 50 m-volts at 1100°C between the screen, the guard and the conductor. This effect may be uncoupled in the amplifier, but an accurate measurement of the isolation resistance above 700°C is difficult. The amplifier operates by oscillating a 16 kHz voltage on the guard, which induces an in-phase voltage on the center wire. With the silica-filled cable above 800°C the phase of the induced voltage on the center wire is shifted, and above 950°C the system ceases to function.

7.10 BLADE TIP TIMING METHODS

Contacting vibration measurement devices for turbine blades such as strain gauges provide better accuracy, but have limited life, rely on complicated telemetry or slip rings, and may interfere with aerodynamic and mechanical properties of the airfoil. Alternative non-contacting measurement techniques such as blade-tip timing measure deflection at the tip using optical probes mounted in the assembly casing, thus eliminating such adverse effects.

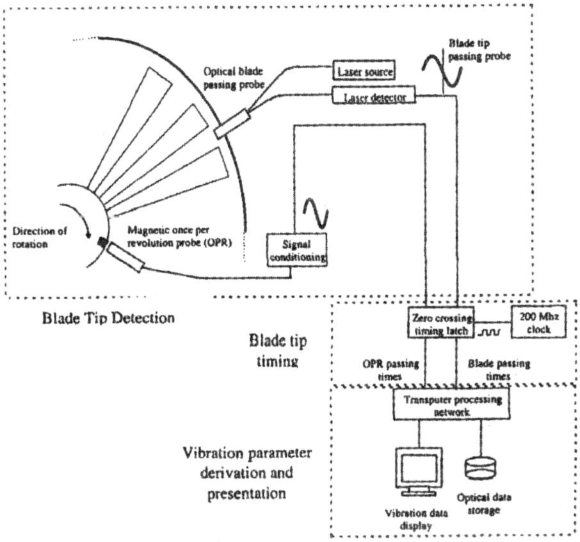

Fig. 7.22. *Elements of Blade Tip Timing System (Heath, 1997)*

When vibrations of the external structure are limited, as in industrial gas turbines, the time for the tip of a probe, the so-called blade arrival time, depends on rotor speed alone. But when the blade is vibrating, blade arrival time also depends on the amplitude and frequency of vibration. In principle, blade motion can be characterized from such data. As with other transducers such as strain gauges and accelerometers, whether a particular mode can be captured by a given laser probe depends on the measurement location. Typically the measurements are taken near the leading edge of the blade to maintain reasonable sensitivity to the motion of the first few bending and torsion modes.

A tip-timing system consists of three main elements: (*i*) a number of probes placed in the casing to acquire raw blade arrival time, (*ii*) derivation of primary vibration parameters (displacement, velocity or acceleration) from the measured data, (*iii*) processing of the parameters to describe the vibration pattern of the bladed disk assembly. Figure 7.22 provides details of the scheme (Heath *et al.*, 1997). The first two items are readily available using data acquisition and derivation technology. The third step plays a major role in determining the success of the concept. For asynchronous vibration the mean value of the measurements over a limited number of samples (or speed points) aids in deleting the steady part of the response. For synchronous vibration the unsteady components rely on location of the blade relative to a non-vibrating rotating datum from a shaft encoder or a once per revolution probe (McCarty *et al.*, 1980).

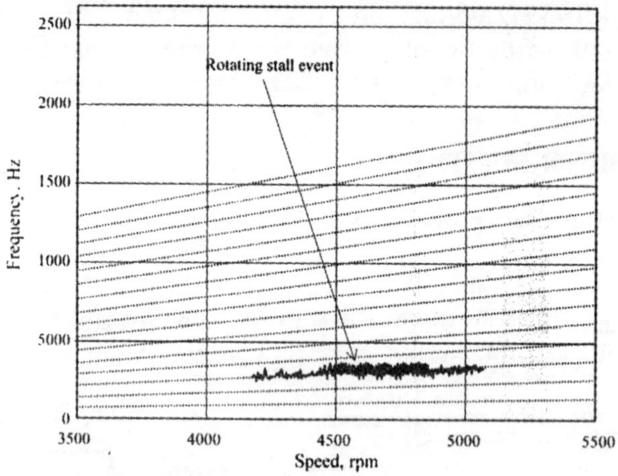

Fig. 7.23. *All-Blade Spectrum Indicating Rotating Stall Event (Heath, 1997)*

The process is illustrated for an industrial turbine experiencing blade failures when operating over extended periods. Strain gauge testing failed to indicate critical vibration levels that would result in a failure. Since the blades are only 20 mm high, the dynamic properties are affected by attaching the stain gauges. Tip timing measurements made under identical operating conditions pinpointed two blades with excessive response levels.

Asynchronous response analysis methods aim at identifying the amplitude and frequency of non-integral order resonances from measurements at the blade tip. One or two measurements made at the same axial plane permit spectral analysis of the response in two ways. Measurements from a single blade on successive rotations can be used to compute individual spectra for each blade in the assembly. Alternatively, the measurements taken in the blade arrival order may be combined to give a single all-blade spectrum. Prior knowledge of the expected assembly resonances is required for both methods. The measured ratio of maximum and minimum amplitudes of a single blade over a finite period is assumed to be equal to the ratio for all blades in the assembly. Resonant frequencies are identified from a Fourier analysis of the response parameters. When the assembly vibration is dominated by a single resonance, the frequency can be recognized with good accuracy if the response is high. The quality of the Fourier spectrum is degraded by noise when the maximum amplitude is below 20 μm, which may correspond to 1 percent of a 10^7 cycle endurance limit.

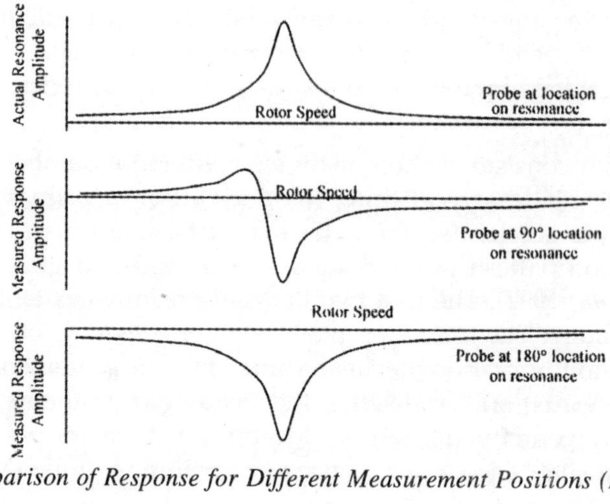

Fig. 7.24. *Comparison of Response for Different Measurement Positions (Heath, 1997)*

An all-blade spectrum analysis assumes the bladed disk assembly vibrates in nodal diameter modes, so the contour of the stationary points describes a certain number of diameters of the disk. Because of circular symmetry the nodal diameter modes will occur in orthogonal pairs that become traveling waves under the effect of rotation. Forward traveling waves correspond to the rotation of the nodal lines in the same direction as assembly rotation, backward waves having their nodal lines rotating in the opposite direction. For each nodal diameter there is constant phase angle between consecutive blades. If the blades are assumed to vibrate at the same frequency, a spatial Fourier transform of the measured response amplitude of successive blades will give the traveling wave response in the stationary measurement response reference frame. The objective of the analysis is to identify the nodal diameter modes associated with the frequency components of significant amplitude. The nodal diameter pattern is inferred by the rather difficult task from expected resonant frequencies. A typical spectrum is shown in the Campbell diagram of fig. 7.23 for an industrial compressor experiencing a rotating stall. The number of the nodal diameter associated with the stall event can be extended to two measurement probes, so the mode can be identified directly from $n = \Delta\phi/\Delta\theta$ where $\Delta\phi$ is the phase difference of the spectral peaks and $\Delta\theta$ is the angle between the probes.

Synchronous response analysis may be divided into direct and indirect methods, based on the number of response samples for each assembly rotation. The direct method needs four samples, and resonance characteristics are identified for each sample set. A least squares sine fit to the measurements from the four probe locations may be used to obtain the amplitude and frequency of an integral order of vibration, although the technique may be susceptible to the presence of noise.

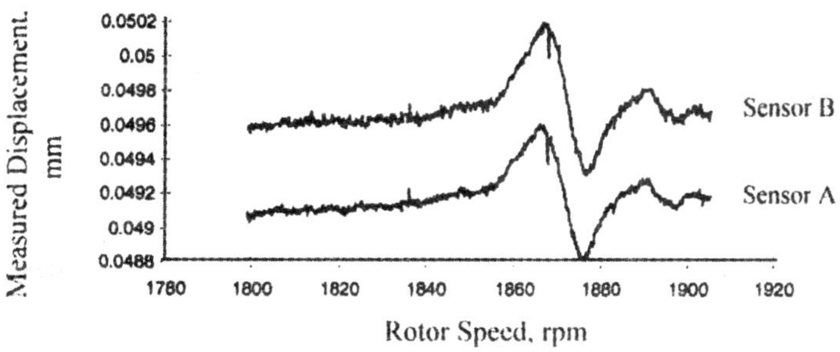

Fig. 7.25. *Displacement versus Speed Characteristic of Bladed Disk Assembly (Heath, 1998)*

The indirect method calls for taking one or two response samples for each turn of the rotor, giving only the resonant frequency (Chi *et al.*, 1988). Displacement, velocity or acceleration signals from the single probe may be used to identify the resonance excited by varying the assembly rotation speed. The form of the measured response depends on the position of the probe, as shown in fig. 7.24.

Errors are caused when the dependence of the blade tip arrival time on the actual blade vibration amplitude is ignored. For synchronous responses this amplitude is a function of angular position and of peak response amplitude, so the response phase with respect to a stationary forcing function will change as the amplitude varies. Hence, the response amplitude will also change. But as the arrival time is dependent on the amplitude, the apparent change in the phase of the response at the measurement point will not be the same as the actual phase change between the forcing term and the response (Zablotsky *et al.*, 1970).

Mistuning, or small blade-to-blade variations, causes blades to have dramatically different response levels, and the assembly will exhibit many close modes because of the splitting double modes with close frequencies and similar but not identical mode shapes. Mistuning will also distort the displacement versus speed characteristic, and hence adversely affect the results of the analysis. The effects are quantified using a numerical simulation of a 12-bladed disk running in the vicinity of a split 3 nodal diameter mode, and subject to an engine order excitation of 3. The finite element model of the assembly is adjusted to give a ± 5 percent variation in the individual blade cantilever frequencies. Maximum response is obtained from the model, and the corresponding displacement versus speed characteristic is obtained from the tip-timing numerical simulator. Variations in the maximum displacement error are to be expected, since mistuning results in a different actual maximum displacement for each blade when traversing the resonance. However, the percentage error has a fixed sign for a fixed probe location. The effects of mistuning can be observed in the displacement versus speed plots, and unless there is a marked mistuning effect, the measured maximum displacement values are probably as accurate as those provided by strain gauges. Figure 7.25 provides the displacement-speed characteristics obtained at two probes while passing through a resonance for a 36-bladed assembly. The objective in this analysis is to determine the engine order of excitation, which is known to be 2 from related sources. The measured data is analyzed for each blade in turn, and the excitation order calculated individually. The results are plotted in fig. 7.26, where the mean value of the predicted engine order excitation is verified.

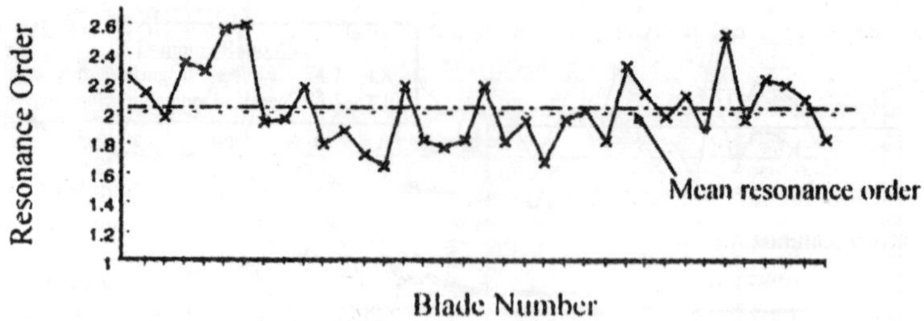

Fig. 7.26. *Identification of Engine Excitation Order (Heath, 1998)*

7.11 TESTING UNDER TRANSIENT CONDITIONS

Loosely defined, ceramics encompass a variety of materials, including monolithic ceramics, ceramic composites, and carbon-carbon composites. Monolithic ceramics have been studied extensively over the last several years, with attention gradually shifting toward composites. Covalently-bonded ceramics offer the prospect of good resistance to heat, to nearly 3000° F, coupled with low density and in some cases excellent oxidation and corrosion resistance. Additional benefits accrue from low cost and easy availability of the principal starting materials, silicon, carbon and nitrogen.

Unfortunately, these ceramics are also brittle, prone to thermal shock and not as thermally conductive as heat-resistant metals, causing significant deterioration in tension load capability. The nature of the inter-atomic bonds controls the physical properties. Mechanical properties heavily rely on preparation techniques, impurities and part shape, and hence vary considerably; the process basically controls the properties of the ceramics. Nevertheless, toughness and thermal shock resistance of SiC and its ability to form a protective coating makes it a good prospect for turbine applications.

Dilzer *et al.*, (1998) present a design for ceramic first stage nozzle vane for a 1400°C, 70 MW stationary gas turbine, taking advantage of the superior high temperature properties while compensating for the deficiencies. The outer shell of the vane is made from sintered silicon carbide (SSiC), with a heat-insulating layer separating it from the metallic core. The insulation material provides a balanced temperature distribution in the outer shell, while compensating for the differential thermal growth. Local thickness of individual layers is adapted according to thermal boundary conditions used in a finite element analysis. The hybrid construction is predicted to reduce induced stresses to half when compared with the metal design it replaces. To ensure most regions of the nozzle vane are accessible for inner cooling, the profile is modified to give optimum cooling, resulting in a slight increase in thickness at the trailing edge. Stress from mechanical load is not of importance, but thermal loads arising from non-uniform growth dominate in the structure. Turbine inlet temperature is 1673 K and coolant temperature is 673 K. Chord length of the vane is 58 mm and height is 60 mm.

Maximum and minimum temperatures and stresses for the full operating cycle in the outer ceramic shell are given in fig. 7.27, with maximum tensile stress staying between 75 MPa and 135 MPa. In comparison, a directly cooled ceramic nozzle vane is predicted to induce maximum stress exceeding 400 MPa. The advantage of using multiple layers to enhance thermal resistance and load carrying capability for continuous and transient conditions is overwhelming. Failure probability calculations due to slow crack growth indicate sufficient reliability for 10,000 hours of continuous engine operation. Loads during engine trip also indicate adequate margin for the fast fracture mode.

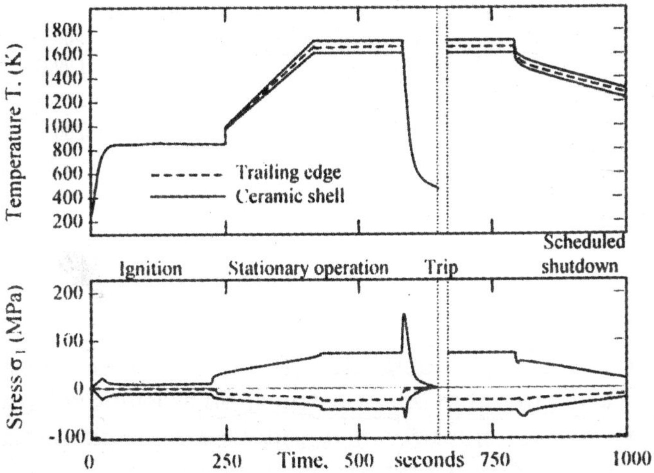

Fig. 7.27. *Maximum and Minimum Temperatures and Stresses (Dilzer, 1998)*

Adequacy of the design is verified in a test setup simulating engine transient and continuous operating conditions, and to examine thermal shock resistance of the ceramic shell. An electric furnace permits raising the temperature of the vane, and internal cooling is achieved by coolant flow through tubes (fig. 7.28). The distance between the tubes can be adjusted to obtain local temperatures according to actual engine conditions. The test specimen is then moved to a cooling down duct, causing temperatures in the ceramic shell to decrease to levels encountered during a trip situation in the real engine. Surface temperatures are measured and recorded by infrared radiation thermography concurrently during the cool down. Three viewing ports are provided to observe the specimen from different perspectives.

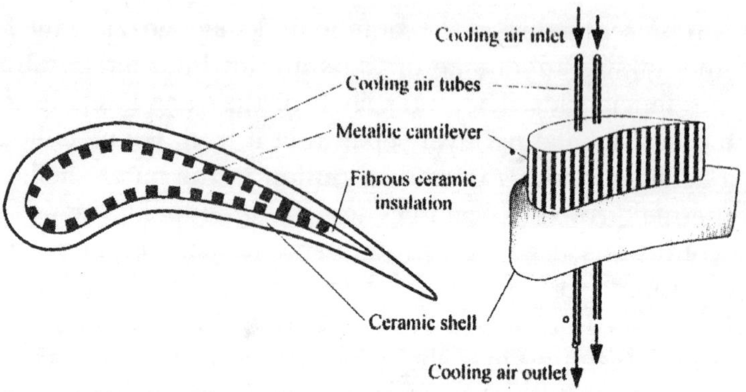

Fig. 7.28. *Internal Cooling of Test Vane (Dilzer, 1998)*

In the event of an engine trip, fuel flow to the combustor is cut-off abruptly. The impact of this incident is to subject the first stage nozzle vanes at a nearly uniform temperature of 1773 K to a flow at 293 K over its surface. This change in temperature of the main engine airflow takes place without an appreciable time lag. During the event the coolant flow is maintained at 293 K. Simulation of the thermal shock in the experimental tests is carried out by holding the temperature in the furnace, until steady temperatures are achieved in the metallic core. The specimen is then moved to the cooling duct. Airflow in the test duct is adjusted to obtain the same condition as in the engine, at which point surface temperatures are determined. A comparison of the maximum and minimum stresses during regular operation and in a tripped situation is shown in fig. 7.29.

In the finite element model a similar time dependent temperature and stress analysis is performed to predict the effects of the shock. Stress contours at fuel cutoff point and 4 seconds after the start of the shock are shown in fig. 7.30. For both cases maximum stress occurs at the mid-height of the trailing edge. Load and stress levels are considerably lower in other parts of the vane. Close correlation of the two sets of results indicates that the thermal shock test is well suited in estimating the adequacy of the three-layered ceramic vane.

The test vane survived a multitude of experiments with maximum induced loads of approximately 150 MPa. Visible damage has not been observed. Another test specimen is subjected to increasing loads, enduring thermal shock tests with induced tensile stress exceeding 250 MPa.

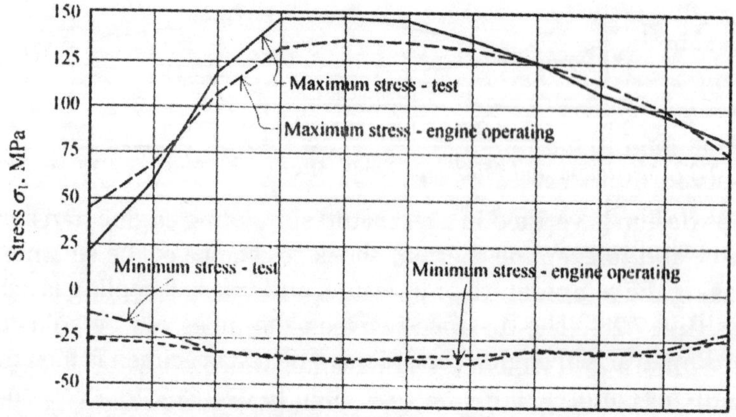

Fig. 7.29. *Time Dependent Stresses During Thermal Shock Test (Dilzer, 1998)*

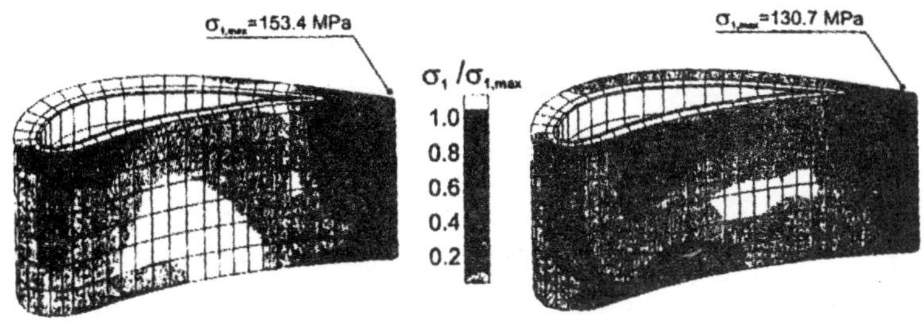

Fig. 7.30. *Stress Distribution Following Fuel Cut-off* (left), *4 Seconds After Cut-off* (right) *(Dilzer, 1998)*

The piece fractured in the fourth test. Prior to the failure the outer shell did not display any distress signs, except for some discoloration. For a final assessment of the design additional experiments are necessary to obtain statistically significant information about the thermal shock behavior of the ceramic stator vane. But the tests affirm that the design method adopted for the part to reduce thermally induced stresses holds promise.

7.12 MODE LOCALIZATION AND FORCED RESPONSE IN BLADED DISK

The design and analysis of a nominally cyclic structure such as a bladed disk is mostly done under the assumption that the structure as a whole is comprised of perfectly identical sectors, but small imperfections arising from manufacturing tolerances and uneven wear negate this assumption. The variations between the sectors can have a large impact on dynamic behavior, leading to a response that is qualitatively different from that of an ideally tuned assembly. Complex modes of vibration occur when individual blades of the assembly are detuned. These modes are distinctly different from the perfectly tuned mode shapes, and are known as *localized modes.*

Free vibration mode shapes of cyclic structures are characterized by the number of nodal diameters and nodal circles. A perfectly tuned-bladed disk has modes that occur in pairs at exactly the same frequency. Both modes in the pair are spatially and circumferentially harmonic around the disk, and the two are orthogonal to one another. But the introduction of even a small amount of mistuning causes the frequencies to differ roughly in proportion to the amount of stiffness mistuning among the blades. If the damping in the system is low enough, the resonant peaks of the pair of modes will be distinct. Peak splitting can be demonstrated using holographic interferometry and strain-gauge measurements to observe dual modes in a nominally tuned and intentionally mistuned bladed disk (Jay *et al.*, 1986), with both standing and travelling wave response observed to a travelling wave form of excitation. Other studies of amplitude magnification (Kruse *et al.*, 1996) demonstrate that while some mistuning enhances maximum response, further mistuning can reverse the effect and result in lower amplitudes. The phenomenon of peak splitting and mode localization can lead to increased resonant amplitudes of blade vibration during operation. The accelerated rate of fatigue can lead to significantly shorter life than predicted by an analysis assuming perfect symmetry.

The focus of an experimental investigation (Judge *et al.*, 2000) of a bladed disk is on a frequency region with stronger structural inter-blade coupling and lower modal density, where the modes are dominated by out of plane bending motion of the blades. The purpose is to validate theoretical predictions about the occurrence of localized modes and increases in forced response amplitudes, and to gain greater insight into their mechanism. A flat, single-piece 12-

bladed disk of simple geometry and held stationary by a clamp at the center is used for the tests. The blisk is 3.24-mm thick, machined from precision ground 4140 steel plate. The disk is 300-mm in diameter, and the blades are 92-mm long and 10-mm wide. Precise tolerances provide the specimen with effects as close as possible to a perfectly tuned assembly. Typical values of damping for the material are a loss factor η between 2×10^{-4} and 1×10^{-3}. Excitation is applied by piezoelectric actuators glued to the surface at the root of each blade. Measurements are taken with an eddy current proximity probe held in a rotating fixture to move the probe over the tip of each blade. At the end of the tuned portion of the experiment, various amounts of lead weight are glued to the underside of the blade tip in order to introduce mistuning to the system.

Since damping in the blisk is low, the resonant peaks are narrow. By exciting the blades one at a time, the resonant response of the structure can be used to determine the natural frequencies and mode shapes. When all piezoelectric actuators are driven simultaneously at the same amplitude with varying phase, excitation in the form of traveling waves corresponding to various engine orders can be simulated. The effect is then similar to that of a rotating bladed disk passing through different numbers of stationary sources of excitation. Phase synchronized signal generators passing through an adjustable gain amplifier may then be used to provide the desired excitation pattern.

Figure 7.31 summarizes the modes of vibration of the tuned system from the finite element computations, characterized according to the frequencies at which they occur and the number of the nodal diameter.

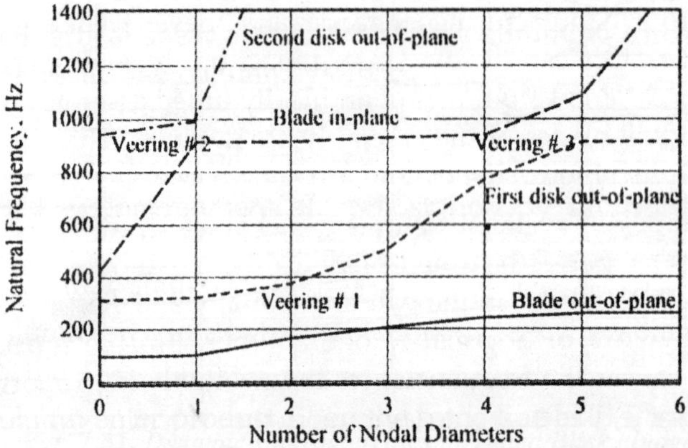

Fig. 7.31. *Dynamic Characteristics of Tuned Blisk (Judge, 2000)*

Note that all modes except those with 0 or 6 nodal diameters are double modes, with orthogonal mode shapes occurring at the same frequency. The regions labeled veering # 2 and veering # 3 pertain to modes that are mostly a combination of in-plane bending of the blades and out of plane bending of the disk, and allow a transfer of energy between the blades and the disk. As a consequence, some amount of coupling between the blades is also present. In these regions of high modal density but with limited inter-blade coupling, strong modal localization may be expected when the system is mistuned. Veering # 1 occurs in the out of plane modes between 200 and 260 Hz. The wider nature of this veering indicates a stronger coupling between the blades and the disk, and the modes are more widely spread in frequency than in the higher frequency regime. This has the tendency to reduce mode localization in the presence of mistuning.

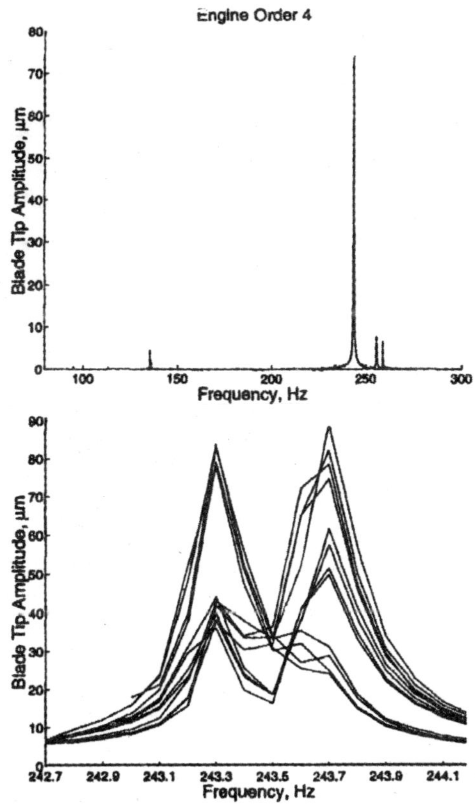

Fig. 7.32. *Measured Frequency Response: Blade 1 Spectrum* (upper), *Response of All Blades* (lower) *(Judge, 2000)*

The excitation of the bladed disk at equal amplitudes but at different phases can be written as:

$$F_n = F_0 \cos[\omega t + (n-1)\phi] \quad (7.3)$$

where F_0 is the force amplitude, ω is the frequency of excitation, n is the number of blade from 1 to 12, and ϕ is the phase difference between adjacent blades, and is defined by $\phi = 2\pi C/12$, where C is the engine order of excitation. The above form can be rewritten in the form:

$$F_n = F_0 \cos[(n-1)\phi]\cos(\omega t) - F_o \sin[(n-1)\phi]\sin(\omega t) \quad (7.4)$$

Each of the terms may be interpreted as a harmonic excitation at the same phase on all blades, but with the amplitude dependent on the blade number in a sinusoidal pattern. Thus, the traveling wave excitation of the form of equation (7.3) is equivalent to the excitation of two standing waves orthogonal to each other spatially and out of phase temporally by 90°. A sign change in the phase shift in (7.3) results only in a change of sign of the second term in (7.4), hence changing the direction of the traveling wave excitation mainly reverses the relative phase of the two standing waves. Because the engine order of excitation determines ϕ, an engine order excitation m can be regarded as the sum of two orthogonal standing waves having m modal diameters. Engine orders 0 and 6 deserve a special mention. In the zero order case, $F_n = F_0 \cos(\omega t)$, so there is no phase difference between the excitation of adjacent blades, and the excitation becomes a standing wave. Engine order 6 excitation in a 12-bladed disk means a 180° phase shift between adjacent blades, so that $F_n = (-1)^{n-1} F_0 \cos(\omega t)$. Therefore, engine order 6 is also a standing wave excitation.

Figure 7.32 plots the experimental frequency response of the tuned disk for engine order 4 excitation. The paired modes are not at the same exact frequency, and the peaks are narrow enough that only one mode shape exhibits significant response at a given frequency. Although the travelling wave excitation decomposed into components aligns with both mode shapes, it drives a response only in one mode or the other. As damping is increased, the two peaks widen and thus merge, so that both modes respond significantly at either resonant frequency.

Mistuning through the application of lead weights at the blade tips is designed to best demonstrate the phenomena being investigated, and is obtained after a number of numerical predictions to obtain high-amplitude magnification from excitation in all engine orders. Monte Carlo simulations using the finite element model incorporating mistuning by the addition of the mass at the blade tips may be used for the selection of the pattern. The simulations may also be based on stiffness mistuning by adjusting the Young's modulus of the blade material. Both methods of mistuning predicted localization in the 10th, 11th and 12th modes, which in the tuned case are shapes characterized by 5 (10th and 11th modes) and 6 (12th mode) nodal diameters. The modes are all blade dominated and quite close to one another in frequency. Lower modes occurring in and around veering # 1 are thus comprised of both blade and disk motion, retaining their spatially extended nature even in the presence of strong mistuning. Table # 7.2 compares the resonant peak frequencies observed experimentally with those predicted numerically.

Table 7.2. *Resonant Frequencies of Mistuned Blisk*

Mode #	*Theoretical Prediction, Hz*	*Experimental Data, Hz*	*Percent Difference*
1	93.02	91.00	2.17
2	94.40	92.00	2.54
3	107.83	109.40	1.46
4	129.69	129.10	0.45
5	130.51	130.20	0.24
6	194.98	194.00	0.50
7	195.81	195.10	0.36
8	219.57	217.80	0.81
9	224.92	224.20	0.32
10	229.60	228.70	0.39
11	237.74	237.30	0.19
12	240.49	239.80	0.29

Modes 1 and 2, 4 and 5 and 6 and 7 are all paired quite closely. Mode 3 is a zero nodal diameter mode, and thus is single. The paired modes are essentially the same as the paired orthogonal nodal diameter modes of the perfectly tuned case and excited most directly by engine order excitation of that same harmonic.

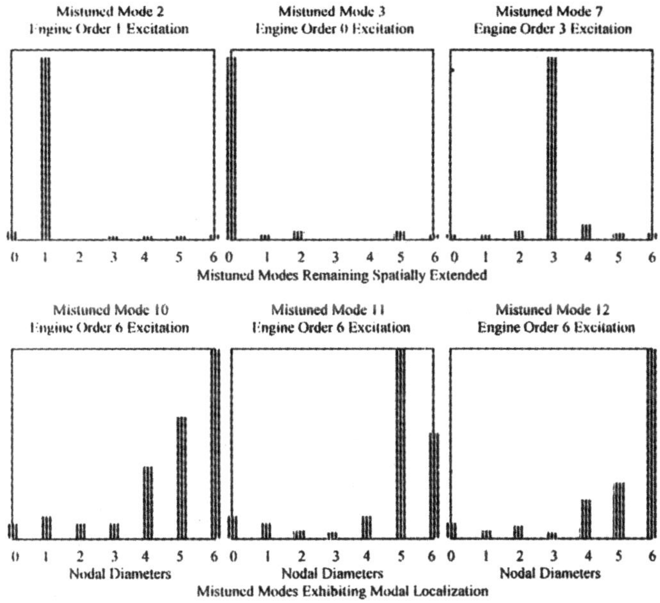

Fig. 7.33. *Nodal Diameter Composition of Mistuned Modes (Judge, 2000)*

But the picture is dramatically different for the higher modes. In the tuned case, modes 10 and 11 pair for the five nodal diameter modes, while mode 12 is the single six nodal diameter mode. The splitting between modes 10 and 11 is greater than the spacing between modes 11 and 12 and between modes 9 and 10, indicating they are no longer spatially extended but are localized. Engine orders 5 and 6 both excite strong responses in these three mode shapes. The measured mode shapes are analyzed by Fourier transformation to determine their composition in terms of nodal diameter shapes. Figure 7.33 shows some examples, with the lower modes comprised overwhelmingly of a single nodal diameter harmonic and the higher modes exhibiting localization of multiple nodal diameter shapes. Modes 11 and 12 result mostly from combinations of 5 and 6 nodal diameter shapes, while mode 10 has strong components in both of these shapes as well as of 4 nodal diameter shape. Engine order 6 excites the strongest response in the 11th and 12th resonant peaks. In the perfectly tuned case this mode consists of all blades vibrating with equal magnitude, albeit at an alternating phase. Response to engine orders 4, 5 and 6 show the highest increase. The specific mistuning pattern predicts an increase in the maximum amplitude of the response to engine order 6, and excites the strongest response in the peaks between 220 and 240 Hz. The highest response is noted in blades 1 and 8.

7.13 LUBRICANT ANALYSIS

Monitoring the debris in bearing lubricants forms the backbone of oil analysis condition monitoring programs, and pertains primarily to the detection of metallic wear particles. The most common devices and techniques include atomic emission spectroscopy, atomic absorption spectroscopy, X-ray fluorescent spectroscopy, ferrography, magnetic plugs, magnetic chip detectors and microscopic examination of filter debris. The process is effective in determining if a system is nearing, or has reached, a failure mode. Further damage can then be averted through immediate shutdown and repair of a turbo-machinery. A modern oil analysis program takes the form shown in fig. 7.34.

A widely applied technique, spectroscopy provides a quantitative analysis to determine the concentration of as many as 20 different elements reported in parts per million. Iron, aluminum, chromium, copper, tin, lead, silver, titanium and nickel are detectable, as well as lubricant additives such as calcium, barium, zinc, phosphorus, magnesium, boron and molybdenum. Concentration trends are routinely monitored to indicate if a continuing wear condition exists, the rate of wear, and as a consequence, the urgency of the problem. The rotating disk arc emission is preferred for applications with large number of oil samples in many different types of environment. An effective program is dependent upon manual or computer-based interpretation of the analytical data, and is based on wear metal guidelines for the type of equipment.

A combination of variables arising from operating conditions and the environment must be taken into account.

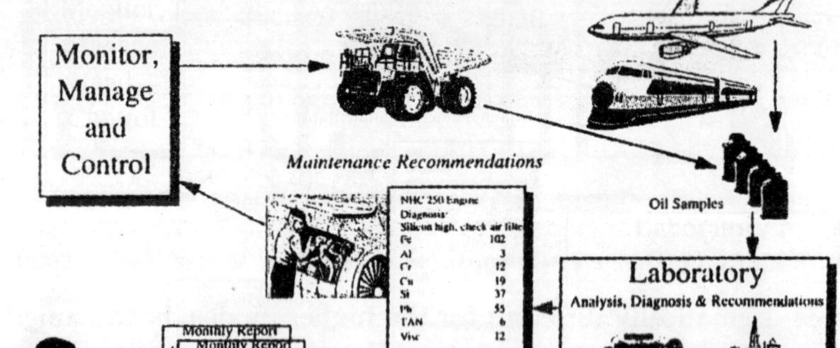

Fig. 7.34. *Lubricant Oil Analysis Program — Flow Diagram*

Thus, an increase in silicon readings and a corresponding rate of wear may be expected in an arid location. Elapsed time since last oil change and oil consumption will affect readings that may possibly mask a problem. The length of time the machine is in service is another important aspect. During the break-in period of an engine, either when new or after an overhaul, wear metal concentrations tend to be abnormally high due to rubbing in the inter-stage seals and other close contact areas, but are no cause for alarm. If equipment is left to stand idle over extended periods of

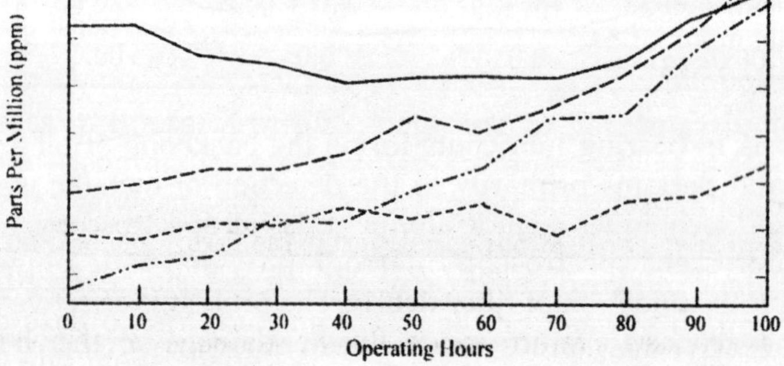

Fig. 7.35. *Spectrometric Trend Indicating Dirt Ingestion (Lukas, 1996)*

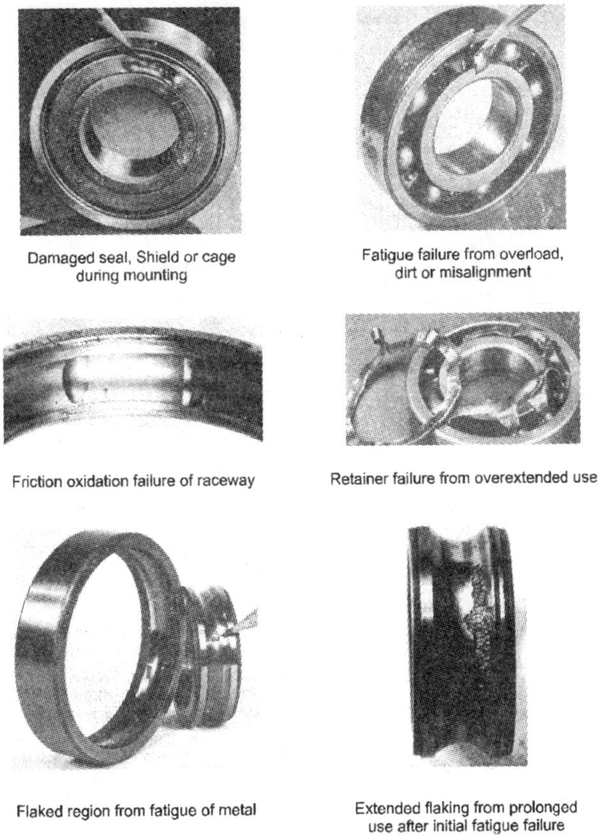

Fig. 7.36. *Examples of Bearing Failure (Morton, 1967)*

time, rust can form and iron readings will increase. Older systems will also generate more wear than fairly new ones of the same model. Load on the engine is also a factor, particularly during load changes. The chemical composition of the oil and coolant are important, with metals appearing not due to wear but from an oil additive or a coolant leak (Lukas *et al.*, 1996).

Since the spectrometer can differentiate between metals present in an oil sample, the readings can be used advantageously in a number of ways. For example, if only iron and aluminum are present, the job is simplified and the inspection can be restricted to components made of the metals. Knowing the relative concentration of the elements will further narrow down their possible source. Increase in levels of nickel and silver is indicative of bearing wear, and a relatively simple replacement of the bearing liners may suffice. Higher levels of silicon in conjunction with a corresponding increase in iron, aluminum and chromium is probably caused by dirt ingestion, and replacement of air filter and oil change may be the only maintenance actions required (fig. 7.35). A trend showing the presence of boron in a water-cooled system would indicate a coolant leak. If left unchecked, the coolant combines with combustion products and forms harmful acids that corrode metal.

Sometimes even a slight increase or presence of an element can be cause for alarm. The failed bearings shown in fig. 7.36 are removed from a number of machines. The presence of only 2 ppm of copper may be sufficient to warrant maintenance action. The source of the copper in this case was the bronze bearing cage.

7.14 BEARING ELEMENT DEFECT DETECTION

Condition monitoring of rolling element bearings offers the benefits of reduced downtime and improved maintenance efficiency, and calls for understanding effective techniques and instrumentation to monitor and diagnose problems. Signature analysis of the data from acceleration or velocity probes mounted on the bearing housing or machine casing permits observation of changes by the root-mean-square method (Monk, 1972). Shock pulse measurement (Boto, 1971) and spectral analysis (Mathew *et al.*, 1984) are also widely used for the diagnosis of bearing conditions.

Direct measurement of the outer race displacement is facilitated by employing an eddy current transducer, and is called the REBAM (rolling element bearing activity monitor) (*ref:* Bentlly, 1982*a*). The probe with its associated procedure permits detection of defects in the rolling elements and the inner and outer races, and is identified by spike signals on the time varying outer race deflection curve along with the known bearing frequencies. Defects on the inner race, moving relative to the probe tip as it contacts the balls or rollers, needs further investigation.

Experimental data from a test rig consisting of a rotor on a support bearing and the test bearing is used to obtain the results (Yu *et al.*, 2001). The dynamically-balanced shaft is 420-mm long with diameters ranging from 30 to 41 mm, and has two disks of 180-mm diameter and 25-mm thick. The assembly is powered by a variable frequency 750 W motor to allow for ramped data collection at a controlled acceleration, and is attached at the other end to a hydraulic pump. Radial load is applied on the test bearing vertically, and ranges between +2200 N and –2200 N. The transducer has a gain of 80 mV/μm. The test rig is mounted on a concrete foundation.

Table 7.3. ***Test Bearing Frequency Ratios***

	Ratio	*Ball Bearing*	*Roller Bearing*
i_{IRBP}	(Inner race ball passing frequency/shaft speed)	4.9	7.2
i_{ORBP}	(Outer race ball passing frequency/shaft speed)	3.1	4.8
i_R	(Rolling element spin frequency/shaft speed)	2.0	2.3
i_C	(Cage frequency/shaft speed)	0.38	0.39

Based on experimental data from good bearings, the approximate ratios of the test bearing frequencies to the shaft speed are given in table 7.3, although the values are susceptible to changes due to the effects of the defects. Thus, if a defect occurs on the inner race, the rolling elements will hit the defects i_{IRBP} times the shaft speed. The sum of the ratios i_{IRBP} and i_{ORBP} is equal to the number of rolling elements, regardless of changes in speed or slip. If there is a defect on the rolling element, the two races will contact the defect at a frequency of i_R times the shaft speed. The frequency ratio of the moving cage holding the rolling elements will be related to the detection of defects in the element.

A good bearing free of flaws is characterized by a smooth deflection curve in the time base. The eddy current transducer is mounted in a hole in the housing to detect elastic deflections of the outer race. The measured deflection has peak-to-peak amplitude of 1.65 μm, with 3.06 and 4.94 dominant periodic waveform cycles and their harmonics appearing for each shaft rotation as the element presses the outer race near the hole for the probe. The synchronous shaft speed signal is practically non-existent. When two balls are equally away from the hole, deflection in the outer race around the hole diminishes. Surface roughness, ball tolerances and ball undulations, however, will cause slight fluctuations on the deflection curve.

Transmission of load from the inner race to the outer race takes place through the rolling elements, hence defects in the inner race are sometimes mistakenly considered to be difficult to detect. An assumed single defective spot on the inner race will contact the rolling elements i_{IRBP} times during each shaft turn. The transducer detects the negative spikes when the moving defective spot is located within an effective zone near the transducer. Figure 7.37 shows the procedure for detecting an artificial spall made on the inner race located at the bottom of the bearing when the reference signal is triggered.

A negative spike occurs in the effective zone, and is marked $\blacklozenge_a$ in the time base. After $(360°/i_{IRBP})$ shaft rotation the defect spot contacts the next ball at $\blacklozenge_b$, but the probe does not detect the negative spike because the contact occurs beyond the effective zone. The same situation occurs at the following contact at $\blacklozenge_c$. The situation continues to exist until the defect spot contacts the *j*th ball, where *j* is a whole number for i_{IRBP} in the effective zone. Thus, two conditions must be satisfied for detecting negative spike for the spot on the inner race. The inner race defect spot must contact the balls, and the defect spot in contact with the balls has to be within the effective zone. The delay angle between the contact spots $\blacklozenge_a$ and Δ_a is given by:

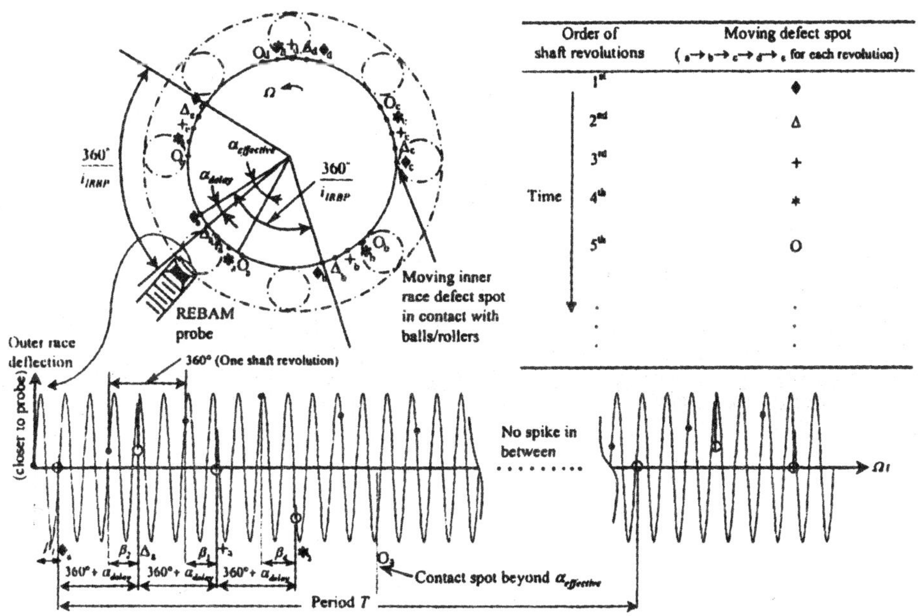

Fig. 7.37. *Detection of Defect on Bearing Inner Race (Yu, 2001)*

$$\alpha_{delay} = [Whole(i_{IRBP}) - i_{IRBP}] \times \frac{360°}{i_{IRBP}} \tag{7.5}$$

where a whole i_{IRBP} is the nearest integral *n*. The time period *T* is obtained by dividing *n* by i_{IRBP}, and the number of spikes is approximately equal to $\alpha_{effective}$ divided by α_{delay}.

Deflection response and spike signals will vary with bearing load and speed. As the speed increases, enhanced centrifugal force on the balls and expansion of the inner race cause a tighter contact between the balls and the races. As the loading increases, higher pressure is manifested between the balls and the races. The combined effect of the increased speed and load is to produce greater deflection and spike amplitude in the outer race, while also smoothening the deflection curve.

A defective spot on the outer race in contact with the rolling elements does not move against the probe when the outer ring is fixed to the casing. Figure 7.38 shows the case, with the rotor speed of 680 rpm and 1400 kPa hydraulic load. As the ball passes the defect spot the sudden release of pressure between the race and the ball leads to a negative spike on the outer race deflection curve, and the number of spikes matches the number of deflection cycles in the time base. The spike interval is given by:

$$\alpha_{OR} = \frac{360^\circ}{i_{ORBP}} \tag{7.6}$$

The fundamental frequency of the spikes is equal to that of the outer race deflection, or i_{ORBP} times shaft speed. The magnitude and direction of the spike are controlled by the orientation of the defect.

The second part of an effective oil analysis program is to determine the effectiveness and remaining life of the lubricant based on degradation and contamination. The number and type of tests performed on an oil sample depend on its characteristics and the machine in question. Physical property tests often rely on modifications of ASTM procedures, and include checks on viscosity, total base number, total acid number, water content, fuel dilution and insolubles. Oxidation by chemical reaction of the oil with oxygen at elevated temperatures of operation results in degradation of the oil. The product has an acidic condition and a sludge that causes the oil to thicken, and enhances the rate of wear of the components. Nitration is similar in some aspects, where nitrogen containing compounds combine with used oil to form oxides. The flow rate of a lubricant in relation to time is referred to as viscosity, and is an important property of the oil. An abnormally low viscosity reduces the strength of the oil film, and may lead to metal-to-metal contact. The ability to carry contaminants away from moving parts and to seal can also cause varying levels of degradation. A very high viscosity, on the other hand, also reduces the effectiveness of the lubricant, and accelerates wear of the components and creates harmful deposits.

Chemical additives are widely used to combat an adverse condition or to enhance lubricating properties. Some additives are required for efficient operation. The presence of an incorrect additive, however, may prove detrimental to the machine, hence the quantity and type must be continuously monitored. Additive depletion testing has also become an important part of oil analysis programs. Water in the oil is generally a result of coolant or anti-freeze leaks, or in the case of marine applications, from seawater. Excessive levels of water will increase wear, attack the bearing metal and affect the lubricating quality of the oil.

Several case histories are documented at the Nellis Air Force Base, Nevada, and are good examples of oil analysis for aircraft turbine engines:

- Spectrometric analysis results for a J79 engine out of an F4C Phantom aircraft indicated an increase in the level of aluminum content from 3 to 10 ppm. A teardown revealed that the inlet gearbox vertical driveshaft shims had torn loose, and metal particles were found in the filter screen.
- A TF30P3 engine out of an F111 aircraft did not show a drastic increase in the wear metal trend. The iron content increased gradually with operation, and oscillated near the upper allowable level threshold. When the iron level crossed the threshold over several hours of operation, the engine was recommended for tear-down. Upon disassembly, the discrepancy appeared in the form of a loose outer case in the tower shaft bearing, causing it to turn in its housing.

- Some spectrometric oil analysis results of the lubricant taken from an F100-100 engine out of an F15 aircraft did not initially indicate a trend. After 430 hours of operation the iron content jumped from 4 to 11 ppm from one flight to the next. A check sample and ground run gave the laboratory sufficient reason to request a tear-down. The number 4 bearing was found to be broken, and pieces were found in the sump. This is a good example of the program's capability to predict an imminent failure and prevent a catastrophic incident.

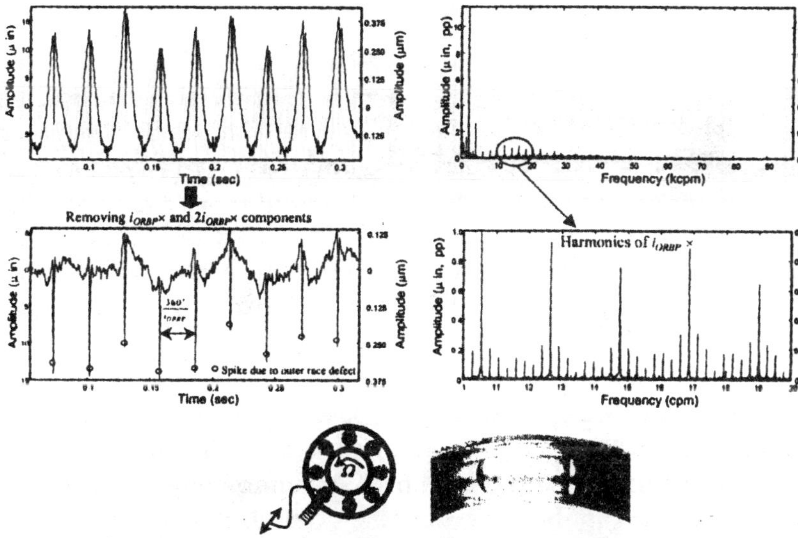

Fig. 7.38. *Detection of Defect on Bearing Outer Race (Yu, 2001)*

If a flaw is present in the ball, or roller, the contact takes place against the races at its spin frequency of $2i_R$ times the shaft speed. The probe detects a negative spike when the defect spot on the element contacts the cage rotating at a frequency of i_C times the shaft speed. The following expression holds when the probe detects the spikes after k contact events:

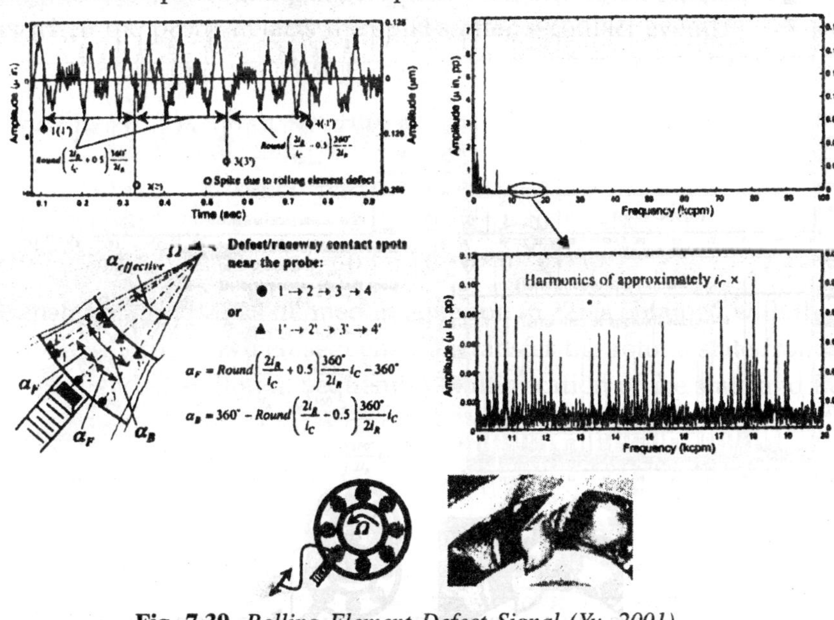

Fig. 7.39. *Rolling Element Defect Signal (Yu, 2001)*

$$\frac{360^\circ}{i_e} \approx k\frac{360^\circ}{2i_R} \tag{7.7}$$

where k is an integer. Figure 7.39 provides details of the signal for a defective roller, with the defect successfully identified in the time domain and in the frequency spectrum.

7.15 DETECTION OF CRACK IN ROTOR

Reliability and durability concerns make the presence of cracks in the rotor one of more than passing interest for the users of turbo-machines, but their early detection is not facilitated by the complexity of the problem. The existence of a 'breathing' crack introduces variations in the transient vibration signals of the rotating system, and that may be used as the basis for their detection. A crack influences the local stiffness and damping of the rotor and changes the dynamic behavior of the rotating system, with the effect similar to that of an external exciting force. For instance, a crack extending to a depth of 5 percent of the radius of the rotor can be detected by an observer-based method, but the procedure requires mathematical definition of the rotor and the crack.

Considerable effort has been expended in the research and understanding of the dynamic characteristics of a cracked shaft with the intention of developing effective monitoring procedures using vibration signals (Gasch, 1993). An interesting feature of the vibration signal of a cracked rotor is the increased amplitude of the second harmonic when compared with a faultless rotor. However, other problems such as shaft misalignment also exhibit similar traits (Imam *et al.*, 1989). By applying periodic axial impulses to a rotating cracked shaft model, Collins *et al.*, (1991) suggest that extra peaks in the spectrum represent a dynamic property of the rotor that may be used to detect the crack as the rotor passes through the critical speed (Plaut *et al.*, 1994).

Analytical evaluation of the composition of the vibration signal of a cracked rotor can provide an approach for detection (Zheng, 1997). The presence or absence of additional time and speed dependent stiffness and damping matrices in the mathematical model acts as a controlling factor in determining the influence of a fatigue crack. The system equation of motion is written as:

$$M(d^2X/dt^2) + C(dX/dt)+KX = \delta\Omega[\Delta K(\Omega t)(dX/dt] + (f(t,\Omega) \tag{7.8}$$

where M, C and K are mass, stiffness and damping matrices, ΔK and ΔC are the additional matrices produced by the crack, t represents time, Ω is the rotor speed and f is a function of exciting forces. $\delta(\Omega t)$ is 0 when the crack is closed and equals 1 when the crack is open. For a linear system this equation may be recast as:

$$dS/dt + [A + \delta\Delta A(\Omega t)]S \;=\; F(t) \tag{7.9}$$

where
$$S(t) = \begin{Bmatrix} X(t) \\ dX(t)/dt \end{Bmatrix},\; F(t)-\{M^{0-1}f\}$$

$$A + \delta\Delta A(\Omega t) = \begin{bmatrix} 0 & I \\ M^{-1}[K - \delta(\Omega t)H\Delta K(\Omega t)] & M^{-1}[C - \delta(\Omega t)H\Delta C(\Omega t)] \end{bmatrix}$$

and H is a matrix function to define the position of the crack. For a small breathing type of crack the equation is piece-wise linear for each state of the crack. If the crack closes after a time period t_o, the vibration signal can be expressed by:

$$S(t) = \sum_{i=1} U_i e^{\lambda_i (t-t_o)} + \sum_{i=1} V_i / (\lambda_i - j\Omega) e^{j\Omega t} + E \qquad (7.10)$$

where $\lambda_i = \zeta_i + j\omega_i$, ζ_i and ω_i are damping coefficient and natural frequency of the system and E represents the complex conjugate portion of the solution. When the crack opens in the time interval between t_o and T, equation (xxxxx13.39) is time varying and there is no solution. Assuming a small crack, the method of multiple scales (Nayfeh *et al.*, 1979) may be used to express the vibration:

$$X(t) = \sum_{i=1} U_i e^{\lambda_i t} + \sum_{k=1} \sum_{i=1} V_{ik} / (\lambda_i - jk\Omega) e^{jk\Omega t} + \sum_{m,n=1, m \neq n} \sum_{k=1} W_{mnk} / (\lambda_m + \lambda_n - jk\Omega) e^{jk\Omega t} + E \qquad (7.11)$$

where λ_i are the eigen-values of the matrix A, U_i, V_{ik} and W_{mnk} are coefficients determined by the system parameters and the initial conditions at the instant the crack begins to open and i, k, m, n = 1, 2, 3... For a large crack, the required matrix is obtained by using the rule of composition for the state transition matrix and by dividing the time interval into a sequence of smaller time intervals where $A + \Delta A(\Omega t)$ is constant. For a linear system, eventhough a steady system, the cracked rotor's vibration signal is made of transient and harmonic components. The transient vibration is determined from the vibrations at the instant when the crack opens or closes. Since the crack eliminates symmetry in the rotor, the initial condition of the transient vibration corresponding to a different 'breathing' pattern of the crack may be different. If the system is sensitive to changes in the initial conditions, the vibration generated by the crack may be chaotic.

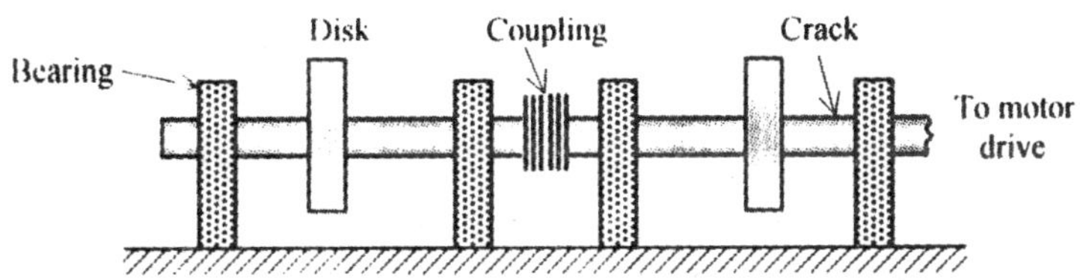

Fig. 7.40. *Test Rig for Cracked Rotor Vibration Evaluation*

Along with the continuous opening and closing, the crack brings energy to the system, causing the amplitude of the cracked rotor to increase (*ref*: Muller *et al.*, 1994). This energy is also dissipated with time, together with the oil film and other damping sources in the system. If the input and dissipating energies are in a state of balance, the vibration will probably be maintained at a certain level. But when the crack is large this balance will be destroyed, and the level of vibration will increase significantly, as occurred in the 963 MW turbo-generator rotor described by Schollhorn *et al.*, (1993).

A number of signal processing techniques are available for detecting transient components. Among them are the short-time Fourier transform, the Gabor transform and the wavelet transform. The transforms may all be considered as a class of windowed Fourier transform because their kernels are composed of a time domain window function and a frequency function. An interpretation of these methods is available in the envelope analysis, but they have the common problem of not having simultaneously high time and frequency resolution. The vibration signal of a turbo-machinery rotor is generally dominated by harmonics of the shaft speed and sub and super-harmonics produced by the non-linearity in the bearings. If these harmonics are not removed, the transient components may be masked because of problems in the resolution. The harmonic components may be removed with the aid of Fourier and inverse Fourier transforms, and the residue is a re-construction of the transient components produced by the crack, if there are no other forces that can excite transient vibrations.

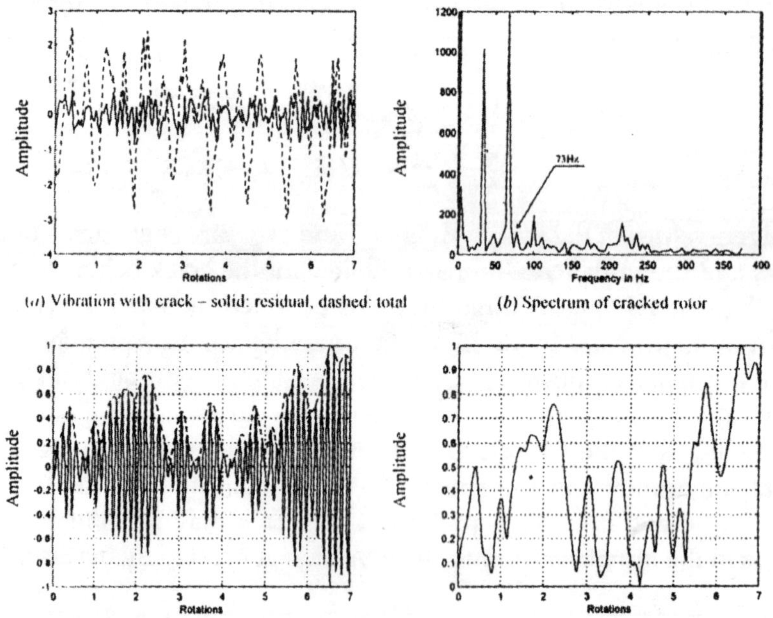

(*a*) Vibration with crack – solid: residual, dashed: total

(*b*) Spectrum of cracked rotor

(*c*) 73 Hz component – solid: frequency, dashed: envelope

(*d*) 73 Hz frequency envelope

Fig. 7.41. *Detection of Crack from Vibration Signature (Zheng, 1997)*

In actual situations, the signal may be contaminated by noise. Since the crack opens and closes mostly in a random manner, the time domain average method commonly used in machinery condition monitoring may not be suitable in eliminating the noise. An effective approach is to let the residual signal pass through a band pass or a low pass filter when most of the noise energy is in the high frequency range. The Gabor transform has an advantage in this respect since it includes the function of a band pass filter, if the parameter determining the width of the time domain is properly selected.

The above discussion is applied to a rotor-bearing test rig shown in fig. 7.40. Shaft diameter is 80 mm. A transverse fatigue crack, about a quarter of the diameter, is created on the shaft after 50 hours of continuous operation. The shaft is then dynamically balanced. Figure 7.41(*a*) shows the vibration signal measured at the far right bearing in the vertical direction, together with the residual signal. The signal contains some transient components in addition to the harmonic signals. The frequency versus amplitude spectrum is given in fig. 7.41(*b*). The system is at a steady 1800 rpm during the measurement. Figures 7.41(*c*) and (*d*) give the 73-Hz component and its envelope. There are two peaks present in each cycle of the envelope. The crack breathing process excites transient vibration in the system, but they are difficult to detect in the early stages of crack formation. The pattern of the envelope is important to the correct detection of a fatigue crack. Having at least two peaks for each rotation on the envelope of a transient component is an important feature of the vibration signal of a cracked rotor, distinguishing it from other vibrations produced by asymmetric shafts or rubbing between the rotor and the casing.

7.16 RADIATION THERMOMETRY METHOD

Surface temperature measurements play a major role in obtaining turbine airfoil heat transfer distributions, but the hot and moving target precludes the use of traditional contacting instruments. A radiometric thermometer offers some distinct advantages in this environment: a high level of working frequency can be achieved using a real optic sensor with an electronic conditioning system, improved sensitivity and accuracy and avoidance of thermal and fluid dynamic interference because of the non-intrusive nature of the scanning system.

Measurement errors arise mainly from the hostile environment in which the radiometric sensor must work, causing optical contamination of the signal reaching the infrared sensor. The constituents of the combustion gases and unburned fuel particles generate considerable numbers of emissive, reflective and scattering spectra. Mono-atomic gases mostly emit ultra-violet wavelengths. Symmetrically bi-atomic molecules (O_2, N_2) are transparent to infrared rays at low and middle temperatures, and the same happens to asymmetric ones (CO, NO), so their emission can be neglected. Tri-atomic molecules such as H_2O, CO_2, SO_2 and N_2O, on the other hand, are characterized by a strong emission in the infrared band, so their impact needs careful evaluation. Emitted and reflected radiation from the hot combustion gases and from associated mirror systems is a substantial source of error. The problem is best overcome by using filters capable of restricting the working wavelength to the field in which the influence of the hot gases is minimized. Interference is sometimes observed during transient events.

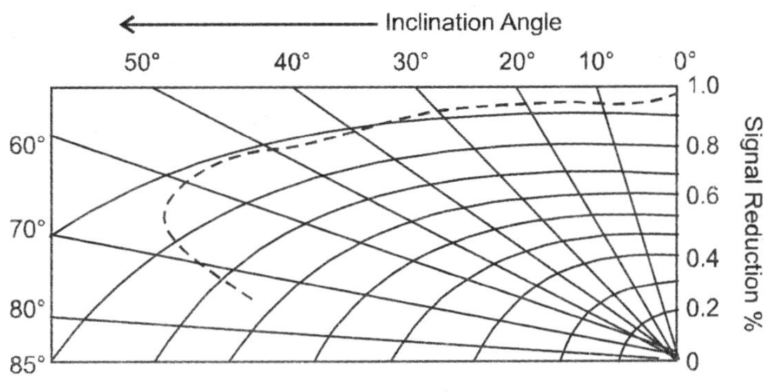

Fig. 7.42. *Signal Reduction Due to Inclination Between Target Surface and Thermometer Axis*

The signal may also be dumped when the target surface is excessively inclined with the thermometer's axis. In a perfect environment, a radiation thermometer measures the target temperature correctly when the surface is exactly perpendicular to the axis of the thermometer. The signal experiences a reduction as the axis is inclined, and at a certain point becomes unacceptable for scanning. A calibration curve of the type shown in fig. 7.42 may be used to obtain the correction, with errors becoming noticeable when the inclination exceeds 30°.

Transient events in radiometric scanning in gas turbines occur due to the sight of the flame coming from the combustion chamber and the carbon particles in the flow stream. The problem is more severe when the unburned particles with a high volume to surface ratio and in the process of finishing the oxidation process act as moving radiation sources. Their influence is significant since they over-evaluate the temperature, appearing as unreasonable and non-periodic peaks in the output signal. The problem can be fully eliminated by filtering out the non-periodic spikes.

When it is necessary to increase the instrument view area for a complete blade radiometric scanning, one or more mirrors can be used to divert the signal. But the presence of reflected signals causes some problems because of radiation emission and reflection by the mirrors themselves. The mirror must be cooled in order to reduce the emission radiation. Experiments indicate that the mirrors must be cooler than the target surface by 100°C to avoid all sensitive influence on the measurements. Figure 7.43 shows a comparison of the views obtainable with and without mirrors. As the rotor turns, the target area is moved along the blade's tip to the hub till a point is reached when the sight thermometer covers the next blade. By changing the angle of inclination of the instrument it is possible to move this point as near to or as far away from the hub as possible. In the passage between adjacent blades the measured temperature does not hold significance.

Incorrect behavior of the cooling system of a blade, superficial irregularity or local oxidation can result in a different temperature distribution between blades. Surface roughness increases the available area for heat transfer, but if the roughness is caused by deposition of low thermal conductivity material, the heat load and blade temperature may be reduced. If the roughness results from erosion, average blade temperature will increase due to the higher gas side heat transfer. The consequences of the surface roughness and its consequences on the heat transfer can be evaluated by obtaining the surface temperatures. An infrared camera system facilitates generation of a temperature map of the airfoil's surface.

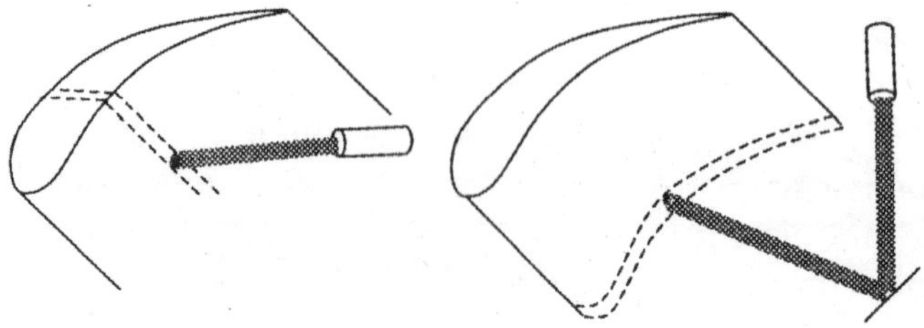

Fig. 7.43. *Sensor View without Mirror* (left), *with Mirror* (right) *(DeLucia, 1999)*

Heat transfer measurements using an infrared camera have been reported by Boyle *et al.*, (2000) for turbine stator vanes. The purpose of the test is to show the effects of surface roughness on heat transfer, and to quantify the results when surface roughness is important. Temperature difference across an insulating layer of unknown thickness at the surface would adversely affect the measurements, especially at higher Reynolds numbers. In experimental test setups accurate flow measurements over a range of Reynolds numbers are required. Consequently, while the physical vane roughness remains constant, the surface goes from being hydraulically smooth to fully rough as the Reynolds number is varied. The lower Reynolds numbers may not be representative of a typical high-pressure turbine application, but they are approached in low power engine applications.

7.17 LIFE PREDICTION FROM POST-SERVICE OXIDATION

The life of a cooled gas turbine blade is primarily limited by fatigue from thermal and mechanical loads and by surface degradation due to oxidation and corrosion. Peak load applications call for frequent startups and shutdowns, so thermal and mechanical load related fatigue is the most likely limiting mode of failure. But in combined cycle and co-generation base loaded plants oxidation and corrosion will govern the life of the turbine blades. In natural gas fired units, oxidation is the dominant mode of environmental attack.

Cooled blade airfoils are generally coated with a diffusion aluminide coating or a MCrAlY overlay coating. The latter consists of a NiCoCrAlY composition in varying amounts plus, in some cases small additions of Hf, Si, Ta or Re. Duplex coatings are also used where a diffused aluminide is applied over a MCrAlY coating or vice versa. Without a suitable coating satisfactory life cannot be achieved at the surface temperatures at which the blades are exposed.

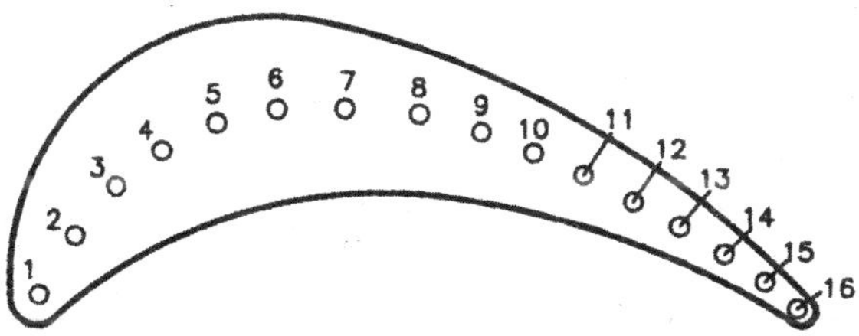

Fig. 7.44. *Turbine Blade Cooling Hole Pattern (Swaminathan, 1996)*

Experience indicates that cooling passages near the substrate surface also oxidize considerably during service. Internal aluminization processes have been developed to address this aspect of the problem. Oxidation of cooling hole passage is not as severe as on external surfaces because of the lower temperatures, but substantial surface layer and grain boundary damage could be caused by internal oxidation and by nitridation. Moreover, higher tensile stresses occur on the cooler surfaces inside the blade. These chemical processes reduce the mechanical properties and increase the risk of failure. Understanding the impact of such damage helps to evaluate structural integrity and the rate of wear (Swaminathan *et al.*, 1996).

Oxidation in the cooling hole of a first stage turbine blade after 32,000 hours of base-loaded service in a W501D5 model engine is examined, and the feasibility of using oxidation measurements to estimate thermal history is explored. Surface cracking noted on some holes is analyzed using elastic stress calculations and Neuber's rule to estimate local inelastic strain around the holes. The blades operate at a firing temperature of 2020°F for the first 16,000 hours, and after re-coat and re-heat treatment, at 2065°F firing temperature for the second 16,000 hours. The blade is made of Udimet 520. The pattern for the cooling holes is shown in fig. 7.44. The holes vary in diameter, and are placed along the camber line, or the line along the mid-thickness, of the airfoil. The holes are electro-chemically milled (ECM) from the tip to the base of the airfoil at one diameter and intersect larger diameter holes milled from the root end.

After the evaluation of the external coating and general condition of the blade, the cooling passages are examined by sectioning the blades at 10, 60 and 90 percent of airfoil height above the platform. Hole diameters are examined with an optical microscope, and their diameters measured at the three airfoil heights. Since the holes are not exactly round their minimum and maximum diameters are noted. The maximum depth of inter-granular oxidation is also measured and located. The holes are also examined for cracking, and the crack depths recorded.

The holes are generally round within 0.001 in. except for the holes at the trailing edge, where the variation is between ± 0.005 in. The holes are somewhat tapered, with the larger diameters occurring at the 10 percent section where the temperature is lowest and the incidence of oxidation is the least compared with the other two sections. Since the cooling holes are cooled to constant nominal diameter values, it may be inferred that the oxidation is primarily internal, and that little loss of metal is experienced. The oxidation at a typical cooling hole appears as uniformly scattered dark aluminum oxide particles and as needle-shaped light gray particles of titanium nitride.

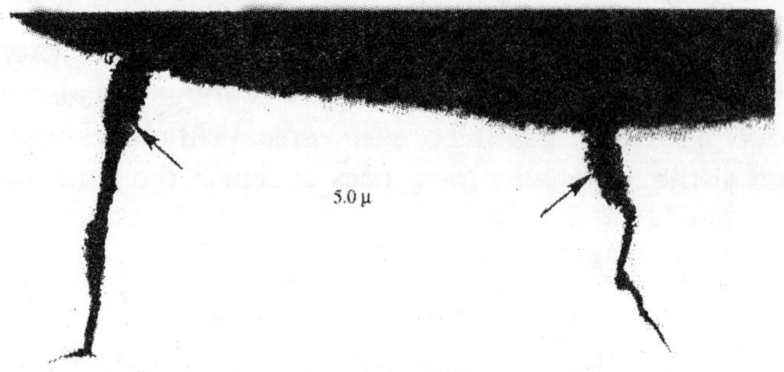

Fig. 7.45. *Inter-granular Cracking at Cooling Hole (Swaminathan, 1996)*

The depth of the oxidation and nitridation is insignificant at the 10 percent height due to the lower metal temperatures. At the 60 and 90 percent height sections the depth of the attack is substantial in the holes at the trailing and leading edges, where they are closest to the surface. At the 60 percent height cracking is observed in holes 3 to 7, with a typical crack shown in fig. 7.45 where the maximum crack depth is 0.009 in. The arrows indicate oxide deposits on the cracks.

It is known that internal oxidation and nitridation can cause embrittlement of super alloys (Chang, 1972), but data on the extent to which blade life is actually reduced is not available. Mechanical properties from similarly oxidized material under simulated turbine operating conditions are needed to answer this question. An important first step is correlation of oxidation measurements with the thermal history of the part. Oxidation measurements made in this study are used to predict surface metal temperature at the cooling holes, and the results compared with what is known about actual temperatures.

Assuming a parabolic depth of oxidation deposit formation with time, the metal temperatures may be predicted by the Arrenhius relationship between time and temperature:

$$[\log C - \log (d^2/t)][T + 460] = Q/R \tag{7.12}$$

where log C = a constant, d = depth of oxidation front, in., t = time, hours, T = temperature, degrees F, Q = activation energy, ft-lb/lb and R = gas constant, ft/R. For blade alloy Rene 80, constants C and Q/R are obtained from static furnace tests, providing the expression:

$$[4.300 - \log (d^2/t)][T + 460] = 25700 \tag{7.13}$$

Since the turbine operates at two different firing temperatures, and hence different metal temperatures, for the two 16,000 hour periods an equivalent operating time referenced to the higher firing temperature is computed using equation (7.6). The equivalent computed time is 26,400 hours.

Estimates of the temperature for the 60 and 90 percent height sections are shown in fig. 7.46, where the results are plotted against camber line distance of the hole from the leading edge hole. Polynomial regression fit of the data is made on the premise that the actual hole temperatures in the blade vary smoothly along this line, the standard deviation being 31.7°F at 60 percent height and 10.7°F at 90 percent height.

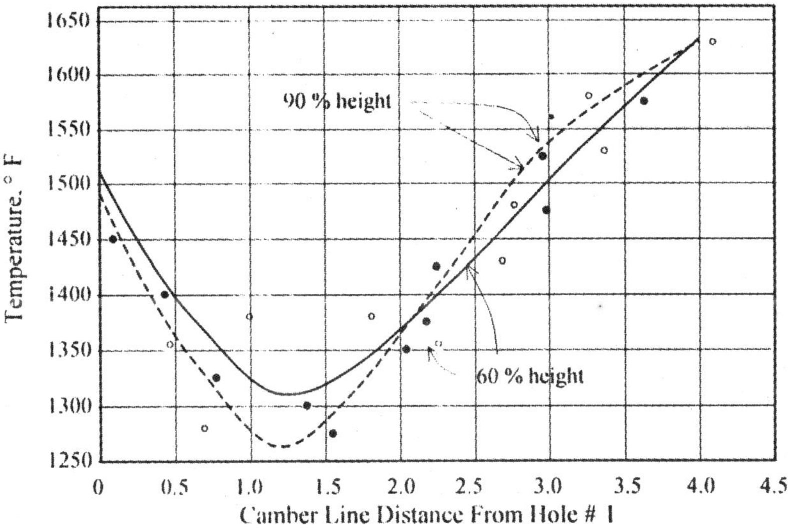

Fig. 7.46. *Predicted Metal Temperatures at Cooling Holes (Swaminathan, 1996)*

The observed hole cracking is evaluated using a nominal elastic stress distribution. Local strain at the edge of the holes is estimated from Neuber's rule and stress, strain and creep properties of the material. Maximum temperature difference through the metal occurs in holes 3 to 7 where high steady state elastic stresses occur in directions perpendicular to the cross-section. High local stresses arise from stress concentration at the holes. Peak values at the location of the cracks is a function of the ratio of spacing between the holes and their diameter and of the in-plane components of nominal stress perpendicular and parallel to the cracks. The consequent stress concentration factor is around 2.5 for the row of holes under bi-axial loading conditions, resulting in a peak elastic stress of 174 ksi.

Nominal stress S versus time at the edge of the holes is obtained from the equation for stress relaxation, assuming creep based on the power law (Timoshenko, 1956). Creep properties are based on a Monkman-Grant relation between creep rate and time to rupture (creep rate × time to rupture = 0.02) and stress rupture data (Udimet, 1999) at a temperature of 1300° F. From Neuber's rule (1961), local strain is given by:

$$\sigma\varepsilon = SeK_t^2 \tag{7.14}$$

where σ and ε are local stress and strain, S = nominal stress, e = nominal constant strain and K_t = elastic stress concentration factor. The power law between stress and strain, $e = kS^n$ and $\varepsilon = k\sigma^n$ where k is a constant and n is the law exponent, will yield the equation between nominal and local strain:

$$\varepsilon = eK_t^{2n/(n+1)} \tag{7.15}$$

To estimate local strain and then stress versus time at the edge of the hole, n = 4.2 when local stress equals yield stress (107 ksi), increasing to 6.3 as inelastic strain approaches total strain. Of the total strain of 1.3 percent the inelastic component is 1.2 percent, which may be split into 0.7 percent plastic strain and 0.5 percent attributable to creep relaxation. Under the given operating conditions of high temperature and observed oxidation, most nickel-based super alloys have inelastic strain components of the rather low value of around 1.2 percent. The cracks run longitudinally along the hole, probably arresting as they propagate into lower stress regions and as the thermal stresses relax from creep. Since they are not normal to the direction of centrifugal stress, the risk of blade failure from these cracks appears to be low.

Because the local strains are large the possibility of low cycle fatigue form of cracking needs to be considered. The blades have been cycled less than 50 times in service. From available material data the strain range required to cause LCF cracks is 3.6 percent. The actual cyclic strain assuming completely elastic behavior during cycling is estimated to be about 0.77 percent, based on 0.67 percent tensile for startup to full load and about 0.10 percent during shutdown. At this level of strain range the material's capability is 3300 cycles. Thus, LCF as a primary cause of failure may be ruled out.

7.18 LEAKAGE IN STEPPED LABYRINTH

Labyrinth seals are advantageous primarily due to their reliability and simplicity, as also because of their forgiveness of radial misalignment between the rotor and stator centerlines. Ease of manufacture and applicability over wide speed, temperature and differential pressure range are also features persuading turbo-machinery manufacturers to use them extensively to control the leakage between high and low pressure regions or to maintain a desired leakage to provide a thermal balance. Each throttling tooth in a labyrinth of any configuration provides a locally high kinetic energy at the throttle, a portion of which is dissipated. The non-dissipated portion is either diffused or constitutes the kinetic energy carry-over to the subsequent tooth.

Rounding of teeth tip occurs in labyrinths designed for a light wear, while heavier wear designs use a hardness coating on an initially rounded tip that changes little with rubbing. Rub grooves are considered representative for aircraft engines, with a nearly rectangular cross-section of 0.010 in. radial depth and width of 0.025 in. at the land surface, narrowing to 0.020 in. at the bottom of the groove.

The effect of wear of tooth tip and rub grooves on abradable labyrinth performance has been demonstrated in tests to substantially increase leakage (Rhode *et al.*, 1999). Current seal design methods do not attempt to account for the effect of rounded tips. This form of wear is commonly found, for example, in non-hardened teeth-on-stator aluminum labyrinths in centrifugal compressors. The rub may be encountered due to centrifugal and thermal growth, and exist in a wide range of pre-rub clearances ranging from 0.002 in. to 0.01 in. A water rig using two large flat plates is developed to visualize and measure the leakage, and is designed to represent 100× enlargement of an actual size seal with a radial clearance of 0.01 in. (fig. 7.47). The rig permits a stepped, interlocking or straight through configuration, with alterations made externally.

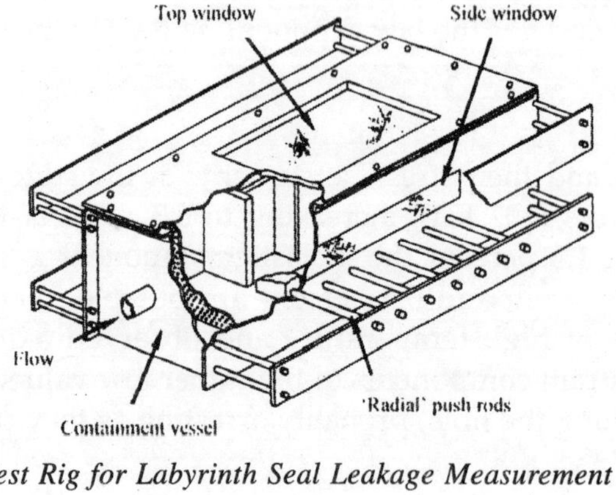

Fig. 7.47. *Test Rig for Labyrinth Seal Leakage Measurement (Rhode, 1999)*

The larger degree of scale up allows small tooth throttling details to be photographed. Relationship between the geometry and leakage is determined by adjusting various aspects of a given seal configuration and comparing the relative levels of turbulence and leakage. Tiny plastic glitter particles, sized less than 5 percent of the tooth clearance, are injected into the water mixed with fluoroscent sodium dye to trace the flow. The dye is illuminated by a thin sheet of light from a 5W laser. The flat plates simulate rotor and stator surfaces to represent circumferential flow around the shaft, and contain several adjustable seal elements to vary step height and radial clearance. Acrylic windows in the containment vessel shell allow visualization of flow.

Understanding of the leakage mechanism in a labyrinth seal is facilitated by using the Reynolds number $Re = 2\rho V_c c/\mu$ to quantify the leakage rate, where V_C is the bulk axial velocity based on the pre-rub tooth radial reference clearance c, and ρ and μ are fluid density and dynamic viscosity. Pressure differential ΔP in the labyrinth is represented by leakage resistance coefficient $K = 2\Delta P/(\rho V_C^2)$, and represents the energy loss due to the reduction in the pressure. Thus, K varies quadratically with c but only linearly with ΔP. Water is used in the investigation for easy visualization of important details.

With a small pre-rub clearance and a fixed step size, almost the same leakage resistance is measured for a medium and wider width rectangular groove and with the rounded medium width groove as the Reynolds number changes from 15,000 to 50,000. The non-grooved teeth, whether worn or not, have a much higher leakage resistance of $K = 1.0$ than do the grooved case, with K between 0.2 and 0.4. The flow pattern for a grooved worn teeth case is shown in the instantaneous digital image of fig. 7.48(a), where the clearance $c = 0.25$ in. and step height $s = 3.25$ in. The heavy dark lines are overlaid to approximate the time-mean through-flow jet boundaries, and the length of a particle trajectory line indicates its velocity. An unavoidable shadow occurs in a portion of the tooth-tip region, so the particles there are essentially invisible.

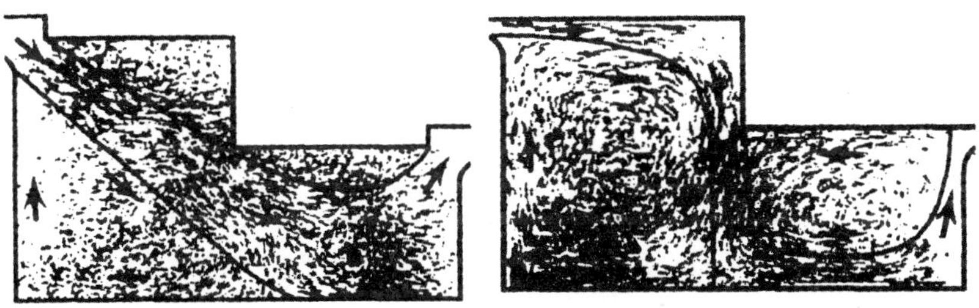

(a) Grooved Teeth (b) Non-Grooved Teeth

Fig. 7.48. *Leakage Flow in Grooved and Non-Grooved Worn Teeth (Rhode, 1999)*

Upon exiting the tooth throttling, the jet is immediately deflected toward the tooth surface, giving a large cavity entrance angle that may be attributed to the considerable stagnation pressure at the downstream end wall of the groove located slightly downstream of the tooth. The jet slightly bypasses the step corner and gives a highly non-perpendicular impingement on the toothed surface. Further, the jet exhibits a minimized kinetic energy carryover coefficient, producing three fairly large re-circulation zones of low turbulence intensity.

The overall flow pattern in a non-grooved worn teeth cavity is shown in fig. 7.48(b) for the same clearance and step height dimensions. The cavity entrance angle is essentially zero, and the jet impacts directly on the vertical surface of the land step. A perpendicular impingement of the jet on the toothed surface is found here. The pattern of flow yields a desirable long distance for traversing a cavity. This results in two large and relatively slow re-circulation zones with lower turbulence intensity than that found in the jet at the cavity entrance.

When the step size s is reduced to 1.75 in., the leakage resistance values are similar to that for the larger-step dimension. A higher resistance for the non-grooved case arises from the smaller effective jet throttling clearance and the altered cavity flow pattern. The grooved worn teeth configuration is similar to currently operating abradable seals, giving $K = 0.25$ at Re = 20,000 versus $K = 0.16$ for the corresponding large step size. If a perfectly sharp tooth tip can be achieved to give a grooved and unworn configuration, the value of K will rise to 0.35 at the same Reynolds number. If a perfect clearance control accounting for thermal and centrifugal effects could be installed at assembly, $K = 0.73$ would be the result.

Figure 7.49 provides the results for the medium clearance pre-rub $c = 0.75$ in. and large step height $s = 3.25$ in. case. The non-grooved cases have higher resistance for the same reasons as for the small clearance categories, and the non-grooved and unworn teeth case has the highest values. At Re = 15,000 the non-grooved worn teeth and the grooved unworn teeth cases exhibit a surprisingly substantial reduction in K. Further, the latter case is worse than the former case. The flow visualization pictures indicate that this finding is attributable to the overall flow pattern. When the step size is reduced to 1.75 in. the non-grooved worn tooth case gives a sharp drop in K at low Re. One distinction is that the expected laminar like rise in K, as found in Moody's pipe friction factor, does not appear here till the Reynolds number falls below 15,000.

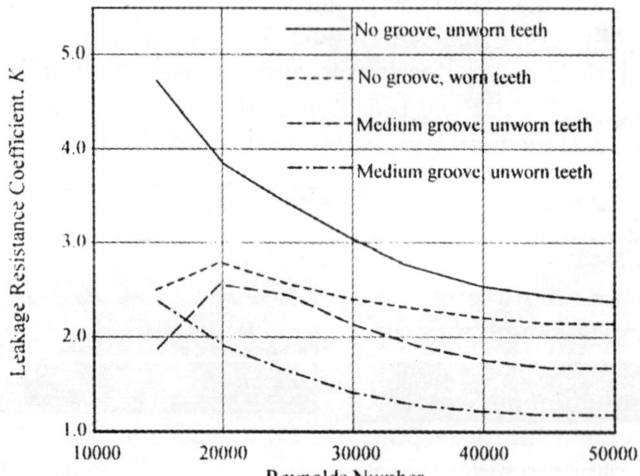

Fig. 7.49. *Leakage Resistance Coefficient versus Reynolds Number (Rhode, 1999)*

With a large pre-rub clearance of $c = 1.0$ in. and large $s = 3.25$ in., the trends are similar to the configurations shown for the medium clearance category. K values are generally higher. The grooved unworn teeth, the non-grooved unworn teeth and the non-grooved worn teeth cases exhibit K values of 2.3, 6.6 and 3.0. When a smaller step is employed the rank ordering and the overall trends remain similar to the medium clearance category case.

7.19 COMPRESSOR SURGE INVESTIGATION USING DIGITAL VELOCIMETRY

Improvement in the efficiency of turbo-machines calls for understanding the flow phenomena in the equipment. Besides the efficiency at design point, compressor stall requires close scrutiny.

Compressor stall is a catastrophic breakdown of the flow in a compressor that can lead to a loss of engine power, large pressure transients in the inlet and nacelle and can result in engine flameout. The distance on a performance map between the operating point of a compressor and its stall point is referred to as the stall margin. This margin is required to offset a loss of stability within the compressor caused by throttle transients and component deterioration with age. Optimal engine designs tend toward minimal stall margins, since modifications to increase the stall margin typically result in a heavier, less efficient and under-loaded compressor.

The instantaneous flow field capture capability of Digital Particle Imaging Velocimetry (DPIV) is well suited to the task of studying the change in flow conditions surrounding the development of stall precursors, stall cell propagation and eventual compressor surge. The technique provides near real-time flow field measurements through the use of refined data processing methods, combined with advances in the Charged Coupled Device (CCD) sensor technology.

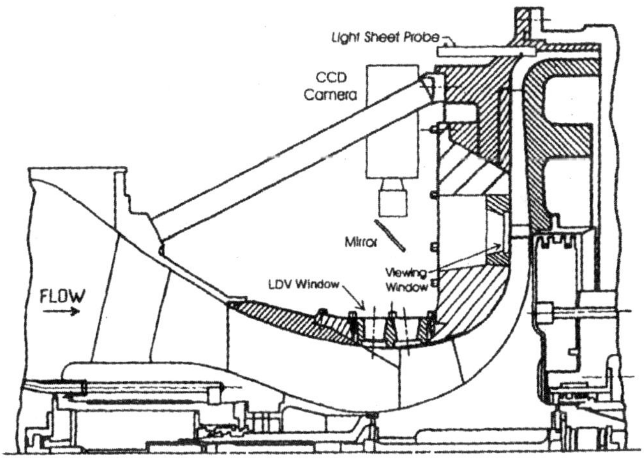

Fig. 7.50. *Centrifugal Compressor Test Facility (Wernet, 2001)*

Planar measurements are used with a pulsed laser light sheet to illuminate a flow field seeded with tracer particles small enough to accurately follow the flow. The position of the particles is recorded on a digital camera at each instant the light sheet is pulsed. In high speed flows pulsed Neodymium:Yttrium-Aluminum-Garnet (Nd:YAG) lasers are needed to provide sufficient light energy (~100 mJ/pulse) in a short time interval, < 10 ns, to record a clear image of the particles entrained in the flow. Data processing calls for obtaining either the average displacement of the particles over a small interrogation region in the image or of individual particles between pulses of the light sheet, from which flow velocity may be derived.

Numerous researchers have employed various velocimetry methods to study the unsteady flow in turbo-machines. Day *et al.*, (1998) studied the rotor-stator interactions in a low speed turbine facility. Braynston *et al.*, (1997) performed particle-tracking measurements of the high-speed flow in the stator trailing edge region of a transonic axial compressor blow-down facility. Dynamic pressure variations occurring in a compressor flow are used to investigate the stall in a scaled Rolls Royce-Allison design centrifugal compressor, schematically shown in fig. 7.50 (Wernet *et al.*, 2001). The casing accepts four window inserts interspersed along the flow path for optical access. The flow size is scaled to 4.54 kg/s from the original 1.66 kg/s. The impeller and vaned diffuser are designed to produce a pressure ratio of 4:1 at the design mass flow. Standard day corrected speed is 21,800 rpm, with an exit speed of 490 m/s. Inlet relative Mach numbers on the suction surface range from 0.9 at the tip to 0.45 at the hub, and absolute Mach number at the exit is 0.87. The impeller has 15 main and splitter blades of 50° back sweep from the radial direction at the discharge.

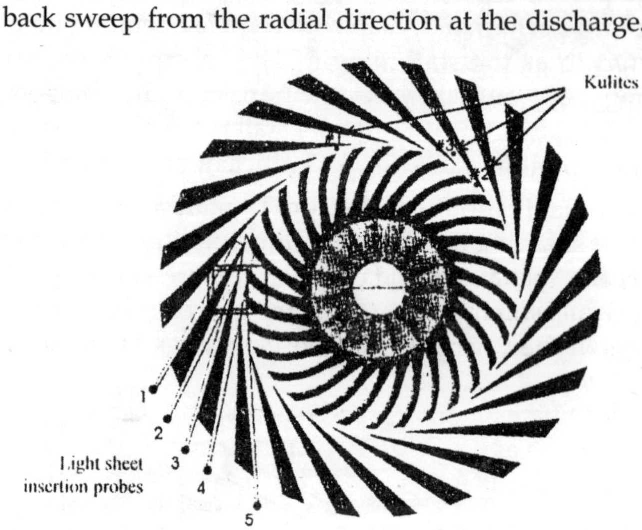

Fig. 7.51. *Vaned Diffuser and Impeller Schematic Drawing (Wernet, 2001)*

The splitter blade's leading edge, located at 30 percent of the main blade chord, is offset slightly toward the main blade suction surface to provide an even flow split. The vaned diffuser consists of 24 two-dimensional wedge vanes, with leading edges located at a radius ratio of 108 percent relative to the impeller exit. The diffuser has an overall area exit ratio of 2.75 with a total divergence angle of 7.8°. The nominal operating impeller exit clearance is 0.4 mm, or 2.4 percent of the passage height. The flow is subsonic under stable operating conditions, and the unit can be driven into stall by back pressuring the compressor through the collector valve at a mass

Fig. 7.52. *Velocimetry Image of Flow during Compressor Surge (Wernet, 2001)*

flow rate of 4.25 kg/s. Figure 7.51 shows a frontal view of the impeller and the diffuser. Four illuminated diffuser vane passages provide access to the suction and pressure sides and the space between the exit of the impeller and the diffuser. The casing is fitted with dynamic pressure transducers at four locations in the vaneless space and in the diffuser throat. To measure the incipient stalling mechanism, dynamic pressure readings over a matrix of mass flows and rotor speed combinations of the stall precursors are collected prior to the velocimetry recordings. The measurements indicate the rotating stalls travel at approximately 25 to 33 percent of impeller speed. The measurements note that below 70 percent speed the stall initiates in the impeller, and at design speed the stall originates in the diffuser.

The light sheet probe is suitable for obtaining the requisite 13 × 1 mm sheet with minimal scattering from the diffuser vanes and impeller blades. The normal seed particle concentration in the diffuser is around 8 particles/mm^3. During a surge incidence, the concentration increases by nearly an order of magnitude. A sample single exposure image illustrating this incidence is shown in fig. 7.52, where the diffuser vanes have been overlaid to highlight their locations. The impeller is located on the right side. The compressor is operating just above stall, and the camera field of view yields a spatial resolution of 61 μm/pixel. The time between laser pulses is 1.8 μs. Particle image displacements range from 0 to 16 pixels. Velocity measurements are obtained at five spans between the hub and the casing on both the suction and pressure sides. The compressor is driven into stall by restricting the flow at the downstream valve. The valve is reopened after the stall initiates to allow the compressor to recover. Between 3 and 5 cycles take place between the initiation and suppression of the surge. The dynamic pressure transducer measurements are used to reconcile the acquired image with the time evolution of the rotating stall and surge events.

A section of the signals from the # 3 and # 4 Kulites preceding and following the surge events is shown in fig. 7.53. The spiked signals represent laser trigger pulses.

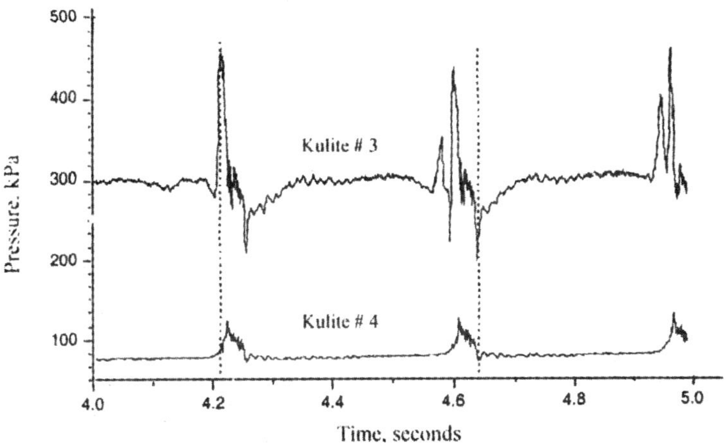

Fig. 7.53. *Digitized Kulite Signals During Surge Event (Wernet, 2001)*

The spacing between the laser firings is a little irregular because of the once per revolution signal from the compressor used to activate it. Surge events are marked by the larger pressure fluctuations. A surge event may be distinguished from a rotating stall by the simultaneous occurrence of the large fluctuation of pressure on all dynamic pressure transducers located in the diffuser. Stable operating point pressure in the diffuser throat is around 280 kPa, while the observed fluctuations during the surge range between a maximum of 540 kPa and a minimum of 170 kPa. After an initial large spike in pressure the increasing back pressure caused by closing the collector valve decreases the mass flow, resulting in flow separation on both the pressure and suction sides of the diffuser passage. Lower momentum fluid on the pressure side occurs downstream of the diffuser throat prior to the surge, reducing the effective flow area. The pressure in the diffuser increases until the impeller can no longer operate against the developed head. At this point, the high-pressure fluid in the diffuser rushes back upstream into the vaneless space, resulting in a sharp drop in pressure in the diffuser. The pressure then gradually recovers from the minimum back to the stable operating point pressure level before the next surge cycle begins. The frequency of the surge cycle is approximately 3 Hz. The impeller makes about 100 revolutions during the cycle. The stable operating pressure level before the surge is higher than following the event, since the valve is opened wider than the pre-stall position to allow quick recovery. Rotating stall cells are directly influenced by the rate at which the collector valve is operated, consequently, not much rotating stall information can be gleaned from the pressure data prior to the surge.

Evidence indicating that the fluid from the diffuser may actually reverse all the way up and out of the impeller is present in fig. 7.53. The traces show a disturbance at the impeller inlet about 10 ms after the pressure spike occurs in the diffuser. Total pressure at this point is atmospheric (100 kPa), but the unsteady static pressure peaking at 120 kPa is higher than the stable plenum static pressure of 83 kPa. If the flow exiting the impeller were to be stopped suddenly, as in the case of a blockage during a surge, then the pressure would not exceed 100 kPa. This is less than the measured peak static pressure of 120 kPa, indicating that the high-pressure fluid built up in the diffuser must have expanded back-out through the impeller.

7.20 FLOW IN ENGINE EXHAUST SYSTEM

With a growing trend towards higher efficiency combined cycle (steam and gas turbine) power plants, attention is focussed on both aerodynamic and thermodynamic management of gas turbine exhaust for improved performance. The diffuser section used to be mainly targeted for most of the pressure recovery, with most of the remaining dynamic pressure lost in the downstream ducting. For a given compressor discharge pressure and nearly ambient pressure at the inlet to the heat recovery steam generator (HRSG) unit in a combined cycle operation, higher pressure recovery in the exhaust system means lower pressure at the turbine outlet plane. This translates into higher overall pressure ratio across the turbine, giving higher power output and efficiency. A high-pressure recovery exhaust system generally also favors a nearly uniform flow velocity (or dynamic pressure) and low turbulence levels into the downstream silencer and HRSG units. Note that a peaky velocity distribution and high turbulence level adversely impact the performance of the silencer. The mass flow rate and flow angle at the inlet varies widely with the turbine load for the full spectrum of operating conditions.

The analysis of the three-dimensional diffusion in an exhaust nozzle is complicated by the presence of a number of geometric and flow parameters. Early work in delineating the diffuser flow regime found that the maximum pressure recovery in a straight walled structure with thin turbulent inlet boundary layer occurs when short duration flow reversal propagates up and down the diffuser walls, a condition known as transitory stall. In practice, exhaust systems are seldom designed for maximum pressure recovery under a transitory stall condition because of the small margin for separation of boundary layer at the wall. The latter, if it occurs, results in a marked deterioration in the diffuser pressure recovery performance. Pressure recovery normally increases when a rectangular diffuser is fitted with a straight constant area tailpipe instead of a plenum chamber (Kelnhofer *et al.*, 1971).

These considerations form the basis for an experimental and CFD investigation into a scale model of General Electric's MS9001E exhaust system (Sultanian *et al.*, 1998). Detailed flow measurements are made in a 1/10-size scale model to validate analytical predictions (fig. 7.54). The model includes an annular diffuser, 10 struts with an airfoil type of cross section located inside the diffuser passage, axi-symmetric turning vanes with curved surfaces to deflect the flow from an axial to a radial direction, a rectangular (310 mm × 526 mm) exhaust plenum and a stack.

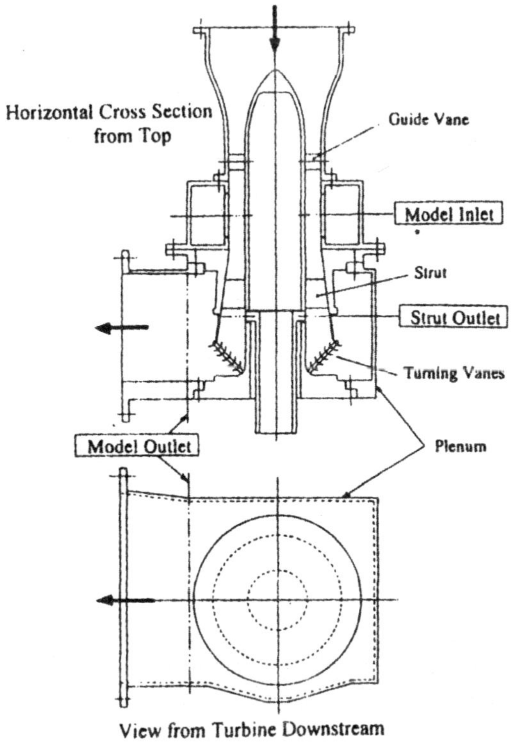

Fig. 7.54. *Exhaust System Test Model (Sultanian, 1998)*

The assembly is held together with 16-tie rods. The side exhaust receives discharge from the last stage turbine flowing axially into the annular diffuser, enters the rectangular plenum through the turning vanes and exits at 90° in the horizontal plane. Sixteen movable guide vanes are installed close to the inlet plane to generate the desired flow swirl. Compressed air (maximum of 3 atmosphere) is cooled in a heat exchanger and supplied to the test section through a control valve. The scale model is designed to simulate the complex flow field under all operating conditions, and includes detailed features of the system.

The tests are carried out for the engine operating at full speed no load, mid load and full load conditions. The mass flow rate varies from 3.9 kg/s at no load to 6.8 kg/s at full load, and the swirl angle at the inlet from – 50.8° at no load to 13.6° at full load. Inlet Mach number and swirl angle simulating the actual turbine outlet conditions are adjusted by controlling the mass flow rate and the upstream vane angle. Inlet air temperature does not simulate turbine exhaust condition since the program is mainly concerned with an aerodynamic investigation and CFD validation.

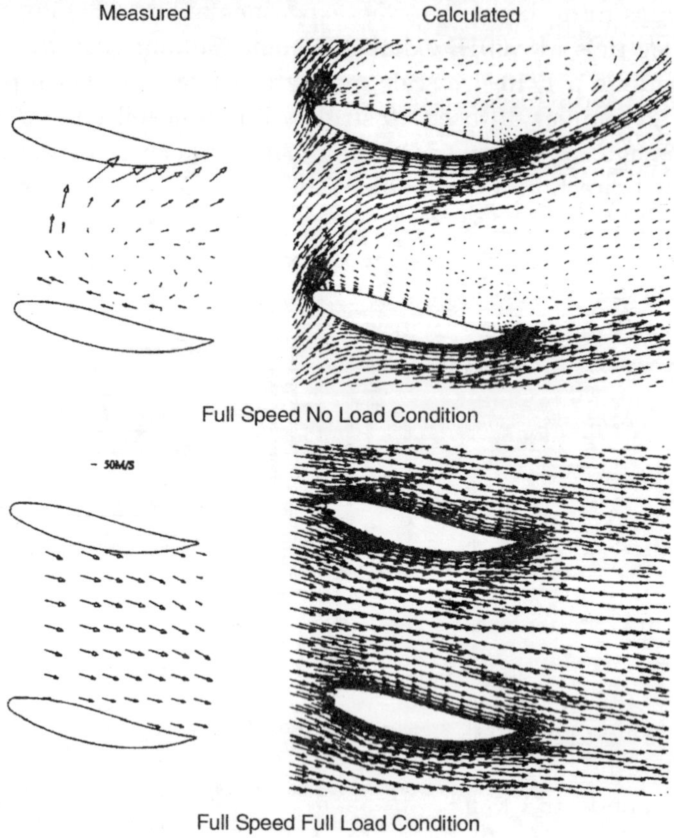

Fig. 7.55. *Flow Velocity Vectors Around Struts (Sultanian, 1998)*

The aerodynamic measurements are made at the inlet, around the struts and at the outlet. Five-hole pressure probes with a head of 2.4 mm are traversed to obtain the distribution of total and static pressures and pitch and yaw angles at the inlet and outlet planes. The pressure at the holes is detected by piezo-resistive transducers, with the probes calibrated in advance in a wind tunnel. Total temperature probes are also traversed in a similar manner. Rake probes with temperature and pressure heads are circumferentially traversed at the strut's outlet. A 5-watt argon-ion Laser Doppler Velocimeter (LDV) is used to measure velocity vectors around the struts and in the region between the struts. Water is sprayed upstream to produce 5 micron droplets for seeding in the velocimeter measurements. A hot film probe with a sensor is traversed at the inlet plane for measuring the distribution of turbulence intensity. Readings from the LDV and the pressure probes are compared to indicate agreement.

Analytical predictions of the CFD technology are based on three key selections: turbulence model, computational grid and a commercial Navier-Stokes code. A standard high Reynolds number k-ε model and an explicit non-linear algebraic stress closure model to predict an S-shaped duct flow are used in the investigation. The geometric complexity of the computational domain and accuracy favor a multi-block high quality structured grid. STAR-CD is selected as the solver. The governing conservation equations for both the mean flow and the k-ε turbulence model are discussed by Lakshminarayana (1996). Boundary conditions are prescribed at the inlet and outlet planes and at the solid walls. Detailed measurements of the mean flow and the turbulence field are used as inlet boundary conditions for all three load conditions. At the outlet plane conditions corresponding to zero axial gradient for all the dependent variables are used. For the internal surface walls, struts and turning vanes, the standard wall functions are employed (*ref*: Launder *et al.*, 1974).

Figure 7.55 provides the velocity vectors around the diffuser strut at mid-span from the analysis and the test. At no load the flow approaches the strut's leading edge with a swirl angle of –50°. This leads to a large flow separation region covering the entire pressure surface and extends further downstream. The position of the center of the secondary flow vortex from the analysis coincides with that obtained from the measurement. At full speed, the incoming flow has negligible angle of incidence with the strut's leading edge, then moderately turns toward the axial direction in the passage without creating a flow separation.

Calculated static pressure distribution on the surface of the struts also agrees with measured values. In comparing total pressure contours in the plane of exit from the struts, at no load full speed condition the total pressure is high downstream of the suction side as the flow is deflected and low downstream of the separated reversed flow region, as seen in the left fig. 7.56. A comparison with the measurements around the circumference at the strut exit plane indicates the flow field is well predicted. The maximum pressure is measured at 123.8 kPa as opposed to 129.0 kPa calculated value, while the minimum measured value is 90.1 kPa and minimum calculated value is 94.3 kPa. For the full load full speed condition the maximum measured and calculated values are 116.3 kPa and 118.1 kPa and the minimum measured and calculated values are 98.8 kPa and 100.3 kPa. At full speed the high-pressure region corresponding to each passage between the struts almost connects circumferentially with that from the adjacent passage, suggesting that the flow velocities are fairly large even in the wakes, except in the root and tip areas.

The calculations indicate two large counter-rotating vortices are present in the upper and lower halves as the flow field exits the rectangular exhaust plenum for the no load condition. The direction of the vortices matches that of the secondary flow generated by the turning from the diffuser to the plenum outflow. The calculated velocity vectors between the vortices differ from the experimental data that exhibit greater symmetry between the upper and lower halves. For the full load case the velocity vectors in this region assume a more horizontal direction in both the predicted and test measurements, and the two vortices come closer to being symmetric.

During no load operation the high-pressure region along the right end wall of the exit plenum spans from the top to the bottom. Weak and broad pressure peaks are noticeable in the low velocity regime from the center to the left. As in the no load case, for the full load condition the flow pattern is reproduced by the calculations. Several large and weak pressure peaks are detected along the right end wall in the upper and lower halves corresponding to the low velocity regions.

The variation in the total pressure loss and static pressure drop as the load changes from 0 to 100 percent is depicted in fig. 7.56. The plots are created for the front portion between the inlet to the strut edge downstream and for the full passage from the inlet to the exit at the exhaust plenum. The total pressure loss and the static pressure drop decrease considerably near the mid operating load condition. Better agreement is seen between the calculations and the experimental values in the front part for all load conditions.

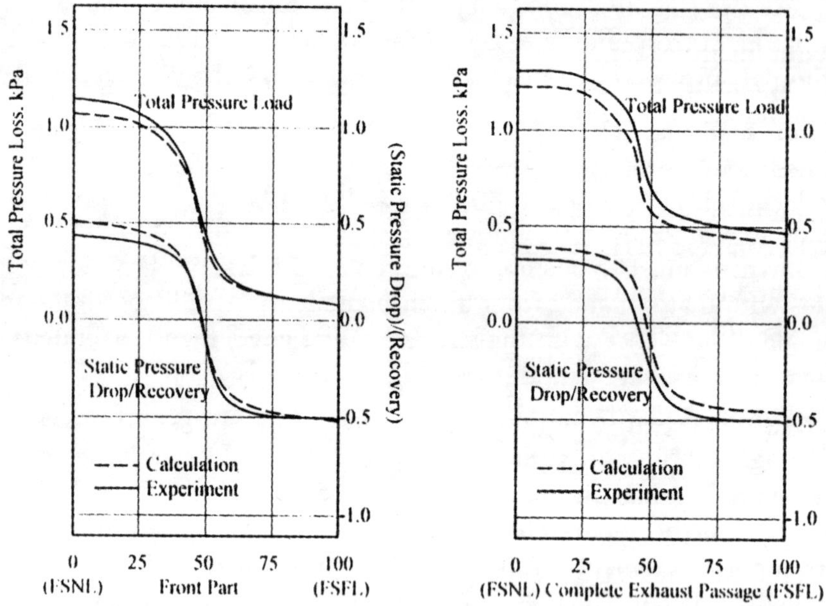

Fig. 7.56. *Variation of Pressure Loss with Load (Sultanian, 1998)*

Even with the larger discrepancy in the exit plane, the calculations tend to under-predict the total pressure losses and static pressure drops when compared with the measurements. A large static pressure drop is encountered in the front part at no load, indicating the diffuser does not function at full efficiency. For the mid and full-load conditions, large negative pressure drop, or positive pressure recovery, is obtained in the diffuser, suggesting a satisfactory diffuser performance. Pressure recovery is maximized when the exhaust system operates at full-load condition.

Most of the total pressure losses in a gas turbine's exhaust system are associated with the secondary flows in the stalled regions. Only a small part of the overall loss can be attributed to friction in the boundary layers at the wall. Capturing the flow behavior away from the wall using a more advanced turbulence model to accurately simulate effects of streamline curvature and turbulence anisotropy due to flow separation and widely varying swirl could further bring predictions closer to measurements.

REFERENCES & BIBLIOGRAPHY

American Petroleum Institute, "Special Purpose Gear Units for Refinery Service", API Std. 613, 4th edition, December, 1994.

American Petroleum Institute, "Centrifugal Compressors for General Refinery Service", API Std. 617, 5th edition, April, 1994.

American Petroleum Institute, "Centrifugal Pumps for General Refinery Service", API Std. 610, 8th edition, December, 1994.

Bentley, D.E., "Polar Plotting Applications for Rotating Machinery", Vibrations Institute, Clarendon Hills, ILL, 1982.

Bently, D.E., "Monitoring Rolling Element Bearings", Orbit 3, No. 3, Pages 2–15, 1982 *a*.

Bishop, R.E.D., Parkinson, A.G., "Second Order Vibration of Flexible Shafts", Proceedings Royal Society of London, 259:1-31, 1966.

Boto, P.A., "Detection of Bearing Damage by Shock Pulse Measurement", Ball Bearing Journal, SKF, 1971.

Boyle, R.J., Spuckler, C.M., Lucci, B.L., Camperchioli, W.P., "Infrared Low Temperature Turbine Vane Rough Surface Heat Transfer Measurements", ASME Paper # 00-GT-216, New York, 2000.

Bryanston-Cross, P.J., Burnett, M., Lee, W.K., Udrea, D.D., Chana, K., Anderson, S.J., "Transonic Particle Image Velocimetry Measurements Made in the Stator Trailing Edge and Rotor Region of the ILPF at Pyestock Farnborough", Proceedings, SPIE Conference on Optical Diagnostics in Fluid and Thermal Flow, Volume 3172, Pages 561–574, 1997.

Chang, W.H., "Tensile Embrittlement of Turbine Blade Alloys After High Temperature Exposure", Super-alloys, Page V-1–V-41, 1972.

Chi, R.M., Jones, H.T., "Demonstration Testing of a Non-interference Technique for Measuring Turbine Engine Rotor Blade Stresses", AIAA/SAE/ASME/ASEE 24th Joint Propulsion Conference, AIAA-88-3143, Boston, 1988.

Collins, K.R., Plaut, R.H., Wauer, J., "Detection of Cracks in Rotating Timoshenko Shafts using Axial Impulses", ASME Journal of Vibration and Acoustics, Volume 113, Pages 74–78, New York, 1991.

Compressed Air and Gas Institute, "In-service Standards for Centrifugal Compressors", Cleveland, OH, 1993.

Cooley, J.W., Tukey, J.W., "An Algorithm for the Machine Calculation of Complex Fourier Series", Mathematics of Computation, 19(90):297–301, 1985.

Day, T.K., Lawless, P.B., "Particle Image Velocimetry of Vane-Rotor Interaction in a Turbine Stage", AIAA Paper # 98-3599, 1998.

DeLucia, M., DeSabato, R., Nava, P., Cioncolini, S., "Temperature Measurements in a Heavy Duty Gas Turbine Using Radiation Thermometry Technique: Error Evaluation", ASME paper # 99-GT-311, New York, 1999.

Den Hartog, J.P., "Mechanical Vibrations", 4th ed., Mc-Graw Hill, New York, 1985.

Dilzer, M., Gutmann, C., Schulz, A., Witting, S., "Testing of a Low Cooled Ceramic Nozzle Vane Under Transient Conditions", ASME paper # 98-GT-116, New York, 1998.

Frarey, J.L., "Pitfalls in the Analysis of Machinery Vibration Measurements", Sound and Vibration, Pages 18–24, September, 2002.

Frischbier, J., Schulze, G., Zielinski, M., Ziller, G., Blaha, C., Hennecke, D.K., "Blade Vibrations of a High Speed Compressor Blisk-Rotor–Numerical Resonance Tuning and Optical Measurements", International Gas Turbine and Aerospace Congress, Birmingham, UK, June, 1996.

Gasch, R, "A Survey of the Dynamic Behavior of a Simple Rotating Shaft with a Transverse Crack", ASME Journal of Vibration and Acoustics, Volume 160, Pages 313–332, New York, 1993.

Griffin, J.H., "Optimizing Instrumentation When Measuring Jet Engine Blade Vibration", Transactions of the ASME, Journal of Engineering for Gas Turbines and Power, Volume 114, Pages 217–221, New York, 1992.

Heath, S., Imregun, M., "A Survey of Blade Tip Timing Measurement Techniques for Turbomachinery Vibration", ASME paper # 97-GT-213, New York, 1997.

Hydraulic Institute, "Acceptable Field Vibration Limits for Horizontal Pumps", 14th edition, Centrifugal Pump Applications, 1994.

Imam, I., Azzaro, S.H., Bankert, R.J., Scheibel, J., "Development of an On-line Rotor Crack Detection and Monitoring System", ASME Journal of Vibration, Acoustics, Stress and Reliability in Design, Volume 111, Pages 241–250, New York, 1989.

ISO 2372, "Mechanical Vibration of Machines with Operating Speeds from 10 to 200 rps–Basis for Specifying Evaluation Standards", 1989.

ISO 2954, "Mechanical Vibration of Rotating and Reciprocating Machinery–Requirements for Instruments for Measuring Vibration Severity", 1984.

ISO 4867, "Guidelines for Evaluation of Vibration in Merchant Ships", after ISO 6954, 1993.

ISO 6945, "Guidelines for Evaluation of Vibration in Merchant Ships", 1987.

Jay, R.L., Burns, D.W., "Characteristics of Diametral Resonant Response of a Shrouded Fan Under Prescribed Distortion", ASME Journal of Vibrations, Acoustic, Stress and Reliability Design, Volume 108, Pages 125 – 131, New York, 1986.

Judge, J., Pierre, C., Mehmed, O., "Experimental Investigation of Mode Localization and Forced Response Amplitude Magnification for a Mistuned Bladed Disk", ASME paper # 00-GT-358, New York, 2000.

Kelnhofer, W.J., Derick, C.T., "Tailpipe Effects on Gas Turbine Diffuser Performance with Fully Devoted Inlet Conditions", ASME Journal of Engineering Power, Volume 93, Pages 57–62, New York, 1971.

Kielb, J.J., Abhari, R.S., "Experimental Study of Aerodynamic and Structural Damping in a Full Scale Rotating Turbine", ASME paper # 2001-GT-0262, New York, 2001.

Kruse, M.J., Pierce, C., "Dynamic Response of an Industrial Turbo-machinery Rotor", Proceedings 32nd AIAA/ASME/SAE/ASEE Joint Propulsion Conference and Exhibit, Lake Buena Vista, FL, 1996.

Lakshminarayana, B., "Fluid Dynamics and Heat Transfer of Turbo-machinery", John Wiley & Sons, New York, 1996.

Launder, B.E., Spalding, D.B., "The Numerical Calculation of Turbulent Flow", Computer Methods in Applied Mechanics and Engineering, Volume 3, Pages 269–289, 1974.

Lukas, M., Anderson, D.P. "Lubricant Analysis for Gas Turbine Condition Monitoring", ASME paper # 96-TA-54, New York, 1996.

Mathew, J., Alfredson, R.J., "The Condition Monitoring of Rolling Element Bearings Using Vibration Analysis", ASME Journal of Vibration and Acoustics, 106, Pages 447–453, New York, 1984.

McCarty, P.E., Thomson, J.W., "Performance Evaluation of a Prototype Non-interference Technique for Measurement of Turbine Compressor Blade Stress", Report # AEDC-TR-80-5, Arnold Air Force Station, TN, 1980.

Military Standard, "Mechanical Vibration of Shipboard Equipment (Type 1 – Environmental and Type 2 – Internally Excited)," MIL-STD-167-1 (SHIPS), May, 1984.

Mignolet, M.P., Choi, B.K., "Robust Optimal Positioning of Strain Gauges on Blades", ASME Turbo Expo paper # GT-2002-30454, Amsterdam, 2002.

Monk, R., "Vibration Measurement Gives Early Warning of Mechanical Fault", Process Engineering, Pages 135–137, November, 1972.

Morton, H.T., "Anti-friction Bearings", Hudson T. Morton Publishing Co., Ann Arbor, MI, 2nd edition, 1967.

Muller, P.C., Bajkowski, J., Soffker, D., "Chaotic Motions and Fault Detection in a Cracked Rotor", Non-linear Dynamics, Volume 5, Pages 223–254, 1994.

Muller, D., Sheard, A.G., Mozumdar, S., Johann, E., "Capacitive Measurement of Compressor and Turbine Blade Tip to Casing Running Clearance", ASME Journal of Engineering for Gas Turbines and Power, Volume 119, Pages 877–884, New York, 1997.

Nayfeh, A.H., Mook, D.T., "Non-linear Oscillations", John Wiley & Sons, New York, 1979.

Neuber, H., "Theory of Stress Concentration for Shear-strained Prismatic Bodies with Arbitrary Stress-Strain Law", ASME Journal of Applied Mechanics, Volume 28, Pages 544 – 550, 1961.

Plaut, R.H., Andruet, R.H., "Behavior of a Cracked Rotating Shaft During Passage Through a Critical Speed", ASME Journal of Vibration and Acoustics, Volume 173, Number 5, Pages 577–589, New York, 1994.

Purcell, T.E., "Dynamic Stress Analysis of Gas Turbine Rotor Airfoil Using Thermo-elastic Techniques", Experimental Techniques, Pages 9–16, May 1996.

Rhode, D.L., Allen, B.F., "Measurement and Visualization of Leakage Effects of Rounded Teeth Tips and Rub Grooves on Stepped Labyrinths", ASME Paper # 99-GT-377, New York, 1999.

Schollhorn, V.K., Ebi, G., Steigleder, K., "Frettinganrisse in Einen 936 MW Generator", VGB Kraftwerkstechnik, Volume 73, No. 4, Pages 340–344, 1993.

Sensmeier, M.D., Nichol, K.L., "Optimum Placement of Sensors for Vibration Measurements on Turbine Engine Blades", AIAA, 98-1849, Tennessee, 1998.

Sheard, A.G., O'Donell, S.G., Stringfellow, J.F., "High Temperature Proximity Measurement in Aero and Industrial Turbo-machinery", ASME paper # 97-GT-198, New York, 1997.

Special Metals Corp., "Udimet 520 Alloy Performance Data", New Hartford, 1999.

Sultanian, B.K., Nagao, S., Sakamoto, T., "Experimental and Three Dimensional CFD Investigation In A Gas Turbine Exhaust System", ASME paper # 98-GT-111, New York, 1998.

Swaminathan, V.P., Allen, J.M., Touchton, G.L., "Temperature Estimation and Life Prediction of Turbine Blades Using Post Service Oxidation Measurements", ASME paper # 96-GT-528, New York, 1996.

Szwedowicz, J., Senn, S. M., Abhari, R.S., "Optimum Strain Gauge Application to Bladed Assemblies", ASME Turbo Expo paper # GT-2002-30306, Amsterdam, 2002.

Timoshenko, S., "Strength of Materials: Part II", D. Van Nostrand, 3rd edition, New York, 1956.

Udimet 520 Alloy Performance Data, Special Metals Corporation, New Hartford, NY, 1999.

Wernet, M.P., Bright, M. M., Skoch, G.J., "An Investigation of Surge in a High Speed Centrifugal Compressor Using Digital PIV", ASME Journal of Turbo-machinery, Volume 123, Pages 418 – 427, New York, 2001.

Winkler, A.F., "High Speed Rotating Unbalance, Coupling or Rotor", Proceedings, Vibration Institute, Nassau Bay Conference, Clarendon Hills, Ill, 1983.

Yu, J. J., Bently, D.E., Goldman, P., Dayton, K.P., Van Slyke, B.G., "Rolling Element Bearing Defect Detection and Diagnostics Using Displacement Transducers", ASME paper # 01-GT-028, New York, 2001.

Zablotsky, I.Y., Korostelev, Y.A., "Measurement of Resonance Vibrations of Turbine Blades with ELURA Device", Energomashinastroneniye, Volume 2, Pages 36–39, 1970.

Zheng, G.T., "Vibration of a Rotor System with a Switching Crack and Detection of the Crack", ASME Journal of Engineering For Gas Turbines and Power, Volume 120, Pages 149–154, New York, 1997.

CHAPTER 8

Balancing of Rotors

8.1 INTRODUCTION

A rotor is considered balanced if it rotates about its principal axis without wobbling, assuming that there is no dynamic motion due to the elasticity of the rotor and the lubricant in the bearings. When the journal axis of rotation does not coincide with its principal axis, a net radial force acts on the rotor due to centrifugal acceleration. The process of effecting essential coincidence between the principal inertia axis of the rotor and the shaft axis is called dynamic balancing. Besides unbalance, there are other sources of vibration in a rotating machine, some of which may be aggravated by unbalance under certain circumstances, and so may be of significance. Balancing of rigid rotors (one that does not experience distortion) is important because it comprises the majority of balancing work done in the industry. When machine operation takes place in a speed regime where bearing support stiffness at both ends is essentially rigid, shaft elasticity also comes into play. The overall problem in which dynamic action and interaction of rotor elasticity, bearing and support elasticity and damping are taken into account is called *flexible rotor*, or *modal, balancing*.

Advanced design capabilities have supported the continuing emphasis to build rotating machines with larger capacities. Larger capacity designs often entail longer and more flexible rotors capable of operating above a number of critical speeds. Safe operation through these critical speeds necessitates a good balance to control the amplitudes of vibration and bearing reacting forces. Even small mass unbalances can result in significant inertia effects because of the high speed at which the machines operate. Under certain conditions, high levels of vibration can impose severe stresses on the moving parts and on the supporting structure. Hence, it is essential that all major rotating components be as fully balanced as possible.

Although a rotor may have been designed and manufactured with great care to ensure proper balance under working conditions, small residual unbalances will almost always remain in the finished product. Inherent non-homogeneity of the material and unavoidable machining tolerances are some culprits of this condition. Such discrepancies can be detected and rectified to a high degree of precision by using balancing machines. If the rotor is short and stubby, static balancing may be acceptable. A comparatively long and slender shaft, on the other hand, will call for dynamic balancing. In its simplest form a balancing machine consists of three main components, namely, a variable speed driving unit, a balancing head and a light platform in the form of a cradle to support the work piece to be balanced. The platform is mounted on springs, and is capable of oscillating about either one or two horizontal axes located in the balancing planes of the rotor. The balancing head is a device mounted on the platform's structure for determining the weight and angular position of the required balancing masses. This type of balancing machines is out of favor, but newer electronic machines work in a similar manner.

Possible causes of rotor unbalance arise because of varied reasons during manufacturing and assembly. Eccentric mounting of a disk on a shaft will cause detectable run-out in the form of a slow roll when the assembly is mounted on knife edges and as the center of gravity seeks the low position. Dimensional inaccuracies lead to a lack of symmetry and a noticeable run-out. Eccentric machining or lack of precision in forming will also cause the same problem. An inadequately tightened bolt can permit slippage between the components, and this can cause the vibrations to reappear because of angular movement. Shrink fitted pairs require heating of one component to obtain thermal growth relative to the mating dimension of the other part, but excessive or improper heating procedures may lead to permanent deformation of some of the features. Non-homogeneous structure of a component in the form of sub-surface voids in a casting can lead to considerable vibrations in the bearings, even if the rotor is machined concentrically.

Once a unit goes into operation and accumulates a number of start and shutdown cycles and operating hours, unbalance develops from many different sources for a number of reasons. Accumulation of foreign particles on compressor and turbine airfoils invariably tends to take an asymmetric pattern, with some sectors experiencing heavier deposits. Non-uniform erosion of compressor blades from sand particles in the incoming air stream can lead to a similar vibration problem. Differential thermal expansion can cause a shaft to bend to throw out the center of gravity, and is a major source of considerable vibrations. Condensing and vaporizing fluids due to changes in the process cycle trapped inside a hollow rotor can also cause the vibrations to reappear after adequate balancing as the whirling fluid causes the center of gravity of the system to shift. Wear of the elements of a rolling element bearing will cause eccentric orbiting of the rotor at a multiple frequency of that of the shaft.

The balancing procedure for correction balance of a rotor after it is placed in service may be considerably different from the initial process employed during its manufacture. Commercial balancing machines using either the soft bearing or the hard bearing concept are used to balance individual rotating components and the total assembly. Field balancing of the machine in place avoids the time consuming job of removing the rotor to be reworked and re-balanced at the factory. The object of trim balancing is to add correction weights in one or more planes along the length of the rotor to maintain acceptable levels of vibration.

Guidelines to classify rotor types and to determine the required quality of balancing are provided by International Standards Organization (ISO 1940, 1973). Depending on the dynamic behavior in a given operating speed range, a rotor is considered rigid if the strain or potential energy in the supporting bearings exceeds 80 percent of the total system's energy. The rotor may then be balanced in two selected planes for satisfactory operation. If the shaft's energy exceeds 20 percent of the total, and if the rotor is operating through one or more critical speeds, then the system may be classified as flexible. A two-plane balancing procedure may then not be adequate, and additional correction weights may be required along the length to control the vibrations in the full operating speed range. The classification also takes into account the quality of balance needed for the specific machine configuration and service application (*ref*: Rieger, 1986). For example, a symmetric multi-stage centrifugal compressor rotor with two-end correction planes and operating speed range between the first and second critical speeds is considered to be a quasi-flexible rotor of ISO class 2. An aircraft engine's axial flow compressor has an asymmetric rotor with a controlled initial unbalance and two transverse correction planes that may be treated in a similar manner. If the components are individually balanced before final assembly as well as at final assembly, then even long and flexible rotors may operate successfully without additional adjustments in the field. But for class 3 generator rotors it is necessary to further balance it at high speed in a spin facility. This is usually the case when the rotor operates through a number of critical speeds, and the balance procedure is elaborate and time consuming.

8.2 ASSEMBLY AND OPERATIONAL PROCEDURES AND ROTOR VIBRATIONS

Most rotors are assembled from a number of components, and the danger for high vibrations exists from many sources. One method of controlling relative concentricity between rotating parts is to use rabbets, sometimes with a shrink fit. Splines and curvic couplings are other methods. In all cases the assembly needs to be locked up with a single tie bolt or multiple individual bolts at some radial location. A normal force is made available at the contact surfaces, and because of friction the parts are prevented from relative motion. Internal forces, such as thermal conditions, fluid flow forces and rotor unbalance act to dislodge the mating parts from their seating. If the tension in the bolts is not large enough to overcome the unseating forces, a permanent shift in the relative position of the parts could result in a shift in unbalance.

Any relative motion at the contact surface results in destabilizing energy dissipation, which can lead to unstable vibration of the rotor. The effect is referred to as hysteretic whirl. Axial asymmetry may be caused when keyways are used to transmit torque between components in an assembly, causing dissimilar bending stiffness about two orthogonal planes. This can result in unstable operation in a rotating machine.

A rotor operating in a temperature varying condition is subject to unbalance.

Thermal conditions can cause problems in two ways. In a rotor assembled with components made of different materials, variance in the coefficient of thermal expansion will cause unequal growth. In addition, if thermal gradients are present along the radial or axial directions in the machine, thermally-induced growth or shrinkage will be even more severe.

Significant thermal distortions are known to occur in steam and gas turbines when the machine is started or when shutdown. Gravitationally forced cooling air tends to cool the lower half of the machine at a faster pace than the half above the shaft centerline. This causes substantial thermal gradients and unequal shrinkage in a rotor, particularly if the span between the bearings is large, and thermal bow is the consequence. To avoid this problem the rotor is kept spinning at half the speed of full-load condition for a period of time through a turning gear mechanism. If a machine is restarted during a cooling period when the machine is not at a uniform temperature, similar thermally-induced problems may be expected.

The design of a rotor must include accommodation for introduction of balance weights, both during manufacturing and in the field. During production, even when individual rotating parts or sub-assemblies are dynamically balanced, the need arises to balance the assembled rotor. The final trim balance is done in several ways. At planes selected in the machine from engineering analysis balance weights are placed, which may take the form of washers under the nut. On turbines with known blade weights and inertia properties, the balance may be executed through exchange of blades at different tangential positions. Disks with substantial outer rims can have selected amounts of material removed at a location determined by the balance machine. This procedure has the drawback that once the material is removed it will be final, hence adequate care must be taken before deciding on this method. Also, the ground edges should not lead to stress risers where a fatigue failure might initiate.

8.3 BALANCE METHODS

At low speeds a shaft supported in flexible bearings does not bend much, but as the lowest natural frequency of vibration is approached the bending distortions cannot be ignored. The theory of balancing is based on the premise of canceling out motions in principal modes, as many of them as necessary, by the use of compensating weights. The stiffness of the bearings is assumed to be independent of frequency, since they are essentially massless springs. The modal approach to balancing has won almost complete acceptance. Although the original modal balancing theory did not require the use of a balancing machine, the newer ones do. Balancing machines can be exceedingly costly devices, especially if they are built to accommodate large rotors.

Balancing machines may be used at the factory where individual components are balanced and the full rotating assembly balanced as a rigid rotor, even when the rotor is of considerable proportions and may operate through a critical speed. Many other rotors, such as for an electrical generator, operate through a number of critical speeds, and may require further balancing. Balancing is often done within the machine's own bearings with displacement measurement probes and strobe lights to record the dynamic motion.

The single-plane, or one-shot, balancing method offers the advantage of simple and quick application on machines in service. The process requires a trial mass to be placed at a known radial location on the rotor. Dividing the change in the measured response by the magnitude of the trial weight provides a useful vector quantity called an *influence coefficient* at the specific speed. The magnitude of the vector represents the amplitude of vibration and the phase angle between the forcing function and the displacement. Phase measurements can be obtained from directly reading non-contact or displacement probes, velocity transducers or accelerometers. An electronic key-phasor or strobe light is also used to obtain the phase angle on the shaft. The application of a single mass in an assumed linear system theoretically will nullify any dynamic motion in the shaft running at that speed. But in the case of a long rotor with a number of disks other points may experience higher vibrations. At other speeds also high vibrations may be encountered, representing a lack of balance.

The vibrations may be recorded at the ends of the shaft after the placement of the trial mass in the selected plane. The readings are then added and subtracted in the vector form to determine the static and dynamic components. If the average of the static components is greater than the average of the dynamic components, the weights are mounted on the rotor in phase. If the dynamic components are greater, the two masses are located 180° out of phase.

Since the single-plane coefficient suffers the disadvantage of smooth operation at only one speed, the least square error procedure may be employed to control the vibrations at other speeds. The vibrations are recorded over a range of speeds. The magnitude of balance weights determined from linear regression is smaller than the value calculated at one speed. The objective here is to achieve reduced vibrations over a range of speed rather than practically suppressing the amplitudes at one speed.

When the single plane balancing is not sufficient, for example in long machines operating at variable speeds, application of two masses in two planes may be considered. After attaching a trial weight in one plane, the response is measured in the same and a second plane. Influence coefficients are then determined. The weight is removed and placed at the second plane, and the procedure repeated. A 2 × 2 matrix of complex influence coefficients, equivalent to a 4 × 4 matrix of real numbers, is thus obtained, which is inverted. The inverted matrix provides the amount of balance weights and relative phase angle locations for the two selected planes.

A more generalized influence coefficient procedure requires measurements at two or more planes at a number of speeds, and does not focus on reduced vibrations at a single speed as in the two plane balancing procedure. A best fit of the data is then determined to obtain the desired balance weight data. The method is preferred in the case of large turbine and generator rotors where knowledge of the rotor dynamic characteristics is not available, and is successfully applied even when passing through a series of critical speeds. The process may be refined by using a weighting function to emphasize the readings from one set and ignoring those from another set. The balance planes must be carefully selected in order to avoid development of excessive and impractical balance magnitudes in the adjacent plane 180° out of phase. Computer programs may also be used for multi-plane balancing to combine the task of least square error method for obtaining the influence coefficients, with a limit placed on the magnitude of the correction weights.

Critical speeds and mode shapes of flexible rotors are generally available from the manufacturer, or may be determined during operation. The weights placed on the rotor are then in proportion to the deflections, and the distribution is used to balance the modes individually. Modes displaying high amplification can then be balanced without upsetting the balance in other modes. The corrections to be applied are calculated by the influence coefficient method from the measured response. For the method to be successful it is assumed the peak dynamic activity can be measured and a balance weight placed at the specific location. Complications may also arise in determining the precise balance weights and their phase angle when two modes are closely spaced in frequency. When an accurate measurement of the phase angle is not possible, the three trial weight, or the four run, method has been suggested (Blake, 1967). As an example, shaft speed does not remain steady in the presence of aerodynamic interference in a fan. Trial weights are attached at three locations. A locus of balance locations may be developed to graphically determine the magnitude and phase angle of the required correction weight. The procedure may also be extended to flexible rotors (Foiles *et al.*, 1982). After placement of the weight distribution, maximum amplitude of vibration is recorded on a spectrum analyzer at each natural frequency. Individual modes are then balanced without disturbing the response at other critical speeds. Flexible rotors experiencing considerable magnification can benefit from this procedure when smooth operation cannot be attained by following other methods. The mode shapes must be known prior to usage, and three runs must be executed for each critical speed to be balanced.

If the influence coefficients of a rotor are available, the Nyquist plot of rotor amplitudes to calculate the amplification factors at a given mode can make trial weights unnecessary. The modal unbalance eccentricity at a particular speed can be obtained directly from the amplification factors, and the polar Nyquist plot helps in calculating the phase angle and modal weight distribution at each specific speed. In this method also the mode shapes and their critical speed, as also the rotor modal mass, must be known. The procedure may be followed by the influence coefficient method to obtain further fine-tuning.

8.4 MOTION OF UNBALANCED ROTORS

Consider two cases of a rotating structure in soft bearings, as shown in fig. 8.1. In one case the rotor has an unbalance mass at its mid plane, causing the rotor to be statically unbalanced. As the shaft turns, centrifugal force on the mass will cause the centerline of the shaft to travel in a circular path. In the other case the rotor has two unbalance masses at opposite ends, and are offset 180° to each other. The two unbalance weights will offset each other statically. Also, as the shaft rotates, centrifugal force on the masses will cancel each other, however, an overturning moment will be present due to the axial offset of the masses. The resulting trajectory of the shaft's centerline will be in the form of two cones, and the apex of the cones will be at mid plane due to the symmetric nature of the rotor and the unbalances. When the two forms of unbalance are combined, the apex of the cones will be moved away from the center of gravity.

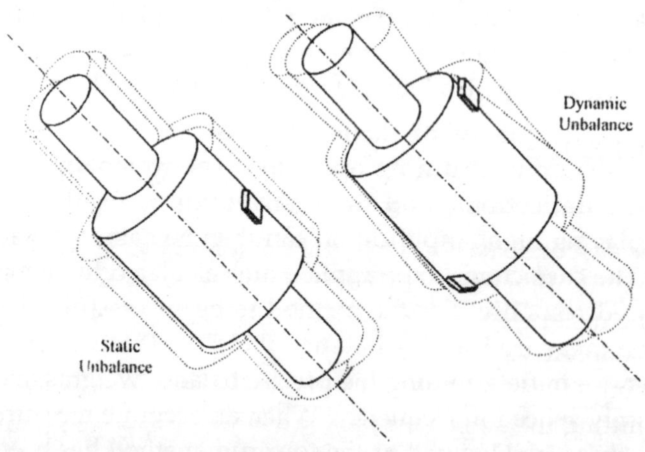

Fig. 8.1. *Static and Couple Forms of Unbalance*

More generally, an apex will not be present, because the shaft will move in a complex combination of motions shown in fig. 8.1. A random combination of axial and radial magnitudes of vibration and phase angle location of the static and couple unbalance will lead to a condition referred to as dynamic unbalance.

Flexibility in the supports of the rotor will add to the motion of the rotor in the lateral directions. When support flexibility at the two ends is not symmetric, the lateral motion will also have differing characteristics, as shown in fig. 8.2.

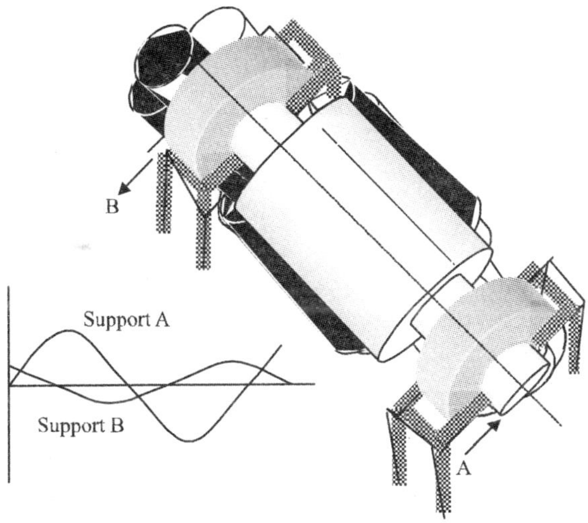

Fig. 8.2. *Rotor Motion in Flexible Bearings and Supports*

Here it is tacitly assumed that the rotor is essentially rigid, with the dynamic activity confined to the supports and no flexure present in the shaft.

To understand the effects of unbalance on a flexible rotor, consider a single mass disk mounted on an elastic shaft. The mass center of the disk is displaced from the center of the elastic shaft, resulting in an unbalanced rotor. The total rotor effective unbalance is equal to the product of the disk's mass and the unbalance eccentricity. The rotating unbalance force on the disk causes the shaft centerline to precess, or orbit, about the geometric centerline. When the bearing supports are rigid, the motion may be referred to as synchronous precession with a circular orbit. If this rotor is mounted on flexible bearing supports, total motion of the disk will include motion of the end supports.

8.5 CORRECTION METHODS

Static balance of a work piece may be performed using gravity. The principle is illustrated in fig. 8.3, where a disk is suspended at its geometric center by a string. Orientation and amount of the disk's tilt gives an indication of the balance mass to be placed to bring it in the horizontal plane. If the weight of the disk precludes this method, it may be mounted loosely at its hub on a vertical mandrel. The balance weight may be split into two equal halves and placed on opposite faces to reduce overturning effects from centrifugal forces during operation. When the disk is attached to a shaft, permitting the shaft to roll on horizontal knife edges gives an indication of unbalance location when it comes to rest.

Balance correction may be done by adding weights or removing material. In special cases the shaft axis may need relocation. The selected procedure must ensure that adequate capacity exists to permit additional correction, because initial correction estimates may not permit reduction within tolerance limits. When carried out carefully, common methods usually result in achieving a 10:1 reduction. Under more closely controlled conditions for weight and angular position, reduction in the range of 20:1 is not uncommon. If the maximum initial unbalance does not bring the rotor within permissible limits by a selected method, a second correction procedure may be required to reduce the residual unbalance.

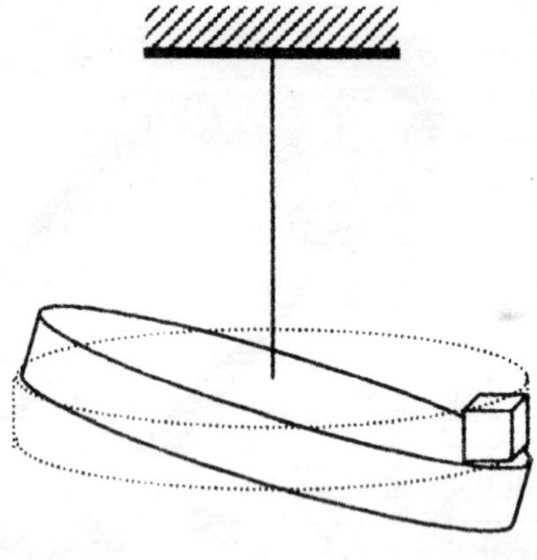

Fig. 8.3. *Static Balance Using Pendulum Principle*

Correction masses may be added in more than one way. It is sometimes difficult to apply wire solder while ensuring that its center of gravity is at the desired location. Errors may also be induced by variations in wire diameter. Bolted or riveted washers are useful if only moderate balance quality is required. Cast iron or lead weights in incremental weights provide large initial unbalance correction, if suitable design features in the rotor permit adequate fastening means. Resistance welding may be used for attaching large correction weights if localized thermal distortion in the rotor from the heat of the welding is kept to a minimum.

Material may be removed from a rotor by drilling a hole to a measured depth for unbalance correction. A depth gauge and switch may be used on the drill spindle to assure intended material removal with a high degree of accuracy. Milling, shaping and fly cutting permit accurate removal if the surface of the rotor from where metal is to be removed is machined, permitting proper measurement of the size of cut. The method is preferable when larger correction is called for. Grinding is essentially a trial and error method of correction, because it is difficult to estimate the amount of metal removed. The method is restricted to design of rotors that do not permit other means of correction. A good example is the shaft connecting the low-pressure turbine with the fan module in aircraft gas turbine engines. The shaft is thin walled, axial spacing between bearing centerlines is large and aircraft operating cruise conditions may call for operation close to the first shaft bending mode of vibration. Since both ends of the shaft are splined, the only provision for balance correction is land areas where metal is removed through grinding.

Location of the two radial planes is not important from mathematical considerations, since the overturning couple does not take into account axial location of the balance planes. Practical considerations call for the planes to be located axially as far as possible to gain greatest leverage, thus minimizing the magnitude of the masses. Balance planes must also be easily accessible for attaching the correction weights, especially if the operation is to be done in the field. A typical approach is the provision of an otherwise non-functional rib to which correction masses may be added, or an annular protrusion in the balance planes from which material can be machined or ground away at the appropriate circumferential location. Since assembly and disassembly is not necessary, the method is advantageous. The process has been adapted to automatic balance correction where material is removed from the rotor by laser melting while rotating in the balance machine. Metal removal procedures are subject to the limitation that balance correction cannot be undone when rebalance of the rotor is required. Also, care must be exercised to avoid introduction of stress risers in highly stressed areas of the rotor in the course of removing the material.

If run-out measurements indicate that a rotor's geometric centerline does not coincide with the axis of rotation, a procedure known as *mass centering* may be useful in reducing unbalance effects. The rotor is mounted in a balanced cage or cradle, which itself is installed in the balancing machine. The rotor is adjusted radially with respect to the cage until the unbalance indication is within limits. The procedure causes the principal inertia axis of the rotor to be brought into essential coincidence with the axis of the balanced cage. Center drills guided along the axis of the cage provide means for establishing an axis in the rotor about which it is in balance. Subsequent machining operations on the rotor, however, may nullify some of the benefits of the mass centering procedure.

An individual turbine stage assembly with blades in place is best balanced by exchanging the location of two blades from their initially assigned dovetail slots. Each blade on the turbine disk is pan weighed before assembly. After a run on the balance machine, vibration indication from the sensor and weight and mass moment of inertia distribution of the blades around the wheel will permit identification of a pair of blades to be switched from their location. More than one such sequence of events may be needed to bring the assembly within tolerance limits.

Balancing of rotors made of two or more individually balanced components or sub-assemblies is often performed to obtain maximum reduction in vibration from unbalance. The complete machine may also be run under service conditions as part of the balancing procedure. Assembly balancing may be necessitated by conditions related to assembly and operation procedures. A balanced flywheel mounted on a balanced crankshaft may not be balanced as an assembly. When connecting rods and pistons are added, more unbalance is likely to be introduced. The unbalance effects can only be reduced by balancing the reciprocating machine as an assembly.

The inner and outer raceways of anti-friction bearings sometimes do not have concentric surfaces. Eccentricity of the races makes it desirable to balance the rotor on its own bearings, with the stator supporting the rolling element bearings. In this respect, bearing alignment and preload also become important factors.

8.6 BALANCING OF RIGID ROTOR

Classification of rotors as rigid or flexible depends on the dynamic behavior of the rotating system in its operating speed range. Rigid rotors substantially operate below their critical speeds, permitting accurate balancing in two axial planes. If analytical methods indicate that potential energy in the bearings is over 80 percent of the system's total strain energy, the rotor is considered to be rigid. Rigid rotors may be balanced through placement of correction weights in two planes in its given operating speed zone. But if the strain energy due to bending in the shaft exceeds 20 percent of the system total, and the rotor operates through one or more critical speeds, then it is regarded as a flexible rotor. A two-plane balancing may not be sufficient for a flexible rotor, and additional weights may be needed along the length of the rotor to control vibration amplitude at speed.

Documentation published by the International Standards Organization (ISO, 11340, 1994) on the subject discusses classification of rotors and balance quality requirements for many different applications. Rieger (1986) also provides a deeper insight on the subject. Five basic groups of rotors from balancing perspective are presented by ISO

1. *Rigid rotors*: Balanced in any two axial planes, staying in balance in the complete speed range.

2. *Quasi-flexible rotors*: Not fully rigid, but can be adequately balanced in a low-speed balancing machine; balance will be maintained in the full speed range.
3. *Flexible rotors*: Cannot be balanced in a slow speed balancing machine, and will require one or more high-speed trim plane corrections.
4. *Flexible-attachment rotors*: Rotors of class 1, 2 or 3 have attached to them, or may contain within them, flexible components.
5. *Single-speed flexible rotors*: Special case of class 3, with balancing of rotor restricted to a single speed.

Class 1 and 2 rotors may be adequately balanced at one speed in one or two planes. Flexible rotors in Class 3 and 4 require the modal technique or the influence coefficient method (*ref*: Ehrich, 1999).

Single-plane balancing of a rotor assumes the mass center to be offset from the spin axis of the shaft. The amount of radial offset, or unbalance eccentricity, is $\boldsymbol{e}_u$. An effective unbalance mass $\boldsymbol{m}_u$ at radius R from the shaft axis is placed on the rotor to move the mass center of the system to coincide with the spin axis. The position vector of the mass from origin is $R \times \boldsymbol{e}_r$. Note that $\boldsymbol{e}_u$, $\boldsymbol{m}_u$ and $\boldsymbol{U} = \boldsymbol{m}_u \times R$ are vectors. If the mass of the rotor is M and spin speed is ω, the unbalance force vector is given by

$$F_u = \boldsymbol{m}_u \times R \times \omega^2 = \boldsymbol{U} \times \omega^2 = \boldsymbol{e}_u \times M \times \omega^2 \tag{8.1}$$

and the unbalance is given by

$$\boldsymbol{m}_u = \boldsymbol{U}/R = (\boldsymbol{e}_u \times M)/R \tag{8.2}$$

If a small balance weight m_b is placed on the rim of the wheel 180° out of phase to the eccentricity vector, total centrifugal load is

$$F_u = \boldsymbol{e}_u \times M \times \omega^2 - m_b \times \boldsymbol{R} \times \omega^2 \tag{8.3}$$

where $\boldsymbol{R}$ now also has angle related to it. In order to obtain a net total load of zero, the value of m_b must be properly selected.

The influence coefficient method requires the relative phase angle of the force, or the displacement at the bearing pedestal, with respect to a timing mark on the shaft to be measured. The force or displacement vectors may be expressed as

$$\boldsymbol{Y} = \boldsymbol{c} \times \boldsymbol{U} \tag{8.4}$$

where $\boldsymbol{c}$ is the influence coefficient. The inverse of this coefficient is the stiffness term, also called the impedance coefficient $\boldsymbol{b}$, and is commonly encountered in rotor dynamics studies. A trial weight U_t is placed on the shaft or the disc, and the new response $\boldsymbol{Y}_t$ is measured. By vector subtraction, the influence coefficient is

$$\boldsymbol{c} = (\boldsymbol{Y}_t - \boldsymbol{Y})/\boldsymbol{U}_t \tag{8.5}$$

To understand the two-plane rigid body balancing method, consider a shaft with two discs of mass M_1 and M_2 rotating in rigid bearings. Unbalance weights m_1 and m_2 are attached to the disks at angles θ_1 and θ_2 from the reference mark. Centrifugal forces due to the weights are

$$\boldsymbol{F}_{u1} = m_1 \times \boldsymbol{R}_1 \times \omega^2 = \boldsymbol{U}_1 \omega^2 \tag{8.6}$$

$$\boldsymbol{F}_{u2} = m_2 \times \boldsymbol{R}_2 \times \omega^2 = \boldsymbol{U}_2 \omega^2 \tag{8.7}$$

The unbalance vectors may be resolved into x and y components in a local coordinate system fixed in the rotor. Ignoring loads from gravitational weights, from D'Alembert's principle the sum of the inertia loads and external bearing reactions must form a system in equilibrium. The measured bearing forces are a function of the rotor unbalance forces. The unbalance forces on the disks, located at axial distance l_1 and l_2 from a fixed point, may be resolved into an equivalent force F_u and a moment M_u acting at the center of the system, and are expressed by:

$$F_u = (U_1 + U_2) \times \omega^2 \tag{8.8}$$

$$M_u = (l_1 \times U_1 + l_2 \times U_2) \times \omega^2 \tag{8.9}$$

The unbalance loads in the two planes may be resolved into equivalent static and dynamic components P_s, P_d (using P to denote force or moment), where:

$$P_1 = P_s - P_d \text{ and } P_2 = P_s + P_d \tag{8.10}$$

Then

$$P_s = (P_1 + P_2)/2 \text{ and } P_d = (P_2 - P_1)/2 \tag{8.11}$$

The influence coefficient method for balancing in two planes may be employed for a rotor spinning at a fixed speed. Initial vibrations at the planes at speed ω are:

$$Y_1(\omega) = Y_1\angle\alpha_1 \text{ and } Y_2(\omega) = Y_2\angle\alpha_2 \tag{8.12}$$

where Y_i is the amplitude and α_i is the phase angle. If the vibrations are a linear combination on the unbalance loads, then:

$$Y_1 = c_{11}(\omega)U_1 + c_{12}(\omega)U_2 \tag{8.13}$$

$$Y_2 = c_{21}(\omega)U_1 + c_{22}(\omega)U_2 \tag{8.14}$$

The speed dependent influence coefficients are obtained by placing a trial mass U_{t1} at each plane and the vibrations recorded. After placing the mass at plane 1 the vibrations are:

$$Y_{11} = c_{11}(U_1 + U_{t1}) + c_{12}\,U_2 \tag{8.15}$$

$$Y_{21} = c_{21}(U_1 + U_{t1}) + c_{22}\,U_2 \tag{8.16}$$

where Y_{11} and Y_{21} are the recordings at planes 1 and 2. The influence coefficients are:

$$c_{11} = (Y_{11} - Y_1)/\,U_{t1} \text{ and } c_{21} = (Y_{21} - Y_2)/\,U_{t1} \tag{8.17}$$

The first trial weight is then removed and a second trial weight U_{t2} placed on the second plane. If Y_{12} and Y_{22} are the recordings at planes 1 and 2, the influence coefficients are:

$$c_{12} = (Y_{12} - Y_1)/\,U_{t2} \text{ and } c_{22} = (Y_{22} - Y_2)/\,U_{t2} \tag{8.18}$$

In matrix notation,

$$\begin{Bmatrix} Y_1 \\ Y_2 \end{Bmatrix} = \begin{bmatrix} c_{11} & c_{12} \\ c_{21} & c_{22} \end{bmatrix} \begin{Bmatrix} U_1 \\ U_2 \end{Bmatrix} \tag{8.19}$$

The balance correction weights are:

$$\begin{Bmatrix} U_{b1} \\ U_{b2} \end{Bmatrix} = -\begin{Bmatrix} U_1 \\ U_2 \end{Bmatrix} = -\begin{bmatrix} c_{11} & c_{12} \\ c_{21} & c_{22} \end{bmatrix}^{-1} \begin{Bmatrix} Y_1 \\ Y_2 \end{Bmatrix} \tag{8.20}$$

So the correction weight for plane 1 is:

$$U_{b1} = -\frac{\begin{vmatrix} Y_1 & c_{12} \\ Y_2 & c_{22} \end{vmatrix}}{\begin{vmatrix} c_{11} & c_{12} \\ c_{21} & c_{22} \end{vmatrix}} \tag{8.21}$$

and the correction weight for plane 2 is:

$$U_{b2} = -\frac{\begin{vmatrix} c_{11} & Y_1 \\ c_{21} & Y_2 \end{vmatrix}}{\begin{vmatrix} c_{11} & c_{12} \\ c_{21} & c_{22} \end{vmatrix}} \tag{8.22}$$

In reality, no rotor is completely rigid. The expression may be applied to specific rotors operating at speeds well below their fundamental flexible natural frequency. The rotor is considered flexible if it does not hold balance after being balanced as a rigid body.

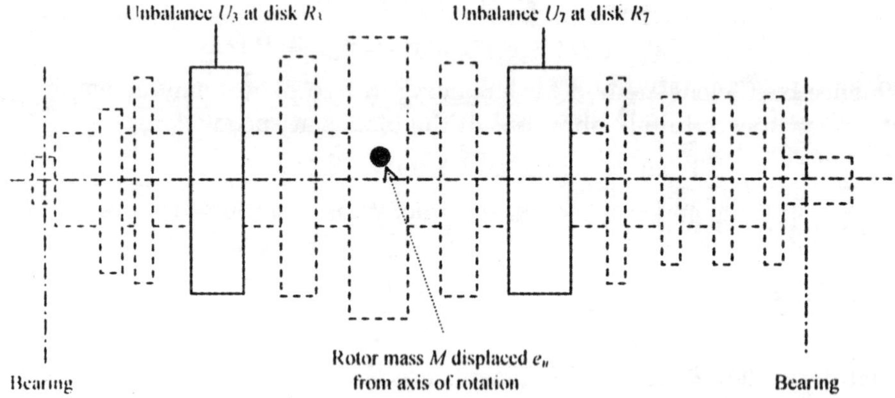

Fig. 8.4. *Rotor with Attached Disks*

Figure 8.4 shows a rotor with attached disks. Disks 3 and 7 have unbalance of U_3 and U_7, while the rotor mass center is displaced from the axis of rotation by distance e_u at disk 5. When this rotor is placed in a balancing machine, either the hard or soft bearing type, a two-plane correction will be indicated. Balance correction weights ω_3 and ω_7 are placed at planes 3 and 7 at radii R_3 and R_7. As a rigid body, the rotor now has the appearance of being balanced. Rotor unbalance is represented by $Me_u = U_5$. The rigid rotor balance condition of $\Sigma(U_3 + U_5 + U_7) = 0$ is satisfied, so the system is in static equilibrium. When operated as shown at a sufficiently low speed, the balance will be maintained. As the speed increases centrifugal forces from the unbalance distribution and elastic effects in the shaft cause the rotor to deflect and bow. The effects become more pronounced as the speed exceeds 70 percent of the critical speed. Additional balancing is required to control vibration amplitudes. Unbalance distribution in the rotor may come in many different variations. Some commonly encountered types include continuous unbalance along the length of the rotor, unbalance due to a bow and moment unbalance from an overhung disk.

8.7 SINGLE PLANE BALANCING OF FLEXIBLE ROTOR

When a rotor operates through multiple natural frequencies, the modal method of balancing is most suitable. Critical speeds and corresponding mode shapes of the system must be known, either from analytical predictions or from experimental measurements. Weights placed on the rotor at an axial location will be proportional to the relative mode shape deflection at that point. Modal distribution for each natural frequency is used to individually balance the rotor for that mode. The objective is to reduce dynamic activity in modes with high amplification factors without upsetting the rotor's balance at other modes. Balance corrections to be applied on the rotor may be computed by using the influence coefficient method to predict modal influence balancing coefficients. Darlow (1989) provides significant contributions on the combined modal and influence coefficient methods, and the procedure is referred to as the unified method of rotor balancing. Note that rotor mode shapes become quite complex above the second critical speed.

Sometimes it is not possible to obtain an accurate measurement of the phase angle. Blake (1967) reports the method of applying trial weights at three different locations to determine the magnitude and location of unbalance. The method has been successfully applied to fans where the speed cannot be held steady due to aerodynamic or other disturbances. Compensation for non-linear effects can be included in this graphical procedure, referred to as the three trial weight method of balancing. The method may be extended to balance flexible rotors through multiple critical speeds (Foiles *et al.*, 1982). A modal distribution of weights is placed on the rotor, and vibration amplitude measured at critical speeds. An FFT analyzer with ability to hold peak values may be used for this purpose. Each mode may then be separately balanced without affecting the balance of the rotor in other modes. The method proves to be advantageous when amplification factors are high and when it is difficult to balance the rotor by the least-square error method. Drawbacks arise from the need to know the rotor's mode shapes, and three runs must be made for each critical speed to be balanced. After this procedure, the generated modal sensitivity may be used for additional improvements by making only two runs.

Two machines may be sometimes individually balanced, but when they are mechanically coupled, high vibration levels ensue. Trim balance at the coupling is required. One procedure calls for rotation of one half of the system by 180° at the coupling, and then taking another set of vibration readings. From the measurements the new position of the coupling may be computed to obtain acceptable vibrations.

Balancing of rotors without trial weights is reported by Palazzolo *et al.*, 1982. The Nyquist's plot of vibration amplitudes is used to determine the amplification factor for various modes, from which modal unbalance eccentricity can be obtained. Phase angle for the modal weight distribution is also determined from the polar plot to balance the rotor at particular critical speeds. Rotor modal mass and mode shapes for the natural frequencies of interest must be known beforehand. Balance predictions from this procedure may be used for initial trial runs, than refined by the influence coefficient method of balancing.

Static single plane balancing of a rotor does not cause coincidence of the shaft's running axis with its principal inertia axis; it only ensures that the two axes have a common point. Thus, when the shaft is turning, dynamic balance is not achieved. This can only be obtained by rotating the principal axis about the center of gravity in the longitudinal plane, characterized by the shaft axis and the principal axis. The rotation can be accomplished by modifying the geometry of the journals (an impractical proposition), or by adding or removing metal from the mass of the rotor in the longitudinal plane. Note, however, that adding or removing material disturbs the static balance already achieved. To avoid this possibility, two masses of equal magnitude are added or removed, one on each side of the principal axis to maintain static balance, and one in each of two radial planes to eliminate overturning effects.

For balancing of machines operating in the field, the rotor is already in place in its own bearings. Suitable instrumentation is placed on the shaft, bearing housing or foundation to monitor the vibrations. A number of procedures are employed, with preference of one method over another depending on geometric configuration of the rotor and bearing systen.

The single plane, single speed balancing procedure is simple. After vibration measurement from an initial run, a trial weight is placed on the shaft. The vibration response of the rotor at the same speed is again recorded. The change in vibration divided by the trial weight provides an influence coefficient, representing rotor response due to the trial weight at the given speed. Note that the influence coefficient is a vector with amplitude and phase, the latter providing the angle between the forcing function and the response. For a linear system application of the balance correction should theoretically reduce amplitude of motion at the given speed to zero. The procedure, called 'one-shot' balancing, is similar to trim balancing and is widely used in many different situations. However, problems in the form of higher vibrations at other speeds and points along the rotor may be encountered.

High vibration at other speeds may be avoided through linear regression, which provides the least-square error method of balancing. Instead of a single speed point, vibration measurements are recorded over a number of points. Linear regression procedure will reduce large responses at other speeds. The magnitude of correction weight is smaller than the single-speed influence coefficient method, and ensures that correction is not limited to a single speed. The method is useful if the shaft has a bow or a run-out.

When the rotor is long, applying correction in one plane may result in high vibrations at another location due to overturning moment effects. Applying correction weights and calculating influence coefficients in two planes at the same time proves more useful. A trial weight is placed on the first plane, and response at first and second planes is recorded. Corresponding influence coefficients are calculated. In the second step the weight is removed from the first plane and placed on the second plane. After measuring the response two more influence coefficients are determined, from which a 2×2 matrix of complex influence coefficients (equivalent to a 4×4 matrix of real numbers) is obtained. Inversion of the matrix will provide magnitude of correction weight and phase angle location for the planes.

As in the case of single plane balancing, vibration readings at multiple speed points may be recorded for the two-plane method. Generalized influence coefficients are then determined using the least-square error procedure. The method provides useful correction weight values even when a rotating system has to pass through a number of critical speeds during operation. Also, the method is preferred when prior knowledge of the system's geometry and rotor dynamic characteristics are not available. Initially developed by Goodman (1974), improvements were reported by Palazzolo *et al.*, (1982). When balance planes are not properly selected, the least-square error method may lead to correction weights of magnitude that are impractical to use. Limitations may not be placed on the size of balance weights in this method.

The influence coefficient method for the single-plane balancing of a flexible rotor does not require prior knowledge of the system's modal characteristics. Synchronous amplitudes of vibration and phase are measured at one or more speeds. Trial weights are attached at a known radial and angular location from a reference mark. The balancing process is facilitated by placing the correction weight by observation of the response charted on a polar plot to decrease the response. As an illustration, consider a disk with its timing indicator lined up with the probe.

Table 8.1. *Vibration Measurements on Disk of Flexible Rotor*

Speed, rpm	*Amplitude, 10^{-3} in.*	*Phase Angle, degrees*
2100	0.25	83
2240	0.33	54
2700	1.15	22
2800	1.70	9
3025	1.95	355
3150	1.80	331
3300	1.77	302
3600	1.55	281
3975	0.95	263
5850	0.35	252
8250	0.20	247

The response is measured as shown in table 8.1. The weight may be placed 67° from the timing indicator opposite the direction of rotation, 180° out of phase with the unbalance in the rotor. The amount of calibration weight may be determined from the guideline that the centrifugal load from the weight is to be approximately 10 percent of the static weight of the rotor.

In the calculations the influence coefficient changes with the speed. For a single mass flexible rotor of stiffness K and damping C with considerable unbalance will require a number of trial runs. The synchronous amplitude Y may be expressed by the complex influence coefficient $\boldsymbol{c}$ and the unbalance vector $\boldsymbol{U}_u$ by:

$$Y = c(\omega)\boldsymbol{U}_u \tag{8.23}$$

where $c(\omega) = \dfrac{f^2}{M(1-f^2+2ipf)}$, $\boldsymbol{U}_u = Me_u e^{-i}\alpha, f = \omega/\omega_{cr}$ and damping p is defined by $2pf = C\omega/K$.

After placing trial weight $\boldsymbol{U}_t$ at a given radius and phase angle the response is measured at the same speed as the initial reading. The new amplitude of vibration is:

$$Y_t = c \times (\boldsymbol{U}_u + \boldsymbol{U}_t) \tag{8.24}$$

The influence coefficient is then:

$$\boldsymbol{c} = (Y_t - Y)/U_t \tag{8.25}$$

The correction weight is:

$$\boldsymbol{U}_b = -\boldsymbol{U}_u = Y \times \boldsymbol{U}_t/(Y - Y_t) \tag{8.26}$$

8.8 OVERHUNG THIN DISK ROTOR

Multi-plane balancing techniques tacitly assume that the transverse axes of the rotor disks are perpendicular to the elastic centerline of the shaft and that the shaft is undistorted. The synchronous excitation is then due only to asymmetric radial mass distributions or disk eccentricities. In practice, the shaft centerline may be bowed and the disks can be skewed, effectively inducing external forces and moments on the shaft. A rotor might appear to be well balanced at a particular design speed when balanced with a technique that considers only the radial unbalance forcing function, but may be considerably out of balance at other operating speeds. These influences can induce large amplitudes of motion on the rotor when it is operating in the vicinity of critical speeds, and lead to a damaged rotor and bearings.

When a skewed disk is present in the system, in addition to the equations of motion of radial unbalance two additional equations are required to represent the disk's angular motion and gyroscopic moments. When the general Euler rotation angles are implemented to develop the dynamic equations of motion for a general precessing and nutating gyroscope, the equations become nonlinear. If the angle of disk skew $\tau = \tau_x + j\tau_y$ is considered small, it can be shown that the product of inertia term is proportional to the product of the difference between the polar and transverse moments of inertia and the skew angle (Salamone *et al.*, 1979). An effect equivalent to disk skew can be generated by two radial out of phase unbalance components $u = u_x + ju_y$, which are separated by a finite axial distance on the disk. The resulting system is mathematically identical to a small disk skew, and the resulting unbalance components produce a bending moment about the disk center.

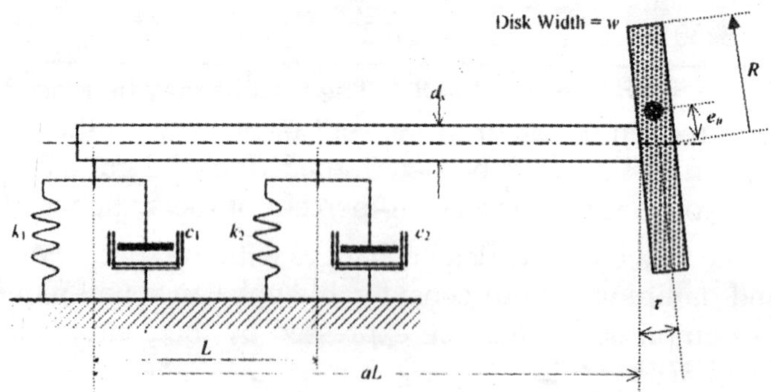

Fig. 8.5. *Overhung Disk with Skew and Unbalance (Salamone, 1979)*

Because of the complexity of the multi-mass rotor behavior with shaft bow and disk skew, a single mass overhung rotor on flexible damped bearings with a skewed disk is examined. Figure 8.5 shows the arrangement. For a thin disk the polar moment of inertia I_p is greater than the transverse moment of inertia I_t, and is assumed to be half the polar moment of inertia. The system will have only one synchronous critical speed of forward precession.

The general transfer matrix equations with isotropic bearings and a massless shaft may be expressed by the flowing four complex equations:

$$\begin{Bmatrix} me_u\omega^2 e^{ja} \\ \tau\omega^2 (I_p - I_t)e^{j\beta} \\ 0 \\ 0 \end{Bmatrix} = \begin{bmatrix} m & 0 & 0 & 0 \\ 0 & I_t & 0 & 0 \\ 0 & 0 & 0 & 0 \\ 0 & 0 & 0 & 0 \end{bmatrix} \begin{Bmatrix} d^2z/dt^2 \\ d^2\theta/dt^2 \\ d^2z_1/dt^2 \\ d^2z_2/dt^2 \end{Bmatrix} + \begin{bmatrix} c & 0 & 0 & 0 \\ 0 & -j\omega I_p & 0 & 0 \\ 0 & 0 & c_1 & 0 \\ 0 & 0 & 0 & c_2 \end{bmatrix} \begin{Bmatrix} dz/dt \\ d\theta/dt \\ dz_1/dt \\ dz_2/dt \end{Bmatrix}$$

$$+ \begin{bmatrix} K_{zz} & K_{Z\theta} & K_{z1} & K_{z2} \\ K_{z\theta} & K_{\theta\theta} & K_{\theta 1} & K_{\theta 2} \\ K_{z1} & K_{\theta 1} & K_{11} & K_{12} \\ K_{z2} & K_{\theta 2} & K_{12} & K_{22} \end{bmatrix} \begin{Bmatrix} z \\ \theta \\ z_1 \\ z_2 \end{Bmatrix} \tag{8.27}$$

Examination of these equations indicates that there are two forcing functions acting on the disk, deflection equation z and slope equation θ that are generated by the radial unbalance and disk skew. α and β represent angular location of unbalance. Also, the sign of the moment forcing term depends on the difference between the polar and transverse moments of inertia. For a thick disk $I_t > I_p$, so the skew gyroscopic moment would be opposite in sign compared to the moment for a thin disk. Hence, the dynamic effects of skew in a thick disk are considerably more than that in a thin disk.

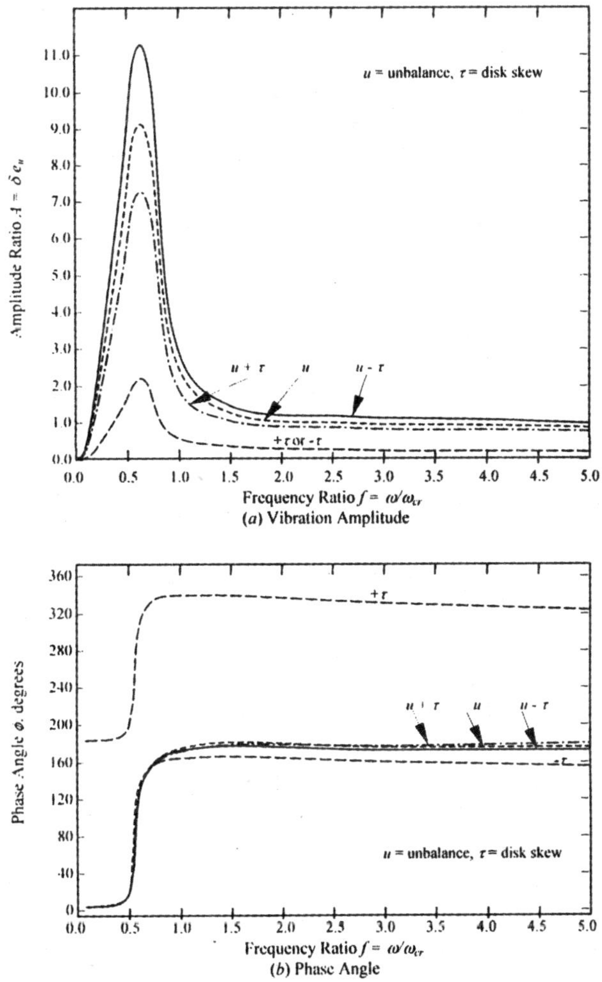

Fig. 8.6. *Vibration Characteristics of Skewed and Unbalanced Disk (Salamone, 1979)*

Synchronous response in the overhung thin disk is calculated for five combinations of unbalance and disk skew angle. The shaft's critical speed is about 60 percent of that obtained with a rigid support system. The reduction from the rigid value arises from the flexibility of the bearings. Figure 8.6*a* plots the disk amplitude against shaft speed. With unbalance alone, peak amplitude ratio at the disk is 9.5 at the critical speed, reducing to 1.0 at higher speeds. Figure 8.6*b* provides the variation of the phase angle. When only unbalance is present the rotor experiences a phase shift of 90° at the critical speed. Above this speed the phase angle increases to 180°, remaining constant at higher speeds.

When only the disk skew is considered the phase angle changes through 90° at the critical speed on to about 180° at the critical speed. If the disk skew is negative, the phase angle change observed is almost identical to the response observed with radial unbalance. For a positive skew angle the phase angle changes are 180° out of phase with the negative disk skew phase angles, and vary from 180° to about 360°. A slight variation in the phase angle behavior, when compared with the response due to pure unbalance, is the phase behavior at supercritical speeds. The phase angle shifts from 180° to 150° as the frequency ratio varies from 1.0 to 5.0. This observation is indicative of the presence of skew in the disk. If a combination of unbalance and disk skew is assumed, the unbalance moment acts in conjunction with the radial unbalance to increase the shaft motion at the disk. The response is then a linear vector superposition of the two items considered individually.

Amplitudes of vibration and phase angle may also be calculated for the far and near bearings to provide useful data when considering techniques for balancing an overhung rotor with a skewed disk. The best way to determine if the disk is skewed is to observe the motion at locations other than the disk. With unbalance alone the maximum amplitude at the near bearing is approximately one-half the amplitude at the disk location. Upon passing through the critical speed the dimensional unbalance response is about 0.7, and remains constant with the speed. But when only the disk is skewed, the maximum amplitude at the near bearing is 1.0, and is again one-half of the amplitude at the disk location. Above the critical speed the amplitude reduces, reaching a minimum at about 0.9 of the ratio f. As the speed increases, the rotor amplitude increases with speed. The associated phase angle with disk skew alone sees little increase in the phase angle at the near bearing. Thus, the build up in the amplitude is due to the skewed disk attempting to straighten out and rotate about its principal inertia axis.

At the far end bearing, with unbalance alone the maximum amplitude ratio is 1.05 at the critical speed. At speeds well above the critical, the amplitude reduces to an asymptotic value of 0.07. The phase angle is 180° out of phase to the disk or the near end bearing. With the disk skew alone, the maximum amplitude ratio is 0.22 at the critical speed. Beyond the critical speed the amplitude reduces, reaching a minimum at f = 0.9, then continues to increase. The corresponding phase angle is even more revealing, changing through the critical by only 150°. After passing through the critical the phase angle reduces over 100° from its maximum value at f = 0.7, gradually increasing again at f = 1.5.

The combination of positive disk skew with positive radial unbalance causes a net reduction in amplitude over the case with unbalance only. Hence, it is logical to assume that if the disk is initially balanced and skewed, a radial correction weight can be placed on the wheel to cancel the disk's motion while passing through the critical speed. The correction weight on the disk may be determined by linear superposition using the amplitude and phase readings from the far bearing probe. After this sequence, the rotor can safely pass through the critical speed. Beyond the critical speed, however, the amplitudes of motion at both bearings increase, with larger amplitudes occurring in the near bearing. For operation at post critical speeds, for example at f = 3.0, a single correction weight may be applied at the far bearing without disturbing the first weight at f = 0.6. Correction in the near bearing at f = 3.0 without disturbing the previous two improvements may be carried out in a similar manner. But a consequence of the third correction will be increased amplitudes at the disk and in the far bearing at the first critical, f = 0.6. Thus, while the single plane balancing technique is sufficient for balancing for a specific rotor location and speed, it is not successful for the overall rotor over the full speed range. Balancing with a couple form of correction weights is then desirable.

A couple correction calls for two correction weights separated by the thickness of the disk and 180° out of phase with each other, producing an equivalent balance moment vector. Starting with the original rotor with positive disk skew only, a trial couple is placed on the disk to balance at the first critical speed $f = 0.6$. The technique is similar to the single plane method, except that a trial couple vector is used instead of a trial weight vector.

8.9 MULTI-PLANE BALANCING OF FLEXIBLE ROTORS

A uniform shaft supported in rigid bearings operating through n critical speeds will have $n + 1$ nodal points at which there is no motion. For effective balancing, ideally the correction weights must be placed where the dynamic motion is a maximum. Balancing of the nth critical speed will thus require n balance planes. The number of measuring probes corresponds to the number of balance planes, and the measurement is made at only one speed. For small amplitudes the vibration for $n = 3$ is expressed by:

$$Y_1 = c_{11}U_1 + c_{12}U_2 + c_{13}U_3 \tag{8.28}$$

$$Y_2 = c_{21}U_1 + c_{22}U_2 + c_{23}U_3 \tag{8.29}$$

$$Y_3 = c_{31}U_1 + c_{32}U_2 + c_{33}U_3 \tag{8.30}$$

In a matrix form the equations are:

$$\{Y\}_n = [c]_{n \times n}\{U\}_n \tag{8.31}$$

The complex inverse of the influence coefficients then provides the balance solution:

$$\{U\}_b = -\{U\}_n = -[c]_{nxn}^{-1}\{Y\}_n \tag{8.32}$$

The coefficients are determined for each speed by placing a trial weight at each disk separately and measuring the response at all the disks.

For the least square error method of balancing, more or less probes may be used than the number of planes n, with $m > n$ readings taken for a number of speed points. The vibration is of the form:

$$Y_m = c_{m1}U_1 + c_{m2}U_2 + \ldots + c_{mn}U_n + \varepsilon_m \tag{8.33}$$

The measurements $Y_t - Y_m$ may be obtained from some of the probes, but must cover a number of speed points. The vibration matrix is given by the expression:

$$\{Y\}_m = [c]_{m \times n}\{U\}_n + \{\varepsilon\}_m \tag{8.34}$$

where $m \geq n$. The matrix of coefficients is generated from the readings after applying the trial weights. From a slow roll of the rotor the terms $\{Y\}_0$ is also recorded before and after the application of the weights to ascertain that shaft run-out is not affected. The transpose of the complex conjugate of the set of influence coefficients is:

$$\{R\}_{n \times m} = [c]^{\mathrm{T}}_{m \times n} \tag{8.35}$$

Combining with (8.33),

$$[R]_{nxm}\{Y\}_m = [R]_{nxm}[c]_{mxn}\{U\}_n + [R]_{nxm}\{\varepsilon\}_m$$

Since $[R][c]$ is a square matrix of order n, it is inverted to give $[H] = [[R][c]]^{-1}$. The least-square-error balancing matrix is given by:

$$\{U_b\}_n = [G]_{nxm}\{Y\}_m \tag{8.36}$$

where $[G]_{nxm} = -[H]_{nxn}^{-1}[R]_{nxm}$. Including the effects of the run-out, the balancing equations using the least square error procedure is:

$$\{U_b\}_n = [G]_{nxm}[\{Y\}_m - \{Y_o\}_m] \tag{8.37}$$

8.10 BALANCING MACHINES

Two types of centrifugal balancing machines are in use to determine the size of the required weight and its angular location in each correction plane, the soft and hard bearing machines. Soft bearing machines permit the idealized free rotor motion, but the activity is usually restricted to the horizontal plane. In addition, the bearings and its attached components move with the rotor, thus adding mass to the rotor at the location. The increased mass of the rotor will reduce the displacement of the bearings, and output from the device sensing the unbalance will correspondingly reduce (fig. 8.7).

Hard bearing balancing machines are basically of a similar construction, except that their bearing supports are significantly stiffer in the horizontal direction. This causes horizontal plane related critical speeds of the machine to be several orders of magnitude greater than that for a comparable soft bearing machine. This type of machine is designed to operate at speeds well below the horizontal critical speed. The output from the sensing elements attached to the bearing supports is directly proportional to the centrifugal force reaction resulting from unbalance in the rotor. Since the output is not affected by bearing mass, rotor weight or inertia, a definite connection between unbalance and sensing element output can be established, hence eliminating the need for calibration of a given rotor.

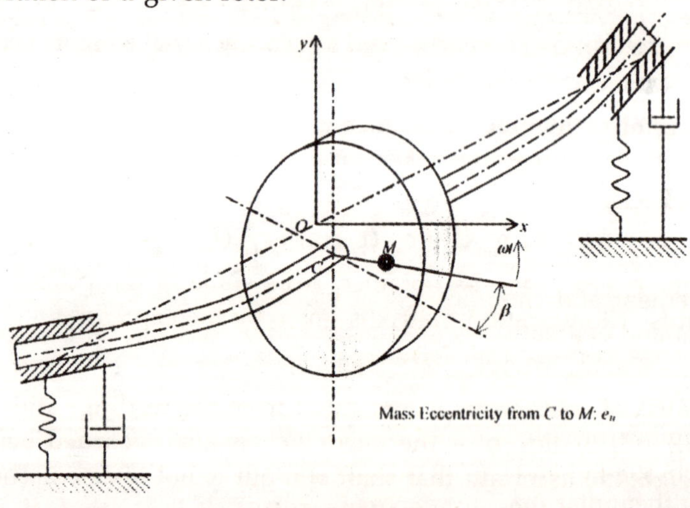

Fig. 8.7. *Deflections of Single Mass Elastic Shaft in Flexible Supports*

Centrifugal balancing machines may also be categorized by the type of unbalance to be indicated (static or dynamic) and the work piece's shaft axis attitude (horizontal or vertical). Trial and error balancing machines are of the soft bearing type, and do not indicate directly the unbalance in weight; they only provide displacement or velocity of shaft lateral motion at the bearings. Balancing of the job calls for a lengthy trial and error procedure, even if it is one of an identical batch. Also, the machine cannot identify balance correction for a specified plane. Field balancing generally does not have access to a microprocessor, so the method falls into this group. Balancing machines that are calibrated for a balanced rotor from an identical group can be instrumented to permit plane separation. However, for the first piece the trial and error procedure cannot be avoided.

Soft bearing machines may also be provided with integral electronic unbalance compensator type of instrumentation, where a balanced master rotor is not needed. Separation of balance planes and calibration is achieved by using precisely weighed calibrating shots temporarily attached in each of two correction planes. Microprocessor-based software determines the influence coefficients. This class of soft bearing balancers may also be provided with electrically driven vibratory shakers fitted to the rotor supports.

Another class of permanently calibrated balancing machines uses hard bearings. Calibration is set up for rotors falling within a range of rotor weight and size in a selected speed range. Unbalance indication is accomplished in the first run by including an analog or digital computer with the instrumentation. Calculations for determining correction weight size and location are based on the following parameters

- Distance from left correction plane to left support.
- Distance between correction planes.
- Distance from right correction plane to right support.
- Radial location of the correction weights in both planes.

The instrumentation then determines the magnitude and angular location of the correction weights in both planes.

Precise definition of the rotor assembly's configuration, capacity and performance is needed in selecting the balancing machine. Details about rotor weight, length, diameter, journal diameters, service speed, application and whether the rotor is rigid or flexible are required to evaluate the suitability of the balancing machine. A manufacturer's selection must be evaluated not only for compliance to the purchase specifications but also on the machine's features, operation and conformance with applicable standards.

8.11 BALANCE CRITERIA

ISO/DIS standard 10816 (ANSI Counterpart S2.12, Part 6-199X) establishes procedures and guidelines for measurement and classification of mechanical vibrations of rotating machines. Vibration measurement and evaluation criteria provided in this standard are recommended for securing safe and reliable operation of machines, and to avoid problems with auxiliary equipment attached to the main structure. In general, the standard calls for vibrations on the main structure of the machine. A number of components are mounted on the machine. Heat exchangers, governors, pumps, filters, gauges and piping components are some examples. Vibration guidelines based on experience with similar machines are useful to ensure damage is not sustained in the mounted attachments. Vibration readings on the main structure, however, are not a true indicator of the state of dynamic stress of the components within the machine.

Vibration of shipboard hull arises from excitation by the propeller shaft and propeller blades. Vibration is normally highest in the aft parts of the ship, with displacement amplitudes sometimes exceeding 0.10 in. At higher shaft speeds hull vibrations tend to be lower. With shipboard machinery mounted on the deck, hull and bulkhead stiffeners, vibration amplitudes at the equipment supports is substantially the same as in the hull. Acceptable values for diesel engine driven merchant ships of 100 meters in length or more are provided in ISO 6945. The shaded area in fig. 8.8 shows acceptable vibration levels for shipboard structures. Additional related information is also available in ISO 2631, ISO 4867 and ISO 4868.

The Hydraulic Institute (1994) specifies acceptable vibration limits for horizontal clear liquid pumps. The limits apply only if the pump is restricted to the following operating conditions: (1) Operation is non-cavitating and non-separating, (2) Suction pipe provides straight uniform flow, (3) Piping arrangement avoids undue stress on the pump and (4) Shaft coupling is aligned according to manufacturer's instructions. If the vibrations in the pump exceed specified values, the machine system should be inspected for possible defects and correction.

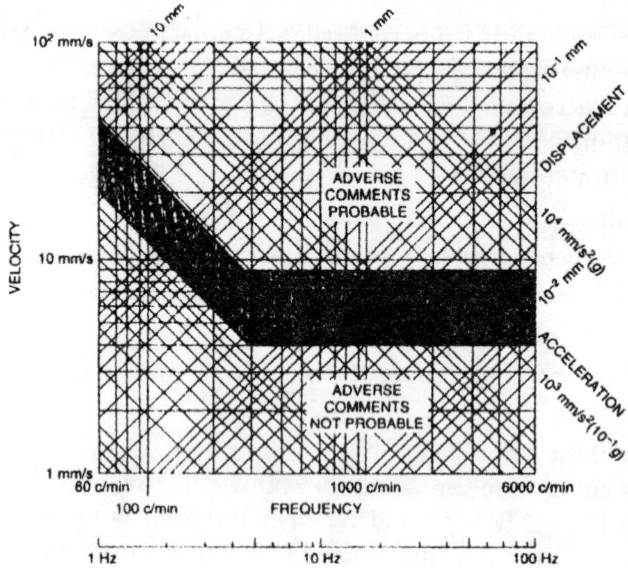

Fig. 8.8. *Guidelines for Peak Vibration in Merchant Ships* (ref. *ISO 4867, 6945*)

Often, changes in the level of dynamic activity over a period of time are more significant than the recorded vibration amplitude. Vibration levels in excess of specified limits may be acceptable if there is no increase over a sustained period of operation. Evidence of damage in the bearings or increased noise level, however, would be reasons for closer investigation.

Vibration criteria are available from the Compressed Air and Gas Institute (1993) for compressors in air and gas service, with operating speeds up to 15,000 rpm. A change in the vibration pattern of an operating compressor, as a rule, is more important than the measured level at any time. Thus, a substantial change in the vibration pattern for the same operating conditions should be promptly examined to determine its cause.

ISO/WD Standard 8579-2 describes the procedure for determining vibrations in enclosed gear reductions units operating within its specified design speed, load and temperature bounds for acceptance testing. Lubrication requirements from the manufacturer must be followed. A set of two vibration measurements is required on units provided with journal bearings, one on the shaft and the other on the bearing housing. Proximity probes may be used to measure shaft vibration relative to the housing for frequencies ranging from 0 to 500 Hz. The transducers must be as close as practical to the bearing, and measurements must be made in three orthogonal directions, one of them parallel to the shaft axis. Housing vibrations may be measured with a seismic type of accelerometer in frequencies ranging from 10 to 10,000 Hz. The standard provides a method for classifying gear boxes based on shaft displacement and housing velocity measurements, together with a subjective assessment of acceptable vibration levels for typical applications.

8.12 EXAMPLE PROBLEMS

Example Problem # 8.1: Consider a shaft of uniform diameter D and length L supported between two bearings, assumed to be ideal rigid supports. The rotating beam has natural frequencies of vibration (called eigen-values), and corresponding to each natural frequency is a mode shape (or eigen-vector), as depicted in fig. 8.9. The displacement mode shape does not represent actual shaft deflection, it represents a normalized shape that the shaft assumes at that frequency. When the shaft accelerates in speed, unbalance or shaft bow will cause the shaft to bow outwards as the speed approaches the shaft critical speed. Actual whirl radius will depend on magnitude and location along shaft axis of unbalance, ratio of rotor speed to critical speed, and amount of damping in the system. Considering a shaft of following dimensions, explain the characteristics and significance of the mode shapes.

Shaft length, L = 75.0 in.

Shaft diameter, D = 3.125 in.

Material Young's Modulus, E = 30.0×10^6 lbs/in^2

Material density, ρ = 0.283 lbs/in^3

Solution: The first three critical speeds of this shaft are $\omega_n = (K_n/2\pi) \times [EIg/wL^4]^{1/2}$, where $K_1 = 9.87$, $K_2 = 39.5$ and $K_3 = 88.8$ and w = weight per unit length. Then

$\omega_1 = 2.648 \times 10^3$ rpm, $\omega_2 = 1.0597 \times 10^4$ rpm and $\omega_3 = 2.3823 \times 10^4$ rpm

Corresponding mode shapes are shown in fig. 8.9.

Critical speeds of a shaft represented by discrete lumped masses will be considered next. Figure 8.10 shows a three-station discrete lumped-mass representation of the uniform shaft. Weights of $W/4$ are located at the ends of the shaft, $W/2$ at the center. Note that this model is valid only for the first rotor critical speed.

The first critical speed of lumped-mass representation gives $\omega_1 = (6.93/2\pi) \times [EIg/WL^3]^{1/2}$ = 2,656 rpm, where $W = wL/2$. Thus, this three-station representation of the uniform shaft provides first critical speed with a deviation of only 0.3%.

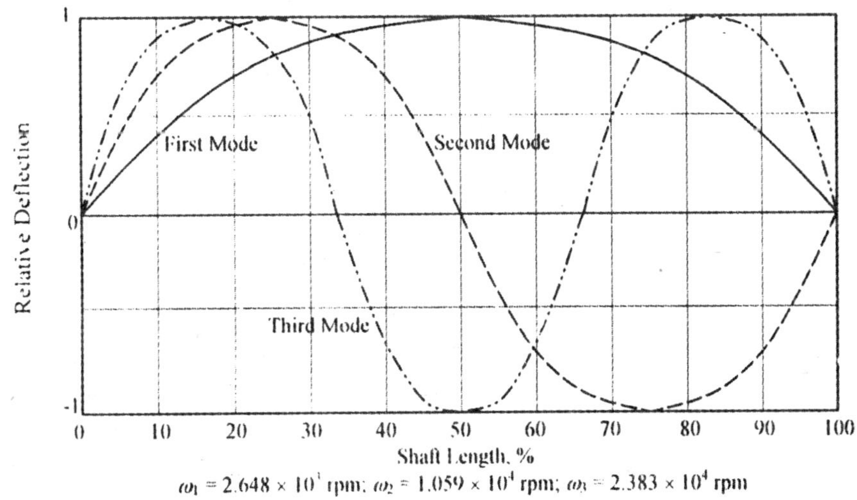

Fig. 8.9. *Mode Shapes of a Uniform Beam*

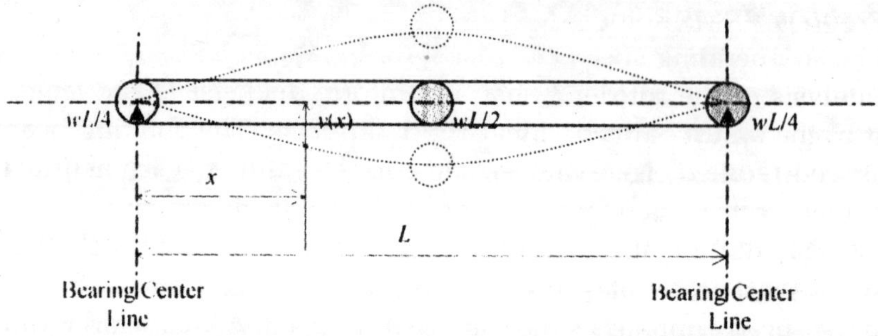

Fig. 8.10. *Three Station First Critical Speed Representation of Uniform Shaft*

The second and higher critical speeds of the lumped-mass representation are considered next. To approximate the characteristics of the uniform beam for the second mode, five mass stations are required (fig. 8.11). The ends have lumped weights of $wL/8$, center stations having lumped weights of $wL/4$. This results in a model that approximates the second rotor critical speed to less than 0.7 percent error. Here it should be noted that closed form solutions for the second and higher-order critical speeds are not available, computerized finite element method is needed for this purpose. Note that the deflected mode shape of the shaft crosses the spin axis twice. The point along the shaft at which zero displacement occurs for a particular critical speed is called a nodal point.

Figures 8.11 and 8.12 show the lumped-mass model and mode shape for the second and third modes.

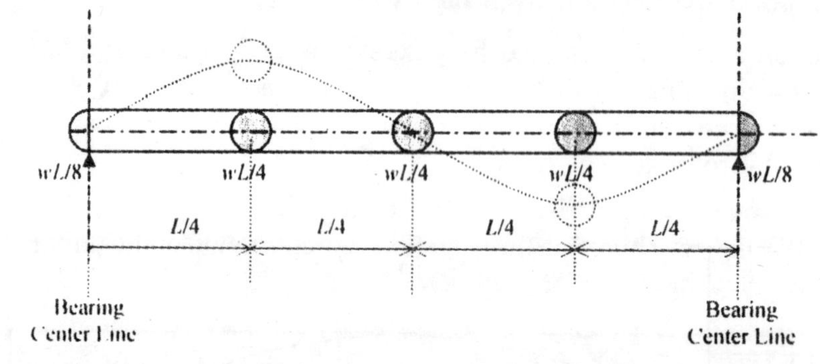

Fig. 8.11. *Five Station Second Critical Speed Representation of Uniform Shaft*

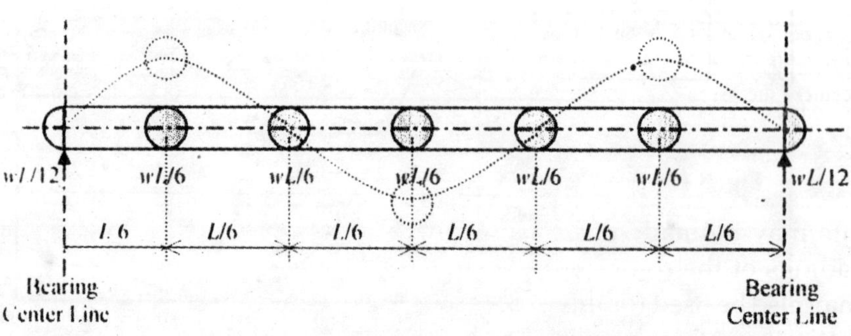

Fig. 8.12. *Seven Station Third Critical Speed Representation of Uniform Shaft*

***Example Problem # 8.2*:** Rotors are classified as either rigid or flexible according to their dynamic behavior at operating speeds. The classification may be determined by performing a critical speed analysis of the rotor-bearing system. If the potential energy in the bearings is over 80 percent of the system's total strain energy, then the rotor may be classified as rigid in that operating speed range. A rigid-body rotor may be balanced in two arbitrary planes. The rotor will then maintain balance throughout its operating speed range. Figure 8.13 shows illustrations of a disk and a cylinder with two-plane couple form of unbalance. Illustrate the two-plane rigid body balancing procedure using the influence coefficient method.

***Solution*:** Practical considerations must be included for proper selection of the planes to balance a rigid-body rotor with unbalance. On long cylindrical rotors the balance planes may be at the ends of the rotor. On a turbocharger shaft assembly balance planes may be opposite disk faces of the impeller.

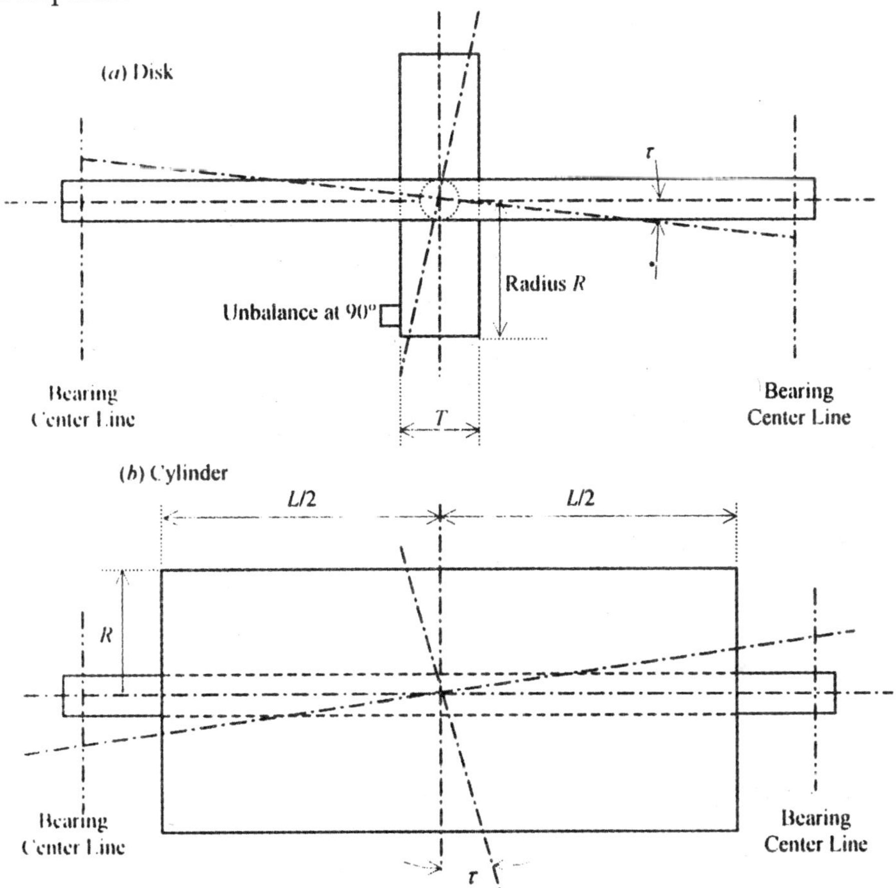

Fig. 8.13. *Two-Plane Couple Unbalance in:* (a) *Disk,* (b) *Cylinder*

Measurements may be made of bearing forces or supports (as in a hard bearing balancing machine); readings of the rotor's displacement, velocity or acceleration, or of the casing or the foundation may also be used for this procedure. A third transducer, called *key phasor,* is required to establish a timing reference mark on the shaft to produce a phase angle reference of peak vibration with respect to the timing mark. Use equations (8.12) to (8.22).

Vibration Readings		Amplitude	Phase Angle
Initial Reading – Plane 1:		$Y_1 = 7.4 \times 10^{-3}$ in.	$\alpha_1 = 83°$
Initial Reading – Plane 2:		$Y_2 = 6.9 \times 10^{-3}$ in.	$\alpha_2 = 234°$
Trial Balance Weight:			
Plane 1:		$U_{t1} = 15$ grams	$\alpha_{t1} = 249°$
Resulting Vibrations:			
Plane 1:		$Y_{11} = 7.8 \times 10^{-3}$ in.	$\alpha_{11} = 138°$
Plane 2:		$Y_{21} = 3.4 \times 10^{-3}$ in.	$\alpha_{21} = 267°$
Trial Balance Weight: (Note – Plane 1 trial weight must be removed)			
Plane 2:		$U_{t2} = 20$ grams	$\alpha_{t2} = 90°$
Resulting Vibrations:			
Plane 1:		$Y_{12} = 8.7 \times 10^{-3}$ in.	$\alpha_{12} = 48°$
Plane 2:		$Y_{22} = 9.8 \times 10^{-3}$ in.	$\alpha_{22} = 138°$
Balance Correction Weight:			
Plane 1:		$U_{b1} = 21.38$ grams	$\alpha_{b1} = 158.5°$
Plane 2:		$U_{b2} = 17.80$ grams	$\alpha_{b2} = 250.1°$

Example Problem # 8.3: The influence coefficient method is used here to balance a flexible, single disk rotor. The rotor (sometimes referred to as a Jeffcott rotor when weight of shaft is comparably small) may be balanced by operating the rotor through various speeds below, at or above the critical speeds and recording the appropriate amplitude and phase angle. It is assumed the shaft is straight with no appreciable run-out or thermal bowing, since the influence coefficients would not be repeatable otherwise. Figure 8.14 shows the relationship between displacement and phase of the disk.

Solution: The procedure involves measurement of the rotor's synchronous amplitude and phase at one or more speeds. The rotor is then stopped and a known trial weight is placed on the rotor at a specific radius and angle measured from a reference mark on the disk, preferably placed so as to reduce rotor response. Also, the magnitude of the trial weight should create a rotating unbalance load of about 10 percent or less of the static rotor weight. A large rotating unbalance force can result in permanent shaft bowing, while also creating non-linear effects in the balancing procedure. Since influence coefficients do change with the level of unbalance, a flexible rotor badly out of balance may require several balancing runs to obtain a low vibration reading.

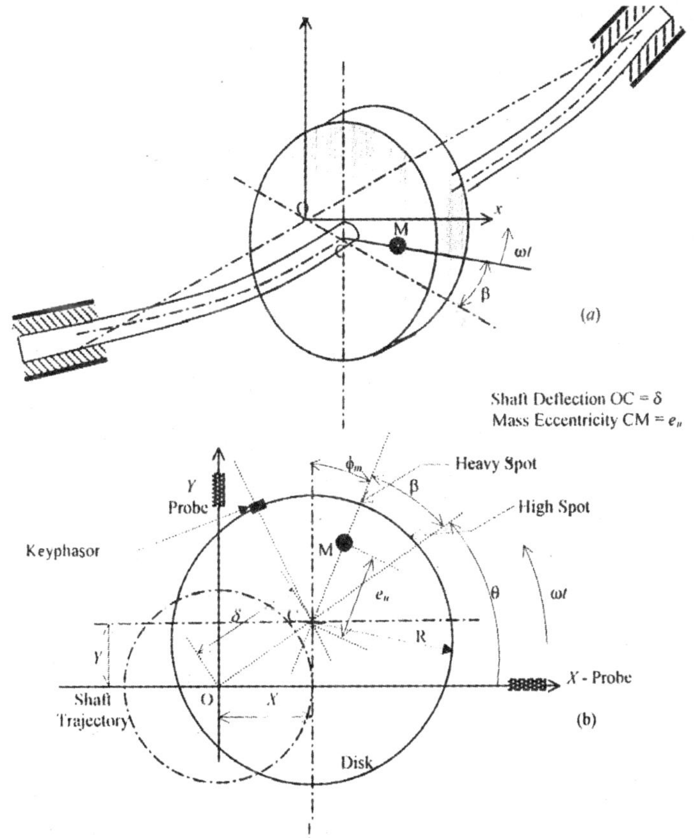

Fig. 8.14. *(a) Single Disk Jeffcott Rotor, (b) Cross Section of Jeffcott Rotor Showing Disk Displacement/Phase Relationship*

Rotor response is then measured at the same speed as the initial peak-to-peak vibration reading. Use equations (8.23) to (8.26).

Rotor Weight: $W = 1475.0$ lbs., rotor speed: $N_1 = 7250$ rpm

Initial Vibration Reading: $Y_1 = 2.75 \times 10^{-3}$ in. at angle: $\phi_1 = 285°$

Trial Weight: $W_{t1} = 0.185$ oz. at radius: $r_{t1} = 11.0$ in. at angle $\phi_{t1} = 45°$ from timing mark against rotation.

New rotor response: $Y_3 = 2.275$ mils at angle: $\phi_3 = 224°$

Balance Correction Weight: $W_{b1} = 2.17$ oz.-in. Angle: $\phi_{b1} = -5.38°$

The following procedure is for trim balancing. When the balance weight is left in the rotor, since the overall rotor amplitude is reduced, the resulting balance correction weight is referred to as a trim balance correction. For the case where a trial weight is welded to the rotor, it may not be convenient to remove this weight. However, if response is much larger than the original reading, the trial weight must be removed and another trial weight, preferably smaller, at another angular location must be attached.

The trim balance correction weight in the example problem above is calculated to be:

Correction: $U_{tr1} = 1.79$ oz.-in. Angle: $\phi_{tr1} = -66.4°$

Example Problem # 8.4: Figure 8.15 shows a typical critical speed map of a uniform shaft for various bearing stiffness values. The first three critical speeds are plotted as a function of bearing flexibility parameter $[\delta_s/(g\omega^2)]^{1/2}$, where ω_1 is the first critical speed with rigid supports and δ_s indicates flexibility of the bearings. The first critical speed $\omega_1 = \pi^2(gEI/WL^3)^{1/2}$, where g = acceleration due to gravity, E = modulus of elasticity, I = moment of inertia and W and L are total span weight and length. For a given value of ω_1 the first and second critical speed ratios decrease continuously with increasing bearing flexibility and the associated mode shapes transforming gradually. For larger values of bearing flexibility the two modes assume the first and second rigid body modes. The ratio between the two critical speeds is 3.5 for large values of bearing flexibility, which is the theoretical value for the rigid body modes.

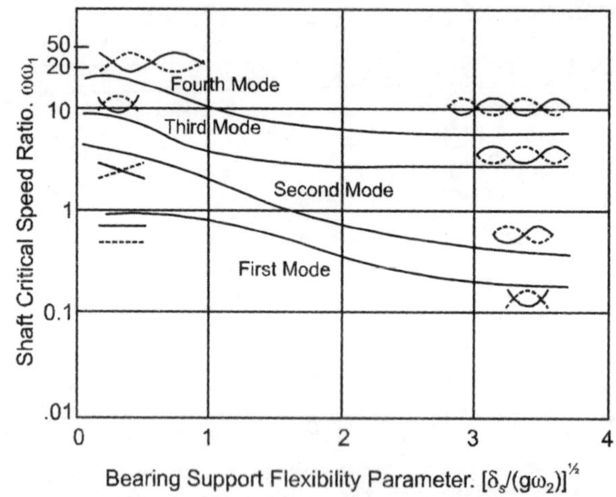

Fig. 8.15. *Critical Speed Map of Uniform Flexible Shaft*

For very large values of bearing flexibility the two rigid body critical speeds both become zero.

To understand the proper method of balancing such a rotor up to the third mode certain concepts such as modal eccentricity, modal weights and rotor response need to be understood. The procedure for balancing a rigid body rotor to take care of the first two rigid body modes has been discussed earlier using the two-plane influence coefficient method and the least-square-error method. The general procedure and method to balance the third mode arising from rotor flexibility is to be explored. Modal eccentricities of a uniform flexible rotor will be considered next.

Consider the uniform shaft of example problem # 8.1.

Shaft Length	L =	75.0 in.
Shaft Diameter	D =	3.125 in.
Material Young's Modulus:		
	E =	30×10^6 lbs/in²
Material Density	ρ =	0.283 lbs/in³ (128.4 gm/in³)
Shaft weight	=	162.8 lbs (73866 gm)

It is required to find the modal unbalance eccentricity vectors for the system using the first and the third modes assuming the following unbalance distribution:

$$U_1 = 125.0 \text{ gm-in at } x_1 = L/4$$
$$U_2 = -250.0 \text{ gm-in at } x_2 = L/2$$
$$U_3 = 125.0 \text{ gm-in at } x_3 = 3L/4$$

Solution: The normal mode vectors for the indicated unbalance locations are:

$$\phi_1 = \begin{bmatrix} 0.707 \\ 1 \\ 0.707 \end{bmatrix} \qquad \phi_2 = \begin{bmatrix} 1.0 \\ 0 \\ -1.0 \end{bmatrix} \qquad \phi_3 = \begin{bmatrix} 0.707 \\ -1 \\ 0.707 \end{bmatrix}$$

Modal weight for the uniform beam for the three modes is $W_1 = W_2 = W_3 = W_{total}/2 = 36933$ gm. Modal unbalance eccentricities e_i are calculated to be:

$$e_1 = (1/36931) \times [.707\ 1\ .707] \times \begin{Bmatrix} 125 \\ -250 \\ 125 \end{Bmatrix} = -0.00198 \text{ in.}$$

$$e_2 = (1/36931) \times [1\ 0\ -1] \times \begin{Bmatrix} 125 \\ -250 \\ 125 \end{Bmatrix} = 0.0 \text{ in.}$$

$$e_3 = (1/36931) \times [.707\ -1\ .707] \times \begin{Bmatrix} 125 \\ -250 \\ 125 \end{Bmatrix} = 0.01156 \text{ in.}$$

Modal unbalance eccentricity for the third mode is nearly 6 times that for the first mode, indicating that rotor response at the third critical speed will be considerable.

Example Problem # 8.5: Calculate the rotor response in example problem # 8.4 due to the specified unbalance for the first and third critical speeds and prescribed amplification factors of $A_1 = 8.0$ and $A_3 = 3.0$.

Solution: Rotor critical speeds were determined earlier. The first three critical speeds are: $\omega_1 = 2.648 \times 10^3$ rpm, $\omega_2 = 1.0597 \times 10^4$ rpm, $\omega_3 = 2.3824 \times 10^4$ rpm. At the first critical speed maximum rotor response at the center of the shaft is

$$\{Y_1\}_{x=L/2} = 2A_1e_1\{\phi_1\}\angle\text{-}90° = 2 \times 8 \times (-0.001983) \times 1\angle -90° = 0.03173\angle -270° \text{ in.}$$

At the third critical speed $\omega_3 = 2.3824 \times 10^4$ rpm the rotor response at the center of the shaft is calculated to be:

$$\{Y_3\}_{x=L/2} = 2A_3e_3\{\phi_3\}\angle -90° = 2 \times 3 \times 0.01156 \times 1\angle -90° = 0.0693\angle -90° \text{ in.}$$

The balance correction weights thus result in considerable dynamic motion in the shaft in the first mode and even greater in the third mode. The balance distribution may be expressed in modal components by:

$$\{U\} = U_{m1}\{\phi_1\} + U_{m2}\{\phi_2\} + U_{m3}\{\phi_3\}.$$

Example Problem # 8.6: Determine the modal unbalance components U_{mn} in example problems # 8.4 and 8.5 due to 150 gm-in unbalance at $x = L/2$ (fig. 8.16) and at $3L/4$ locations on the shaft. This single unbalance may be represented as a modal distribution located at three planes (three balance planes will be used).

Solution: For 150 gm-in unbalance at $x = L/2$ at $\angle 0°$:

$$U\} = \begin{Bmatrix} 0 \\ 150 \\ 0 \end{Bmatrix} = U_{m1}\begin{Bmatrix} 0.707 \\ 1 \\ 0.707 \end{Bmatrix} + U_{m2}\begin{Bmatrix} -1 \\ 0 \\ 1 \end{Bmatrix} + U_{m3}\begin{Bmatrix} -0.707 \\ 1 \\ -0.707 \end{Bmatrix}$$

Solving the simultaneous equations gives $U_{m1} = 75.0$ gm-in, $U_{m2} = 0$, $U_{m3} = 75.0$ gm-in, and the corresponding modal unbalance eccentricities are $e_1 = 0.00406$ in., $e_2 = 0.0$ in. and $e_3 = -0.00406$ in.

When the 150 gm-in unbalance is located at $x = 3L/4$ at $\angle 0°$, then:

$$\{U\} = \begin{Bmatrix} 0 \\ 0 \\ 150 \end{Bmatrix} = U_{m1}\begin{Bmatrix} 0.707 \\ 1 \\ 0.707 \end{Bmatrix} + U_{m2}\begin{Bmatrix} -1 \\ 0 \\ 1 \end{Bmatrix} + U_{m3}\begin{Bmatrix} -0.707 \\ 1 \\ -0.707 \end{Bmatrix}$$

resulting in $U_{m1} = 53.04$ gm-in, $U_{m2} = 75.0$, $U_{m3} = -53.04$ gm-in.

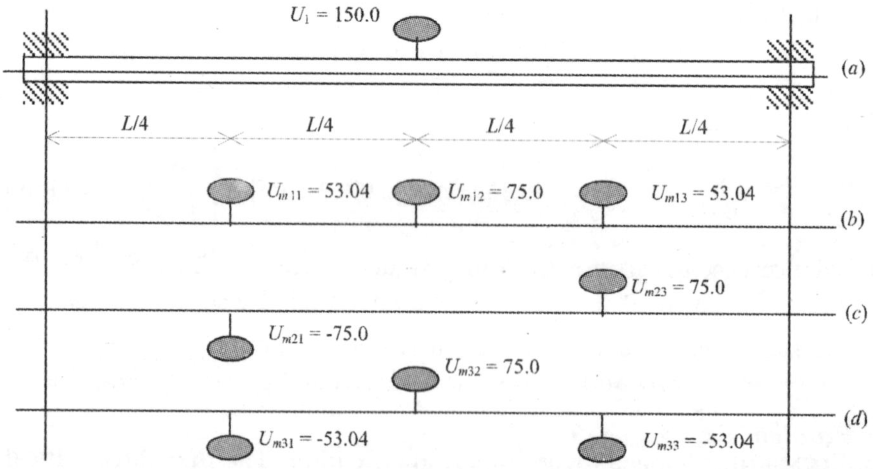

Fig. 8.16. *Single-Plane Unbalance of Flexible Shaft Represented as Modal Distribution*

Example Problem # 8.7: The condition for modal balancing a rotor at the first critical

(*a*) Uniform Rotor with 150 gm-in at Shaft Length $x = L/2$ at $\angle 0°$

(*b*) Equivalent First Mode Unbalance Distribution

(*c*) Equivalent Second Mode Unbalance Distribution

(*d*) Equivalent Third Mode Unbalance Distribution

speed is given by the expression $[\phi_1]^T[\{U\} + \{U_{bj}\}\{\phi_j\}] = M_1 e_1 = 0$. The mode shapes are orthogonal for a uniform beam if a symmetric axial mass distribution in the rotor is assumed. The condition for the balance of the first mode will then reduce to the form

$\phi_1^T\phi_1 [\{U_{m1}\} + \{U_{b1}\}] = 0$, or $\{U_b\}_1 = U_{b1}\phi_1 = -U_{m1}\{\phi_1\}$. The flexible rotor in example problems 8.4 to 8.6 has been balanced rigidly. Calculate the balancing components for the first and third modes.

Solution: Components for the balancing of the first mode are: $\phi_1^T[\{U\} + U_{b1}\phi_1] = 0$, or

$$[0.707 \quad 1.0 \quad 0.707]\times\left[\begin{Bmatrix}125\\-250\\125\end{Bmatrix}+U_{b1}\begin{Bmatrix}0.707\\1.0\\0.707\end{Bmatrix}\right]=0$$

Thus, $-73.25 + 2.414U_{b1} = 0$, or $U_{b1} = 30.343$ gm-in. Hence, balance distribution for first mode is $\{U_{b1}\} = 30.343\{\phi_1\} = \{21.45\ 30.34\ 21.45\}^T$ gm-in. The new unbalance distribution now becomes

$$\{U\}_{current} = \begin{Bmatrix}125\\-250\\125\end{Bmatrix}+\begin{Bmatrix}21.45\\30.34\\21.45\end{Bmatrix}=\begin{Bmatrix}146.45\\-219.66\\146.45\end{Bmatrix}$$

But this configuration is not balanced as a rigid rotor, even then it holds the balance for operation through the first two critical frequencies. For operation through the third mode the balancing component, the condition is $\phi_3^T[\{U\} + U_{b3}\phi_3] = 0$, or:

$$[0.707 \quad -1 \quad 0.707]\times\left[\begin{Bmatrix}146.45\\-219.66\\146.45\end{Bmatrix}+U_{b3}\begin{Bmatrix}0.707\\-1.0\\0.707\end{Bmatrix}\right]=0$$

Hence, $426.80 + 2U_{b3} = 0$, or $U_{b3} = 213.40\angle 180°$ gm-in. For the third mode the balance components are $\{U_{b3}\} = 213.40\{\phi_3\} = \{-146.45\ 219.66\ -146.45\}^T$ gm-in. The final unbalance distribution then becomes:

$$\{U\}_{final} = \begin{Bmatrix}146.45\\-219.66\\146.45\end{Bmatrix}+\begin{Bmatrix}-146.45\\219.66\\-146.45\end{Bmatrix}=\begin{Bmatrix}0\\0\\0\end{Bmatrix}$$

Thus, after the balancing of the third mode the rotor is fully balanced. This procedure is recommended for balancing of flexible rotors in a specified speed range.

***Example Problem # 8.8*:** Figure 8.17 shows a thin disk mounted on a uniform shaft. The shaft has a set bow. When the rotor is placed on a pair of knife-edges and slowly rotated, the disk centerline describes an orbit. The bow is a vector quantity requiring radial bow displacement and phase lag angle from the shaft timing mark, producing an equivalent unbalance whose magnitude is proportional to weight of the rotor and radius of the orbit. Rotor data is as follows:

Rotor weight:	P	=	13.5 lbs
Distance between bearing center-lines:	L	=	64.375 in.
Shaft outer diameter	Od	=	1.375 in.
Shaft inner diameter	Id	=	0.75 in.
Shaft material Young's modulus	E	=	28.5×10^6 lbs/in^2
Damping coefficient	C	=	0.525 lbs-second/in.
Eccentricity of unbalance	e_{u1}	=	0.0015 in.
Unbalance phase angle	ϕ_{m1}	=	15.0° (lag)
Shaft bow vector	δ_{r1}	=	0.001 in.
Shaft bow phase angle	ϕ_{m2}	=	100.0° (lag)
Balancing speed	N_{b1}	=	1600 rpm.

It is required to balance the rotor using the standard influence coefficients method, the run-out subtractions method, the bow vector compensation method and the least square error method (*ref*: Nicholas *et al.*, 1976).

***Solution*:** The equivalent unbalance due to the bow in the shaft is: $U_u = (13.5 \times .001 \times 453) =$ 6.12 gm-in. For a simply supported uniform shaft the stiffness is given by:

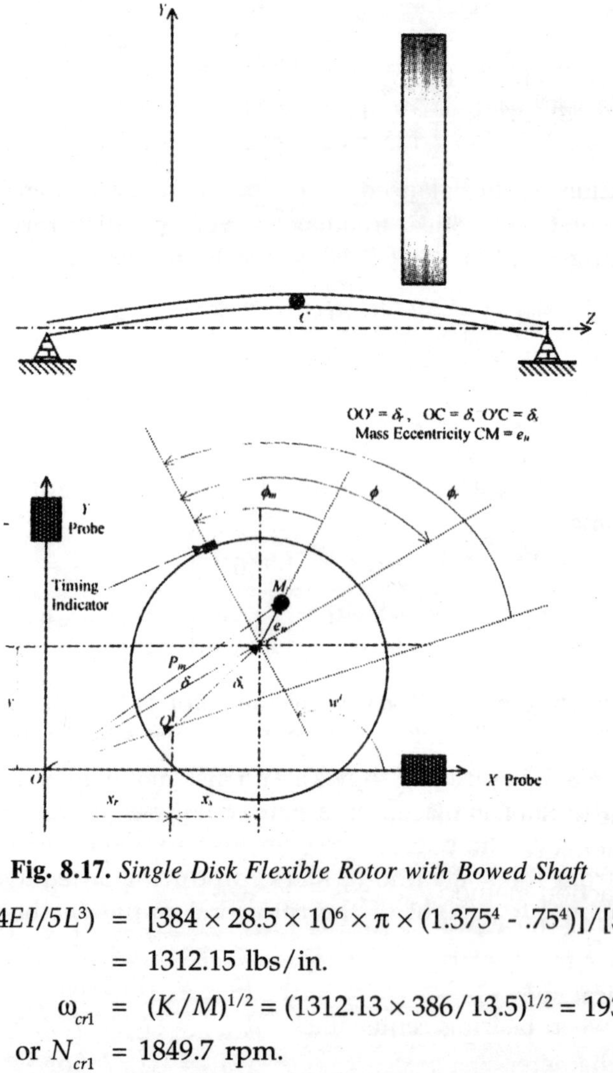

Fig. 8.17. *Single Disk Flexible Rotor with Bowed Shaft*

$$K = (384EI/5L^3) = [384 \times 28.5 \times 10^6 \times \pi \times (1.375^4 - .75^4)]/[5 \times 64.375^3 \times 64]$$
$$= 1312.15 \text{ lbs/in.}$$

Then
$$\omega_{cr1} = (K/M)^{1/2} = (1312.13 \times 386/13.5)^{1/2} = 193.69 \text{ rad/sec,}$$
$$\text{or } N_{cr1} = 1849.7 \text{ rpm.}$$

Critical damping $C_c = 2M\omega_{cr} = 2 \times 13.5 \times 193.69/386 = 13.55$ lbs-sec/in, damping ratio $\xi = C/C_c = 0.525/13.55 = 0.0387$ and amplification $A_{cr} = 1/2\xi = 12.903$. In complex notation shaft unbalance $e_{u1} = 0.0015e^{-i15°}$ in. $= 0.038e^{-i15°}$mm and shaft bow is $\delta_{r1} = 0.001e^{-i100°}$ in. $= 0.025e^{-i100°}$ mm. The balancing speed ratio $f = 1600/1849.7 = 86.5$ percent of the critical speed. Phase angle $\phi = \tan^{-1}(2\xi f)/(1 - f^2) = 14.9°$.

In the method of standard influence coefficients, the coefficients for unbalance are

$$c_u = (f^2 e^{-i\phi})/[(1 - f^2)^2 + (2\xi f)^2]^{1/2} = 2.872\angle{-14.9°}$$
$$c_r = e^{-i\phi}/[(1 - f^2)^2 + (2\xi f)^2]^{1/2} = 3.839\angle{-14.9°}$$

Total initial response at the given balance speed is

$$Y_1 = c_u e_u + c_r \delta_r = (2.872\angle{-14.9°}) \times (0.0015\angle{-15°}) + (3.838\angle{-14.9°}) \times (0.001\angle{-100°})$$

$$= 0.00431\angle{-29.9°} + 0.00383\angle{-114.9°}$$

$$= 0.006015\angle{-69.4°} \text{ in. or, } 0.153\angle{-69.4°} \text{ mm}$$

Attaching a trial weight on the rotor results in response of $Y_{tr} = c_u(e_u + e_{tr}) + c_r\delta_r$. The corresponding influence coefficient is obtained by subtracting the initial response Y_1 from the response with the measured trial weight (2.5∠180° oz.) response of 0.0024∠–85.5° in. due to e_{tr}. Then: $c_u = (Y_{tr} - Y_1)/e_{tr} = 1.508\angle{-59.2°}$.

The standard influence coefficient method assumes that the rotor responds only to mechanical unbalance. Balance correction is obtained from:

$$e_b = -(Y_1/c_u) = -(6.014\angle{-69.4°})/(1.507\angle{-59.2°}) = 0.00399\angle{-190.2°}$$

in.

If the balance radius is 15 mm (0.59 in.), the balance correction weight is:

$$U_b = w\,e_b = w_b R = 13.5 \times .00399 = 0.0539 \text{ lb–in} = 24.40 \text{ gm–in.}$$

Hence, $w_b = 24.4/0.59 = 41.3$ gm at an angle of 190.2° from the timing mark.

But the rotor is balanced at 1600 rpm, and will have high response at the critical speed. At the critical speed $c_u = c_r = A_{cr}\angle{-90°} = 12.903\angle{-90°}$, and the amplitude is:

$$Y_{cr} = (e_u + \delta_r)A_{cr} = (1.5\angle{-15°} + 1.0\angle{-100°}) \times 12.905\angle{-90°}$$

$$= 24.18\angle{42.9°}$$

Response at the critical speed will be:

$$Y_{cr,b} = c_u(e_u + e_b) + c_r\delta_r = 12.905\angle{-90°} \times (1.5\angle{-15°} + 3.99\angle{-190.2°}) + 1.0\angle{-100°} \times 3.838\angle{-14.9°}$$

$$= 51.86\angle{102.9°} \text{ mils} = 1.317\angle{102.9°} \text{ mm.}$$

Total response due to the factors is high and is out of phase with the bow vector, leading to the high response. Therefore, the influence coefficient method is not acceptable.

The run out subtraction method assumes rotor response of the form $\boldsymbol{Y} \approx \boldsymbol{c_u e_u} + \delta_r$. So $\boldsymbol{e_b} = -\boldsymbol{e} = -(\boldsymbol{Y} - \delta_r)/\boldsymbol{c_u} = -(6.014\angle{-69.4°} - 1.0\angle{-100°})/(2.872\angle{-14.9°}) = 2.40\angle{121.3°}$ gm-in. The balance correction is then $\boldsymbol{U_b} = \boldsymbol{e_b} w_b = (4.80\angle{121.3°})$ and the correction weight $w_b = 4.80/0.59 = 8.129$ gm. The amplitude at critical speed is

$$\boldsymbol{Y_{cr,b}} = 12.905\angle{-90°} \times (1.5\angle{-15°} + 2.40\angle{121.3°}) = 21.55\angle{173.1°} \text{ mils}$$

$$= 0.547 \text{ mm}$$

In the bow vector compensation method the influence coefficient for the bow vector is obtained from the expression $\boldsymbol{c_r} = \boldsymbol{c_u}/f_b^2$, where $f_b = N_b/N_{cr} = 1.0$ at the critical speed. With this procedure, since $\boldsymbol{Y_{cr,b}} = \boldsymbol{c_u}(\boldsymbol{e_u} + \boldsymbol{e_b}) + \boldsymbol{c_r}\delta_r = \boldsymbol{c_{cr}}(\boldsymbol{e_u} + \boldsymbol{e_b} + \delta_r) = 0$, the coefficients for shaft bow and unbalance become equal. The balance correction then becomes $\boldsymbol{e_b} = -(\boldsymbol{e_u} + \delta_r)$. For speeds other than the critical speed ($f_b = 0.865$), $\boldsymbol{c_r} = 2.872/(.865^2) = 3.839\angle{-14.9°}$. The amplitude is obtained from $\boldsymbol{Y_b} = \boldsymbol{c_u}(\boldsymbol{e_u} + \boldsymbol{e_b} + \delta_r/f_b^2) \neq 0$, and the balance eccentricity vector then takes the form of $\boldsymbol{e_b} = [\boldsymbol{Y_1}/\boldsymbol{c_u} + \delta_r(1 - 1/f_b^2)]$. For the given values of shaft bow measured at the lower speed and the ratio f_b, the value of the balancing eccentricity is $\boldsymbol{e_b} = 2.78\angle{-112.7°}$ mils, or $0.0706\angle{-112.7°}$ mm.

It is difficult to distinguish between true shaft bow and a constant shaft residual run-out vector. To overcome this, multi-speed balancing based on minimization of rotor response over a speed range has been developed using the least-square-error method. In this procedure the rotor low-speed run-out vector is measured. Rotor response is then measured at several speed points. A trial weight is next attached to the rotor, and new response is then measured at the same speeds. The single mass Jeffcott rotor with shaft bow and unbalance considered earlier will be balanced by the Least-Squared-Error method by measuring vibration amplitude below and above rotor critical speed, at 1600 and 2000 rpm.

Vibration amplitude and the corresponding phase angle at three speeds points are:

$$Y_{1600} = 4.84 \times 10^{-3}\angle{-83°} \text{ in.}$$
$$Y_{1850} = 14.32 \times 10^{-3}\angle{-147°} \text{ in.}$$
$$Y_{2000} = 7.84 \times 10^{-3}\angle{-179°} \text{ in.}$$

Compensation by the slow roll vector from constant shaft run-out is $\delta_r = 0.001\angle{-100°}$ in. Subtracting it from the vibration amplitudes gives:

$$Y_{1c} = 4.84 \times 10^{-3}\angle{-83°} - 0.001\angle{-100°} = 3.895 \times 10^{-3}\angle{-78.7°}$$
$$Y_{2c} = 7.84 \times 10^{-3}\angle{-179°} - 0.001\angle{-100°} = 7.712 \times 10^{-3}\angle{-6.3°}$$

The trial weight is 0.826 gm at a radius of 1.0 in., or $U_{tr} = 1.824 \times 10^{-3}\angle 167°$ in-lbs. Vibration readings after the placement of trial weight are:

$$Y_{tr1600} = 3.89 \times 10^{-3}\angle{-75°} \text{ in.}$$
$$Y_{tr1850} = 9.47 \times 10^{-3}\angle{-148°} \text{ in.}$$
$$Y_{tr2000} = 3.72 \times 10^{-3}\angle{-182°} \text{ in.}$$

Influence coefficients at 1600 and 2000 rpm are:

$$c_1 = 4.71\angle{-245.7°} \text{ and } c_2 = 9.33\angle{-173.3°} \text{ mils/gm.}$$

Complex conjugates of these influence coefficients are:

$$c_1^* = 4.71\angle 245.7° \text{ and } c_2^* = 9.33\angle 173.3° \text{ mils/gm.}$$

Balance correction by the least square error method in a single plane for two speeds is:

$$U_b = -(Y_{1c}c_1^* + Y_{2c}c_2^*)/(c_1c_1^* + c_2c_2^*) = -(49.767\angle 167°)/33.169\angle 180°$$
$$= 0.826\angle{-193°} \text{ gm.}$$

Example Problem # 8.9: Figure 8.5 shows an overhung rotor with radial unbalance and skewed disk. The plane of the disk is inclined from the vertical by a small angle. Balancing this rotor is more complex than a conventional Jeffcott rotor symmetrically situated between bearings due to gyroscopic moments generated by the skew. If the disk skew is out of phase with the unbalance the effects combine to increase the amplitude of motion at critical speed. On the other hand if disk skew is in phase with disk unbalance, the gyroscopic moment will help to offset the radial unbalance force, thus reducing overall vibration amplitude. Provide a procedure for balancing the rotor.

Solution: The recommended procedure for balancing an overhung rotor in the first attempt is to single-plane balance the disk by using influence coefficients and vibration amplitudes obtained from the near and far bearings. This will produce a compromise balance that will prevent excessive vibrations in one bearing. If the resulting vibration is still not low, a trial balance couple using two weights 180 degrees out of phase to each other and separated by an axial distance should be attached to the disk.

Similar balancing can also be achieved by the two-plane balancing procedure in which trial weights are placed on the two flat faces of the disk and vibrations for the near and far bearings recorded for two trial runs.

Vibration monitored at the near and far planes is assumed to be a function of two unknown values of unbalance. A trial balance is placed in each plane and resulting amplitudes of motion are measured. In this calculation positive sign represents lag angles for both vibration and balance weight measures against the direction of rotation.

Measurements for the slow roll vector (from constant shaft run-out) are:

$$\mathbf{Y}_{srn} = 0.8 \times 10^{-3}\angle 256° \text{ in.}$$
$$\mathbf{Y}_{srf} = 0.95 \times 10^{-3}\angle 238° \text{ in.}$$

Note that vibration and trial weight angles are measured against rotation from the vibration probe. Initial vibration readings at near and far locations are measured to be:

$$\mathbf{Y}_{in} = 1.95 \times 10^{-3}\angle 175° \text{ in.}$$
$$\mathbf{Y}_{if} = 3.25 \times 10^{-3}\angle 142° \text{ in.}$$

Compensated near and far vibration amplitudes are obtained by subtracting the slow roll vectors:

$$\mathbf{Y}_{n} = 1.99 \times 10^{-3}\angle 151.6° \text{ in.}$$
$$\mathbf{Y}_{f} = 3.48 \times 10^{-3}\angle 126.2° \text{ in.}$$

The vibrations at the two planes are assumed to be functions of the two unknown unbalances $\mathbf{U}_1$ and $\mathbf{U}_2$:

$$\mathbf{Y}_{n} = c_{n1}\mathbf{U}_1 + c_{n2}\mathbf{U}_2$$
$$\mathbf{Y}_{f} = c_{f1}\mathbf{U}_1 + c_{f2}\mathbf{U}_2$$

A trial weight is first attached in the near plane and the amplitude of motion is measured. Leaving it in place, attach a second weight in the far plane and measure the resulting amplitude of motion.

First trial weights in the two planes are:

$$\mathbf{U}_{1t} = 7.5\angle 83° \text{ gm}$$
$$\mathbf{U}_{2t} = 5.2\angle 180° \text{ gm}$$

Vibration readings at near and far locations after placing the trial weights in the two planes are measured to be:

$$\mathbf{Y}_{n1} = 2.42 \times 10^{-3}\angle 179° \text{ in.}$$
$$\mathbf{Y}_{f1} = 2.29 \times 10^{-3}\angle 356° \text{ in.}$$
$$\mathbf{Y}_{n2} = 3.19 \times 10^{-3}\angle 248° \text{ in.}$$
$$\mathbf{Y}_{f2} = 2.87 \times 10^{-3}\angle 169° \text{ in.}$$

The set of four influence coefficients may be expressed in the following form:

$$c_{n1} = (Y_{n1} - Y_n)/U_{1t} = 0.150\angle 150.4^\circ$$
$$c_{f1} = (Y_{f1} - Y_f)/U_{1t} = 0.701\angle 62.7^\circ$$
$$c_{n2} = (Y_{n2} - Y_n)/U_{2t} = 0.758\angle -81.9^\circ$$
$$c_{f2} = (Y_{f2} - Y_f)/U_{2t} = 0.458\angle 71.4^\circ$$

The balance correction weights are given by:

$$\begin{Bmatrix} U_{b1} \\ U_{b2} \end{Bmatrix} = -\begin{vmatrix} U_1 & U_{1t} \\ U_2 & U_{2t} \end{vmatrix} = -\begin{vmatrix} c_{n1} & c_{n2} \\ c_{f1} & c_{f2} \end{vmatrix}^{-1} \begin{Bmatrix} Y_{n2} \\ Y_{f2} \end{Bmatrix}$$

The balance correction weights are:

$$U_{b1} = (Y_{f2}c_{n2} - Y_{n2}c_{f2})/(c_{n1}c_{f2} - c_{f1}c_{n2})$$
$$U_{b2} = (Y_{n2}c_{f1} - Y_{f2}c_{n1})/(c_{n1}c_{f2} - c_{f1}c_{n2})$$

Correction weight and location at near plane is calculated to be:

$$U_{b1} = 5.77\angle 120.9^\circ \text{ gm against rotation}$$

and the correction weight and location at far plane is:

$$U_{b2} = 3.19\angle 141.76^\circ \text{ gm against rotation.}$$

Vibration readings after placing correction weight in near plane are measured to be:

$$Y_{n1} = 0.6 \times 10^{-3}\angle 271^\circ \text{ in.}$$
$$Y_{f1} = 0.55 \times 10^{-3}\angle 4^\circ \text{ in.}$$

Vibration readings after placing correction weight in far plane are measured to be:

$$Y_{n2} = 2.7 \times 10^{-3}\angle 298^\circ \text{ in.}$$
$$Y_{f2} = 1.5 \times 10^{-3}\angle 139^\circ \text{ in.}$$

REFERENCES & BIBLIOGRAPHY

American Petroleum Institute, "Special Purpose Gear Units for Refinery Service", API Std. 613, 4th edition, December, 1994.

American Petroleum Institute, "Centrifugal Compressors for General Refinery Service", API Std. 617, 5th edition, April, 1994.

American Petroleum Institute, "Centrifugal Pumps for General Refinery Service", API Std. 610, 8th edition, December, 1994.

Balda, M., "Dynamic Properties of Turbo-set Rotors", Symposium on Dynamics of Rotors, IUTAM, Lyngby, Denmark, 1975.

Barrett, L.E., Allaire, P.E., Gunter, E.J., "Optimum Bearing and Support Damping for Unbalance and Response and Stability of Rotating Machinery", Journal of Engineering for Power, 100(1):89-94, 1978.

Bentley, D.E., "Polar Plotting Applications for Rotating Machinery", Vibrations Institute, Clarendon Hills, ILL, 1982.

Bishop, R.E.D., Parkinson, A.G., "On the Isolation of Modes in the Balancing of Flexible Shafts", Proceedings of the Institute of Mechanical Engineers, 177:407-423, 1963.

Bishop, R.E.D., Parkinson, A.G., "Second Order Vibration of Flexible Shafts", Proceedings Royal Society of London, 259:1-31, 1966.

Blake, M.P., "Use Phase Measuring to Balance Rotors in Place", Hydrocarbon Processing, August, 1967.

Compressed Air and Gas Institute, "In-service Standards for Centrifugal Compressors", Cleveland, OH, 1993.

Cooley, J.W., Tukey, J.W., "An Algorithm for the Machine Calculation of Complex Fourier Series", Mathematics of Computation, 19(90):297 - 301, 1985.

Darlow, M., "Flexible Rotor Balancing by the Unified Balancing Approach", Rensselaer Polytechnic Institute, Troy, NY., 1989.

Den Hartog, J.P., "Mechanical Vibrations", 4th edition, Mc-Graw Hill, New York, 1985.

Ehrich, F.F., "Handbook of Rotor Dynamics", Krieger Publishing Co, Malabar, FL, 1999.

Foiles, W., Gunter, E. J., "Balancing a Three Mass Rotor with Shaft Bow", University of Virginia, Charlottesville, 1982.

Goodman, T.P., "A Least-Squares Method for Computing Balance Correction", Journal of Engineering for Industry, 86(3), Pages 273 – 279, 1974.

Hydraulic Institute, "Acceptable Field Vibration Limits for Horizontal Pumps", 14th edition, Centrifugal Pump Applications, 1994.

ISO 11340, "Mechanical Vibration – Methods and Criteria for the Mechanical Balancing of Flexible Rotors", 1994.

ISO 4867, ISO 4868, "Guidelines for Evaluation of Vibration in Merchant Ships", after 6954, 1993.

ISO 1925 and ANSI S2.7, "Balancing – Vocabulary", 1990.

ISO 2372, "Mechanical Vibration of Machines with Operating Speeds from 10 to 200 rps – Basis for Specifying Evaluation Standards", 1989.

ISO 2631, "Balancing of Rotors", 1979.

ISO 6945, "Guidelines for Evaluation of Vibration in Merchant Ships", 1987.

ISO 1940 and ANSI S2.19, "Balancing Quality of Rotating Rigid Bodies", 1986.

ISO 2954, "Mechanical Vibration of Rotating and Reciprocating Machinery – Requirements for Instruments for Measuring Vibration Severity", 1984.

ISO/DIS 2953, "Balancing Machines — Description and Evaluation", 1984.

ISO/DIS 10816, "Measurement and Classification of Vibrations in Rotating Machines", ANSI counterpart S2.12 part 6-199X, 1993.

ISO/WD Standard 8579-2, "Vibrations in Enclosed Gear Reduction Units", 1981.

ISO 2371 and ANSI S2.38, "Field Balancing Equipment – Description and Evaluation", 1982.

ISO 1940, "Balance Quality of Rotating Rigid Bodies", 1973.

Military Standard, "Mechanical Vibration of Shipboard Equipment (Type 1 – Environmental and Type 2 – Internally Excited)," MIL-STD-167-1 (SHIPS), May, 1984.

Nicholas, J.C., Gunter, E.J., Allaire, P.E., "Effect of Residual Shaft Bow on Unbalance Response and Balancing of a Single Mass Flexible Rotor: Part I – Unbalance Response, Part II – Balancing", Journal of Engineering and Power, Volume 98(2), Pages 171 – 189, 1976.

Palazzolo, A.B., Gunter, E.J., "Multimass Flexible Rotor Balancing by the Least-Squares Error Method", Vibration Institute, Clarendon Hills, Ill., 1982.

Rieger, N., "Balancing of Rigid and Flexible Rotors", SVN 12, The Shock and Vibration Center, 1986.

Salamone, D.J., Gunter, E.J., "Synchronous Unbalance Response of an Overhung Rotor with Disk Skew", ASME paper # 79-GT-135, New York, 1979.

Schneider, H., "Balancing Technology", 4th edition, Carl Schenck, AG, Darmstadt, Germany, 1991.

Stadelbauer, D.G., "Fundamentals of Balancing", 3rd edition, Schenck Trebel Corp., Deer Park, NY., 1990.

Winkler, A.F., "High Speed Rotating Unbalance, Coupling or Rotor", Proceedings, Vibration Institute, Nassau Bay Conference, Clarendon Hills, Ill, 1983.

CHAPTER 9

Turbo-Machinery Noise

9.1 INTRODUCTION

Mechanical power produced by gas turbine engines, compressors and pumps has been responsible for a great deal of prosperity and abundance over the last century. Agricultural machinery has increased grain production several folds, electric power generators illuminate the remotest corners of the world and motorboats and snowmobiles provide leisure time activity for everyone. Turbo-machines have been at the forefront to lead the transformation. But this progress has come at a price. Vast amounts of pollutants are spewed out into the air by the engines, depleted energy resources and created noise pollution. Noise is now no longer considered to be only an annoying irritant. Medical science has proved that prolonged exposure to high levels of noise results in hearing impairment, sometimes even in a total loss of hearing. Excessive noise can have severe psychological and physiological effects on humans. Cardiovascular, muscular and fatigue effects can be directly correlated to the intensity of exposure to sound levels. Hence, the impetus to make the machines operate quietly certainly exists.

The connection between mechanical vibration of a component and noise caused by it is easily recognized. Man has used sound for a long time to discriminate between properly functioning machines and those where failure of a component is imminent. Before the advent of accelerometers and spectrum analyzers production line inspectors merely listened to an operating machine as it was run for final inspection. This type of a quality evaluation, however, had several problems. The ambient noise level plays a significant role. A particularly noisy machine can be readily identified when the background noise is low. But if the noise in the background is high it becomes more difficult to hear the telltale signature of an unacceptable component. In addition, after a long exposure to the same sound the human faculty of hearing loses its ability to identify discrete noise sources. Individual mood changes may also play a role in applying the acceptance criteria. More recently, sound and vibration meters are extensively used together with signal processing equipment for production line quality control testing of machines and components. The analyses can be made more objectively and definitively, and the data can be stored for further evaluation at a later date.

Direct combustion noise arises when a volume of gas expands at constant pressure as it is heated during combustion. The surrounding gas expands and produces a sound wave as it travels outside the boundary of the flame. The intensity of the sound, controlled by the pressure in the acoustic wave, depends on the rate of volume generation from the combustion. The size distribution of eddies in a turbulent flow regime plays a role in the sound generation mechanism since it controls the turbulent mixing lengths throughout the combustion zone. The radiated sound power and the heat release in the combustion are of importance in the description of direct combustion noise. Interference within the source region causes only a fraction of the total thermal power to be emitted in the form of acoustic power during combustion in a gas turbine. The radiated sound power covers a broad spectrum of frequencies from around 100 to 2000 Hz, reaching a blunt peak between 300 and 500 Hz. Although geometric features, engine power and flame temperature play a role in affecting the radiated sound power level, the general shape of the sound power/frequency curve remains consistent. General Electric Company and Pratt and Whitney have used the combustion noise theory, supplemented by sound data recordings to derive mathematical formulations to calculate overall sound power level (combination of direct and indirect combustion noise) based on combustor and turbine operating parameters. The procedures are outlined in publications by Ho *et al.,* (1979) and Matthews *et al.,* (1977).

Combustion in a flowing turbulent fuel-air stream is accompanied by noise due to the process itself and by a more insidious form of noise that arises when combustion instabilities couple with the acoustic modes in the chamber. The combustor system has natural frequencies that may be excited by the combustion process to produce oscillations in the longitudinal, radial or circumferential or a combination of these modes. In the axially oriented modes the resonant pressure waves travel along the length of the chamber much as in an organ pipe, and in the circumferential mode the waves slosh tangentially to the primary flow direction. Combustion instability has figured prominently in aircraft engines provided with an afterburner (*ref*: Lefebvre, 1999).

Analysis of dynamic excitation, structural response and sound radiation characteristics of machines helps to identify dominant sound radiating surfaces, and to determine important dynamic features of the internal and external components related to sound transmission and radiation. Evaluation of the data then permits selection of various treatments and modifications for controlling the noise at its source. The analyses could be performed experimentally or analytically. However, complexities and non-linearities make analytical studies time consuming and generally more expensive than experimental methods. Digital signal processing techniques using Fourier analysis type of instrumentation make many of the measurements practical. Data acquired during operating and static tests provide noise source identification, dynamic force generation and structural response. Frequency response of the structure can be measured by applying a controlled force when the machinery is shutdown. Resonant frequencies, their corresponding mode shapes and damping values can then be obtained. The task of defining sound emission characteristics of the machine may appear at first sight to be straightforward, but it requires ingenuity and difficult work that is rarely complete. After understanding the noise generation and radiation process, a cure has to be found for the diagnosis. Finite element studies come in useful at this point. Once the cure is found, it has to be acceptable from production and cost considerations, otherwise one may be faced with a terminal case!

9.2 OPERATING PARAMETERS

The amplitude of the root-mean-square of the acoustic pressure wave gives sound pressure. At the threshold of hearing the magnitude of sound pressure $p_o = 2 \times 10^{-4}$ μbar, or 2×10^{-5} N/m². Sound power per unit area defines sound intensity, and can be computed from sound pressure:

$$I = p^2/\rho c, \text{ W/m}^2 \tag{9.1}$$

where ρ = density of air, kg/m³ and c = acoustic speed of air, m/second.

At the threshold of hearing the sound intensity $I_o = 10^{-12}$ W/m². Sound intensity ratio is the ratio of sound intensity I over the reference intensity I_o. The ratio varies from 10^0, the threshold of hearing, to 10^{14}, the threshold of pain. For convenience the exponent of 10 is used as the unit bel. Therefore, the thresholds of hearing and pain are 0 and 14 Bels, or 0 and 140 decibels (dB). Sound intensity and pressure levels in dB are defined as:

$$SIL = 10\log\left(\frac{I}{I_o}\right) \tag{9.2}$$

$$SPL = 20\log\left(\frac{p}{p_o}\right) \tag{9.3}$$

A decibel is a logarithmic unit used to measure the relative loudness of sounds. A ten unit increase in decibels, for example, from 70 to 80 dB, corresponds to a ten-fold increase in the sound energy and a doubling in the perceived loudness. Since sound pressure level is a physical quantity, it does not consider human response to noise. A weighted average, dBA, approximately simulates the frequency response of the human ear to noise. The dBA level may be referred to as the noise of the source. The advantage of considering changes in noise levels is that changes can be calculated from sound pressure level, provided measurement conditions are maintained consistently. Hence:

$$\Delta L = \Delta\,(\text{dBA}) = \Delta\,(SPL) \tag{9.4}$$

where L signifies noise level.

Since auditory perception of the magnitude of noise is generally subjective, other methods for determining its effects on humans and the environment exist. Composed of a number of frequency bands of rapidly changing acoustic intensity, each of the bandwidths cause a varying response in the auditory sense. Hence, broadband noise may be split into selected bands to determine the overall impact on the auditory senses to the loudness of a sound. One procedure expresses the subjective effects in the form of perceived noise level (PNL), defined by the non-dimensional logarithmic parameter PNL = 10 × {(log PN)/(log 2)} + 40, where PN is the perceived noisiness. The constant value of 40 indicates the auditory level of a standard noise source of PN = 1, such as normal speech. Perceived noisiness is also expressed in decibels, and is expressed as PNdB.

Industrial turbo-machinery noise is often measured at a distance of 1 m from the machine. Only the bare machine is used for noise evaluation. The machine is equipped only with built-in accessories essential for operation, such as fuel, oil and water pumps and manifolds. Sound from the manifolds, test stand and dynamometer must be minimized to measure the sound level of the basic, or bare, machine. Sound radiation occurs at machine surfaces. These include valve covers, oil pan and other non-structural load carrying panels. Radiation of surface noise is affected by factors like force transmissibility from machine structure to support, area and radiation efficiency, damping and modal density of the machine's radiating surfaces. Hence, the basic machine noise depends on exciting force characteristics, structural response, and vibration transmissibility and radiation efficiency of the machine's external surfaces.

Sound transmission loss is 10 times the common logarithm of the ratio of the airborne sound power transmitted by a surface and radiated on the opposite side of the surface. Used to designate the reduction provided by various partitions, the transmission loss achieved by a given construction depends on physical parameters of the partition, such as mass, stiffness and damping. The loss of sound transmission is frequency dependent and is expressed in decibels.

Noise reduction is denoted by the numerical difference in dB between the average sound pressure levels in the source area and the receiving area. The combined effect of transmission loss of the component separating the areas plus the absorptive effect present in the receiving and source areas is included in the difference. Insertion loss is the numerical difference in dB between two sound pressure levels measured at the same point before and after the insertion of a silencer between the measuring point and the noise source. Flow velocity, temperature and direction with respect to the sound source affect the performance of the silencer.

Divergence provides the reduction of a sound pressure level as a function of the distance from the source. The reduction is proportional to the inverse square of the distance, since it is based on the spreading of sound energy from the source area to a target area some distance away.

9.3 JET ENGINE NOISE

Jet engine noise is an unequivocal characteristic of its design, coupled strongly by operating conditions. A smaller aircraft engine with reduced air mass flow rate but a high exhaust velocity (2000 ft/second) will be characterized by the exhaust jet and it's mixing with the ambient air. A low bypass ratio turbofan engine permits increased airflow rate at the intake and a lower exhaust flow velocity, permitting a reduction in the noise level from the exhaust jet. At the same time, however, noise from the turbo-machine tends to increase. High bypass ratios (in the range of 5.0 to 7.0) direct more of the airflow into the fan duct of the engine. Flow velocity at the exit end of the fan duct maybe around 1000 feet/second and around 1300 feet/second from the core engine, relegating the exhaust flow noise to a lower level than the turbo-machinery noise.

Newer airliners are required to comply with noise regulations provided in Federal Aviation Regulation (FAR) part 36 in the US and in Joint Airworthiness Requirements in Europe, both of which are consistent with rules set by the International Civil Aviation Organization. All turbofan powered transport aircraft must be certified with effective perceived noise level (EPNL) limits based on gross weight of the aircraft during takeoff and landing for three measuring stations: under the approach path 2000 m before touchdown, under the takeoff path 6500 m from the start of the takeoff roll and at sideline points of maximum noise located 450 m from the runway centerline at takeoff. The sideline points are more indicative of the intrinsic noise pattern of the engine during takeoff because the engine is running at full throttle. Maximum noise at the designated measuring line generally is produced when the aircraft reaches some altitude. The noise limits do not depend on the number of engines on approach path (normally 3°) and on the sideline. On takeoff, however, aircraft with fewer engines climb out faster and thus are higher above the measuring point. Hence, two and three engine aircraft have lower noise levels at the takeoff station than aircraft with four engines. FAR 36 noise limits as a function of gross weight at takeoff are shown in fig. 9.1.

Internal and external noise sources prevail in turbo-machines. Internally generated noise sources are the fan, compressor, combustor and turbine. In most high bypass engines the fan ranks the highest in terms of noisiness. The fan noise propagates out of both the inlet and discharge ends of the fan duct. Compressor noise also comes out of the inlet, while combustion and turbine noise is released from the turbine exhaust end. External noise is mostly due to the exhaust jet.

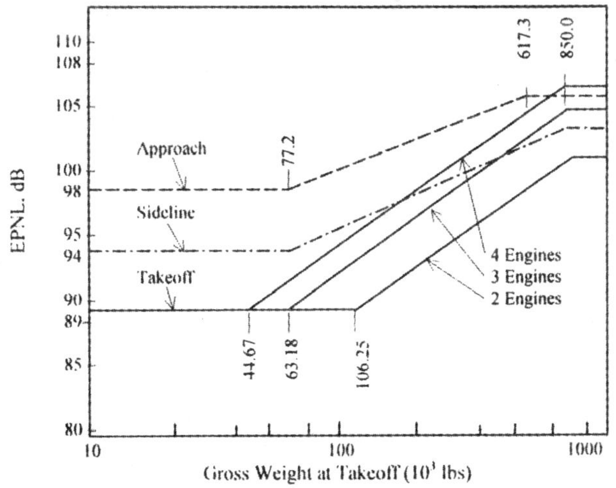

Fig. 9.1. *FAR 36 Noise Limits for Aircraft*

9.4 UNSTEADY FLOW ASPECTS AND NOISE

Noise is a consequence of unsteadiness in the fluid flow. Aircraft engines have three major sources due to this time dependence: velocity of the blades, which if above Mach 1 gives rise to weak shocks causing a buzz-saw type of noise, relative motion between the airfoils that lead to a pure tone siren form of noise and turbulence and eddies in the flow that radiate sound either through interaction with the airfoils and stationary surfaces or from the instabilities themselves to generate jet noise.

The phenomena can be mathematically formulated by the compressible fluid theory, although in practice the description can be complicated by many factors. Starting with equations for mass and momentum conservation, the fluid is assumed to be non heat conducting and inviscid, so the entropy stays constant along the flow. Quadratic terms in small disturbances may be ignored to obtain a linear connection between entropy, pressure and temperature, leading to expressions for pressure and wave velocity. The latter quantity is identified as the sound velocity. One-dimensional plane waves may be identified by pressures varying with distance traversed by the wave over a time period, after specification of boundary conditions. The energy transferred across a plane of unit area by this wave is given by pressure multiplied by velocity integrated over a cycle. For a spherical wave, such as would emanate from a point source, the total power radiated over the sphere's surface surrounding the noise source dies off with $(1/\text{radius})^2$. The pressure may be thought of as caused by harmonic expansion and contraction of a small sphere of radius r_o, and the amplitude of pressure fluctuation at its surface is P. In the limit as $r_o \to 0$ with Pr_o held constant, the solution can be obtained for a point source, or a monopole, with acoustic power $\mathbf{P}_m \sim (Pr_o)^2$.

From this fundamental solution other solutions can be derived by superposition. A dipole is made of two monopoles of opposite sign, and hence shifted by π when harmonic, separated by a distance d. Equality in magnitude and opposite sign leads to no net mass flow, but as they oscillate the fluid moves back and forth between the two. This feature causes an oscillating force on the blade, leading to vibrations in it. Adding two monopole solutions and neglecting its directional aspects find the sound radiated by the dipole. The power of a dipole $\mathbf{P}_d \sim (Pr_o)^2 \times (k\,d)^2$, where $k = 2\pi/\lambda$, λ being the wavelength. If now two dipoles are combined so that force exerted on the fluid cancels, the disturbance also lacks mass injection, with only pressures and accelerations remaining. Such a disturbance can still radiate sound, as indicated by repeating the argument in constructing a dipole by separating the dipoles by a distance d to form a quadrupole. The radiated acoustic power of a quadrupole is $\mathbf{P}_q \sim (Pr_o)^2 \times (k\,d)^4$. Hence, the dipole radiate radiates $(k\,d)^2$ times, and a quadrupole radiates $(k\,d)^4$, times the acoustic power of a monopole.

Shear occurs between the moving and stationary flows when a jet stream enters into a nearly stagnant background of fluid, causing mechanical instability that leads to a breakup of the interface into vortices (*ref*: Dimotakis, 1989). The vortices travel downstream at a rate in between the high and low speed flows. The noise generated by the jet is influenced by whether this velocity of the vortices is subsonic or supersonic relative to the outside flow. For subsonic jets the turbulent stream in the jet may be considered as a distribution of quadrupoles (*ref*: Lighthill, 1952 and 1963), since no other acoustical sources or bodies are present to create forces of a lower order. Radiated sound intensity according to Lighthill's theory is proportional to air density, second power nozzle exit diameter and a variable power of exhaust velocity, or $\mathbf{P} \propto \rho D^2 u^n$. The magnitude of the noise may be estimated from the scale of the turbulent fluctuations, and hence of the quadrupoles, from the jet diameter D and the pressure fluctuations. Since the pressure fluctuations are also dependent on D, the acoustic power of the jet is then proportional to D^2. Directional aspects characterize jet noise intensity, with noise magnitude reaching a peak value in directions 45° from the jet. This may be explained by the fact that the quadrupoles are transported along by the flow at about half the jet's Mach number and by the effects of refraction as the sound propagates out of the stream into the stagnant air. This results in a quiet zone directly behind the jet. The frequency distribution of the sound depends on the ratio of the local jet velocity and the local height of the mixing region. The initial segment of the jet has an inviscid core, combined with a linearly growing mixing region where the local jet velocity u is half that at the exit from the nozzle, u_e. In the fully developed part of the jet the velocity decreases so that $ux \approx u_e D$ where x is the distance traveled, and the frequency $\omega \approx u_e D/x^2$. Thus, the mixing layer contributes to the high frequencies and the fully developed jet to the low frequencies. The frequency spectrum may be estimated from $dP/d\omega = (dP/dx)(dx/d\omega)$. For low frequencies $dP/d\omega$ increases with ω^2 and for high frequencies it decreases as ω^{-2}. The peak occurs at a frequency given by $\omega/2\pi \approx \mathrm{Sr}(u_e D)$, where Strouhal number Sr is between 0.15 and 0.20 for subsonic jets.

For supersonic jets experimental works indicate the acoustic power varies with $u_e^{\,3}$. Two modes are present to generate noise. One is from large vortices arising from the supersonic shearing layer, which radiate waves when the rate of convection of eddies exceeds the speed of sound in the ambient air. The other source is due to incomplete expansion of the jet, resulting in shocks as the pressure adjusts to the ambient. The shocks are formed and compounded by the deflection of the unsteady jet boundary, adding considerably to the noise, to the extent of 3 to 5 dB.

9.5 ROTATING AND STATIONARY BLADE INTERACTION AND NOISE

Time-dependent pressure fluctuations brought about by the passing of rotating blades past the stationary vanes may be considered to approximate as dipoles. The blades then radiate noise with acoustic power $\mathbf{P}_{\text{blade}} \sim (\rho u^6 c^2)/a^3$, where u is the blade tangential speed, c is a characteristic blade dimension (such as chord) and a is the speed of sound. The blade tangential Mach number $M_T = u/a$. The power exchanged by the blade with the fluid varies with M_T^2, so the blade's acoustic efficiency $\eta \propto M_T$.

But the rotating blade radiates sound to an outside observer after the sound waves propagate out of the engine's inlet or exhaust duct, hence passage through the duct is of interest. The acoustic excitation has a harmonic pressure pattern, and travels in the annular duct at a speed $\omega = m\Omega_R$ where m is the number of rotor blades and Ω_R is the rotating speed. Axial variation of the acoustical amplitude as the wave propagates along the length of the duct is discussed in section 6.11. Determination of the behavior of the pressure pattern is of interest, and in particular whether it propagates or is attenuated. Assuming a uniform and purely axial velocity, or alternatively, the Mach number M facilitates this. The solution must satisfy the boundary conditions of zero radial velocity perturbation at the inner and outer walls of the duct, $r = r_H$ and r_T. If the number of blades m is large enough, the condition for propagation takes the form $M^2 + M_T^2 > 1$, or the relative Mach number to the blade tip must exceed 1. Cutoff tangential Mach number values are shown as a function of ratio of hub and tip radii r_H/r_T and number of blades m for $M = 0$ in fig. 9.2. If a mode is appreciably below cutoff and the number of blades is large, the mode attenuates by $e^{-m(x/r)}$.

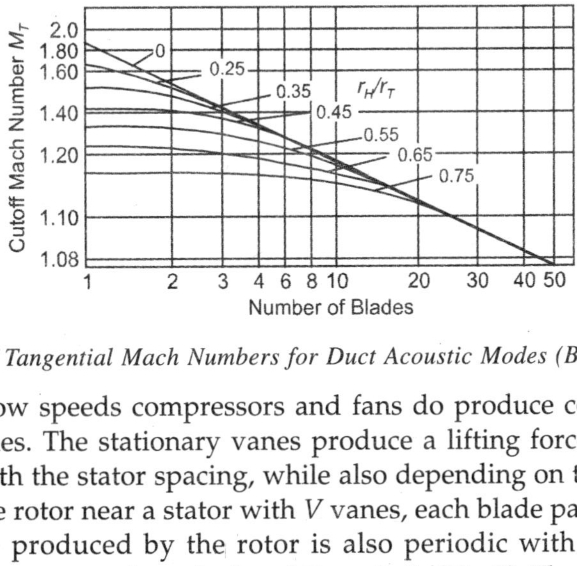

Fig. 9.2. *Cutoff Tangential Mach Numbers for Duct Acoustic Modes (Benzakein, 1972)*

However, even at low speeds compressors and fans do produce considerable noise at the blade passing frequencies. The stationary vanes produce a lifting force, and hence a pressure field, that is periodic with the stator spacing, while also depending on the angle of incidence. If there are B blades on the rotor near a stator with V vanes, each blade passes through V pressure fields. The disturbance produced by the rotor is also periodic with the rotor spacing. The pressure pattern rotates at an angular velocity of $\Omega_{eff} = \Omega_R B/(V - B)$. Thus, the combined rotating and stationary airfoils generate pressure patterns rotating faster than the rotor by the factor $B(V - B)$. Newer turbofan engines generally have $V > 2B$, so the interaction between the blades is reduced substantially and less likely to excite acoustic modes than the rotor by itself. Elimination of inlet guide vanes also means the rotor does not experience their wake. A large axial spacing between the rotor and the downstream stator permits the rotor wakes to decay, and diminishes the interaction between the two.

A more or less random excitation of most of the harmonics of the shaft natural frequency up to and above the blade passing frequency is usually observed in the noise spectrum of a single rotor operating at a tip speed above the cutoff. The low frequencies are responsible for the buzz saw form of noise in high bypass turbofans during takeoff. The lower frequencies are in a way beneficial because if they approached the blade passing frequencies, the combined effect would not be tolerable since the blade passing tones are near the upper limit of sensitivity of the ear. Blade tips generate weak shocks that spread forward into the duct. When dissimilar shocks from the neighboring blades due to geometric variations interfere, they coalesce to form a reinforced shock containing a number of harmonics of the shaft frequency.

9.6 NOISE GENERATION IN INDUSTRIAL TURBINES

Noise generated in industrial steam and gas turbines and compressors may be classified into three groups: aerodynamic noise, combustion noise and mechanical noise. Aerodynamic noise includes exhaust gas and intake air noise, as also the noise arising from cooling and auxiliary fans and other airflow. Combustion noise in engines refers to noise emitted from the vibrating surfaces of the machine's structure, components and accessories as a direct consequence of the combustion process. Mechanical noise is generated by the vibrating surfaces of the overall machine structure, inlet and exhaust manifolds, components and accessories after excitation from a host of sources.

Noise may also be classified by the source where it originates. Exhaust system noise includes noise from exhaust gases leaving the exhaust system, together with noise emitted by the mechanically vibrating surfaces of exhaust system components. Exhaust system components are also excited by the engine structure's vibration energy. Intake system components generate noise in a similar manner by the flow of fluid or air in the machine's inlet manifold.

A number of components require cooling during operation of the machine system, with an axial flow fan forcing cooler air through the heat exchanger. Air-cooling is accomplished generally by a centrifugal fan with its shrouding to direct the air across the machine's surfaces. Fan noise consists of discrete frequency tones and broadband noise. Broadband components of fan noise are caused by the shedding of vortices from the rotating fan blades and by turbulence in the fan's air stream. Discrete frequency components arise from pressure fluctuations when a fan blade passes an obstruction in the pressure field. For dynamic balance fan blades are spaced at equal angular intervals, hence the fundamental discrete tone will occur at the fan's blade passing frequency, given by:

$$f_o = \text{Fan speed, rpm)} \times (\text{Number of fan blades})/60$$

where the frequency is in cycles/sec. Noise intensity of the fundamental tones, as also the strength of its harmonics, will depend on the nature of the pressure impulses generated by the obstruction.

In axial flow fans blade tip speed is the dominating factor affecting the sound emission level. In many applications the cooling fan is driven directly by the main drive shaft at a fixed ratio of engine speed. The speed ratio is selected usually for the most adverse condition (load or temperature). During part load or idle operation, excessive power may be needed to drive the fan, resulting in higher sound generation. Mechanical vibration of blades and shrouding may also add to the noise level.

Higher temperatures in the exhaust system require more exacting materials to provide for extended life of the equipment. In the designing for noise control requirements of various criteria, two major categories need special attention: the near field and the far field noise codes. A set limit of 90 dBA for an 8-hour exposure of the workers 3 feet from the perimeter of the unit is generally followed by most governmental agencies. Far field noise criteria relate to noise measurements at the property lines of the site facility operating the unit. Other authorities call for noise levels between 35 and 45 dBA for the exterior of residential premises.

9.7 INLET AND EXHAUST SYSTEMS

Compressor inlet noise, casing radiated noise and the noise emitted by the exhaust are the primary sources in gas turbines, as shown in fig. 9.3. Secondary noise sources that also deserve attention are enclosure ventilation inlet and exhaust ducting, fans used to move the cooling air, regenerated noise caused by exhaust gases as they leave the silencer, scavenge air fans for the filters and penetrations through the enclosure. The driven equipment must also be included in the evaluation as a potential noise source. If a compressor is employed, casing radiation is present. For an air-cooled generator the inlet and exhaust cooling air system and radiation from the casing generator must be checked.

The noise emanated from all silenced and unsilenced sources of the complete gas turbine system must be combined to determine compliance with the stated criteria. In contemplating the design of the silencing system, the total assemblage of the components must be taken into account.

The turbine inlet system consists of an inlet plenum, silencer and a filtering system. The orientation of the major components can vary depending on the site conditions. Each configuration has a differing effect on the acoustical and aerodynamic properties of the inlet system. A vertical inlet flow system has the advantage of better directivity, thus requiring less attenuation of the silencer (fig. 9.4). Also, the walls of the plenum can be acoustically lined so that there is additional attenuation obtained, causing still further reductions in the inlet silencer requirements. Should louvers be added to the front of the filters to keep moisture from entering, the noise coming out of the filter face and the louvers tends to be randomized. The positive beaming effect of the 0° directivity is then partially offset. Depending on the characteristics of the filtration elements, some noise attenuation primarily at the higher frequencies may also be achieved in the filters.

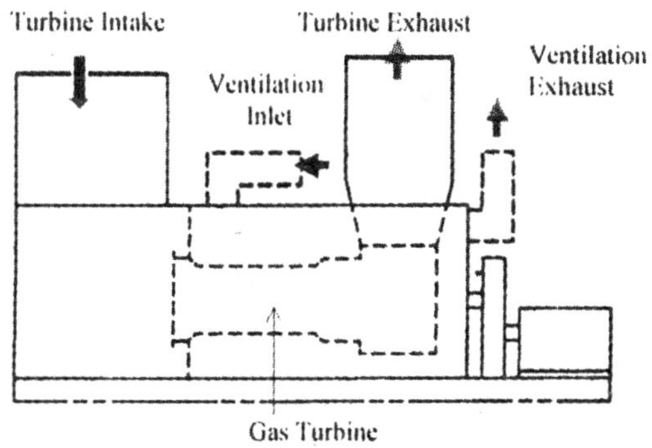

Fig. 9.3. *Primary Noise Sources in Gas Turbine System*

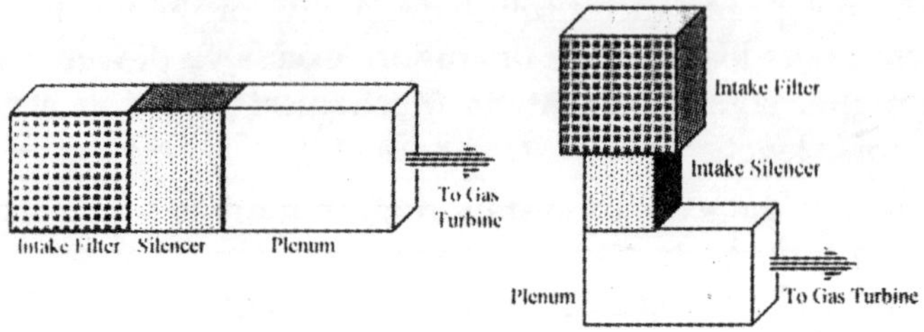

Fig. 9.4. *Horizontal* (left) *and Vertical* (right) *Gas Turbine Intake with Filtration System*

The dual, or either side, horizontal inlet scheme al1ows for the use of a lined bend in the plenum area similar to that used for the vertical scheme, and this again permits the reduction of silencer treatment. But the 0° directivity effect of beaming the noise causes the arrangement to be less efficient in terms of noise than its vertical counterpart. The straight inline horizontal system is the most difficult from noise considerations.

In addition to the acoustical requirements, the silencer must take into account aerodynamic and geometric considerations. Baffle plates in the silencer must be aerodynamically appropriate. Thus, a baffle with no aerodynamic nose or tail will experience a high-pressure drop. A baffle plate with a round nose and a blunt tail has a slightly better aerodynamic performance. A round nose and tail increase the aerodynamic effectiveness of the baffle even further. Baffles provided with an aerodynamic nose and tapered tail has the best flow efficiency. In this regard, baffle plate thickness and orientation with the centerline of the turbine play a role. The angle of the tapered tail must prevent turbulence in the engine inlet. The key to an effective design for the intake silencer system is an optimum combination of meeting acoustical and aerodynamics requirements. The walls of the plenum can be lined with fibrous material held in place with perforated sheet. Lagging or lining the silencer shell may also be considered if the distant field criteria are stringent. With the use of self-cleaning filters care should be exercised that the noise created by the air pulse used to clean the filters is within limits. When an anti-ice system is required with the inlet, the noise from the system is of interest. Re-circulatory form of systems use exhaust gas bled from the turbine and fed into the anti-icing system. The duct carrying this air can radiate noise through its walls.

The exhaust system from the turbine can likewise be config.d in many ways, although the vertical form is widely used because of superior acoustic and aerodynamic features. A silencer is most beneficial if it is located as close to the exit from the turbine as good aerodynamic considerations will permit. Internal components in the silencer need to be configured to obtain maximized acoustic performance. Figure 9.5 shows construction details for an exhaust silencer. Most of the silencing is accomplished by using parallel type baffles. If high frequency attenuation is desired the diamond shaped baffle offers the advantage of blocking the line of traverse of the noise coming from the exhaust system. The exhaust system, transition shell and the silencer shell must be analyzed for noise radiating from them. The expansion joint between the exhaust casing and the exhaust silencer plenum is also a possible source of noise, especially if the joint is placed outside the gas turbine enclosure. Enclosure of the joint with an acoustical labyrinth may then be required to reduce the radiated noise.

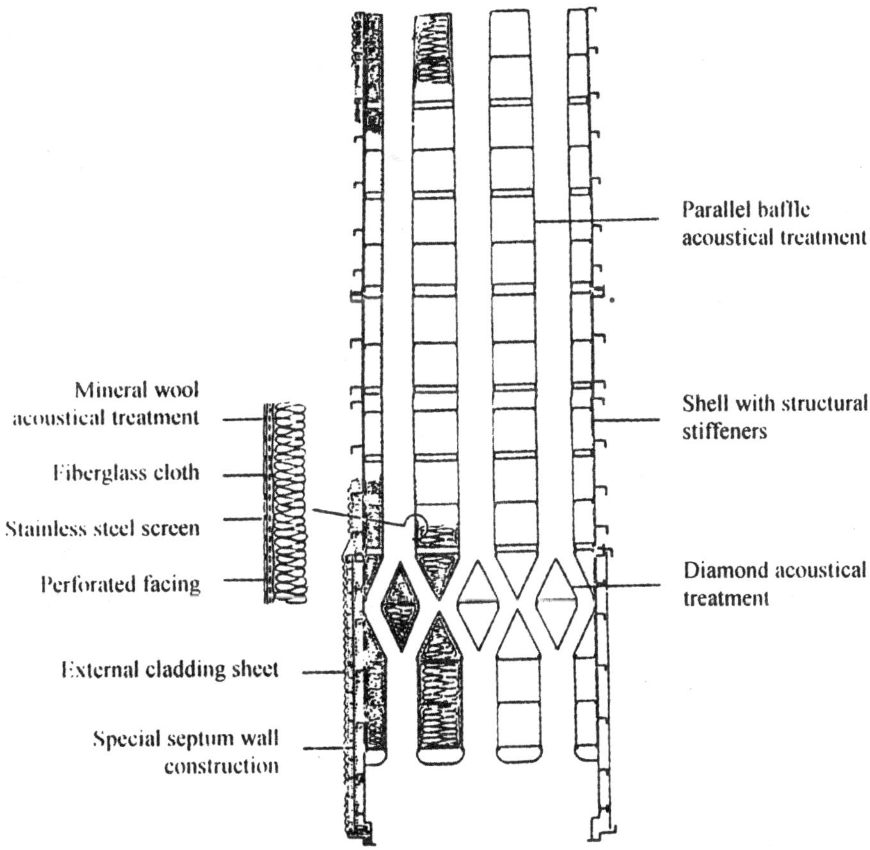

Fig. 9.5. *Silencer Layout at Gas Turbine Exhaust (Sawyer, 1982)*

9.8 SHAFT VIBRATION AND NOISE

Multiple ignition centers at hot deposits coating the combustor of an engine account for rapid and excessive pressure build up, and give rise to a noise sometimes described as 'thud'. 'Rumble' is associated with high compression turbo-machinery running for long periods at part load. Noise emanating from main bearings is also of considerable interest. Rotor acceleration due to gas pressure and flexural elastic oscillations of its elements cause the bearings and main journal to strike the bearings, generating vibration and noise. Many names have been given to this phenomenon, including 'roughness', 'bearing thrash' and, unfortunately, 'rumble'.

In a diesel engine noise generation follows the following sequence:

- Variable inertial forces act on the reciprocating mass, pulling the crank pin.
- Variable gas pressure loads are imposed on the piston. A sudden increase in the gas pressure takes place after the ignition and combustion phase of the cycle. The connecting rod transmits this load and strikes the crank pin bearing.

The crankshaft accelerates and strikes the main journal bearings. Shock waves transmit to the cylinder block through the main bearings, and noise is radiated from the cylinder block surface. In a 4-cycle machine the process is repeated $n/2$ times, where n is the number of cylinders.

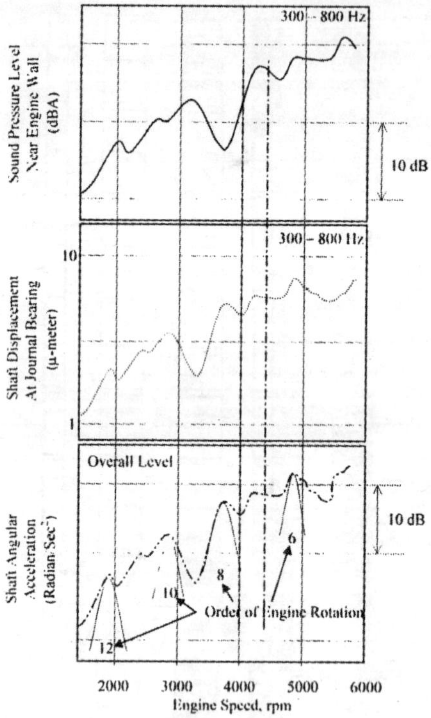

Fig. 9.6. *Rumbling Noise and Shaft Vibrations in Automotive Engine* (ref. *Aoki, 1986*)

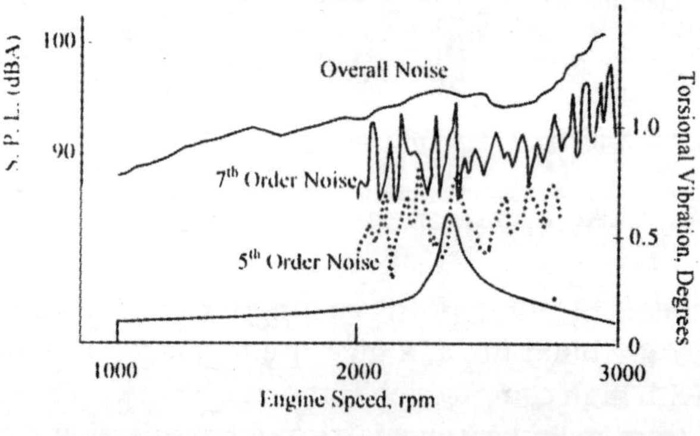

Fig. 9.7. *Noise and Shaft Torsional Vibrations of Diesel Engine (Ide, 1990)*

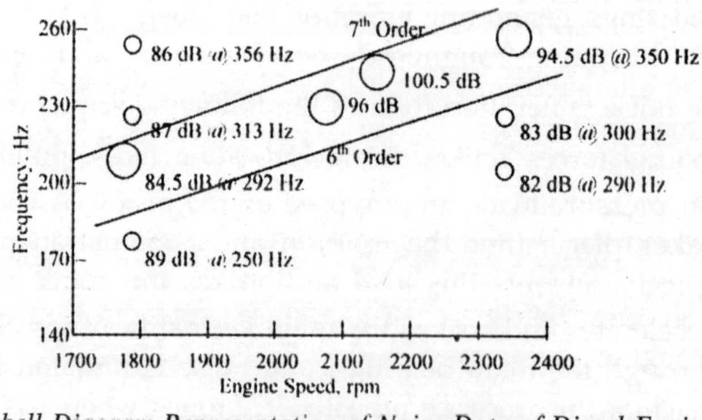

Fig. 9.8. *Campbell Diagram Representation of Noise Data of Diesel Engine (Ide, 1990)*

Aoki (1986) reported analytical investigation of the phenomenon, combined with measurements of torsional oscillations of the crankshaft and sound pressure levels. Measurements are made on a water-cooled 4-cycle, 4-cylinder, in-line, gasoline driven automotive engine. Engine displacement is 1.958 liters (81 mm bore × 95 mm stroke), output 144 horsepower at 6,200 rpm, torque 172 N-m at 4,000 rpm. Using displacement sensors, accelerometers and proximity probes, continuous and simultaneous measurements were made of: (*i*) vertical and lateral crankshaft displacements at each journal, (*ii*) rotational and axial acceleration at each throw of crankshaft, (*iii*) flywheel displacement in all three directions, (*iv*) acceleration of crankshaft pulley in all three directions. Figure 9.6 shows sound pressure levels measured near the engine wall and angular acceleration of the crankshaft at the pulley. Rumbling noise peaks in the area around 4,100 rpm, indicating a link between the crankshaft's torsional vibration amplitude and sound pressure level.

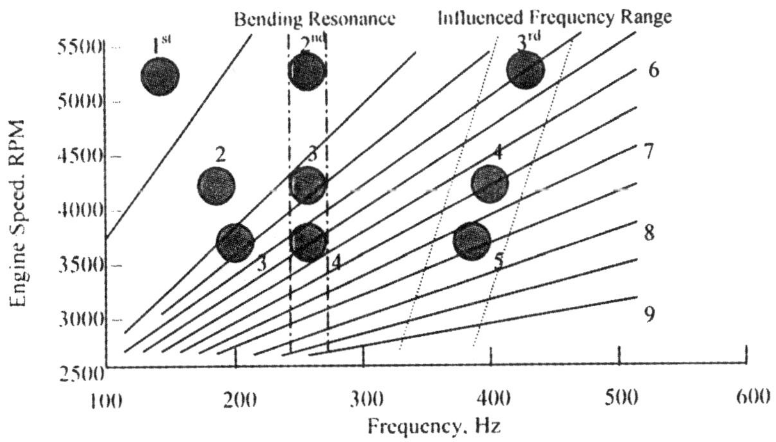

Fig. 9.9. *Shaft-Flywheel Bending Resonance and Noise (Sasaki, 1984)*

Interestingly, the author predicted resonance at the sixth order of engine speed of 4,100 rpm using the finite element approach.

Torsional vibrations of a 6-cylinder in-line 4-cycle diesel engine are investigated by Ide, 1990. Noise spectrum of the engine operating at 2,800 rpm at full load revealed order 7 to be of interest. Figure 9.7 shows engine noise and crankshaft torsional vibration signatures, where a peak occurs at order 6 of engine speed. Exciting forces caused by torsional vibration of order n has components of $(n + 1)$ and $(n - 1)$ orders; this explains the presence of orders 7 and 5 components of engine noise peaks when order 6 in torsional vibration appears. Figure 9.8 shows noise speed characteristics of the diesel engine in Campbell diagram format.

Discomforting noise often accompanies bending vibrations of a shaft and flywheel system operating at or near maximum load condition. Bending in the crank pin results in angular misalignment with the bearing. Noise level and bending stress increase at high operating speeds, as shown in fig. 9.9 for an internal combustion engine (*ref*: Sasaki, 1984). Bending vibration at crank pin # 4 is shown in fig. 9.10, and engine noise spectrum in fig. 9.11. Since the noise frequency band is composed of harmonic components of engine revolution in the 200 to 500 Hz frequency range, it may be concluded that noise problem is caused by solid propagation of vibration from the engine.

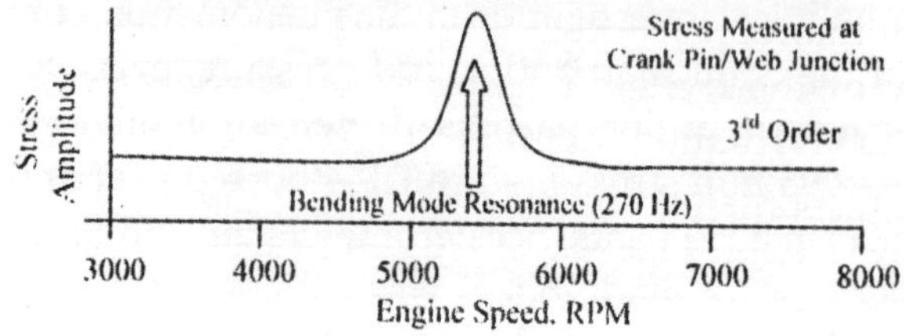

Fig. 9.10. *Bending Vibration of Shaft at # 4 Crank Pin (Sasaki, 1988)*

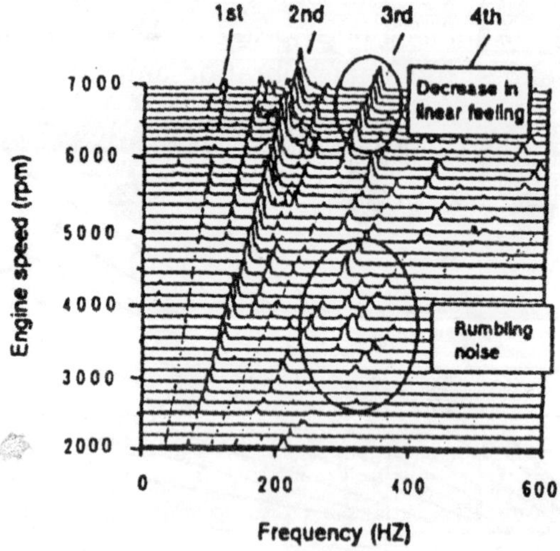

Fig. 9.11. *Noise Spectrum of Engine in Figs 9.9 and 9.10* (ref. *Sasaki, 1984)*

As noted in the example of diesel engine, an order n of crankshaft vibration appears as $(n + 1)$ and $(n - 1)$ order of vibration under static condition. Bending stress in crankpin number 4 shows a resonant peak at 270 Hz, the third order of 5,400 rpm rotor speed.

9.9 CORRELATION BETWEEN DYNAMICS AND NOISE

The intensity of a machine's radiated noise level depends on the characteristics of exciting forces and response of its structure. An understanding of the exciting force mechanism, the machine structure and their interaction is the first step towards the goal of reducing the noise. Exciting forces may be expressed in the time domain as a force-time diagram or in the frequency domain as a force-frequency spectrum. Figure 9.12 displays these characteristics for some representative machine types and their associated components. Force spectral data can be obtained directly from time data by Fourier or harmonic analysis methods. In the time domain as the force function's initial ascent becomes more rapid, high frequency components in the frequency domain will increase. In the frequency range above 240 Hz the forcing function can be approximated by a straight line, with the slope depending on the rate or impulsiveness of the forcing function. Frequency domain data has the advantage that the effect on the output can be estimated, as seen in the noise frequency spectrum system setup of fig. 9.13. Evaluation of data from a transducer such as a microphone, a load cell, an accelerometer or a strain gauge begins with analog time domain signals, and can be reviewed in the time domain on an oscilloscope or a strip chart. Complexity of time domain signals and sensitivity of the dynamic response of physical systems to frequency call for studying the signals in the frequency domain.

Surface noise emits from the vibrating surfaces of the machine. Excitation forces come from combustion, gas or fluid pressure and from impact caused by the rotating components. Rapid increase in the combustion chamber pressure of engines following ignition generates large

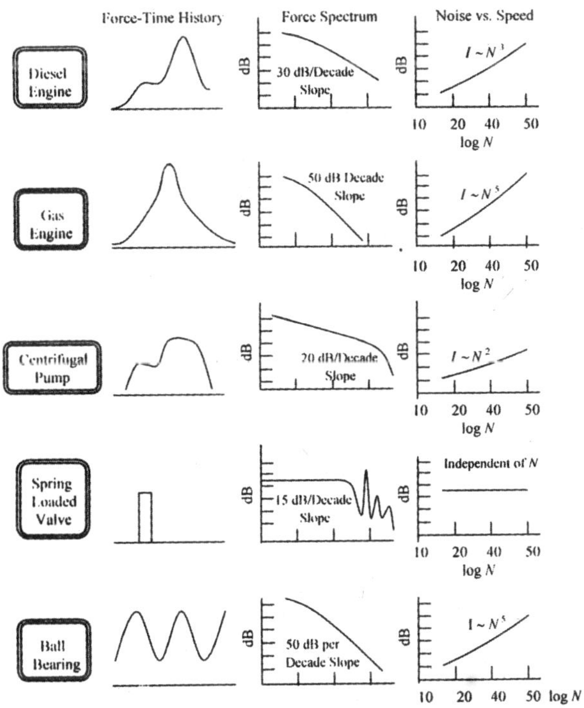

Fig. 9.12. *Examples of Force-Time History, Force and Noise-Frequency Spectra (Clarkson, 1982)*

excitation forces. Cylinder gas or fluid pressure applied to the casing and diffuser walls due to combustion and expansion is referred to as combustion noise.

Mechanical noise is also a direct consequence of metal-to-metal impact between moving parts. Since a shaft slides in the bearings, clearance must be available between the diameter of the journal and the bearing.

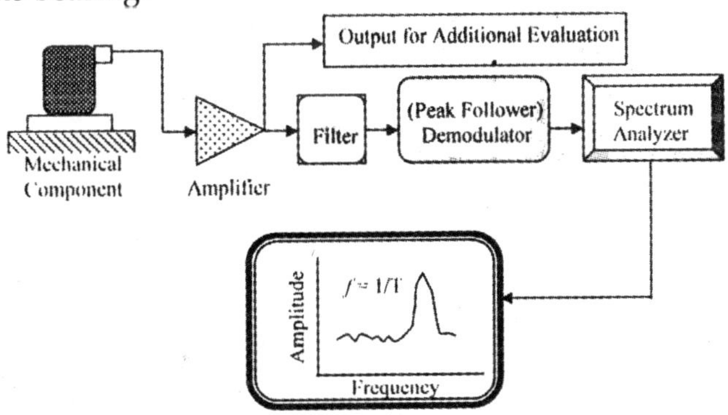

Fig. 9.13. *Demodulation and Analysis of Resonance Signal (Harting, 1984)*

Also, manufacturing tolerances in component dimensions have to be provided to facilitate production. If a journal with a smaller than normal diameter is assembled into a bearing with a greater than normal bore diameter, the resulting clearance will be greater than the nominal design value. Besides motion in the radial direction, the shaft will be able to rock about the transverse axis within the clearance envelope. Ungar *et al.*, (1985) determined that the noise generated by piston slap in reciprocating machines is affected by the impact distance, the rate of change of impact force, machine speed and machine load. The following expression provides the intensity of piston slap:

$$I_{P-Slap} \propto \frac{\rho_{piston}}{\rho_{cyl.block}} \left[\frac{Z}{S} \frac{B}{t_b} \frac{B}{t_l} \right] \left(\frac{\delta}{B} \right)^{4/3} C_m^{\ 3} \tag{9.5}$$

where:

ρ = material density

B = bore diameter

S = stroke

Z = number of cylinders

t_b = thickness of cylinder block

t_l = thickness of liner

δ = clearance between piston and cylinder bore diameters

C_m = mean piston speed

Impact between gear teeth due to the clearance between meshing teeth is responsible for noise in the same manner. Other examples are crankshaft journal and bearing, crank pin and connecting rod bearing and valve and valve seating. As the operating speed increases, impact forces between neighboring components goes up, as also the noise level.

9.10 PROCEDURES FOR NOISE EVALUATION

Identification of the source of noise in a rotating machine is a principal objective. In the partial exposure method noise contribution of a given source is measured by blocking sound radiated by all other sources through the use of sound barrier wrapping material. Since sound pressure is measured directly, noise from other sources must be quantified or reduced to negligible levels. Generally, this calls for acoustically treating intake and exhaust manifolds, dynamometer, test support structure and any other noise sources in the test chamber. Vibration isolators between the foundation and the rotating machine and dynamometer may be necessary to ensure structure-borne noise does not interfere. Sound barrier materials consist of layers of acoustical foam, fiberglass or ceramic fiber, with an outer layer of a dense material. The inner softer layer isolates the outer layer from the component vibrations. Such sound barrier treatments generally provide an insertion loss of 10 to 20 dBA. The insertion loss is mostly due to acoustic leaks at high frequencies and poor isolation at low frequencies. With the treatment in place, sound pressure level L_1 at any position equals background noise plus unwrapped engine noise minus transmission loss of the barrier. Expressed acoustically:

$$L_1 = L_b + L_e^{\ w} \tag{9.6}$$

where L_b = background noise and L_e^w = sound level of wrapped engine. If each source is assumed to radiate sound independently of that radiated by other sources, then with one source exposed the sound level L_2 is given by:

$$L_2 = L_b + L_s + L_e^w - L_s^w \tag{9.7}$$

where L_s = sound level of exposed source and L_s^w = sound level of exposed source when wrapped. The two equations can be solved only by assuming a value for L_s^w; it may be neglected if it is much less than L_s. The advantage of this method is that it does not require sophisticated instruments or highly trained personnel. However, the procedure is time consuming and requires costly sound barrier material.

Acoustic intensity measurement method relies on finite sum and difference approximations of acoustic pressure and particle velocity, respectively, of two closely spaced sound pressure measurements. The measuring apparatus consists of two microphones separated by a distance Δr and obtaining two sound measurements $p_1(t)$ and $p_2(t)$. Particle velocity can be approximated from the pressure gradient. Cross- spectral density of $p_1(t)$ and $p_2(t)$ can be computed on a two channel Fourier analyzer. Sound intensity is then obtained from the expected value function of the cross spectral densities. Additional treatment of the mathematical theory has been done by a number of writers (*ref*: Potter, 1994, Baxa, 1982). Supplementary information about the frequency content of the acoustic intensity and variation of acoustic intensity over the surface improves the accuracy of this method to identify the noise source. Acoustic intensity data may be plotted using line contour plots, raster plots and three-dimensional contour plots. Figure 9.14 shows contour plot results of measurements made on one side of a turbine engine using the same interpolation algorithm.

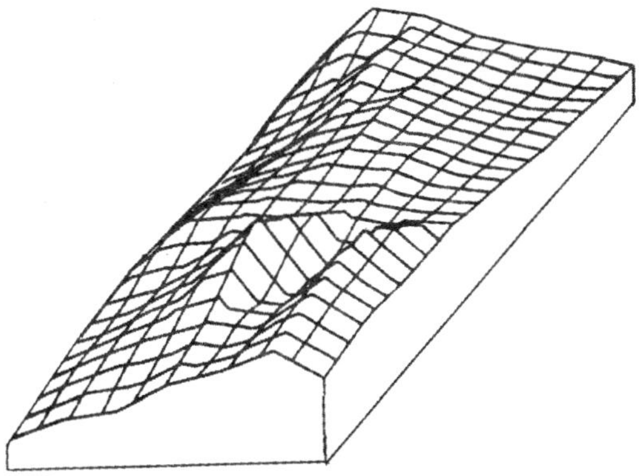

Fig. 9.14. *Acoustic Intensity Three-Dimensional Contour Map of Turbine Engine*

Surface vibration method of sound measurement is available for a few simple radiators, such as a rigid circular piston in an infinite baffle. Combining the closed form analytical solution of the source field with the concept of radiation efficiency permit calculation of sound power radiated by the source. Other techniques that hold promise use the partial coherence function to relate a given input to the output by removing contributions from partially coherent inputs. The phased-array microphone method calls for use of a number of microphones in the far field of the noise generating system. Varying the relative phase of individual microphones in the array focuses on a specific area of the noise generating system. Incremental variation in the phasing of the array permits scanning all areas of the system, and yields relative contributions of the various areas.

9.11 STATIC TESTS

The purpose of static testing is to obtain frequency response information on the structural response elements of the machine. Objectives of the test relate to identifying resonant frequencies, measuring structural damping values, determining resonant mode shapes and measuring the associated frequency response functions. Operational exciting forces are applied at locations where vibration response is of interest. In a noise control task it is often found that relatively low frequency and sufficiently well spaced resonance modes dominate the transmission of structure borne energy and the radiation of airborne sound.

Frequency response is basically a determination of the output of a system in terms of the input to the system. For structural dynamics work this usually takes the form of vibration response in terms of displacement, velocity or acceleration at some location on the structure relative to a force applied at the same or another point in the structure. Complete description of a frequency response function requires both magnitude and phase information, although often only the magnitude of the response is of interest. Force may be applied to the machine structure in several ways. A piezoelectric or an electromagnetic exciter is attached to the structure through a load cell for measuring the applied force. An impulsive force may be applied by a hand held hammer with a load cell attached to it. When measurements are to be made on a weak or lightly damped structural component, attachment of an exciter can significantly alter the structure's characteristics. If the impact technique is not suitable in such a case, a non-contacting acoustic or magnetic exciter may be useful.

Before the advent of digital equipment the swept sine wave method was widely used for frequency response testing. With the traditional analog system the exciter signal is a sine wave swept through the frequency range of interest; the rate must be low relative to the response time of the structure. Thus, in structures where damping level is low measurement time is high. Another disadvantage of the method is that on non-linear systems frequency response is also non-linear. A major advantage derives from the fact that energy is concentrated at a single frequency. Hence, signal-to-noise ratio is relatively high, so high force levels can be maintained at a constant level through a feedback loop.

The random excitation method has become popular with the availability of digital Fourier analyzers. The signal applied to the exciter is a band-limited random signal with a period much greater than the analysis sample time. The technique is much faster than the swept sine method, because information is being obtained simultaneously throughout the frequency band. Also, it produces the best linear approximation, in the least-square sense, of a non-linear frequency response function. However, the spectrum shape of the signal to the exciter cannot be controlled easily.

Problems frequently arise in the testing of rotating machines due to the presence of non-linearities. Clearance at bearings and joints is a prime example. The usual approach is to pre-load the gap with flexible springs. This will ensure that mass-controlled compliance associated with the modes introduced by the flexible springs will not contribute substantially to the overall response in the frequency range of interest. Problems also occur because of the basic non-linear aspects of a machine using the slider-crank mechanism – inertia characteristics vary with the crank angle, so frequency response also varies with it. In internal combustion engines primary gas forces from combustion occur over a small crank angle, so a dedicated piston and crank arrangement will not have a serious problem. However, multi-cylinder engines pose a different problem. In a 90° V configuration, for example, the connecting rods of the opposite banks are attached to the same crank, hence ignition will take place after the crank turns a corresponding angle. A host of other factors also come into play. Non-linear damping characteristics may be encountered internally among machine components. Since damping force varies with gap size and velocity across a joint, impact testing may be ill suited. A high ratio of peak to root-mean-square level of the excitation signal leads to substantial errors in the measurement.

9.12 MONITORING FOR QUALITY CONTROL

Among other measurable parameters (gas pressure, temperature, flow, torque, horsepower, speed) the information obtained from a noise and vibration control program of a machine system provides useful insight on the machine's operation. In measuring the machine's sound spectrum, however, it has to be made in a location where background noise is minimal. Such a situation does not commonly occur on a production line. If the sound level is selected as a measuring parameter for the quality control testing, a low ambient anechoic test chamber must be available, and the machines must pass through the chamber one by one for sound test. Most manufacturers now routinely carry out this procedure as part of the overall quality evaluation of the engines. As an alternative to sound testing, a machine's vibration spectrum can be evaluated.

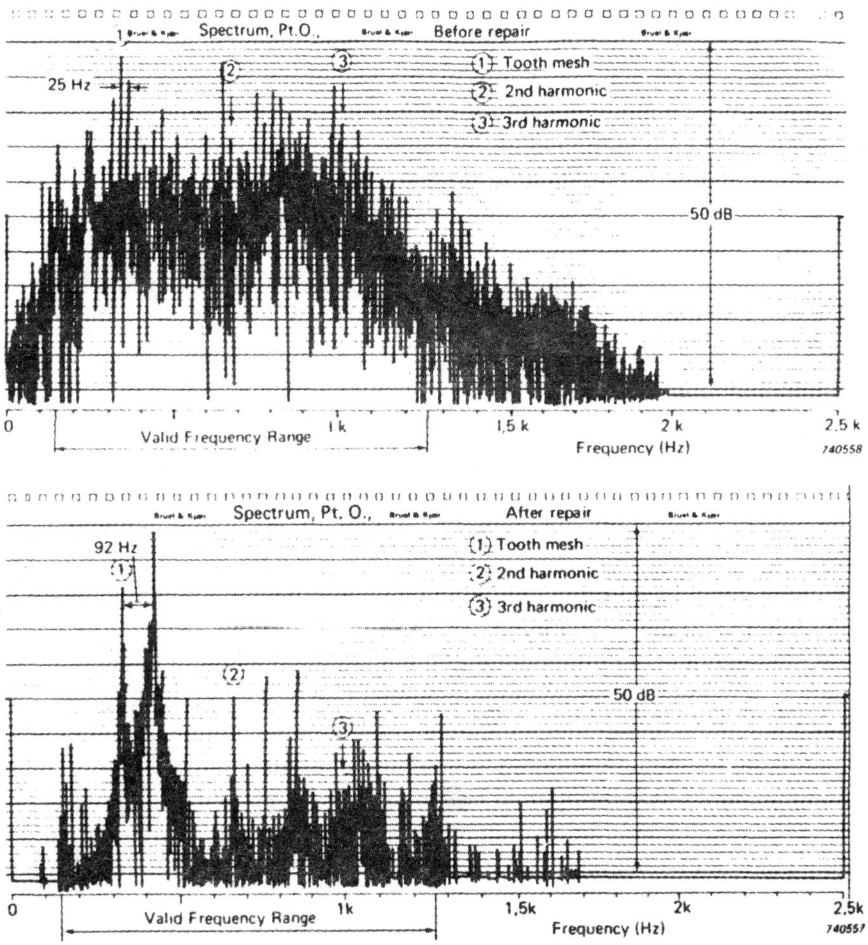

Fig. 9.15. *Low Speed Gearbox Vibration Spectra Before* (upper) *and After* (lower) *Repair*

The correspondence between sound emitted by a defective machine and its vibration levels at certain selected points in the structure must be determined. Evaluation of the machine's vibration signature can then be used for quality control. Although vibration may be measured in terms of displacement, velocity or acceleration, the amplitude and frequency range of interest of many high speed rotating machine's vibration spectra make measurement of acceleration most convenient. If the vibration measurement is performed on a production line, the machine must be vibration isolated from the assembly line. This will ensure that interference between assembly line fixtures and the machine structure is eliminated, and only the machine structure's vibration spectrum will then be considered in determining the machine's acceptance. Local structural resonances and its effect on the level of noise and vibration must be taken into account. Since resonant frequencies tend to change from unit to unit, means of detecting and eliminating this parameter is vital. Selection of accelerometer type and its location should take into account the frequency range of interest.

Selected signals under various operating conditions must be carefully checked, keeping in mind the analyzer's frequency range, resolution and need for tracking for information. Figure 9.15 shows the vibration signature of a badly worn gearbox before and after repair.

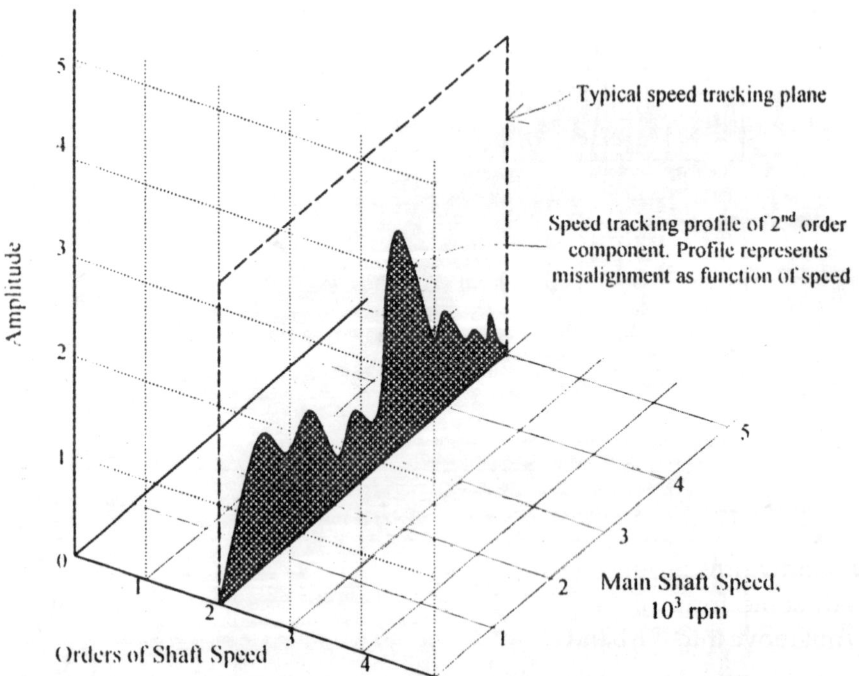

Fig. 9.16. *Harmonic Order Analysis*

For normal operation a machine's characteristic sound and vibration signature will vary around some average spectrum. With limits placed around the average spectrum the procedure can be used as a quality control tool. Many machines in a given operating mode will generate harmonically related components, which can yield meaningful data. In a simple frequency analysis, however, separation of the machine's harmonic contents from random noise or vibration may be difficult. Also, small variations in the machine's speed will result in larger variations in the frequency of higher order harmonics.

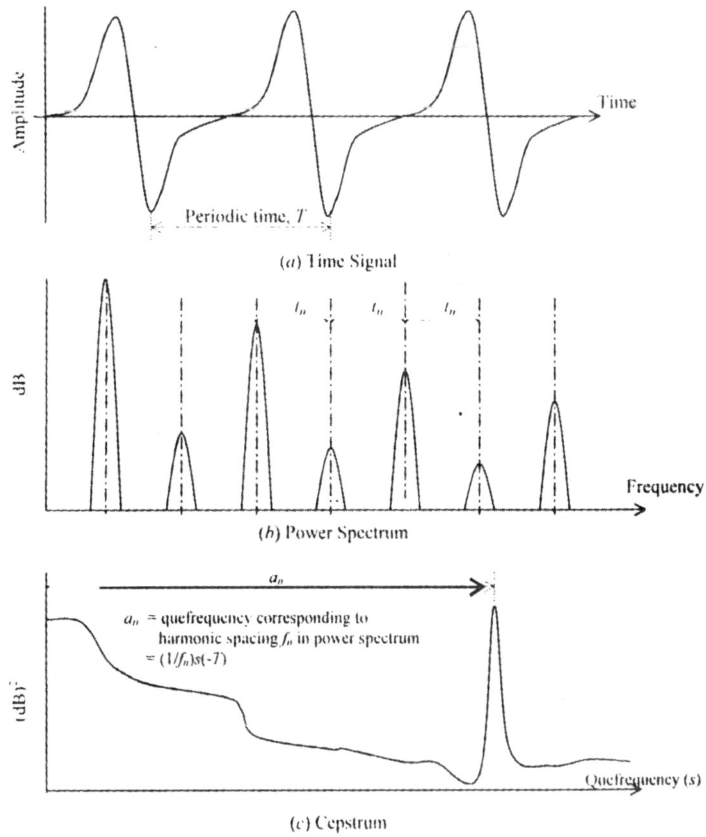

Fig. 9.17. *Time Signal, Log Spectrum and Cepstrum of a Periodic Signal*

If narrow bandwidths around the harmonic content of a machine's frequency spectrum are established, some form of frequency tracking may be used to ensure that higher order harmonics do not move into the bandwidth when operating speed changes. Figure 9.16 illustrates this concept.

Cepstrum analysis method is used to enhance information made available from an engine's harmonic contents. The cepstrum is defined as the power spectrum of the logarithm of a signal's power spectrum. The unit of amplitude is dB, and the dimension of time is called quefrequency, meaning spectrum periodicity. Measures of harmonic content and strength of side bands are provided by peak amplitudes. Figure 9.17 shows the concept. In addition to processing of signals by a single channel analyzer, a cross spectrum analysis technique between two accelerometer signals improves the forcing function by minimizing the effects of local resonances.

Instantaneous acoustic intensity at two different locations on an object may be used for cross spectrum between the two locations. Another technique suggested towards the same objective is meant for high frequency vibration monitoring of gears and bearings. Since a defect causes a broadband frequency spectrum, the resonant frequency will be amplitude modulated at the defect's impact rate (fig. 9.18). To determine a problem the carrier resonant frequency must be filtered, demodulated and spectrum analyzed as illustrated in the arrangement of fig. 9.13.

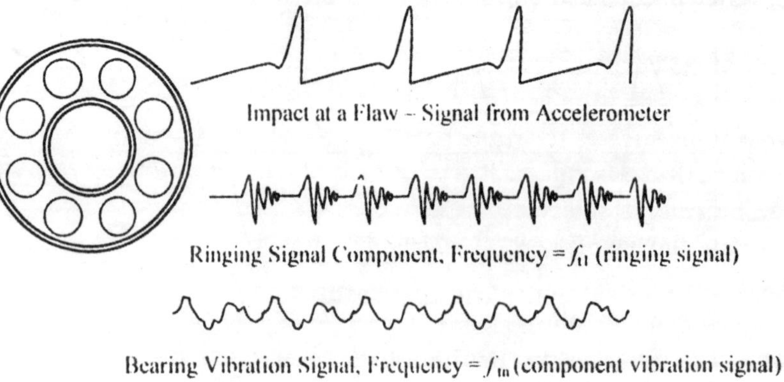

Fig. 9.18. *Amplitude Modulation of Transducer Signal due to Flaw in Bearing (Harting, 1984)*

9.13 SUPPRESSION OF JET NOISE

Jet noise can be suppressed by exploiting the characteristics of the jet and of the human ear, the latter being less sensitive to high frequencies. By replacing one large nozzle with a number of small ones, a shift to higher frequencies can be obtained because the frequency varies with the diameter of the jet, *D*. Reducing the jet velocity can substantially lower the acoustic power **P**, since P varies with the jet velocity raised to the power between 8 and 3, depending on the magnitude of exit velocity.

The concept of a turbofan engine itself has a significant effect in lowering the noise level by distributing the airflow over a larger area. But its impact on thrust, system weight and fuel consumption requires evaluation. Thrust $F \propto (dm_e/dt)u_e$, and if $M_e < 2$ as in turbofans, then jet acoustic power $\mathbf{P}_j \propto (dm_e/dt)u_e^7$, where u_e = exhaust velocity. In a turbofan engine if dm_o/dt represents mass flow at F_o, then $(dm/dt)/(dm_o/dt) = (F/F_o)^2 = 1 + \alpha$, where α is the bypass ratio. For $\alpha = 2$ the reduction in jet noise is about 10 dB, which is the gain obtained in Pratt and Whitney's JT-3D turbofan relative to the JT-3 turbojet engine. In smaller engines with the same jet velocity, $\mathbf{P} \propto (dm/dt) \propto F$, so a similar advantage is not possible.

Most noise suppressors, including elaborate multi-tube ejectors studied for supersonic transports, cost about 1 percent loss in thrust. A number of such schemes have been studied for turbofan engines. These include inverted temperature profile nozzles, where a hot outer flow surrounds a cooler core flow. Disturbances propagating at the interface between the cold inner and hot outer jets are subsonic with respect to the outer hot gas, and this can bring reduction in the overall noise level when compared with a conventional turbofan where the outer jet is cooler than the inner one. Use of mixer-ejector nozzles relies on reducing the mean jet velocity by ingesting additional airflow through a combination of ejector nozzles.

Duct linings in the form of acoustic panels are widely employed to modify the walls in order to cause pressure fluctuations at the walls. This results in a normal velocity component in the region and extraction of energy from the sound field, thereby achieving attenuation in the axial direction. The acoustic panels line the duct with a porous layer to provide a series of small cavities (see fig. 9.19). The pressure in each small cavity acts as a spring upon which the flow through the orifice oscillates in response to pressure fluctuations outside the orifice, thus functioning as a Helmhlotz form of resonator. The pressure in the cavity, the effective length L, area A of the orifice, velocity of sound in air a and flow velocity V determine the resonant frequency of the cavity according to the expression $\omega_n^2 \approx a^2A/VL$. Pressure p is harmonic of the form $p = Pe^i\omega^t$. Attenuation length at the resonant frequency is established from design and operating conditions such as the number of blades, tip radius, tip tangential Mach number, rotor angular velocity and viscous drag coefficient of the cavities. If ε is the ratio of orifice area to wall area, and if A is selected to approximate $\varepsilon V/L$, then $\omega_n^2 \approx a^2\varepsilon/L^2$. Then for a rotor with B blades and angular velocity Ω_R, $L/r_T \approx \sqrt{\varepsilon}/(BM_T)$. A honeycomb structure covered by a perforated panel is often preferred. Detailed calculations and extensive experimentation are necessary to establish the attenuation properties of the acoustic liners for a specific engine type. A summary of the data is provided in NASA SP 311 (1972). A review of aircraft engine and flight vehicle noise is also available in publications by Little *et al.*, (1968) and NASA Publication 1258 (1990).

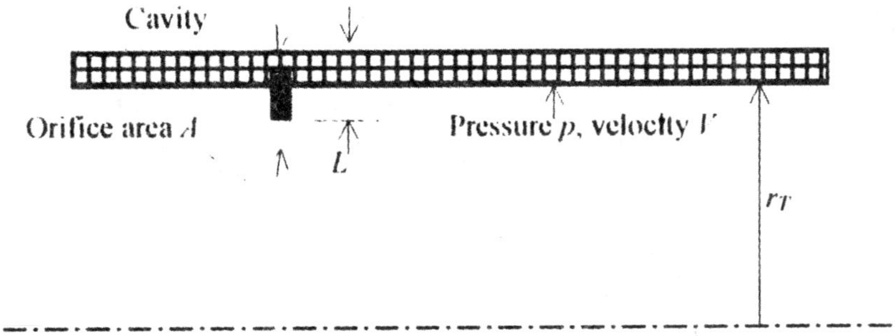

Fig. 9.19. *Schematic of Acoustic Panel Damping Cavities in Engine Duct*

9.14 DAMPING TECHNIQUES

Damping materials have been used effectively to reduce forced vibration amplitudes in structures, increase fatigue life and to reduce structure borne noise. Damping may not be the ideal solution in many circumstances, but it provides an alternative solution for controlling vibration and associated noise problems in a cost effective manner. Damping becomes advantageous when:

- A component has a number of natural frequencies of vibration in a given frequency band, and the source is exciting many of these vibration modes.
- Modifications to the source or to the load-carrying path cannot be readily made.
- Little improvement is expected by stiffening or reducing the mass.
- Structure's resonances are important to the problem.

An understanding of the basic characteristics of the visco-elastic damping material, as also its configuration, is necessary for a successful application of the technique. The energy dissipated by the damper is proportional to the Young's modulus and the loss factor of the material. The product of the two parameters is called the loss modulus, a mostly uniform basic material property.

Two types of damping treatments are commonly used, surface and tuned damping devices. Surface damping is used to solve vibration and noise problems on flat or curved sheet metal structures, and provides high damping capability over wide temperature and frequency ranges. Surface dampers may be subjected to extensional or shear deformation.

Extensional damping, also referred to as free-layer or unconstrained damping, is more commonly used. It calls for application on one or both sides of a structure, so the material is subjected to tension and compression deformation as the structure undergoes cyclic loading. The performance of any visco-elastic damping material is affected by temperature, thickness of material, frequency, structural damping and method of bonding. The loss modulus of most commercially available damping treatment materials reaches the maximum value in the 70° to 110° F region, hence the effectiveness of the material to reduce vibration and noise also is at a peak in this range. Increasing the thickness of the damping material achieves higher damping up to an extent. If the thickness of the material is much greater than that of the base structural material, the composite system damping asymptotically approaches that of the material itself. For small thickness the increase in the damping is nearly linear. The structural damping coefficient of the base material in some cases is negligible, but in many other cases it is significant and should be included in the evaluation. This is particularly the case in assemblies with joints; some diesel engine sub-assemblies have modal damping values approaching 0.05. Addition of unconstrained damping treatment in small amounts will not help in such structures. The effect of frequency on the performance of damping is secondary to that of temperature. For most applications the effect is small and can be ignored, unless a wide frequency range is of interest.

It is to be expected that the performance of the damping material to control both vibration and sound is as good as the method of attachment to the structure. Some commercially available materials come with their own adhesive, and generally provide easier and better attachments if the component surface permits it. However, a majority of unconstrained damping materials require a structural epoxy or an adhesive layer in order to reach complex surface contours. The adhesive layer should be as stiff and thin as possible. A softer epoxy layer between the damping treatment and base structure will yield to shear deformation, and will detract from the performance of the damper.

Energy dissipation in the damping material is a function of deformation, hence damper efficiency increases when located in areas of the structure that are subjected to large deformations. Detailed knowledge of the vibration mode shapes of the structure is necessary in determining locations where maximum amplitudes and bending stresses occur. Location at a nodal point in the structure will have minimal effects. However, when vibration and associated noise protection is required in a large frequency range, many modes are likely to be present, and partial coverage of the component based on modal activity may not be adequate.

Attempts have been made to broaden the damping performance of damping materials with temperature. Limited success has been achieved by varying the composition of the visco-elastic material and filler type. A more effective approach is to use multiple materials with different properties. For example, if the operating temperature range is -50° to 150° F, it may be necessary to select one material with an optimum damping value around 0° F and another material at about 100° F. The materials may then be placed on opposite sides of the component wall, or one on top of another, to obtain a wider temperature range of protection.

Shear type of damping treatment (also called sandwich damper) is more efficient than unconstrained layer damping treatment (*ref*: Collacott, 1987). Since a greater analytical effort is required, more precautions are necessary in its application. The rubber like visco-elastic material is constrained by a metal layer. During cyclic bending the metal layer constrains the damping material to deform in shear, and energy is dissipated through shear deformation. Performance of the constrained layer damper depends considerably on the geometry of thc component. Maximum shear strain is present in the visco-elastic material only if the constraining layer is sufficiently stiff. At the same time, however, the layer's stiffness should not exceed that of the structure. Hence, the material of the constraining layer as well as the geometry of the structure to be damped play a role in determining the effectiveness of noise damping.

9.15 TUNED DAMPERS

Tuned dampers have the advantage of low weight, and can be used in structures responding to a number of resonant frequencies. Tuned dampers differ from classical vibration absorbers in that it uses a rubber like material to provide stiffness, which also dissipates dynamic energy in the form of heat.

For simple cross sections the stiffness of the rubber like tuned damper will depend on the mode of deformation. When it is subjected to shear deformation its stiffness is expressed by $k_G = GA/H$, and when it is subjected to tension or compression loads the expression for stiffness is:

$$k_E = \frac{EA}{H}\left[1+\beta\left(\frac{A}{A'}\right)^2\right] \quad (9.8)$$

where
G = shear modulus
E = Young's modulus
H = height
A = loaded area
A' = unloaded area
β = 1.5 for filled rubber like materials
= 2.0 for unfilled rubber like materials

A correction factor based on the shape of the damper is necessary in (9.8); it arises because Poisson's effects can introduce a bi-axial state of stress in the material. Properties of a typical damper material are shown in fig. 9.20. Performance of the damper depends on the operating temperature, since its properties vary with it. If the operating temperature is going to be over a wide range, it may be necessary to use the properties at the higher temperature end, since both the loss factor and the modulus vary at a slower pace in the region.

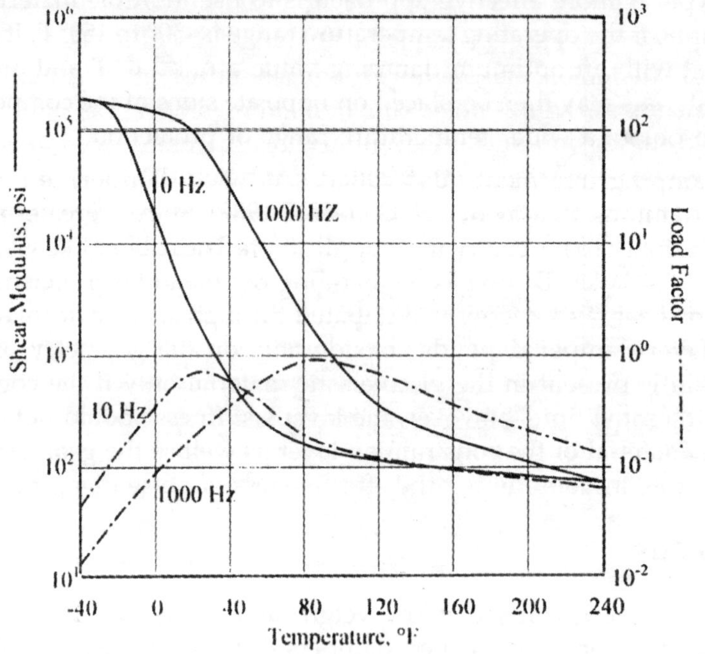

Fig. 9.20. *Material Properties of a Typical Tuned Damper*

Besides damping, reduction in noise level may also be obtained through proper evaluation of the source. For example, reduction in the exhaust system of an industrial turbine is generally achieved through proper silencer selection, as also of piping lengths and diameters (see fig. 9.5), mounting of exhaust system components and positioning of the exhaust outlet. On the air intake side an engine's air filtration system will provide significant attenuation of intake air noise. Additional attenuation requires an intake air silence and adequate design, selection and mounting of the components to take care of surface radiated noise. Noise due to the fan of an aircraft engine depends on fan tip speed (see fig. 9.21), blade passing frequency and aerodynamic design of the fan blade. A thermostatically controlled variable speed fan drive for the enclosure of an industrial gas turbine will add to the cost, but fan-cooling output can be matched to machine cooling requirements. At lighter loads a lower airflow rate may prove to be adequate, and the fan speed can be reduced to decrease driving horsepower and output sound level.

Reduction in machine surface radiated noise requires smaller running clearances and machining tolerances of components. Partial or complete machine enclosures can be used to reduce noise transmitted from machine surfaces to the environment, but will add to the cost of the assembly. An effective enclosure incorporating sound absorption materials, sound barrier materials and vibration damping material can attenuate surface radiated noise by 20 dBA. The enclosure will require vibration isolation from the machine and its support structure to prevent the enclosure panels also from becoming noise sources.

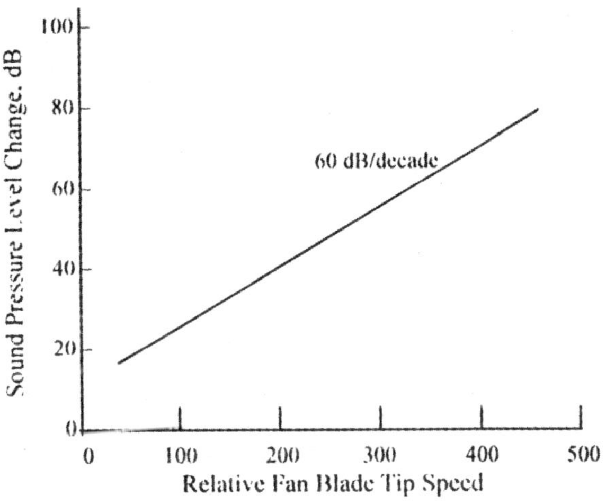

Fig. 9.21. *Effect of Fan Speed on Sound Generation*

9.16 EXAMPLE PROBLEMS

Example Problem # 9.1: A gas turbine has a mass flow 370 lbs/second and an exhaust temperature of 950° F. Unsilenced noise levels from the turbine and the ISO/NR 45 criteria are shown below. System configuration is similar to the vertical arrangement shown in figs. 9.3 and 9.4, but without an intake filter. The intake and exhaust systems require silencers. The noise criterion is ISO/NR 45 at a distance of 400 ft from the site for the far field (requirements shown on next page). The near-field requirement is 90 dBA at 3 ft from the perimeter of the unit and 5 ft from the ground. Assume suitable values for: (*i*) directivity corrections for various orientations of silencers, (*ii*) attenuation associated with the lined bend effect of inlet plenum between the engine bellmouth and silencer, (*iii*) typical attenuation values for silencers, (*iv*) other additive sources of noise in the system. Determine the requirements for the intake silencer, casing enclosure and exhaust silencer to satisfy the far-field criteria.

Octave Band #	*Center Frequency, Hz*	*Sound Pressure Level, dB*			*ISO/NR 45*
		Inlet	*Casing*	*Exhaust*	*Criteria*
–	31.5	119	122	132	86.0
1	63	120	125	135	71.0
2	125	120	129	135	61.1
3	250	120	130	134	53.6
4	500	121	130	133	48.6
5	1000	124	132	132	45.0
6	2000	147	130	125	42.2
7	4000	151	133	124	40.0
8	8000	145	132	115	38.3

Solution: Directivity corrections for various orientation of the silencer are shown below:

Octave Band	–	1	2	3	4	5	6	7	8
Center Frequency, Hz	31.5	63	125	250	500	1000	2000	4000	8000
0°	1	2	3	4	5	5	5	5	5
90°	-1	-2	-3	-5	-8	-10	-11	-11	-11
110°	-1	-3	-5	-8	-12	-15	-18	-18	-18

Assumed values for attenuation associated with the lined bend effect of the inlet plenum between the engine bellmouth and the silencer are as follows:

Octave Band	–	1	2	3	4	5	6	7	8
Center Frequency, Hz	31.5	63	125	250	500	1000	2000	4000	8000
Lined Bend Attenuation	0	-2	-4	-6	-6	-8	-9	-9	-9

Typical values for silencer attenuation and losses from enclosure walls are:

Octave Band	–	1	2	3	4	5	6	7	8
Center Frequency, Hz	31.5	63	125	250	500	1000	2000	4000	8000
Intake silencer	3	8	13	27	52	53	55	54	44
Enclosure wall loss	18	24	26	32	47	48	49	53	54
Exhaust silencer	10	19	34	45	46	42	37	31	24

For the intake silencer the calculation sequence is shown in the table 9.1. The unsilenced sound power level is provided in line # 1. Divergence at 127 m, or 415 ft,

Table 9.1. ***Intake System Calculations***

Line	*Octave Band*	–	*1*	*2*	*3*	*4*	*5*	*6*	*7*	*8*
#	Center Frequency, Hz	31.5	63	125	250	500	1000	2000	4000	8000
1	Unsilenced Noise level	119	120	120	120	121	124	147	151	145
2	Divergence at 127 m.	-50	-50	-50	-50	-50	-50	-50	-50	-50
3	Directivity, 90°	-1	-2	-3	-5	-8	-10	-11	-11	-11
4	Lined Bend Attenuation	0	-2	-4	-6	-6	-8	-9	-9	-9
5	Σ(Line 2+ 3+ 4)	-51	-54	-57	-61	-64	-68	-70	-70	-70
6	Unsilenced Σ(Line1+ 5)	68	66	63	59	57	56	77	81	75
7	Criteria less 7 dB	79	64	54	47	42	38	35	33	31
8	Attenuation Δ(L 6-7)	-11	2	9	12	15	18	42	48	44
9	Silencer attenuation	3	8	13	27	52	53	55	54	44
10	Silenced noise Δ(L 6-9)	65	58	50	32	5	3	22	27	31

radius arises from the reduction in energy as it spreads over a hemisphere of an equal radius r, and is calculated from the equation: Hemispherical Divergence = $10\{\log_{10}(2\pi r^2)\}$ and the calculated values are shown in line # 2. Directivity corrections for 90° orientation are given in line # 3. Attenuation due to the lined bend effect of the inlet plenum depends on the number of acoustically lined sides, thickness of lining and the length of the associated ductwork, and is given in line # 4. Algebraic sum of the correction factors of lines 2, 3 and 4 is given in line # 5, and line # 6 provides the unsilenced noise levels at a distance of 127 m from the unit. If it is assumed that there are five additional contributory noise mechanisms, the expression $10\{\log_{10}(N)\}$ where N is the number of expected additive sources equals 7.0, and determines the extent to which individual sources must be designed below the criterion to obtain a level at or below this target. The computed value is shown in line # 7. Line # 8 determines the necessary silencer attenuation. Line # 9 gives typical values for the intake silencer attenuation and line # 10

Table 9.2. ***Exhaust System Calculations***

Line	*Octave Band*	–	*1*	*2*	*3*	*4*	*5*	*6*	*7*	*8*
#	Center Frequency, Hz	31.5	63	125	250	500	1000	2000	4000	8000
1	Unsilenced Noise level	132	135	135	134	133	132	125	124	115
2	Divergence at 127 m.	-50	-50	-50	-50	-50	-50	-50	-50	-50
3	Directivity, 0°	1	2	3	4	5	5	5	5	5
4	Σ(Line 2+ 3)	-49	-48	-47	-46	-45	-45	-45	-45	-45
5	Unsilenced Σ(L 1+ 4)	83	87	88	88	88	87	80	79	70
6	Criteria less 7 dB	79	64	54	47	42	38	35	33	31
7	Attenuation: Δ(L 5-6)	4	23	34	41	46	49	45	46	39
8	Silencer attenuation	10	19	34	45	46	42	37	31	24
9	Silenced noise Δ(L5-8)	73	68	54	43	42	45	43	48	46

provides the silenced sound pressure level of the total silenced noise emitted from the gas turbine's intake. The sequence of calculations for the intake system is shown on the previous page.

Calculations for the exhaust system and casing enclosure follow a similar pattern, and are shown in tables 9.2 and 9.3. Calculations for the gas turbine casing and enclosure require correction to convert sound power L_w to sound pressure level L_p inside the enclosure using the expression: $L_p = L_w + 10\{\log_{10}(Q/4\pi r^2 + 4S\alpha)\}$, where S = total surface area of the enclosure, α = average absorption coefficient of the walls, r = distance from source to wall and Q = directivity factor of the source. Assume $Q = 1$, $r = 3$ ft, $S = 75$ m^2 and $\alpha = 0.001$. Sound pressure level inside the enclosure is offset by the enclosure loss, given in line # 3. Line # 4 is correction for projection from enclosure surface to the criterion point at 127 m, given by the expression $10\{\log_{10}(A_1/A_2)\}$, where A_1 = surface area of enclosure and A_2 = surface area of 127 m (415 ft) radius hemisphere. Line # 5 gives the unsilenced noise L_p at 415 ft, or 127 m, obtained by adding lines 3 and 4. Line 6 gives the criteria less 7 dB for 5 additional noise sources. Line # 7 is the transmission loss required of the walls (line # 5 minus line # 6), line # 8 is actual wall loss and line # 9 is the silenced noise level (line # 5 minus line # 8). Detailed calculations are shown in table 9.3.

Table 9.3. *Turbine Casing and Enclosure Calculations*

Line Octave Band	–	1	2	3	4	5	6	7	8
#Center Frequency, Hz	31.5	63	125	250	500	1000	2000	4000	8000
1 Unsilenced Noise level	122	125	129	130	130	132	130	133	132
2 Convert L_w to L_p	-5	-5	-5	-5	-5	-5	-5	-5	-5
3 L_p inside Σ(L 1+2)	117	120	124	125	125	127	125	128	127
4 $10\log_{10}(A_1/A_2)$	-31	-31	-31	-31	-31	-31	-31	-31	-31
5 Unsilenced ,415′ Σ(L3+4)	86	89	93	94	94	96	94	97	96
6 Criteria less 7 dB	79	64	54	47	42	38	35	33	31
7 Wall loss needed Δ(L5-6)	7	25	39	47	52	58	59	64	65
8 Silenced, at 415′ Δ(L5-8)	18	24	26	32	47	48	49	53	54
9 Silenced noise, Line 5 – 8	68	65	67	62	47	48	45	44	42

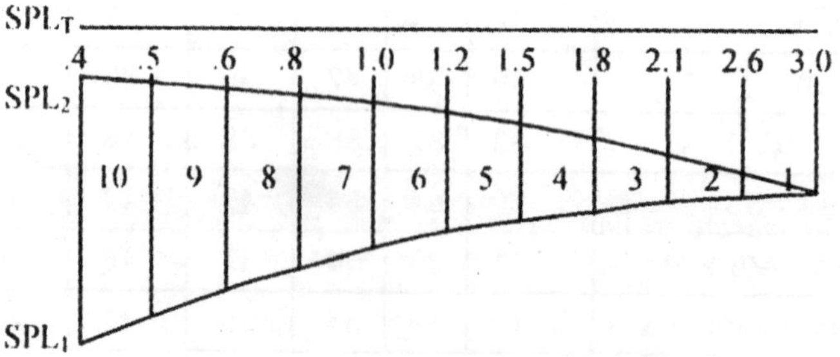

Fig. 9.22. *Summation of Noise Levels*

The procedure for summation of the major noise sources in the gas turbine is shown with the aid of a diagram (fig. 9.22). Line # 1, 2 and 3 need to be combined to give the overall sound level on line # 4, and is obtained from the fig.. First determine ($SPL_2 - SPL_1$), find the corresponding difference ($SPL_T - SPL_2$), then $SPL_T = SPL_2 + (SPL_T - SPL_2)$. Line # 4 is the overall noise level in each octave band, and line # 5 is the actual site criteria to be met. The calculation sequence is provided in table 9.4 shown below.

Table 9.4. *Summation of Noise Levels*

Line Octave Band	–	1	2	3	4	5	6	7	8
#Center Frequency, Hz	31.5	63	125	250	500	1000	2000	4000	8000
1 Silenced Intake Noise	65	58	50	32	5	3	22	27	31
2 Silenced Casing noise	68	65	67	62	47	48	45	44	42
3 Silenced Exhaust Noise	73	68	54	43	42	45	43	48	46
4 Overall noise, L	75	71	67	62	49	50	47	51	50
5 Criteria	86	71	61	54	49	45	42	40	38

REFERENCES & BIBLIOGRAPHY

Alpini, A., Busso, M., Rsupa, G., Turino, G., "Analysis Techniques of Combustion Noise and Vibrations in Diesel Engines", SAE paper # 800406, Warrendale, Pa, 1980.

Anderson, D., Grover, E.C., Lalor, N., Priede, T., "Origins of Reciprocating Engine Noise – Its Characteristics, Prediction and Control", ASME Paper # 82-WA/DGP-3, New York, 1982.

Anderton, D., "Relation Between Combustion System and Engine Noise", paper # 790270, SAE Proceedings, Page 80, Warrendale, Pa, 1980.

Andrews, S.A., "The Analysis and Mechanism of Engine Crank Rumble", Institute of Mechanical Engineers, ISVR, Pages 99-109, 1989.

Aoki, H., "Effects of Power Plant Vibration on the Quality of Sound Inside a Passenger Compartment During Acceleration", Nissan Technical Review, Nissan Motor Co., 1986.

Atkins, K.A., Challen, B.J., "A Practical Approach to Truck Noise Reduction", paper C131/79 in Noise and Vibration of Engines and Transmissions, Institute of Mechanical Engineers Conference Publication, July, 1988.

Baxa, D.E., "Noise Control In Internal Combustion Engines", John Wiley & Sons, New York, 1982.

Beranek, L.L., "Noise and Vibration Control", Mc-Graw Hill, New York, 1991.

Benzakein, M.J., "Research on Fan Noise Generation", Journal of Acoustical Society of America, 51, Number 5, Part I, Pages 1427 – 1438, 1972.

Booker, J., "Dynamically Loaded Journal Bearings: Numerical Application of Mobility Method", Institution of Lubrication Technology, Volume 93, No. 2, 1983.

Booker, J., "Design of Dynamically Loaded Journal Bearing", ASME Design Engineering Conference, Fundamentals of the Design of Fluid Film Bearings, New York, Pages 31-44, 1984.

Broch, J.T., "Application of Bruel & Kjaer Measuring Instruments to Mechanical Vibration and Shock Measurements", Bruel & Kjaer Instruments, Inc., Cleveland, 1994.

Clarkson, B.L., "The Social Consequences of Noise", Nominated Lecture, Proceedings of Institute of Mechanical Engineers, Volume 189, August, 1982.

Collacott, R.A., "Mechanical Fault Diagnosis", Chapman & Hall, London, 1987.

DeJong, R.G., Manning, "Modeling of Vibration Transmission in Engines to Achieve Noise Reduction", SAE paper # 890360, 1989.

Dimotakis, P.E., "Turbulent Free Shear Layer Mixing and Combustion", AIAA paper # 89-0262, 1989.

Haddad, S.D., "Simulating Piston Slap by an Analog Computer", Journal of Sound & Vibration, Volume 72, # 1, Pages 79-93, 1987.

Harting, D. R., Taylor, J.W., "Demodulated Resonance Analysis System", US Patent # 3,842,663, October, 1984.

Hodgetts, D., "Vibration of A Crankshaft", Institute of Mechanical Engineers, paper # C99/87, 1987.

Ho, P. Y., Doyle, V.L., "Combustion Noise Prediction Update", AIAA paper # 79-0588, 1979.

Hunecke, K., "Jet Engines – Fundamentals of Theory, Design and Operation", Motorbooks International Publishers, Osceola, WI, 1997.

Ide, S., "Improvement of Engine Sound Quality Through A New Flywheel System Flexibly Mounted to The Crankshaft", SAE paper # 900391, 1990.

Kerrebrock, J.L., "Aircraft Engines and Gas Turbines", The MIT Press, Cambridge, MA, 1992.

Lalor, N., Petyt, M., "Modes of Engine Structure Vibration as a Source of Noise", SAE paper # 850833, Warrendale, Pa, 1985.

Law, B., "Prediction of Main Bearing and Crankshaft Loading in Reciprocating Engines", CIMAC paper # 6358C/424A, 1983.

Law, B., "Prediction of Crankshaft and Flywheel Dynamics", Institute of Mechanical Engineers, Paper # C382/046, Pages 487-498, Perkins Technology, 1989.

Lefebvre, A.H., "Gas Turbine Combustion", Taylor and Francis, Philadelphia, 1999.

Lighthill, M.J., "On Sound Generated Aerodynamically, I: General Theory ", Proceedings of the Royal Society, Series A, 211, Pages 564 – 587, 1952.

Lighthill, M.J., "Jet Noise", AIAA Journal, 1, Pages 1507 – 1517, 1963.

Little, J.W., Mabry, J.E., "Human Reaction to Aircraft Engine Noise", AIAA paper # 68-548, 1968.

Maas, H., "Calculation of Crankshaft Plain Bearings", 9th CIMAC Conference, paper # A-22, Stockholm, 1984.

Maas, H., "Calculation of Crankshaft Plain Bearings", 11th CIMAC Conference, Pages 43-68, Volume 1, Barcelona, 1986.

Mathews, D. C., Rekos, N. F., "Prediction and Measurement of Direct Combustion Noise in Turbo Propulsion Systems", Journal of Aircraft, Volume 14, Number 9, Pages 850 – 859, 1977.

Morse, P. M., Ingard, K. U., " Theoretical Acoustics", Mc-Graw Hill, New York, 1992.

NASA SP 1258, "Aero Acoustics of Flight Vehicles" and WRDC Technical Report 90-3052, Volume I – Noise Sources, Volume 2 – Noise Control, 1990.

NASA SP 311, "Aircraft Engine Noise Reduction", NASA Lewis Research Center, 1972.

Nashif, A. D., "Application of Damping for Noise Control in Diesel Engine Components", Damping Application for Vibration Control, ASME, AMD Volume 38, New York, 1980.

Pischinger, F. F., Schmillen, K. P., Leipold, F. W., SAE Paper # P-88, Proceedings Diesel Engine Noise Conference, Page 39, March, 1987.

Potter, R. W., Mathematical Work, Unpublished, September, 1994.

Randall, R. B., "Cepstrum Analysis and Gearbox Fault Diagnosis", Application Notes, Bruel & Kjaer Instruments, Inc., Cleveland, OH, 1995.

Russell, M. F., Cavanagh, J. E., "Establishing a Target for Control of Diesel Engine Combustion Noise", Paper # 790271, SAE Proceedings, Page 80, Warrendale, Pa, 1980.

SAE Engine Sound Level Measurement Procedure: J-1074, SAE Handbook, Vol. 2, Pages 35-39, Warrendale, Pa, 1984.

Sasaki, Y., "A Study of Engine Caused Abnormal Noise of Automobiles at Acceleration", MHI, Technical Review, Vol. 18, No. 1, 1984.

Sawyer, T., "Gas Turbines – Volumes I, II and III", International Gas Turbine Institute, Atlanta, ASME, 1984.

Snowdon, J. E., "Vibration and Shock in Damped Mechanical Systems", John Wiley & Sons, New York, 1988.

Society of Automotive Engineers, "Engine Noise", Warrendale, Pa, 1984.

Spectral Dynamics Corporation, "Machinery Analysis – Why Signature Ratio?", Technical Publication # RTA-4, San Diego, September, 1987.

Suzuke, T., "Analysis and Mechanism of Engine Intake Rumble Noise", SAE of Japan paper # 902246, Toyota Motor Co., 1990.

Tsuge, K., "Study of Noise in Vehicle Passenger Compartment During Acceleration", Journal of Society of Automotive Engineers of Japan, Vol. 39, No. 12, 1985.

Tyrell, R. J., "Engine Noise: Practicalities and Prediction, Part 2 – Finite Element Analysis", SAE paper # 870978, Ricardo, 1987.

Ungar, E. E., Ross, D., "Vibrations and Noise due to Piston Slap in Reciprocating Machines", Journal of Sound and Vibration, Vol. 2, Pages 132-146, 1985.

Wu, T., "Discussion of SAE Papers 880405 and 880407", SAE Transactions volume 77, Pages 1408 - 1409, 1988.

Mathews, D.C., Rekos, N.F., "Prediction and Measurement of Direct Combustion Noise in Turbo Propulsion Systems", Journal of Aircraft, Volume 14, Number 9, Pages 850 – 859, 1977.

Morse, P.M., Ingard, K.U., " Theoretical Acoustics", Mc-Graw Hill, New York, 1992.

NASA SP 1258, "Aero Acoustics of Flight Vehicles" and WRDC Technical Report 90-3052, Volume I – Noise Sources, Volume 2 – Noise Control, 1990.

NASA SP 311, "Aircraft Engine Noise Reduction", NASA Lewis Research Center, 1972.

Nashif, A.D., "Application of Damping for Noise Control in Diesel Engine Components", Damping Application for Vibration Control, ASME, AMD Volume 38, New York, 1980.

Pischinger, F.F., Schmillen, K.P., Leipold, F.W., SAE Paper # P-88, Proceedings Diesel Engine Noise Conference, Page 39, March, 1987.

Potter, R.W., Mathematical Work, Unpublished, September, 1994.

Randall, R.B., "Cepstrum Analysis and Gearbox Fault Diagnosis", Application Notes, Bruel & Kjaer Instruments, Inc., Cleveland, OH, 1995.

Russell, M.F., Cavanagh, J.E., "Establishing a Target for Control of Diesel Engine Combustion Noise", Paper # 790271, SAE Proceedings, Page 80, Warrendale, Pa, 1980.

SAE Engine Sound Level Measurement Procedure: J-1074, SAE Handbook, Vol. 2, Pages 35-39, Warrendale, Pa, 1984.

Sasaki, Y., "A Study of Engine Caused Abnormal Noise of Automobiles at Acceleration", MHI, Technical Review, Vol. 18, No. 1, 1984.

Sawyer, T., "Gas Turbines – Volumes I, II and III", International Gas Turbine Institute, Atlanta, ASME, 1984.

Snowdon, J.E., "Vibration and Shock in Damped Mechanical Systems", John Wiley & Sons, New York, 1988.

Society of Automotive Engineers, "Engine Noise", Warrendale, Pa, 1984.

Spectral Dynamics Corporation, "Machinery Analysis – Why Signature Ratio?", Technical Publication # RTA-4, San Diego, September, 1987.

Suzuke, T., "Analysis and Mechanism of Engine Intake Rumble Noise", SAE of Japan paper # 902246, Toyota Motor Co., 1990.

Tsuge, K., "Study of Noise in Vehicle Passenger Compartment During Acceleration", Journal of Society of Automotive Engineers of Japan, Vol. 39, No. 12, 1985.

Tyrell, R.J., "Engine Noise: Practicalities and Prediction, Part 2 – Finite Element Analysis", SAE paper # 870978, Ricardo, 1987.

Ungar, E.E., Ross, D., "Vibrations and Noise due to Piston Slap in Reciprocating Machines", Journal of Sound and Vibration, Vol. 2, Pages 132-146, 1985.

Wu, T., "Discussion of SAE Papers 880405 and 880407", SAE Transactions volume 77, Pages 1408 - 1409, 1988.